STUDENT STUDY GUIDE

BIOLOGY

CONCEPTS & CONNECTIONS

Sixth Edition

Campbell • Reece • Taylor • Simon • Dickey

Richard M. Liebaert

Linn-Benton Community College
Albany, Oregon

PEARSON

Benjamin
Cummings

San Francisco   Boston   New York
Cape Town   Hong Kong   London   Madrid   Mexico City
Montreal   Munich   Paris   Singapore   Sydney   Tokyo   Toronto

Editor-in-Chief: Beth Wilbur
Senior Acquisitions Editor: Chalon Bridges
Project Editor: Carol Whitney, nSight, Inc.
Executive Managing Editor: Erin Gregg
Managing Editor: Michael Early
Production Supervisor: Camille Herrera
Production Management and Composition: Progressive Publishing Alternatives
Cover Production: Richard Whitaker
Illustrators: Progressive Publishing Alternatives
Manufacturing Buyer: Michael Penne
Senior Marketing Manager: Jay Jenkins
Text and Cover printer: Courier Kendallville Plant

Cover Photo Credit: Leopard (Panthera pardus) in grass, looking up, in Sabi Sands Conservancy,
Mpumalanga Province, South Africa; Heinrich van den Berg, Getty Images

Copyright © **2009** Pearson Education, Inc., publishing as Pearson Benjamin, 1301 Sansome St., San Francisco, CA 94111.
All rights reserved. Manufactured in the United States of America. This publication is protected by Copyright and permission should be obtained from the publisher prior to any prohibited reproduction, storage in a retrieval system, or transmission in any form or by any means, electronic, mechanical, photocopying, recording, or likewise. To obtain permission(s) to use material from this work, please submit a written request to Pearson Education, Inc., Permissions Department, 1900 E. Lake Ave., Glenview, IL 60025. For information regarding permissions, call (847) 486-2635.

Many of the designations used by manufacturers and sellers to distinguish their products are claimed as trademarks. Where those designations appear in this book, and the publisher was aware of a trademark claim, the designations have been printed in initial caps or all caps.

ISBN 10-digit: **0-321-54825-6**
13-digit: 978-0-3215-4825-2

Contents

Note to Students

I once heard a teacher remark that he likes to see a textbook that is tattered and torn, because then he knows it is being used. I could think of no better end for this book than for it to be worn out—blanks filled in, notes and sketches in the margins, phrases highlighted, and corners of pages turned down. It is a book that is meant to be used.

How to Use This Book

This Study Guide is intended to do just what the title says—to *guide* you to a real understanding of biology. It will help you learn about important biological concepts and connections—in the textbook, lecture and lab sessions, the website, and podcasts that accompany the text, and in the biological environment around you.

Each Study Guide chapter accompanies a chapter in the sixth edition of *BIOLOGY: Concepts & Connections*, and has six parts:

1. *Focus on the Concepts.* This section summarizes the main concepts and connections that you should concentrate on as you study. Essentially, it is a brief overview of what you need to know.

2. *Review the Concepts.* This section, which is the heart of the Study Guide, contains activities such as questions, diagrams, crossword puzzles, problems, and fill-in-the-blank stories that will help you master the concepts in the text. Some exercises in this section relate to single modules in the textbook; others connect the information and concepts of several modules. (Module numbers are given at the beginning of each exercise.) If you get stuck on a question, refer to the text and your lecture notes. There are also Web questions and activities for almost every module; often they will help you work out the exercises in this Study Guide. Check your work by looking in the Answers section at the back of the book. Try to resist the desire to "peek" at the answers; the value in doing the exercises is in actually thinking about them. If you answer questions incorrectly, you can review the relevant modules.

3. *Test Your Knowledge.* These multiple choice and essay questions require you to recall information from the chapter. When doing the multiple choice questions, look at all the possible answers and choose the best one. Only one correct answer is given. Some answers are tempting but wrong; others are only partially correct. The essay questions are worth your effort, as exams often include essay questions. Check your answers with those at the back of the Study Guide, but again, do not simply look up the answers.

4. *Apply the Concepts.* The multiple choice and essay questions in this section go beyond recall of facts, and test the depth of your understanding by asking you to apply the material from the chapter to new situations, make comparisons, and put ideas together to solve biological problems. Again, there are suggested answers at the back of the Study Guide.

5. *Put Words to Work.* This vocabulary list will help you learn and remember key biological terms. Don't try to memorize definitions; it is more important to be able to use the words than define them. And don't stop with learning the words; it is much more important to understand the concepts they describe.

6. *Use the Web.* Don't forget to use this book with all the resources available to you. The website(s) that accompany the textbook are excellent, and there are even podcasts available to help you review difficult topics.

Studying for Success in Biology

You will learn more in your biology course and enjoy it more if you have an effective study plan. First, it is important to have **a comfortable, well-lit place to study.** Try to **get enough sleep, good food, regular exercise, and time to play and relax.**

It is important to study regularly. Every day you have class you should study for an hour or two. This book will guide you and help you make the most of your study time, so you can build a more sturdy and usable understanding of biology. Each chapter, each lecture, and each concept connects with those you have already covered. If you wait until a day or two before an exam and try to "cram," you don't have time to really understand the concepts, to explore the connections. Facts are easier to cram than concepts, but not as important and useful; if you fill up on mental junk food, there is no room for learning important stuff that is good for you (and your grade in the course!). Develop a study routine.

Studying starts with class lectures and reading, but note that **reading and studying are not the same thing.** Reading is only one of the things you do to study. Simply reading a chapter in the text or listening to a lecture does not guarantee you will learn anything, or even remember it. Don't just try to swallow whole what is stated in your textbook or lecture notes. You need to take a more active role, to get your brain engaged. A good way to do this is to **ask questions as you read and listen.** Make the information your own by asking "What? Where? When? How? How much?" And especially **ask "Why?"** Why do we need to breathe? Why are cells so small? Why are saturated fats harmful? Why do plants bend toward the light? Scientists **look for cause and effect;** they **ask how things work.** If you want to understand biology and succeed in your biology course, this is the most productive approach. Many of the exercises in this study guide will show you how.

When reading and studying, note that **facts are different from concepts,** and **not all facts are equally valuable.** Facts are names, statistics, and simple relationships: The upper arm bone is called the humerus. A human cell contains 46 chromosomes. Sugars are carbohydrates. Concepts are generalizations, rules, explanations: Living things naturally vary, and those variants best-adapted to their environments produce the most offspring and become more numerous. A certain hormone affects some body organs and not others because only certain organs have receptors that enable them to receive the hormone message. Simply **recalling facts is only a starting point** for learning biology. **Your goal should be understanding and explaining concepts—how things work, how they fit together.** To reach this goal, **focus on the big picture.** Don't simply make a list of everything you think you might have to know, from one end of a chapter to the other. It works better to stand back and look at all the main ideas first. Start by reading the *Focus on the Concepts* portion of the Study Guide. Then scan the textbook module headings and illustrations to **get an overview of the main concepts and how they connect.** Then dig for a few key facts that support each concept. If you run out of time, at least you'll have a useful framework. Use your lecture notes to help you focus on what your professor considers the most important material. How does your teacher want you to use the text and Study Guide? Are the books to be used as your primary information source for learning on your own, as preparation for lecture, as background material, or simply as references?

Some students find biology intimidating because it seems to involve so much vocabulary. Although you have to learn some terminology in order to discuss biology intelligently and do well on exams, you don't need to name and define every part to understand how a cell, or an enzyme, or a muscle works. Rather than starting with names, it is better to **first figure out how things work, then learn the terms you need to describe them.** If you read a description of the heart, and look at the illustrations, and perhaps see an animation on the Web, you can comprehend how the heart works and then learn the names of its most important parts. In fact, the names will come easily as you need to use them. The important thing is to **concentrate on the cause-and-effect relationships** among the parts—how they function and interact—not just the parts themselves. Incidentally, **it is more important (and easier) to be able to use biological terms than define them.** You are much more likely to have to say "Oxygen diffuses out of the air in the lung and into the blood" than "Diffusion is the movement of molecules from an area of higher concentration to an area of lower concentration."

This brings us to another important point: **Don't memorize anything unless you have to.** Remembering and understanding—"knowing"—are two different things. For example, you don't "remember" how to drive, you just "know" how! And you can remember something without understanding it! It is more important to know diffusion when you see it and clearly understand what make it happen than to state a definition from memory. If you find yourself spending most of your study time memorizing—simply trying to remember the definition of diffusion without thinking about and understanding what diffusion really is—you are probably focusing too much on facts and vocabulary and not enough on concepts and connections. (I wonder why the authors chose that title?) An efficient way to organize facts and concepts is to **write things down in your own words,** and if possible, **summarize them in an outline, concept map, or diagram.** (Concept maps are discussed in Chapter 3 of the Study Guide.) To translate the material into your own words, you can't just repeat what it says in your notes or in the text. Instead, you have to **think about what the text or lecture really means.** The process of writing helps you to organize your thoughts, and looking at what you have written helps you evaluate your understanding. Outlining, mapping, and making sketches help you to see how ideas connect.

Can you describe how a bicycle works? You can do it because you remember what a bicycle looks like, not because you have memorized a verbal description of how a bicycle works. This same approach works in biology. Biology is a very concrete, tangible subject. **You can really see most of the things biologists talk about,** even though some of them are very small (molecules) or take a long time (evolution). A lot of effort has gone into making the textbook illustrations clear and complete. Study the illustrations carefully, and **try to learn in visual images,** not just in words. In fact, usually it is easier to learn in pictures. Note that the Web animations are immensely helpful in this regard. Remember that even tiny things like molecules and genes are real, just as real as a bicycle, and thinking in pictures is an effective way to understand how they work.

Even if you have never taken a biology course, you already know a lot about the subject. I find that students who enjoy biology and do well in the course are curious about how things work. They devote sufficient time to studying, look at the big picture, and manage to avoid getting bogged down in unnecessary details. And they realize that many of the things covered in the text and in the course are things they already know about. Don't set your experience aside because it seems "unscientific." Instead, embrace it; trust your experience and the knowledge that you bring to the subject of biology. Biology is not some esoteric subject "out there." In fact, there are many points of contact between biology and the rest of your life. Consider the function of your body (breathing, seeing, walking) and your connection with other organisms and the environment (the food you eat, the air you breathe, the microorganisms that make you sick). In addition, a multitude of important issues, including environmental pollution, global climate change, stem cell research, health care issues, and population growth, have biological implications.

Many students find it useful to review that material together in pairs or small groups. Study on your own to get the basics, then get together, talk about it, and quiz each other. Discussing biology aloud gets another "learning channel" involved, and other members of your study group can evaluate your understanding. Compare your answers to the end-of-chapter and Study Guide questions. Make up questions of your own. **Don't just recite what you know, but make up new questions and problems. See if you can anticipate the kinds of questions that might show up on an exam.** If you disagree on an answer, draw in another student to break the deadlock. What you need is a reality check. Do you really understand how blood circulates? What is the connection between genes, DNA, and chromosomes? When my students miss exam questions, they seldom complain that the material was too difficult to understand. More often, they didn't realize certain topics were important, and neglected to go over them before the exam. Your study group is a great place to fill in knowledge gaps so they don't slow you down. A study group is also a great place to **get different points of view, and new ways to study and approach the subject.**

Finally, **if you have questions or difficulties, ask for help.** Make an appointment to talk with your professor during office hours. Don't be embarrassed if you don't understand something. If you knew everything about biology, you wouldn't have to take the course! Teaching assistants and laboratory instructors are also valuable resources. Many colleges and universities have tutorial assistance, learning assistance or study skills centers, and/or learning labs that provide additional help. Use all the resources that are available to you.

I hope this Study Guide contributes to your understanding of biology and to success in your biology course. If you have comments or suggestions, I would like to hear from you. You can e-mail me at: rilie1@comcast.net.

Acknowledgments

I would like to thank the staff at Benjamin Cummings—especially Editor-in-Chief Beth Wilbur, Senior Acquisitions Editor Chalon Bridges, Assistant Editor Rebecca Johnson, and Production Supervisor Camille Herrera. A special thank you to Project Manager Carol Whitney, of nSight, Inc., who expertly steered the Study Guide through its sixth revision. Thank you, Carol! Thank you also to Project Manager Heather Meledin and the folks at Progressive Publishing Alternatives for turning the manuscript into printer-ready pages. Many thanks to the authors of *BIOLOGY: Concepts & Connections*, who had confidence in my ability to write the Study Guide and whose writing sets a high standard. I would also like to thank my friend and mentor Bob Thornton, my most inspiring teacher, and Cat Shelby, my co-conspirator and accomplice. A number of reviewers made important suggestions; I hope the Study Guide meets their expectations. Thanks also to my students and colleagues at Linn-Benton Community College, in Albany, Oregon, who have made teaching and learning a joy.

Richard Liebaert

Reviewers of the first through sixth editions

Jane Aloi Horlings, *Saddleback College*

Loren K. Ammerman, *University of Texas, Arlington*

M.C. Barnhart, *Southwest Missouri State University*

William E. Barstow, *University of Georgia, Athens*

Paul Decelles, *Johnson County Community College*

Stephen Dina, *Saint Louis University*

Charles Duggins, Jr., *University of South Carolina*

Gregory Farley, *Chesapeake College*

John Griffis, *Joliet Junior College*

Martin Hahn, *William Peterson College*

Lazlo Hanzley, *Northern Illinois University*

Yvonne Harris, *Truman College*

Jean Helgeson, *Collin County Community College*

George A. Hudock, *Indiana University*

Kris Hueftle, *Pensacola Junior College*

Ursula Jander, *Washburn University of Topeka*

Alan Jaworski, *University of Georgia, Athens*

Jennifer Katcher, *Pima Community College*

Steve Lebsack, *Linn-Benton Community College*

Melanie Loo, *California State University, Sacramento*

Joseph Marshall, *West Virginia University*

Presley F. Martin, *Drexel University*

James E. Mickle, *North Carolina State University*

Henry L. Mulcahy, *Suffolk University*

Kathryn Stanley Podwall, *Nassau Community College*

John E. Rinehart, *Eastern Oregon University*

Frank Romano, *Jacksonville State University*

Robert Salisbury, *State University of New York College at Oswego*

Roger Sauterer, *Jacksonville State University*

Gary Smith, *Tarrant County Junior College*

Mary Colavito, *Santa Monica College*

Gerald Summers, *University of Missouri, Columbia*

Hilda Taylor, *Acadia University*

Sandra Winicur, *Indiana University, South Bend*

Mary Wisgirda, *San Jacinto College*

Edmund Wodehouse, *Skyline College*

Uko Zylstra, *Calvin College*

Biology: Exploring Life

Focus on the Concepts

This chapter is an introduction to the study of biology—its major themes, methods, and relevance to our lives. As you study this chapter, focus on these major concepts:

- Life encompasses a hierarchy of organization, from atoms and molecules to the entire biosphere. Novel properties emerge at each level. The cell is the fundamental building block where the properties of life appear.

- Living things interact with their environments, exchanging matter and energy in their roles as producers, consumers, or decomposers.

- All organisms share common features, such as the DNA genetic code, regulation, and reproduction, and all species of living things can be grouped into three domains.

- The idea of evolution by natural selection is the core theme of biology; it explains the common features of living things and their diversity.

- Science seeks natural causes for natural phenomena. Discovery science uses inductive reasoning to derive general principles from observations and measurements. Hypothesis-based science seeks causes via deductive reasoning, proposing and testing hypotheses as tentative explanations. Ultimately, scientists build theories, which are broad explanations backed by substantial evidence.

- While science seeks to understand nature, technology applies scientific knowledge to specific needs and purposes. Many issues facing society are related to biology, and it is important for citizens to achieve some understanding of scientific ideas such as evolution.

Review the Concepts

Work through the following exercises to review the concepts in this chapter. For additional review, check out the activities at www.mybiology.com. The website offers a pre-test that will help you plan your studies.

Exercise 1 (Module 1.1)

This module discusses the hierarchy of structural levels into which life is organized. Each of these levels has unique emergent properties, which arise from the organization of its component parts. Review this structural hierarchy by completing the chart below.

Level	*Description*
1.	All environments on Earth that support life
Ecosystem	2.
3.	All the organisms in a particular ecosystem
Population	4.
5.	An individual living thing
6.	Organs that work together to perform particular functions
Organ	7.
8.	A group of similar cells with a specific function
9.	A living unit, separated from its environment by a membrane
Organelle	10.
11.	A cluster of atoms held together by bonds

This module describes how organisms interact with their environments. Review the web of interactions by completing this crossword puzzle.

Across

2. ____ make food that supports the ecosystem.
4. ____ are the main producers in a forest.
6. Energy moves ____ an ecosystem, while nutrients are recycled.
7. Chemical ____ cycle within the ecosystem's web of species.
8. The dynamics of an ____ involve flow of energy and recycling of nutrients.
10. Decomposers ____ chemical nutrients.
11. ____ comes into an ecosystem in the form of sunlight.
12. Living things interact with their ____ .
14. Plants carry out ____, producing food for other organisms.
15. ____ are consumers: they eat plants and each other.

Down

1. Rabbits function as ____ in an ecosystem.
3. ____ such as bacteria and fungi convert dead matter to simpler nutrients.
5. The exchange of matter and energy in an ecosystem involves a complex web of ____.
9. In an ecosystem, there are many interactions among organisms, and between organisms and their ____ environment.
13. Plants absorb mineral nutrients from the ____ .

Exercise 3   (Modules 1.3 – 1.4)

All living things are built from complex systems called cells. Cells are the smallest units that display the properties of life. All cells are controlled by the genetic information in DNA molecules. These two modules briefly discuss the structure and function of cells and DNA and other features shared by all living things. Write a paragraph describing how an organism familiar to you—a college student—displays these features of life.

Exercise 4   (Module 1.5)

Review the three domains of life by matching each statement on the right with the correct domain. Write your answer in the first column. In addition, name the kingdom for each of the organisms in Domain Eukarya, and write your answer in the second column. Choose from:

Domain Bacteria
Domain Archaea
Domain Eukarya
      protists (several kingdoms)
      Kingdom Plantae
      Kingdom Fungi
      Kingdom Animalia

Domain   *Kingdom*

Domain	Kingdom	
_____	_____	1. Tree and fern
_____	_____	2. Prokaryotes
_____	_____	3. Another domain of prokaryotes
_____	_____	4. Multicellular eukaryotes that ingest (eat) other organisms
_____	_____	5. Molds, yeasts, and mushrooms
_____	_____	6. Algae and protozoa
_____	_____	7. Organisms whose cells lack a nucleus
_____	_____	8. Brown pelican, sloth, and spider
_____	_____	9. Multicellular photosynthetic organisms with rigid cellulose cell walls
_____	_____	10. Single-celled eukaryotes such as an amoeba

Evolution is biology's core theme. It explains the diversity of life, the relatedness of all living things, and the adaptation of living things to their environments. Fill in the blanks in the following story to review the concepts of evolution.

While investigating the insect life of the rainforest canopy, a zoologist captured several specimens of a previously unknown species of butterfly. The butterfly was mostly black but had conspicuous red and yellow stripes on its wings. It rested on bare tree limbs in plain view; the zoologist was surprised she had not seen it before. The butterfly was very similar in structure to that of a much less conspicuous all-black species found in the same general area, so the zoologist figured that the two species were closely

1_____ members of the same family.

Biologists have long marveled at the diversity of insect life in the tropics.
2_____, the English biologist who wrote *The Origin of Species*, was surprised by the large number of insect species he encountered in the rain forests of South America. In fact, biologists estimate that most species of living things are rain forest insects.

Like Darwin, the zoologist concluded that the black butterfly and the new species looked alike because they were both descended from a common
3_____ species. But why the difference in color pattern? When she first encountered the striped butterflies, she speculated that the red and yellow stripes were an evolutionary adaptation, a beneficial feature that evolved by means of natural selection. But how could a bright color pattern be of any possible benefit? Wouldn't brightly colored butterflies be attacked by predators?

Her suspicions intensified when the zoologist saw the red and yellow winged butterfly resting on a tree limb. A predatory bird landed nearby and peered at the butterfly. The butterfly responded by rapidly flapping its wings, displaying their striped pattern, and the bird flew off.

This is what first caused the zoologist to suspect that the bright wing pattern was an example of "warning coloration," often seen in harmful or bad-tasting animals—for example, the conspicuous yellow and black stripes of bees and wasps. How could such a color pattern have evolved in this species of butterfly? The zoologist speculated that at one time a 4_____ of black butterflies existed in this area, breeding among themselves but not with other members of their species. These butterflies exhibited
5_____ traits—slightly different wing shapes, sizes, behaviors, and so on. They also may have tasted different. Perhaps some were able to make a bad-tasting substance or store a bad-tasting substance obtained from food plants. Just as Darwin reasoned, the zoologist realized that 6_____ variation must be present in the population for natural selection to operate. Also like Darwin, she realized that there is great overproduction of offspring—many more butterflies are produced each year than can possibly survive. In this population, it appeared that among this abundance of butterfly prey, the good-tasting butterflies were more likely to be eaten by
7_____ than bad-tasting ones. The surviving bad-tasting butterflies were more likely to survive to 8_____, and they passed their ability to make the bad-tasting chemical on to their 9_____. Over time, this

[10]_____ trait accumulated in the population—the bad-tasting butterflies became more numerous.

What about the difference in color pattern? The zoologist speculated that among the bad-tasting butterflies, there may have been variation in wing coloration. Butterflies with bright colors on their wings were easier for predators to remember and avoid. The colorful butterflies had more offspring than less-conspicuous individuals—they had [11]_____ reproductive success. Eventually bright-colored, bad-tasting butterflies became the norm in the population. In this situation, as in others explained by Charles Darwin, [12]_____ occurs as heritable [13]_____ are exposed to [14]_____ factors that favor the [15]_____ success of some individuals over others.

The zoologist speculated that, over a long period, the changes in palatability, wing pattern, and other characteristics must have combined, and a whole new [16]_____ of butterfly came into existence. According to Darwin, the [17]_____ of new species results from the accumulation of minute changes resulting from natural selection over [18]_____. This short story is just one illustration of evolution, biology's core [19]_____. Evolution is an important idea, because it explains both the [20]_____ of life (descent from a common ancestor) and [21]_____ of life (modification as species diverged from their ancestors).

Exercise 6   (Modules 1.7 – 1.8)

Review the scientific method by filling in the blanks in the following story. Choose from **factor, hypothesis, question, inductive reasoning, control, prediction, deductive reasoning, discovery, observation, experiment, experimental,** and **scientific method.** Answers may be used more than once.

In Exercise 5, you read how a zoologist identified a previously unknown species of butterfly. This species was mostly black, but had conspicuous red and yellow stripes on its wings. It was very similar in appearance and structure to an all-black species found in the same area. Carefully comparing the two species, the zoologist concluded that they were closely related—members of the same family. Looking at many examples (of butterflies) and deriving a general principle (the characteristics of the butterfly family) employs a kind of logic called [1]_____. This kind of thinking is often involved when a scientist does [2]_____ science, which is mostly concerned with describing nature.

A different process is followed when a scientist seeks to develop an explanation for natural events. This second method of inquiry is called [3]_____-based science. It employs a process of inquiry sometimes called the [4]_____. Although it is not a single method, its key element is the kind of logic called [5]_____.

In our example, like many examples of hypothesis-based science, the process started with a simple [6]_____: The zoologist noticed that predatory birds avoided the brightly colored butterflies even though they rested in tree branches in plain sight. This evoked a [7]_____: Is there something about the butterflies that the birds don't like? The researcher had a hunch; she suspected that the striped butter-flies tasted bad and that their bright colors acted as a sort of "warning" to predators to stay away. This kind of tentative explanation is called a [8]_____.

The zoologist decided to test this in the laboratory, under conditions that she could manipulate and monitor. Such a test is called a controlled [9]_____. She captured insect-eating birds native to the area and put them in cages at a nearby research station. Then she netted a number of brightly striped butterflies and their black cousins. For her first experiment, she allowed the birds to choose between a black butter-fly and a striped one. The birds invariably chose the black butterflies and avoided the striped ones. This confirmed her field observations.

But did the striped butterflies taste bad? The researcher set up another controlled experiment, designed to compare an [10]_____ group—striped butterflies with their wings painted black—with a [11]_____ group of "normal" striped butterflies. (Actually, the "normal" butterflies were also handled and painted with clear paint, so that only one [12]_____ would differ between the two groups.) Her hypothesis led the zoologist to make a [13]_____ about how she thought the experiment would turn out: *If* the stripes really acted as a warning, *then* the birds would be fooled and eat the butterflies when the stripes were covered—and that *if* the striped butterflies tasted bad, *then* the birds would spit them out. Such "if-then" thinking is called [14]_____ and is an important feature of hypothesis-driven science.

Just as the researcher hypothesized, the birds chose the black-painted butterflies in every trial. Also, most of the birds quickly spat out the black-painted butterflies, and those that swallowed the butterflies became ill. Just to cover things, the zoologist per-formed another experiment in which she painted the wings of the edible black-winged butterflies. The birds ate them with gusto, demonstrating that the paint itself was not dis-tasteful and produced no ill effects.

After repeating the experiments several times, the researcher wrote a paper de-scribing her hypothesis, experiments, results, and conclusions. It was published in the *Journal of Tropical Entomology*. There other scientists could read about the experiments, repeat and expand upon them, even challenge the results—all part of the process of science.

Exercise 7   (Modules 1.9 – 1.10)

After reading these modules, and without referring back to the text, list six ways in which you think the science of biology and its technological applications may affect society in the next decade. Which of these are primarily scientific? Which are technological? Are any of them related to evolution, the core theme of biology? Which do you think will have the most effect on you personally in the coming years?

Test Your Knowledge

Multiple Choice

1. All organisms have which of the following in common?
   a. They respond to stimuli.
   b. They store genetic information in DNA molecules.
   c. They utilize energy.
   d. They reproduce.
   e. all of the above

2. Extrapolating from general premises to specific results is a kind of logic called
   a. inductive reasoning.
   b. synthesis.
   c. deductive reasoning.
   d. experimentation.
   e. observation.

3. Biologists group living things into ___ domains.
   a. 2
   b. 3
   c. 4
   d. 5
   e. about 10

4. A bacterium and an amoeba are placed in different domains because
   a. a bacterium is single-celled.
   b. an amoeba is photosynthetic.
   c. an amoeba can move.
   d. a bacterial cell is much simpler.
   e. an amoeba is single-celled.

5. Almost all of the organisms in Kingdom ____ are photosynthetic.
   a. Animalia
   b. Archaea
   c. Plantae
   d. Fungi
   e. Eukarya

6. At the most fundamental level in life's hierarchy, all living things contain the same basic kinds of
   a. cells.
   b. organs.
   c. molecules.
   d. tissues.
   e. systems.

7. An educated guess posed as a tentative explanation is called a
   a. theory.
   b. control.
   c. variable.
   d. prediction.
   e. hypothesis.

8. The information in ____ underlies all of the properties that distinguish life from nonlife.
   a. carbon
   b. DNA
   c. proteins
   d. populations
   e. chemical nutrients

9. The ____ is the highest level in life's structural hierarchy.
   a. ecosystem
   b. cell
   c. organism
   d. population
   e. biosphere

10. There are many interdependencies in an ecosystem. Prokaryotes and fungi play an important role in the ecosystem primarily because they
    a. cause diseases that keep populations in check.
    b. trap water, which is then used by other organisms.
    c. decompose the remains of dead organisms.
    d. are responsible for producing energy.
    e. carry out photosynthesis, which makes food for other species.

Essay

1. Name the three domains of life and briefly describe their distinguishing characteristics.

2. Name the kingdoms of Domain Eukarya, identify the organisms in each, and briefly describe the criteria that separate each kingdom from the others.

3. What kinds of questions can be answered and what kinds of problems can be solved by science? What kinds of questions are outside the realm of science?

4. Explain how the information in DNA relates to the common features that characterize life.

Applying The Concepts

Multiple Choice

1. A crop scientist noted that, over a period of 10 years, a beetle species that feeds on rice gradually became resistant to insecticide. Which of the following best explains this in terms of natural selection?
   a. The insecticide mutated the beetles exposed to the biggest doses.
   b. Some beetles learned to tolerate the insecticide and passed this ability to their offspring.
   c. Beetles learned to avoid the spray and passed the knowledge to their offspring.
   d. The insecticide caused the beetles to reproduce more quickly than normal.
   e. Those beetles with natural resistance to the insecticide had the most offspring.

2. Researchers testing new drugs usually give the drug to one group of people and give placebos, "sugar pills," to another group. The group receiving the sugar pills
   a. constitutes the experimental group.
   b. is needed so that the test will be repeated enough times.
   c. is the control group.
   d. is a backup in case some of the people getting the drug drop out of the test.
   e. is the experimental variable.

3. ____ has characteristics that result from the organization of its component ____ .
   a. A population . . . ecosystems
   b. A tissue . . . organs
   c. A cell . . . tissues
   d. An organism . . . organ systems
   e. A molecule . . . cells

4. An ecologist studied the effect of nutrients and predators on the population growth of bacteria on the bottom of a pond. His study of bacteria could not involve which of the following levels in life's structural hierarchy?
   a. ecosystem
   b. organ
   c. organism
   d. population
   e. molecule

5. A rain forest primate called an aye-aye has a long middle finger that it uses to probe for insects in cracks and crevices in tree bark. This connection between structure and function developed gradually as a result of
   a. reproduction.
   b. population growth.
   c. natural selection.
   d. DNA replication.
   e. energy exchange.

6. Which of the following does *not* illustrate technology?
   a. DNA research is used to cure an inherited disease.
   b. Scientists develop a bacterium that destroys toxic wastes.
   c. A biologist identifies a new species of monkey.
   d. A chemical that slows cell division is used to treat cancer.
   e. Biologists breed a disease-resistant kind of corn.

7. Which of the following illustrates *discovery* science?
   a. cataloging all the species in a tide pool
   b. trying to find a connection between dietary fat and heart disease
   c. testing a new blood-pressure drug
   d. determining whether higher temperatures slow or speed plant growth
   e. all of the above

8. You have probably seen pictures of the diverse communities flourishing in the dark depths around hydrothermal vents on the ocean floor. Scientists were at first surprised to find that the vents support such thriving ecosystems, because in such an environment, you wouldn't think there could be very many
   a. consumers
   b. bacteria
   c. producers
   d. decomposers
   e. nutrients

9. Which of the following questions might best be answered by the methods of hypothesis-based science?
   a. How many kinds of tissues make up the kidney?
   b. Do human beings have free will?
   c. How many leopards inhabit the Sabi Sands Reserve of southern Africa?
   d. Are there diseases that are caused by evil spirits?
   e. Does the vitamin biotin prevent birth defects?

Essay

1. Choose a familiar wild animal—a squirrel, a toad, or a duck, for example—and describe some of the web of relationships that connect it with other organisms in its ecosystem.

2. Beavers are descendants of land-dwelling rodents similar to rats. Explain how natural selection could have shaped the beaver's flat tail and webbed feet, which it uses for swimming.

3. Jason tried a new fertilizer called MegaGro on his garden. He said, "I used it on all my tomato plants this year, and they grew much better than they did last year! MegaGro is fantastic!" Was Jason's test of MegaGro scientifically valid? Why or why not?

4. Tropical birds called oilbirds nest in caves and emerge at night to forage for seeds. Biologists think that oilbirds might be able to avoid obstacles in their caves and in the tangled growth of the rain forest much the way bats do—by making sounds and listening to the echoes. Describe a controlled experiment to test this hypothesis.

5. Explain how each of the following shows the connection between biological structure and function: your hand, a leaf, a hawk's beak, a frog's hind legs.

6. Describe the hierarchy of structural levels of which your body is composed. Give a specific example of a feature at one level that is not seen in the parts that make it up.

7. A camera sends back pictures of purple gelatinous blobs in near-boiling water near volcanic vents on the ocean floor. What properties should scientists look for to determine whether the blobs represent life?

8. How would you expect human DNA to be different from the DNA of a chimpanzee? From the DNA of a goldfish?

Put Words to Work

Correctly use as many of the following words as possible when reading, talking, and writing about biology:

Archaea, Bacteria, biology, biosphere, cell, community, consumer, controlled experiment, domain, ecosystem, emergent properties, Eukarya, eukaryotic cell, gene, hypothesis, molecule, natural selection, organ, organism, organelle, organ system, population, producer, prokaryotic cell, species, systems biology, technology, theory, tissue

Use the Web

There is more material related to the topics in Chapter 1 at www.mybiology.com.

The Chemical Basis of Life

Focus on the Concepts

This chapter explores the chemicals of life—the ordering of atoms into molecules and the interactions of molecules in shaping and affecting living things. As you study the chapter, focus on the following concepts:

- Living things are made of about 25 chemical elements. Atoms are the smallest particles of an element, and atoms of each element are made of a characteristic number of protons, neutrons, and electrons.

- Arrangement of electrons determines the chemical properties of an atom— how it will bond to other atoms. Sometimes atoms gain and lose electrons and form ionic bonds. Sometimes atoms share electrons, forming covalent bonds and making a molecule. Atoms of different elements combine in specific arrangements and ratios to form compounds.

- Two atoms of hydrogen and one of oxygen bond covalently to form a water molecule. Unequal sharing of electrons makes a water molecule polar. The polarity of water molecules causes them to form weak hydrogen bonds, which give water many of its peculiar and important properties, such as cohesion

- Water molecules can break apart to form hydrogen (H^+) and hydroxide (OH^-) ions. Acids are compounds that add hydrogen ions, and bases remove hydrogen ions. The pH scale describes how acidic or basic a solution is.

- In a chemical reaction, chemical bonds break and reform, changing products into reactants. Living things carry out a myriad of chemical reactions, changing matter in numerous ways.

Review the Concepts

Work through the following exercises to review the concepts in this chapter. For additional review, check out the activities at www.mybiology.com. The website offers a pre-test that will help you plan your studies.

Write the chemical symbol for each of the following elements, and state whether it is one of the four elements used by living things in large amounts (L), whether it is used in moderate amounts (M), or whether it is a trace element (T) required in small amounts.

Symbol	Amount	Element	Symbol	Amount	Element
		1. Magnesium			7. Carbon
		2. Oxygen			8. Calcium
		3. Zinc			9. Phosphorus
		4. Hydrogen			10. Nitrogen
		5. Copper			11. Sodium
		6. Iodine			12. Iron

A compound is a substance that contains two or more elements in a fixed ratio. Indicate with a check mark which of the following are elements and which are compounds. (You will have to guess on some!)

	Element	Compound
1. Table salt		
2. Calcium		
3. Water		
4. Vitamin A		
5. Carbon		
6. Sulfur		
7. Carbon dioxide (CO_2)		
8. DNA		
9. Iodine		
10. Protein		

These modules introduce atoms. It is most important to know what the subatomic particles are, where they are located in an atom, and that atoms of different elements differ because they contain different numbers of protons. Some atoms not covered in these modules are compared below. You can figure out the subatomic particles they contain based on the concepts in the modules. First, fill in the blanks. Then sketch each atom, labeling and coloring **protons** red, **neutrons** gray, and **electrons** blue. (Coloring will help you focus on and remember which is which.)

Element	Symbol	Atomic Number	Mass Number	Number of Protons	Number of Neutrons	Number of Electrons
1. Carbon-12	C	6	12	6	6	6
2. Nitrogen-14	___	7	14	___	___	___
3. Chlorine-35	___	___	35	17	___	___
4. Oxygen-16	___	___	___	___	___	8
5. Oxygen-17	___	___	___	___	___	___

Atoms of four elements important to life are diagrammed below. Pay particular attention to their electron shells. Remember that atoms with incomplete outer electron shells participate in chemical reactions that allow them to attain complete outer shells: 2 electrons for a hydrogen atom, 8 electrons for most other elements important to life.

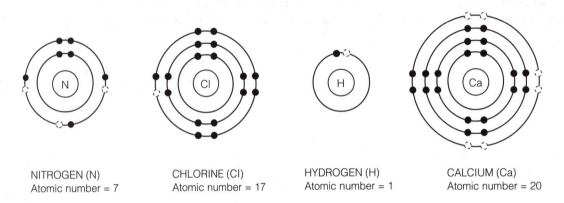

NITROGEN (N)
Atomic number = 7

CHLORINE (Cl)
Atomic number = 17

HYDROGEN (H)
Atomic number = 1

CALCIUM (Ca)
Atomic number = 20

1. Given the information and diagrams above, show how electrons would be transferred between calcium and chlorine atoms to form calcium and chloride ions, which would then attract each other to form calcium chloride, $CaCl_2$. (Hint: An atom can gain or lose more than one electron.)

2. Using the information and diagrams above, show how nitrogen could form covalent bonds with several hydrogen atoms, forming a molecule of ammonia. What would be the molecular formula for ammonia?

Review the properties of water by filling in the blanks in the following story.

When Amy came through the door, she found Liz poised over a glass of water, ready to drop a needle into the glass. Amy asked, "Liz, what are you trying to do? Trying to kill your roommate with a poisoned needle? Or is this another of your 'experiments'?"

"We're studying the [1]_____ basis of life in my biology class," Liz replied. "I don't get some of the stuff she's teaching us, so I need to do some experiments to figure it out."

Amy scrolled through her text messages and rolled her eyes. "Another experiment. Welcome to Geek-O-Rama."

Liz gently placed the needle on the water surface. "Watch this," she said. The needle rested in a dimple on the surface of the liquid.

"Whoa—How'd you do that?"

"I didn't. The water did. Water molecules have a tendency to stick together, which is called [2]_____. The water molecules are stuck together so tightly at the surface that they form a film that can support the weight of the needle. Bugs can walk on it. It's called [3]_____."

"O.K. You got me. At the risk of getting too much information, how do the water molecules do it? What's so special about water?"

Liz explained, "A water molecule is H_2O, right? It is made up of one [4]_____ atom and two [5]_____ atoms. The atoms stay together because they [6]_____ electrons. This holds them together. A shared pair of electrons forms a chemical bond called a [7]_____ bond between each hydrogen atom and the oxygen atom. Now, if the electrons were shared evenly, the bond would be called a [8]_____ covalent bond. But they are not shared evenly. The oxygen tends to 'hog' the electrons away from the hydrogens. It has a greater attraction for electrons; it is more [9]_____ than hydrogen—"

"TMI! TMI! Just tell me what this has to do with floating needles."

"Well, because the oxygen atom attracts the electrons more strongly, the shared electrons are closer to the oxygen than to the hydrogens, giving the oxygen a slight [10]_____ charge. Because the electrons are pulled away from the protons in the nuclei of the hydrogen atoms, the hydrogens are left with slight [11]_____ charges. So the bonding electrons are shared unevenly, producing a [12]_____ covalent bond between each hydrogen atom and the oxygen atom. In fact, the whole water molecule is polar, even though the molecule as a whole is electrically [13]_____."

Amy was getting impatient. "So what does that have to do with surface tension? And what's the biology connection?"

Liz went on, "Well, it is their polarity that causes water molecules to stick together. The [14]_____ charged oxygen of one water molecule is attracted to the [15]_____ charged hydrogens of other water molecules. These special bonds between water molecules are called [16]_____ bonds. These bonds form a network at the water's surface, creating surface tension strong enough to support the needle. Each water molecule can connect with [17]_____ others. Hydrogen bonds are weak, but important. For example, they are responsible for holding the two strands of a [18]_____ molecule together, and for keeping [19]_____ molecules folded.

Hydrogen bonds give water some peculiar properties. For example, water is the only common substance on Earth that naturally exists in all three states of matter [20]_____, [21]_____, and [22]_____. And lots of things will dissolve in water; it is a versatile [23]_____. Blood plasma, for example, is an [24]_____ solution containing many different [25]_____, or dissolved substances, such as salt and blood sugar."

Amy got up and opened the bathroom door, looked inside, and said, "It's steamy in there. Are you going to take a bath?"

Liz replied, "No, that's just another experiment. I'm trying to figure out the difference between heat and temperature."

"Are they different?"

"Yes. [26]_____ is the total amount of energy resulting from the movement of molecules in a body of matter, like a bathtub full of water. [27]_____ measures the intensity of movement. I compared the amount of heat in a cup of water at 98°C and a bathtub of water at 45°C. In the [28]_____, the intensity of movement of water molecules was greater, but the [29]_____ held more heat energy. I knew it did because the bathtub of water added more heat to the room as it cooled, warming up the room more than the cup of hot water did."

"It doesn't take a genius to figure that out."

"I just wanted to see it for myself. Water has a great capacity to store heat, by the way. When water is heated, a lot of the energy goes into breaking the [30]_____ between water molecules before the molecules can move faster. For instance, if you had a kilogram of water and a kilogram of rock, the same amount of heat would raise the temperature of the water [31]_____ than the temperature of the rock. This means water can soak up a lot of heat, and its temperature will go up only a few degrees."

"And when water cools a few degrees, it [32]_____ a lot of heat."

"Correct. And since animals are mostly water, this helps us control our body temperature. It also stabilizes the temperatures of the ocean and coastal areas. In the summer, the ocean [33]_____ heat, and in the winter, it [34]_____ heat."

Amy's eyes narrowed. "So why do we sweat when we are hot? Wouldn't we want to hang onto all that good water?"

Liz was ready with an answer. "No, not necessarily. Because of their strong hydrogen bonds, it takes a lot of heat energy to get a water molecule moving [35]_____ enough to [36]_____, to separate from its neighbors. This gives water an unusually high [37]_____, but it also makes [38]_____ cooling possible. The hottest—or fastest moving—water molecules evaporate first, taking a lot of heat energy with them and leaving the cooler—slower—molecules behind. So sweating cools you off on a hot day."

Amy looked at the clock and said, "Arrgh—It's 3:30. I told Sara I'd meet her at the ice rink at 3:30. Thanks to the science lesson, I'm gonna be late!"

Liz rambled on. "Ice. Now, ice is very interesting. In ice the water molecules are locked into a crystal, linked by hydrogen bonds, but farther apart than they are in liquid water. This means that ice is [39]_____ dense than liquid water, so it [40]_____. This is important to life, because . . ."

But Amy was already out the door. Liz had a puzzled expression on her face as she got up and slid an ice tray out of the freezer.

Exercise   (Modules   )

Practice using the pH scale by giving the approximate pH of each of the following. Some are listed in the modules; others you can estimate from the information given.

_____ 1. Tomato juice

_____ 2. Human blood

_____ 3. Vinegar (moderately acidic)

_____ 4. Pure water

_____ 5. Cola (moderately acidic)

_____ 6. Household ammonia

_____ 7. Concentrated nitric acid (very acidic)

_____ 8. Acid precipitation

_____ 9. Drain cleaner (very basic)

_____10. Antacid pills (mildly basic)

_____11. Urine

_____12. Gastric juice

Exercise   (Module   )

This module introduces chemical reactions, chemical processes that change matter. A common chemical reaction in many cells is one that changes hydrogen peroxide (H_2O_2) into water and oxygen gas:

$$2H_2O_2 \longrightarrow \underline{\quad} H_2O + O_2$$

_____   _____

Hydrogen peroxide is a harmful by-product of many reactions. Cells get rid of it by carrying out the above reaction, converting it to harmless water and oxygen. What are the reactants in this reaction? What are the products? Label them in the blanks below the equation. Note that the equation for a chemical reaction must be "balanced." Since atoms cannot be created or destroyed in a chemical reaction—only rearranged—the numbers of atoms on both sides must be equal. In this example, there are four hydrogen atoms in the two hydrogen peroxide molecules on the left. After the reaction occurs, the hydrogen atoms reappear in the water on the right. Similarly, the four oxygen atoms in the hydrogen peroxide molecules on the left reappear in the water and oxygen molecule on the right. How many water molecules must be formed to account for all the atoms in the H_2O_2 molecules? Write the correct number in the small blank in front of H_2O.

Review basic chemical terminology by completing this crossword puzzle.

Across

2. ____ is the energy due to movement of molecules in a body of matter.

5. An ____ is a sub-atomic particle that circles an atom's nucleus.

7. The smallest particle of an element is called an ____.

9. An ____ is a charged atom or molecule.

12. Acid ____ is caused by pollutants that combine with water in the air.

16. ____ is anything that occupies space and has mass.

17. Two or more atoms held together by covalent bonds form a ____.

19. Neutrons and protons are found in an atom's ____.

22. The cohesion of water molecules is responsible for surface ____.

23. Variant forms of an element with different numbers of neutrons are called ____.

24. ____ is the tendency of water molecules to stick together.

Down

1. A ____ is a subatomic particle with no electrical charge.

3. ____ measures the intensity of heat.

4. Electrons are shared unequally in a ____ covalent bond.

5. There are 92 naturally occurring ____.

6. A ____ contains two or more elements in a fixed ratio.

8. Weak bonds between water molecules are called ____ bonds.

10. An ____ donates H^+ ions to solutions.

11. A ____ bond is formed when two atoms share electrons.

13. A ____ is a positively charged particle from the nucleus of an atom.

14. In a solution, the dissolving agent is called the ____.

15. A ____ is a liquid containing a uniform mixture of substances.

18. When two ions of opposite charges attract each other, an ____ bond forms.

20. The ____ is the substance dissolved in a solution.

21. A ____ accepts H^+ ions and removes them from solution.

Test Your Knowledge

Multiple Choice

1. Which of the following is a trace element, required only in small amounts by most living things?
   a. oxygen
   b. iron
   c. nitrogen
   d. carbon
   e. hydrogen

2. An acid is a substance that
   a. dissolves in water.
   b. forms covalent bonds with other substances.
   c. donates hydrogen ions to solutions.
   d. is a versatile solvent.
   e. removes hydrogen ions from solutions.

3. How an atom behaves when it comes into contact with other atoms is determined by its
   a. nucleus.
   b. size.
   c. protons.
   d. neutrons.
   e. electrons.

4. Most of water's unique properties result from the fact that water molecules
   a. are very small.
   b. tend to repel each other.
   c. are extremely large.
   d. tend to stick together.
   e. are in constant motion.

5. Atoms of different phosphorus isotopes
   a. have different atomic numbers.
   b. have different numbers of neutrons.
   c. react differently with other atoms.
   d. have different numbers of electrons.
   e. have different numbers of protons.

6. An ion is formed when an atom
   a. forms a covalent bond with another atom.
   b. gains or loses an electron.
   c. becomes part of a molecule.
   d. gains or loses a proton.
   e. gains or loses a neutron.

7. The smallest particle of water is
   a. an atom.
   b. a crystal.
   c. an element.
   d. a compound.
   e. a molecule.

8. Why are biologists so interested in chemistry?
   a. Chemicals are the fundamental parts of all living things.
   b. Most chemicals are harmful to living things.
   c. They know little about life except the chemicals it is made from.
   d. If you understand the chemistry of life, you can make a lot of money.
   e. Everything about life can be known by understanding its chemistry.

9. Molecules are always moving. Some molecules move faster than others; _____ is a measure of their average velocity of movement.
   a. polarity
   b. heat
   c. temperature
   d. electronegativity
   e. density

10. Which of the following holds atoms together in a molecule?
    a. ionic bonds between atoms
    b. transfer of protons from one atom to another
    c. sharing of electrons between atoms
    d. loss of neutrons by atoms
    e. sharing of protons between atoms

11. Ice floats because
    a. it is colder than liquid water.
    b. its molecules are moving faster than in liquid water.
    c. it is more dense than liquid water.
    d. its hydrogen molecules bond to the water surface film.
    e. its water molecules are farther apart than in liquid water.

12. Adding acid tends to ____ of a solution.
    a. increase the hydrogen ion concentration and raise the pH
    b. increase the hydrogen ion concentration and lower the pH
    c. decrease the hydrogen ion concentration and raise the pH
    d. decrease the hydrogen ion concentration and lower the pH
    e. c or d, depending on the original acidity

13. Scientists are excited by evidence that water once flowed on Mars, because
    a. it shows that all planets share the same chemical processes.
    b. there will be water to drink when humans explore Mars.
    c. they thought it was impossible for water molecules to exist in the Martian environment.
    d. it means that Mars might support some form of life.
    e. it suggests that Mars once orbited Earth.

Essay

1. List the four elements needed by living things in large amounts, two others needed in moderate amounts, and two elements needed in trace amounts.

2. Explain why the smallest particle of iron is an atom, but the smallest particle of water is a molecule.

3. Explain the following statement: The temperature of the water in a teakettle is higher than the temperature of water in a swimming pool, but the swimming pool contains more heat.

4. How does acid precipitation form? How does it injure animals? Plants?

5. Explain why water molecules are polar, how this makes them tend to bond to each other, and how this causes water to have a large heat-storage capacity.

6. Explain how evaporation of water from your skin cools you on a hot day.

Apply the Concepts

Multiple Choice

1. An atom that normally has _____ in its outer shell would tend *not* to form chemical bonds with other atoms.
   a. 1 electron
   b. 3 electrons
   c. 4 electrons
   d. 6 electrons
   e. 8 electrons

2. You can use a product such as "Jet Dry" in your dishwasher to keep water from clinging to dishes and causing spots. Jet Dry must work by interfering with
   a. cohesion
   b. covalent bonding
   c. evaporation
   d. adhesion
   e. ionic bonding

3. Researchers studying the effects of toxic wastes knew that animals were poisoned by the heavy metal cadmium, but they wanted to know where cadmium accumulated in the body. They could find out by
   a. tracing the movement of cadmium isotopes in test animals.
   b. measuring the size of cadmium atoms.
   c. finding out whether cadmium atoms form ionic or covalent bonds.
   d. finding out whether cadmium is acidic in water.
   e. determining the number of bonds formed by cadmium atoms.

4. Changing the number of _____ would change it into an atom of a different element.
   a. bonds formed by an atom
   b. electrons circling the nucleus of an atom
   c. protons in an atom
   d. particles in the nucleus of an atom
   e. neutrons in an atom

5. A glass of grapefruit juice, at pH 3, contains _____ H^+ as a glass of tomato juice, at pH 4.
   a. one-tenth as much
   b. half as much
   c. twice as much
   d. three times as much
   e. ten times as much

6. Fluorine atoms tend to take electrons from any atoms that come near. As a result, fluorine atoms
   a. tend to become positively charged.
   b. are nonpolar.
   c. do not react readily with other atoms.
   d. tend to form ionic bonds.
   e. are not very electronegative.

7. Sean added 10 milliliters (mL) of hydrochloric acid and 10 mL of water (pH 7) to a beaker containing 100 mL of water. The pH of the resulting solution was 4. Next he is going to add 10 mL of hydrochloric acid and 10 mL of pH 7 buffer to a different beaker containing 100 mL of water. What do you think will happen?
   a. The resulting pH will be less than 4.
   b. The resulting pH will be between 4 and 7.
   c. The resulting pH will be 7.
   d. The resulting pH will be between 7 and 11.
   e. The resulting pH will be greater than 11.

8. A sodium atom has a mass number of 23. Its atomic number is 11. How many electrons does it have (if it is not an ion)?
   a. 11
   b. 12
   c. 22
   d. 23
   e. 34

9. Which of the following is the smallest in volume?
   a. nucleus of an oxygen atom
   b. water molecule
   c. proton
   d. ice crystal
   e. electron cloud of an oxygen atom

10. Potassium chloride consists of potassium ions (K^+) and chloride ions (Cl^-) in a crystal. If potassium chloride is placed in water, what do you think happens?
    a. The K^+ ions are attracted to the oxygen atoms of water molecules.
    b. It will not dissolve.
    c. The Cl^- ions are attracted to the oxygen atoms of water molecules.
    d. It acts as an acid.
    e. The K^+ ions are attracted to the hydrogen atoms of water molecules.

Essay

1. A sulfur atom has 6 electrons in its outer shell. How many covalent bonds is it likely to form with other atoms? Why? What do you think the formula for hydrogen sulfide would be?

2. Plants carry out chemical reactions that make sugars, which contain carbon, hydrogen, and oxygen, from carbon dioxide (CO_2) and water (H_2O). Researchers want to know whether the oxygen atoms in sugar come from the carbon dioxide or the water. How could they use radioactive tracers to find out? What would they look for in terms of results?

3. If you drop a hot 10-kg rock into 10 kg of cold water, the rock cools off and the water warms up, but the final temperature of both is much closer to the starting temperature of the water than that of the rock. Why?

4. A sugar molecule contains carbon, hydrogen, and oxygen atoms. The oxygen atoms tend to steal electrons from the other atoms in the molecule, giving the oxygen atoms negative charges and leaving other parts of the sugar molecule positively charged. The molecules in oil, however, consist only of hydrogen and carbon, which share electrons equally. Oil molecules do not have areas of positive and negative charge. Using this information and what you know about water, explain why sugar mixes with water but oil does not.

5. Why are the ratios of elements in molecular formulas always fixed—H_2O instead of H_5O, and CH_4 instead of CH_6? Explain this in terms of the characteristics of the atoms making up the molecules.

6. On a typical July day in Seattle, Washington, the high temperature is around 75°F, the low around 55°F. In January, the average high is around 45°F, the low 35°F. In Minneapolis, Minnesota (at roughly the same latitude as Seattle), July highs average around 83°F, lows around 60°F; January highs average around 20°F, lows close to 0°F. Explain the differences between temperatures in Seattle and Minneapolis in terms of the concepts discussed in this chapter.

7. If you heat up a mixture of water and alcohol, the alcohol evaporates first, leaving most of the water behind. Why do you think this happens?

Put Words to Work

Correctly use as many of the following words as possible when reading, talking, and writing about biology:

acid, acid precipitation, adhesion, aqueous solution, atom, atomic mass, atomic number, base, buffer, chemical bond, chemical reaction, cohesion, compound, covalent bond, double bond, electron, electron shell electronegativity, element, heat, hydrogen bond, ion, ionic bond, isotope, mass number, matter, molecule, neutron, nonpolar covalent bond, nucleus, pH scale, polar covalent bond; polar molecule, product, proton, radioactive isotope, reactant, salt, solute, solution, solvent, surface tension, temperature, trace element

Use the Web

There is more on the chemicals of life at www.mybiology.com.

The Molecules of Cells

3

Focus on the Concepts

Life is based on the structure and function of carbon-based molecules. This chapter introduces the chemistry of carbon and the four groups of large organic molecules—carbohydrates, lipids, proteins, and nucleic acids. As you study this chapter, focus on these major concepts:

- A carbon atom can form four covalent bonds, so carbons atoms can form a variety of organic molecules containing chains, branches, rings, and double bonds. Certain functional groups of atoms help to determine the emergent properties of a molecule, such as the way it bonds or reacts. Many large organic molecules are polymers—chains of smaller molecular subunits.

- Carbohydrates have formulas that are multiples of CH_2O. The simplest carbohydrates are sugars such as glucose, called monosaccharides, used for energy and to build larger carbohydrates or other organic molecules. Two monosaccharides can pair to form a disaccharide such as fructose. Polysaccharides are polymers of many sugars. Starch is an energy storage polysaccharide, and cellulose forms plant cell walls.

- Lipids are molecules such as fats, waxes, oils, and steroids that possess hydrocarbon regions that do not mix well with water. A fat is an energy-storage molecule consisting of three long fatty acid units attached to glycerol. Fats can be saturated or unsaturated. Phospholipids are modified fats important in membrane structure.

- Proteins are complex polymers of amino acids linked via peptide bonds between their acid groups and nitrogen-containing amino groups. There are 20 kinds of amino acids, each with different projecting side chains, so there is an immense variety of possible amino acid sequences. Beyond this primary structure, different amino acids and their side-chains can bond and interact in various ways, twisting, folding, and combining polymers in secondary, tertiary, and quaternary levels of structure. The specific 3-D shapes of proteins enable them to carry out complex life functions, such as transport, communication, and movement.

- Protein structure is determined by the information stored in genes, consisting of DNA, a nucleic acid. To shape proteins and body characteristics, DNA works though an intermediary nucleic acid called RNA. Each nucleic acid is a polymer of nucleotides, subunits consisting of a sugar, a phosphate group, and a nitrogen-containing base. The information stored in the long, twisted double helix of DNA resides in the sequence of its nucleotides.

Review the Concepts

Work through the following exercises to review the concepts in this chapter. For additional review, check out the activities at www.mybiology.com. The website offers a pre-test that will help you plan your studies.

The great variety of organic compounds results from the ability of carbon atoms to form four bonds, creating branching chains of different lengths. Several hydrocarbon molecules, consisting only of carbon and hydrogen, are shown in Module 3.1. Practice seeing the versatility of carbon by sketching some hydrocarbon molecules of your own, as suggested below.

1. Sketch a hydrocarbon molecule that is a straight chain, containing 5 carbon atoms and 12 hydrogen atoms, molecular formula C_5H_{12}:

   Question: Why does each carbon bond to 4 other atoms?

2. Now sketch a shorter hydrocarbon chain, with only four carbon atoms:

   Question: What is the molecular formula ($C_?H_?$) of the above molecule?

3. Sketch another five-carbon hydrocarbon, but this time include one double bond:

   Question: What is the molecular formula of this molecule?

4. Sketch a five-carbon hydrocarbon molecule that is branched (and contains no double bonds):

   Question: What is the molecular formula of this molecule? What is the term for its relationship to molecule 1 (in this exercise)?

5. Sketch two five-carbon hydrocarbon molecules in the form of rings, one without double bonds and one with one double bond.

   Question: How many hydrogen atoms are in each of these molecules?

Functional groups participate in chemical changes and give each molecule unique properties. Circle the functional groups that are discussed in this module in the molecules below. Label an example of each of the following: **hydroxyl group, carbonyl group, carboxyl group, amino group,** and **phosphate group, methyl group.** There are a total of ____ hydroxyl group(s), ____ carbonyl group(s), ____ carboxyl group(s), ____ amino group(s), ____ phosphate group(s), and ____ methyl group(s). (The properties of the molecules are described at the right.)

Formaldehyde is the starting point for making many chemicals.

Formic acid gives ant venom its sting.

 Thymine is one of the "bases" that make up the DNA genetic code.

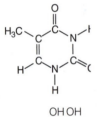 Ethylene glycol is in automobile antifreeze.

Acrolein is produced when meat is heated; it is the barbecue smell.

 Serine is part of many protein molecules.

Urea is a waste product in urine.

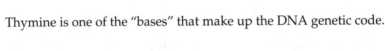 Putrescene's name is descriptive; it is produced in rotting flesh.

 G3P is an intermediate step in plants' production of sugar.

Exercise 3   (Module 3.3)

There are four main classes of macromolecules. Most are polymers, assembled from smaller monomers in a process called a dehydration reaction. Hydrolysis breaks polymers back down to monomers. State whether each of the following relates to dehydration (D) or hydrolysis (H).

_____   1. Connects monomers to form a polymer.
_____   2. Produces water as a by-product.
_____   3. Breaks up polymers, forming monomers.
_____   4. Water is used to break bonds between monomers.
_____   5. Joins amino acids to form a protein.
_____   6. Glycerol and fatty acids combine this way to form a fat.
_____   7. Occurs when polysaccharides are digested to form monosaccharides.
_____   8. —H and —OH groups form water.
_____   9. Nucleic acid breaks up to form nucleotides.
_____ 10. Water breaks up, forming —H and —OH groups on separate monomers.

Exercise 4   (Modules 3.3–3.7)

Review carbohydrates by filling in the blanks in the following story.

Carbohydrates are a class of molecules ranging from the simplest sugars, called ^{1}_____ , to giant molecules called ^{2}_____ , built of many sugars. Carbohydrates are the main fuel molecules for cellular work.

Plants make their own carbohydrates, but humans, like all animals, must obtain them from plants or other animals. Imagine eating a piece of whole-wheat bread spread with strawberry jam. It contains a mixture of carbohydrates, along with other macromolecules such as ^{3}_____ and ^{4}_____ . Much of the carbohydrate in the bread itself is in the form of a polysaccharide called ^{5}_____ , which is simply a chain of ^{6}_____ monomers. The monomers were linked together in the wheat plant in a process called a ^{7}_____ reaction. As the glucose units joined, ^{8}_____ was produced as a by-product. When you swallow a bite of bread, digestive juices in the intestine separate the monomers in the opposite reaction, called ^{9}_____ . In the intestine, this is actually a two-step process. Secretions from the pancreas first break the starch down to maltose, a type of carbohydrate called a ^{10}_____ , which consists of two glucose monomers. Secretions from the walls of the intestine complete the process, breaking each maltose molecule down to two individual glucose molecules. Each glucose is a ^{11}_____ -shaped molecule, containing ^{12}_____ carbon atoms.

There are other carbohydrates in the bread and jam. Whole-wheat flour contains the tough coats of the wheat seeds. These contain a lot of ^{13}_____ , the fibrous polysaccharide that makes up plant cell walls. Like starch, it is made of glucose monomers, but these monomers are ^{14}_____ in a different orientation. The human digestive tract is not capable of ^{15}_____ cellulose, so it passes through the digestive tract unchanged, in the form of ^{16}_____ . Sucrose, a ^{17}_____ refined from sugar cane or sugar beets, may be used to sweeten the strawberry jam. Each sucrose molecule is hydrolyzed in the small intestine to form one molecule of [18]_____ and one molecule of [19]_____. This homemade jam naturally also contains a small amount of fructose, a [20]_____ that is naturally produced by strawberries and is considerably sweeter than sucrose. (High-fructose corn syrup, or HFCS, used to sweeten many processed foods, is produced by hydyrolyzing [21]_____ and using enzymes to convert the resulting [22]_____ to fructose. The increase in use of HFCS may be linked to the recent increase in [23]_____ , [24]_____ , and other chronic diseases.)

Once all the carbohydrates have been hydrolyzed to small monosaccharides, they can be absorbed by the body. Glucose and fructose pass through the wall of the intestine and into the bloodstream, which carries them to the liver. Like all carbohydrate molecules, these sugars are [25]_____ , so they easily dissolve in the water of blood plasma. In the liver, the fructose is converted to glucose. This process is relatively easy because glucose and fructose are [26]_____ , having the same molecular formula, written [27]_____ , but slightly different structures. Glucose circulates around the body as "blood sugar" and is taken up by the cells for fuel as needed. Extra glucose molecules are taken up by liver and muscle cells and linked together by [28]_____ synthesis to form a polysaccharide called [29]_____. This molecule is similar to plant [30]_____ , except it is more branched. Later the glycogen can be hydrolyzed to release [31]_____ into the blood.

Review the structures and functions of lipids by completing the following crossword puzzle.

Across

3. ____ means that hydrogen has been added to unsaturated fats.

4. ____ is a steroid common in cell membranes.

6. A ____ is similar to a fat; found in cell membranes.

9. A fat molecule is composed of ____ and three fatty acids.

10. Glycerol and three____ acids make a triglyceride.

11. ____ is another name for "fat."

12. A ____ forms a waterproof coat that keeps a fruit or insect from drying out.

16. Olive and corn ____ are examples of unsaturated fats.

17. Fats with double bonds are said to be ____.

18. ____ is a lipid-containing deposit in a blood vessel.

19. ____ are grouped together because they do not mix well with water.

Down

1. ____ is a condition where lipid-containing deposits build up in blood vessels.

2. Female and male sex hormones are examples of ____.

5. ____ is an illegal steroid recently banned by the Olympic Committee, FDA, and professional sports.

7. Animal fats are said to be ____.

8. Lipids are water-avoiding, or ____ substances.

13. Unsaturated fats contain more ____ bonds than saturated fats.

14. A ____ is a large molecule whose main function is energy storage.

15. ____ steroids are dangerous synthetic variants of testosterone.

Exercise 6 (Module 3.11)

Everything a cell does involves proteins. Eight functions of proteins are discussed in Module 3.11 (enzymes, transport proteins, etc). Match each of the functional types with one of the descriptions below.

_____ 1. Hemoglobin carries oxygen in the blood.
_____ 2. A protein in muscle cells enables them to move.
_____ 3. Antibodies fight disease-causing bacteria.
_____ 4. Collagen gives bone strength and flexibility.
_____ 5. Insulin signals cells to take in and use sugar.
_____ 6. A protein in a cell receives the insulin signal.
_____ 7. Proteins in seeds provide food for plant embryos.
_____ 8. A protein called sucrase promotes the chemical conversion of sucrose into monosaccharides.

Exercise 7 (Modules 3.12–3.13)

Three amino acids not shown in the modules are diagrammed below.

1. Draw a box around the unique R group of each, and label it **R group**.

2. Draw a red circle around the amino group of each, and label it **amino group**.

3. Draw a blue triangle around the acid group of each, and label it **acid group**.

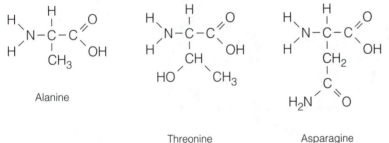

Alanine

Threonine   Asparagine

4. In the space below, sketch the three amino acids to show how they would join to form a tripeptide. What is this chemical reaction called? How many molecules of water would be formed? Show where the water would come from.

Identify each of the levels of protein structure in the diagrams. Then choose the descriptions from the list below that go with each of the levels.

1._____ Structure ⟶

Descriptions: _____

2._____ Structure ⟶

Descriptions: _____

3._____ Structure ⟶

Descriptions: _____

4._____ Structure ⟶

Descriptions: _____

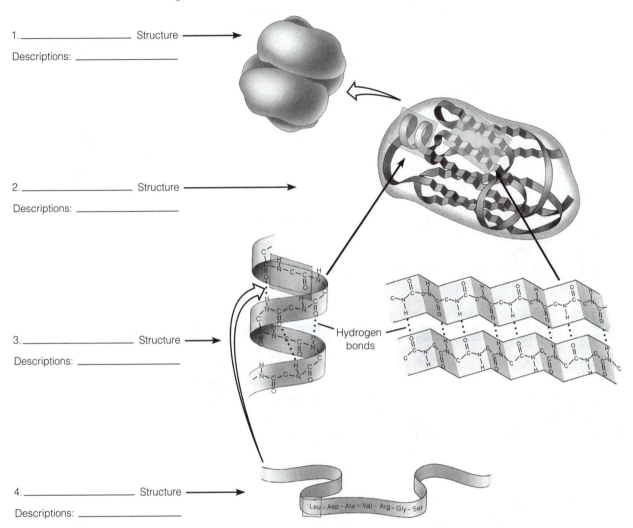

Hydrogen bonds

Leu – Asp – Ala – Val – Arg – Gly – Ser

Choose from these descriptions:

A. Overall three-dimensional shape
B. Amino acid sequence
C. Even a slight change in this can alter tertiary structure.
D. This level occurs in proteins with more than one polypeptide subunit.
E. Coiling and folding produced by hydrogen bonds between —NH and C=O groups
F. Not present in all proteins
G. Level of structure that is held together by peptide bonds
H. Alpha helix and pleated sheet
 I. Stabilized by clustering of hydrophobic R groups, hydrogen bonds, ionic bonds, and sometimes even covalent bonds
J. "Globular" or "fibrous" might describe this level of structure.
K. Folding that results from interactions among R groups of amino acids in the polypeptide chain.

Nucleic acids are the fourth group of biological molecules discussed in this chapter. Review their structures and functions by matching each of the phrases on the left with a word or phrase from the list on the right. Answers may be used more than once.

_____ 1. Sugar in RNA
_____ 2. Overall structure of DNA
_____ 3. Short for "ribonucleic acid"
_____ 4. Molecule passed on from parent to offspring
_____ 5. Nitrogenous bases of RNA
_____ 6. The stuff that genes are made of
_____ 7. Nitrogenous bases of DNA
_____ 8. Programs the amino acid sequence of a polypeptide
_____ 9. DNA works through this intermediary.
_____ 10. A nucleotide consists of sugar, phosphate, and this.
_____ 11. Sugar of one nucleotide bonds to this in next nucleotide.
_____ 12. Monomer of nucleic acids
_____ 13. Sugar in DNA
_____ 14. A mutation in this results in lactose tolerance.

A. Phosphate group
B. Deoxyribose
C. A, T, C, G
D. DNA
E. Nucleotide
F. A, U, C, G
G. Double helix
H. Ribose
I. Nitrogenous base
J. RNA
K. Gene

You may find that making a concept map is a useful way to organize your knowledge. Such a map for the topic of carbohydrates is shown at the top of the next page. A concept map shows how key ideas are connected. Making a concept map can help you learn because it causes you to focus on main concepts and how they are related. It helps you to sort out what is important from unimportant details, and helps you tie your knowledge together into a more meaningful and useful whole.

To make a concept map, you must first decide which ideas are most important. Place the biggest, or most inclusive, concept at the top of the page. Just a word or phrase is enough. Cluster subconcepts around it, and cluster sub-subconcepts around them. Draw lines between the concepts to show how they are connected, and describe these connections next to the lines. Again, use only a word or two.

If the topics or connections are not clear, perhaps they are unimportant, or perhaps you are not clear on how they connect, or perhaps they do not really connect. Remember, clarifying relationships is the purpose of making the map. Generally, maps that are more "branched" are more useful than ones with many long straight "chains" of boxes, but there is no one "correct" map for a particular topic.

Focus on the process of making the map, rather than on the map itself. More learning will take place while you are making the map than when you look at the finished product. You might want to "tune up" your maps by comparing them with maps made by other students. After reviewing the following concept maps, on separate paper try making your own concept maps for proteins and nucleic acids. Keep them simple at first. Remember, practice makes perfect! Also, keep the concept map idea in mind for upcoming chapters.

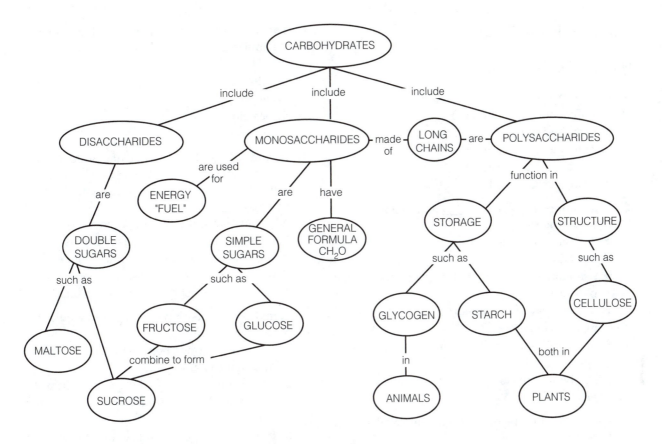

Practice working with a concept map by filling in the blanks on this map for lipids.

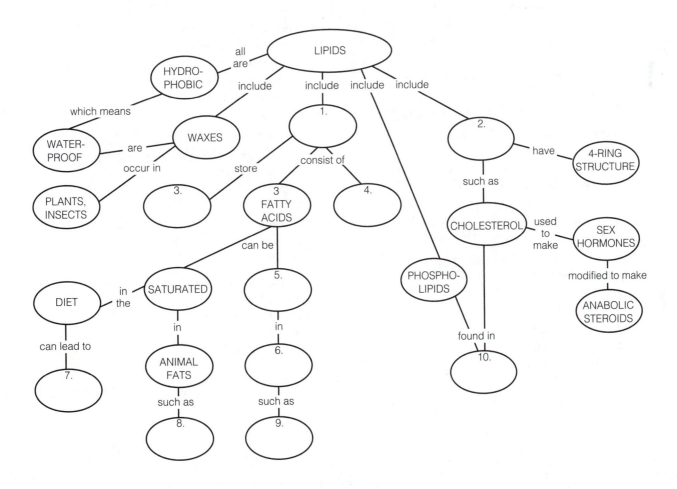

Test Your Knowledge

Multiple Choice

1. Cellulose is a _____ made of many _____.
   a. polypeptide . . . monomers
   b. carbohydrate . . . fatty acids
   c. polymer . . . glucose molecules
   d. protein . . . amino acids
   e. lipid . . . triglycerides

2. In a hydrolysis reaction _____, and in this process water is _____.
   a. a polymer breaks up to form monomers . . . consumed
   b. a monomer breaks up to form polymers . . . produced
   c. monomers are assembled to produce a polymer . . . consumed
   d. monomers are assembled to produce a polymer . . . produced
   e. a polymer breaks up to form monomers . . . produced

3. The four main categories of macromolecules in a cell are
   a. proteins, DNA, RNA, and steroids.
   b. monosaccharides, lipids, polysaccharides, and proteins.
   c. proteins, nucleic acids, carbohydrates, and lipids.
   d. nucleic acids, carbohydrates, monosaccharides, and proteins.
   e. RNA, DNA, proteins, and carbohydrates.

4. A major characteristic that all lipids have in common is
   a. they are all made of fatty acids and glycerol.
   b. they all contain nitrogen.
   c. none of them is very high in energy content.
   d. they are all acidic when mixed with water.
   e. they don't dissolve well in water.

5. A flower's color is determined by the genetic instructions in its
   a. proteins.
   b. lipids.
   c. carbohydrates.
   d. nucleic acids.
   e. all of the above.

6. The most concentrated source of stored energy is a molecule of
   a. DNA.
   b. cellulose.
   c. fat.
   d. protein.
   e. glucose.

7. In some places the backbone of a protein molecule may twist or fold back on itself. This is called _____, and the coils or folds are held in place by _____.
   a. tertiary structure . . . hydrogen bonds
   b. primary structure . . . covalent bonds
   c. secondary structure . . . peptide bonds
   d. tertiary structure . . . covalent bonds
   e. secondary structure . . . hydrogen bonds

8. A hydrophobic amino acid R group would be found where in a protein?
   a. forming a peptide bond with the next amino acid in the chain
   b. on the outside of the folded chain, in the water
   c. on the inside of the folded chain, away from water
   d. forming hydrogen bonds with other R groups
   e. only at one end of a protein chain

9. The overall three-dimensional shape of a polypeptide is called the
   a. double helix.
   b. primary structure.
   c. secondary structure.
   d. tertiary structure.
   e. quaternary structure.

10. How many different *kinds* of protein molecules are there in a typical cell?
    a. four
    b. twenty
    c. about a hundred
    d. thousands
    e. billions

11. Estrogen, cholesterol, and other steroids are examples of
    a. polysaccharides.
    b. lipids.
    c. polypeptides.
    d. triglycerides.
    e. fats.

12. Functional groups called _____ groups are often used to transfer energy between organic molecules.
    a. hydroxyl
    b. amino
    c. carboxyl
    d. carbonyl
    e. phosphate

13. The "building blocks" of nucleic acid molecules are called
    a. polysaccharides.
    b. amino acids.
    c. fatty acids.
    d. nucleotides.
    e. DNA and RNA.

Essay

1. Briefly describe the various functions of proteins in the cell and body.

2. Animal fats tend to be solid at room temperature, plant oils more liquid. Explain how a difference in the chemical structure of their molecules causes this physical difference.

3. What forces and bonds maintain the three-dimensional folded shape of a protein molecule? How does this relate to the sensitivity of proteins to changes in their environment?

4. Using circles to represent monosaccharides, show the difference between glucose, maltose, and starch. Maltose is an example of what kind of carbohydrate? Starch is an example of what kind of carbohydrate?

5. Sketch a protein molecule, using squares connected by lines to represent amino acids connected by peptide bonds. Does your protein display primary, secondary, and tertiary structure? Where?

Apply the Concepts

Multiple Choice

1. Citric acid makes lemons taste sour. Which of the following is a functional group that would cause a molecule such as citric acid to be acidic?
    a. hydroxyl
    b. hydrocarbon
    c. amino
    d. carbonyl
    e. carboxyl

2. Which of the following do nucleic acids and proteins have in common?
    a. They are both made of amino acids.
    b. Their structures contain sugars.
    c. They are hydrophobic.
    d. They are large polymers.
    e. They each consist of four basic kinds of subunits.

3. A biochemist is analyzing a potato plant for the disaccharide sucrose. Where would he be most likely to find it?
    a. in cell membranes
    b. in grains in the cells of underground tubers (potatoes)
    c. in the nuclei of potato cells
    d. in the sap of the potato plant
    e. in the walls of the potato plant cells

4. Which of the following ranks the molecules in the correct order by size?
    a. water . . . sucrose . . . glucose . . . protein
    b. protein . . . water . . . glucose . . . sucrose
    c. water . . . protein . . . sucrose . . . glucose
    d. protein . . . sucrose . . . glucose . . . water
    e. glucose . . . water . . . sucrose . . . protein

5. How does glucose differ from sucrose, cellulose, and starch?
    a. It is a carbohydrate.
    b. It is larger.
    c. The others are polysaccharides.
    d. It is a monosaccharide.
    e. It contains carbon, hydrogen, and oxygen.

6. Seth noticed that his friend Jon had gained a little weight during the holidays. He commented, "Storing up some _____ for the winter, I see."
    a. polysaccharides
    b. triglycerides
    c. nucleotides
    d. polypeptides
    e. steroids

7. How does DNA differ from RNA?
    a. DNA is larger.
    b. One of their nitrogenous bases is different.
    c. They contain different sugars.
    d. DNA consists of two strands in a double helix.
    e. All of the above are differences.

8. Certain fatty acids are said to be essential because the body cannot make them itself; they must be obtained in the diet. If your diet were deficient in these essential fatty acids, you would not be able to make certain
   a. fats.
   b. glycerol molecules.
   c. monosaccharides.
   d. proteins.
   e. You would not be able to make any of the above.

9. Hydrolysis of a protein would produce
   a. amino acids.
   b. monosaccharides.
   c. polysaccharides.
   d. peptide bonds.
   e. nucleotides.

10. Glucose and hexanoic acid each contain six carbon atoms, but they have completely different properties. Glucose is necessary in food; hexanoic acid is poisonous. Their differences must be due to different
    a. monomers.
    b. macromolecules.
    c. hydrolysis.
    d. quaternary structures.
    e. functional groups.

11. Which of the following would probably *not* be affected when a protein is denatured?
    a. primary structure
    b. secondary structure
    c. hydrogen bonds
    d. tertiary structure
    e. All of the above must be affected for the protein to be denatured.

12. Palm oil and coconut oil are more like animal fats than other plant oils. Because they ____ than other plant oils, they can contribute to cardiovascular disease.
    a. contain fewer double bonds
    b. are less saturated
    c. contain more sodium
    d. are less soluble in water
    e. contain less hydrogen

13. A shampoo contains "hydroxypropyl methylcellulose and hydrolyzed soy protein." These substances are
    a. a polysaccharide with added functional groups and broken-down polypeptides
    b. broken-down carbohydrates and broken-down proteins
    c. a broken-down carbohydrate and proteins with added functional groups
    d. polysaccharides and proteins with added functional groups
    e. a lipid with added functional groups and broken-down polypeptides

Essay

1. Briefly explain why all starch molecules are pretty much the same, but there are millions of kinds of protein molecules.

2. Specific enzymes in your intestine enable you to break down starch and use the glucose molecules produced by this process. But you cannot break down cellulose. Explain why, in terms of both carbohydrate structure and protein shape.

3. Heating slightly is often enough to render a protein nonfunctional. But a polysaccharide such as starch must literally be boiled in acid before it is significantly affected. Explain why.

4. Sketch the structural formulas of two hydrocarbon molecules that are isomers. Be sure the C and H atoms form the correct numbers of bonds. What are the molecular formulas of the molecules? What is identical about the molecules? How do the molecules differ?

5. Fred suffers from a disease that makes it difficult for his cells to produce glycogen. For him, three meals a day are not enough; he needs to snack constantly. Explain why.

6. A tripeptide, a molecule consisting of three amino acids, is a very small protein. Yet a huge variety of tripeptides is possible. Assume that the first, second, and third amino acids can be any of 20 choices. How many different tripeptides could there be? (Hint: Imagine you are stringing beads. If you have 20 colors to choose from, how many three-bead sequences are possible?) How could you calculate the number of possible polypeptides 100 amino acids long? (You probably won't want to actually work it out—it's a *very* large number.)

Bacteria and archaea consist of small, simple prokaryotic cells. Label the following on this diagram of a prokaryotic cell: **capsule, cell wall, plasma membrane, nucleoid, ribosome, prokaryotic flagella, pili, bacterial chromosome.** Briefly state the function of each structure next to its label.

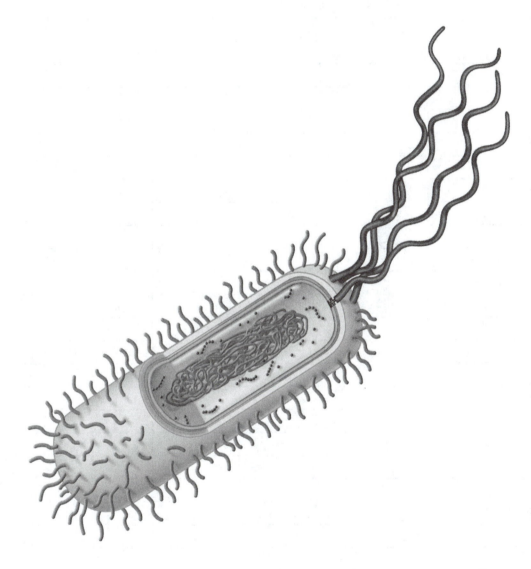

Examine the diagrams and text, and then compare the structures of the cells of prokaryotes, plants, and animals by checking off their characteristics below. You may want to revise or refer to this checklist as you complete the chapter.

Characteristic	Prokaryotic Cell	Plant Cell	Animal Cell
Prokaryotic structure			
Eukaryotic structure			
Relatively large size			
Relatively small size			
Membranous organelles			
Plasma membrane			
Cell wall			
Cytoplasm			
Ribosomes			
Bacterial flagellum			
Nucleus			
Nuclear envelope			
Rough endoplasmic reticulum			
Smooth endoplasmic reticulum			
Golgi apparatus			
Lysosome			
Peroxisome			
Central vacuole			
Mitochondrion			
Chloroplast			
Cytoskeleton			
Flagellum			
Centriole			
Extracellular matrix			
Plasmodesmata			
Cell junctions			

Exercise 5 (Modules 4.5–4.13)

Review membranes, the nucleus, and the various structures that make up the endomembrane system by matching each phrase on the left with a structure from the list on the right. Answers can be used more than once.

_____ 1. Lipids manufactured here

_____ 2. Small structure that makes protein

_____ 3. Contains chromatin

_____ 4. Sac of digestive enzymes

_____ 5. Carries secretions for export from cell

_____ 6. Breaks down drugs and toxins in liver

_____ 7. Makes cell membranes

_____ 8. Cell control center

_____ 9. Numerous ribosomes give it its name

_____ 10. "Ships" products to plasma membrane, outside, or other organelles

_____ 11. May store water, needed chemicals, wastes, pigments in plant cell

_____ 12. Buds off from Golgi apparatus

_____ 13. Defective in Tay-Sachs disease

_____ 14. Proteins made and modified here for secretion from cell

_____ 15. Pumps out excess water from some cells

_____ 16. Nonmembranous organelle

_____ 17. Takes in transport vesicles from ER and modifies their contents

_____ 18. Boundary between the cell and its surroundings

_____ 19. Surrounded by double layer of membrane with pores

_____ 20. How proteins, other substances get from ER to Golgi apparatus

_____ 21. Stores calcium in muscle cells

_____ 22. Marks and sorts molecules to be sent to different destinations

_____ 23. Buds from here become lysosomes

_____ 24. Digests food, wastes, foreign substances

_____ 25. Phosopholipid bilayer with proteins controls flow in and out of cell

_____ 26. Drug tolerance results from enlarging this

_____ 27. Fuses with food vacuole and digests contents

A. Nucleus

B. Transport vesicle

C. Central vacuole

D. Smooth ER

E. Lysosome

F. Golgi apparatus

G. Rough ER

H. Contractile vacuole

I. Ribosome

J. Plasma membrane

Exercise 6   (Modules 4.6–4.13)

Sketch and label the endomembrane system on this diagram. Include **rough ER, smooth ER, ribosomes, Golgi apparatus, lysosome, nuclear envelope, transport vesicles,** and **plasma membrane.** (1) Trace the path of a protein from its site of manufacture to the outside of the cell with a red arrow. (2) Trace the path of a protein incorporated into a lysosome in blue. (3) Trace the path of a protein incorporated into the plasma membrane in green. (4) Trace the path of a lipid secreted from the cell in yellow.

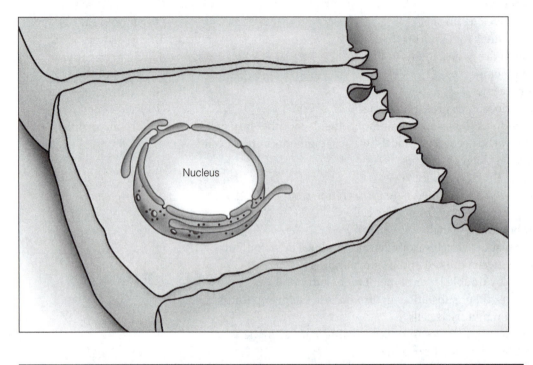

Exercise 7   (Modules 4.14–4.16)

Both mitochondria and chloroplasts are energy converters, but their functions are quite different. Compare them by filling in the chart below.

	Chloroplast	*Mitochondrion*
Found in the following organisms . . .		
Carries out process of . . .		
Converts energy of . . .		
Into chemical energy in . . .		
Evolved by endosymbiosis from . . .		

Exercise 8 (Modules 4.17–4.19)

The cytoskeleton is important in cell shape, organization, and movement. Compare the components of the cytoskeleton by checking which of the following are characteristics of microfilaments, intermediate filaments, or microtubules.

	Microfilaments	*Intermediate Filaments*	*Microtubules*
Hollow tubes			
Solid rods			
Ropelike structure			
Made of tubulin			
Made of actin			
Made of fibrous proteins			
Work with myosin to cell change shape			
Reinforce cell shape, anchor organelles			
Act in muscle cell contraction			
Act as tracks for organelle movement			
Shape and support cell			
In cilia			
In flagella			
In centrioles			
9 + 2 pattern			
Dynein arms cause bending movement			
Extend from basal body			
Don't function in the disease PCD			

Exercise 9 (Modules 4.20–4.22)

The surfaces of cells have some important structural features. Match each of the cell surface characteristics or structures on the left with a phrase on the right. Answers may be used more than once.

____ 1. Channel between animal cells
____ 2. Rigid covering of a plant cell
____ 3. Link animal cells in leakproof sheet
____ 4. Channel between plant cells
____ 5. Connects animal cells into a strong sheet
____ 6. Sticky layer holds animal cells together
____ 7. Integrins in membrane link this to cytoskeleton
____ 8. Helps support plant against gravity
____ 9. Connects cytoplasm of adjacent plant cells
____ 10. Acts like a "rivet" between animal cells
____ 11. Made mostly of cellulose
____ 12. Made mostly of collagen

A. Tight junction
B. Plasmodesma
C. Anchoring junction
D. Cell wall
E. Gap junction
F. Extracellular matrix

Exercise 10 (Module 4.23 and Summary)

Label the organelles listed in Module 4.23 on these diagrams of animal and plant cells. (If you get stuck, refer to Module 4.4.) Try to group your labels according to the four functional categories in Module 4.23 so that you can also circle and label each category. Complete your diagrams by underlining in red the names of structures found in animal cells but not in most plant cells. Underline in green the names of structures found in plant cells but not in animal cells.

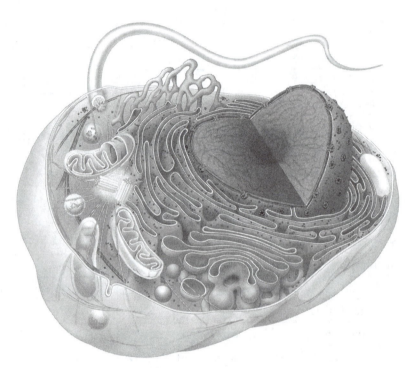

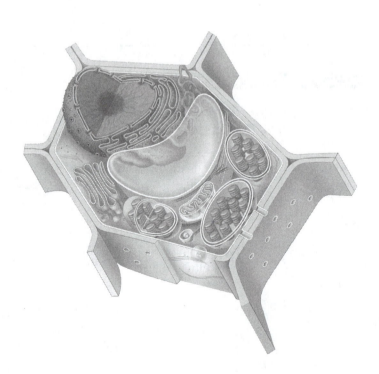

Test Your Knowledge

Multiple Choice

1. To enter or leave a cell, substances must pass through
   a. a microtubule.
   b. the Golgi apparatus.
   c. a ribosome.
   d. the nucleus.
   e. the plasma membrane.

2. Which of the following would *not* be considered part of a cell's cytoplasm?
   a. a ribosome
   b. the nucleus
   c. a mitochondrion
   d. a microtubule
   e. fluid between the organelles

3. Which of the following consist of prokaryotic cells?
   a. plants and animals
   b. bacteria and archaea
   c. plants, fungi, bacteria, and archaea
   d. animals
   e. plants, bacteria, and archaea

4. Organelles involved in energy conversion are the
   a. rough ER and Golgi apparatus.
   b. nucleus and smooth ER.
   c. nucleus and chloroplast.
   d. lysosome and ribosome.
   e. mitochondrion and chloroplast.

5. The maximum size of a cell is limited by
   a. its need for enough surface area for exchange with its environment.
   b. the number of organelles that can be packed inside.
   c. the materials needed to build it.
   d. the amount of flexibility it needs to be able to move.
   e. the amount of food it needs to survive.

6. You would expect a cell with an extensive Golgi apparatus to
   a. make a lot of ATP.
   b. secrete a lot of material.
   c. move actively.
   d. perform photosynthesis.
   e. store large quantities of food.

7. Which of the following correctly matches a structure with its function?
   a. mitochondrion—photosynthesis
   b. nucleus—cellular respiration
   c. ribosome—manufacture of lipids
   d. lysosome—movement
   e. central vacuole—storage

8. Cellular metabolism is
   a. a type of cell division.
   b. the process by which certain parts cause a cell to "self-destruct."
   c. the chemical activity of a cell.
   d. movement of a cell.
   e. control of the cell by the nucleus.

9. Which of the following stores calcium, important in muscle contraction?
   a. mitochondria
   b. smooth ER
   c. the Golgi apparatus
   d. contractile vacuoles
   e. rough ER

10. Which group below is involved in manufacturing substances needed by the cell?
    a. lysosome, vacuole, ribosome
    b. ribosome, rough ER, smooth ER
    c. vacuole, rough ER, smooth ER
    d. smooth ER, ribosome, microtubule
    e. rough ER, lysosome, peroxisome

11. The internal skeleton of a cell is composed of
    a. microtubules, intermediate filaments, and microfilaments.
    b. cellulose and intermediate filaments.
    c. cellulose, microtubules, and centrioles.
    d. microfilaments and collagen.
    e. microfilaments and cellulose.

Essay

1. What are the advantages of an electron microscope over a light microscope? For what tasks would it be preferable to use a light microscope?

2. Briefly describe the major differences between prokaryotic and eukaryotic cells.

3. Name the structures present in plant cells but lacking in animal cells, and describe their functions.

4. Explain the advantages eukaryotic cells derive from being compartmentalized by many internal membranes.

5. Compare the functions of chloroplasts and mitochondria in a plant cell.

Apply the Concepts

Multiple Choice

1. A cell has mitochondria, ribosomes, smooth ER, and other parts. Based on this information, it could *not* be
   a. a cell from a pine tree.
   b. a grasshopper cell.
   c. a yeast (fungus) cell.
   d. a bacterium.
   e. Actually, it could be any of the above.

2. Dye injected into a cell might be able to enter an adjacent cell through a
   a. tight junction.
   b. microtubule.
   c. vacuole.
   d. plasmodesma.
   e. lysosome.

3. If a cell's chromatin were damaged, the cell would
   a. swell up and burst.
   b. run out of energy needed for its activities.
   c. go out of control.
   d. not be able to absorb light.
   e. divide immediately.

4. A researcher made an interesting observation about a protein made by the rough ER and eventually used to build a cell's plasma membrane. The protein in the membrane was actually slightly different from the protein made in the ER. The protein was probably changed in the
   a. Golgi apparatus.
   b. smooth ER.
   c. mitochondrion.
   d. nucleus.
   e. chloroplast.

5. If the nucleus is a cell's "control center," and chloroplasts its "solar collectors," which of the following might be called the cell's combination "food processor" and "garbage disposer"?
   a. lysosome
   b. Golgi apparatus

c. flagellum
d. ribosome
e. nucleolus

6. When elongated, tube-shaped cells from the lining of the intestine are treated with a certain chemical, the cells sag and become round blobs. The internal structures disrupted by this chemical are probably
   a. cell junctions.
   b. microtubules.
   c. smooth and rough ER.
   d. mitochondria.
   e. microfilaments.

7. The electron microscope has been particularly useful in studying prokaryotes, because
   a. electrons can penetrate tough prokaryotic cell walls.
   b. prokaryotes are so small.
   c. prokaryotes move so quickly they are hard to photograph.
   d. they aren't really alive, so it doesn't hurt to "kill" them for viewing.
   e. their organelles are small and tightly packed together.

8. A cell possesses ribosomes, a plasma membrane, a cell wall, and other parts. It could *not* be
   a. a bacterium.
   b. a cell from a fungus.
   c. a cell from a mouse.
   d. an oak tree cell.
   e. a bacterium or a plant cell.

9. A mutant plant cell unable to manufacture cellulose would be unable to
   a. build a cell wall.
   b. divide.
   c. capture sunlight.
   d. move.
   e. store food.

10. A plant cell was grown in a test tube containing radioactive nucleotides, the parts from which DNA is built. Later examination of the cell showed the radioactivity to be concentrated in the
    a. rough ER.
    b. Golgi apparatus.
    c. smooth ER.
    d. central vacuole.
    e. nucleus.

Essay

1. Explain whether you would use a light microscope, a transmission electron microscope (TEM), or a scanning electron microscope (SEM) to perform each of the following tasks, and explain why: examining fine structural details within cell organelles, observing how a cell changes shape as it moves, studying tiny bumps on the cell surface, filming changes in the shape of the nucleus as a cell prepares to divide.

2. Imagine a cell shaped like a cube, 5 μm on each side. (Cells are not perfect cubes, but this assumption simplifies the question.) What is the surface area of the cell, in $μm^2$? What is its volume, in $μm^3$? What is the ratio of surface area to volume for this cell? (Sketches may help.) Now imagine a second cell, this one 10 μm on each side. What are its surface area, volume, and surface-to-volume ratio? Compare the surface-to-volume ratios of the two cells. How is this comparison significant to the functioning of cells? How could the surface-to-volume ratio of the larger cell be increased?

3. An enzyme (a type of protein) called salivary amylase is manufactured in the cells of your salivary glands and secreted as part of saliva. Explain how these parts of the cell cooperate to produce and secrete salivary amylase: transport vesicles, rough ER, plasma membrane, nucleus, Golgi apparatus, ribosomes.

4. A poison that acts specifically on mitochondria was found to interfere with the movement of cilia, slow down protein synthesis, reduce the frequency of cell division, and slow down the manufacture of lipids. Explain how one chemical could affect so many different cell activities.

5. When you work harder, your muscle cells work harder and increase in size. How might various organelles in a muscle cell increase in size, number, or activity to respond to the challenge of an increased workload?

Put Words to Work

Correctly use as many of the following words as possible when reading, talking, and writing about biology:

basal body, cell theory, cell wall, cellular metabolism, central vacuole, chloroplast, chromatin, chromosome, cilia, cytoskeleton, cytoplasm, electron microscope (EM), endomembrane system, endoplasmic reticulum (ER), eukaryotic cell, extracellular matrix, flagellum, glycoprotein, Golgi apparatus, integrins, intermediate filament, light microscope (LM), lysosome, microfilament, micrograph, microtubule, mitochondrion, nucleoid, nucleus, nucleolus, organelle, peroxisome, plasma membrane, plasmodesma, prokaryotic cell, ribosome, rough endoplasmic reticulum, scanning electron microscope (SEM), transmission electron microscope (TEM), transport vesicle, vacuole, vesicle

Use the Web

To best understand and remember how cells are built and how they function, see the activities, questions, and animations at www.mybiology.com.

The Working Cell

Focus on the Concepts

This chapter discusses the roles of membranes, energy transformations, and enzymes in the functioning of cells. As you study the chapter, focus on the following concepts:

- Membranes are fluid mosaics of phospholipids and proteins. The phospholipids form a hydrophobic bilayer that is selectively permeable. Membrane proteins carry out many functions.

- Molecules may diffuse across a membrane in a process called passive transport. Diffusion of water is called osmosis; osmosis and water balance are important to living things.

- Transport proteins may facilitate diffusion of a solute across a membrane or transport a solute against its concentration gradient.

- Chemical reactions may release or store energy. ATP shuttles energy from energy-producing to energy-consuming processes and drives cellular work.

- An enzyme is a protein that speeds up a chemical reaction by lowering activation energy needed to start the reaction. The enzyme's active site fits a specific substrate. Environmental conditions affect the enzyme. Inhibitors may block it and cofactors may assist it.

Review the Concepts

Work through the following exercises to review the concepts in this chapter. For additional review, check out the activities at www.mybiology.com. The website offers a pre-test that will help you plan your studies.

Exercise 1   (Module 5.1)

Review fluid mosaic membrane structure by coloring and labeling the diagram on the next page. (Coloring will help you focus on each specific part.) The drawing is a composite based on the figures in Module 5.1. Label the items in **boldface** type: Start with the **cytoplasm**, **extracellular fluid**, and a **fiber of the extracellular matrix**. In the membrane, color **phospholipids** gray, protein molecules purple, **carbohydrate I.D. tags** on **glycoprotein** and **glycolipid molecules** green, and **cholesterol** molecules yellow. Also show the functions of certain proteins by labeling them **enzyme**, **receptor protein**, and **transport protein**.

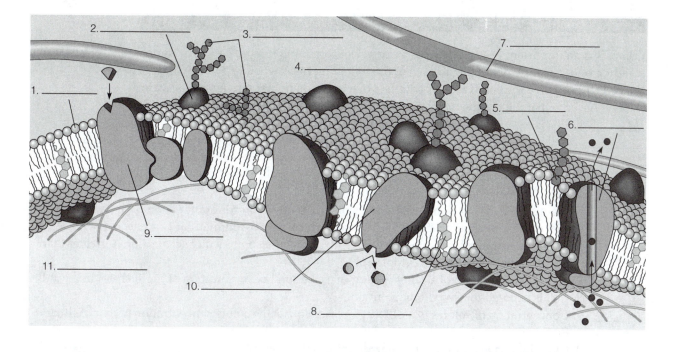

Review diffusion and the function of cell membranes by matching each of the phrases on the right with the appropriate mechanisms from the list on the left. Two questions require more than one answer.

_____ 1. Diffusion across a biological membrane

_____ 2. Moves solutes against concentration gradient

_____ 3. Any spread of molecules from area of higher concentration to area of lower concentration

_____ 4. Diffusion with the help of a transport protein

_____ 5. Three types of endocytosis

_____ 6. Engulfing of fluid in membrane vesicles

_____ 7. Diffusion of water across selectively permeable membrane, from hypotonic to hypertonic solution

_____ 8. Transport molecules need ATP to function

_____ 9. Enables cell to engulf bulk quantities of specific large molecules

_____ 10. How oxygen and carbon dioxide enter and leave cells

_____ 11. Two types of passive transport

_____ 12. Engulfing of particle in membrane vesicle

_____ 13. Fusion of membrane-bound vesicle with membrane, and dumping of contents outside cell

_____ 14. How a cell might capture a bacterium

_____ 15. Helped by aquaporins

A. Diffusion
B. Active transport
C. Osmosis
D. Phagocytosis
E. Passive transport
F. Facilitated diffusion
G. Pinocytosis
H. Receptor-mediated endocytosis
I. Exocytosis

Exercise 3   (Modules 5.4–5.5)

Osmosis is an important process that has many effects on living things. Test your understanding of osmosis by predicting in each of the following cases whether water will enter the cell (In) or leave the cell (Out), or whether there will be no net movement of water (None). Assume that the plasma membrane is permeable to water but not solutes.

_____ 1. Cell is exposed to a hypertonic solution.

_____ 2. Cell is placed in a salt solution whose concentration is greater than that of the cell contents.

_____ 3. Due to disease, the solute concentration of the body fluid outside a cell is less than the solute concentration inside cells.

_____ 4. Cell is immersed in an isotonic solution.

_____ 5. A single-celled organism is placed in a drop of pure water for examination under a microscope.

_____ 6. Cell is immersed in solution of sucrose and glucose whose individual concentrations are less than concentration of solutes in cytoplasm, but whose combined concentration is greater than concentration of solutes in cytoplasm.

_____ 7. Solute concentration of a cell is greater than the solute concentration of the surrounding fluid.

_____ 8. Cell is exposed to a hypotonic solution.

_____ 9. Concentration of solutes in a cell's cytoplasm equals the solute concentration of extracellular fluid.

_____ 10. Cytoplasm is more dilute than surrounding solution.

Exercise 4   (Modules 5.1–5.9)

Try to picture membranes and their functions close up by completing the following story.

Your first mission as a Bionaut requires you to enter a blood vessel and observe the structure and functions of cell membranes. You step into the water-filled chamber of the Microtron, which quickly shrinks you to a size much smaller than a red blood cell.

You tumble through the tunnel-like needle and into a blood vessel in the arm of a volunteer. Huge, rubbery red blood cells slowly glide past. Floating in the clear, yellowish blood plasma, you switch on your headlamp and examine the epithelial cells of the vessel wall. Their plasma membranes seem made of millions of small balloons. These are the hydrophilic "heads" of the [1]_____ molecules that make up most of the membrane surface. Through the transparent surface, you can see their flexible, [2]_____ tails projecting inward toward the interior of the membrane and beyond them an inner layer of [3]_____ molecules with their tails pointing toward you. Here and there are globular [4]_____molecules embedded in the membrane; some rest lightly on the surface, but most project all the way into the interior of the cell. The membrane is indeed a [5]_____ mosaic; the proteins are embedded like the pieces of a picture, but you can see that they are free to move around. You push on one of the proteins, and it bobs like an iceberg. Some of the phospholipids and proteins have chains of sugar molecules attached to them, forming [6]_____ and [7]_____. These are the molecules that act as cell [8]_____ tags. You notice that one of the proteins has a dimple in its surface. Just then a small, round molecule floating in the plasma nestles in the dimple. The molecule is a hormone, a chemical signal, and the dimpled protein is the [9]_____ that enables the cell to respond to it.

In your light beam, you can see the sparkle and shimmer of many molecules, large and small, in the blood and passing through the cell membrane. Oxygen is moving from the plasma, where it is more concentrated, to the cell interior, where it is less concentrated. This movement is [10]_____; when it occurs through a biological membrane, it is called [11]_____ transport. Similarly, carbon dioxide is flowing out of the cell, down its [12]_____ gradient, from the cell interior, where it is [13]_____ concentrated, to the blood, where it is [14]_____ concentrated.

You note that water molecules are passing through the membrane equally in both directions. The total concentration of solutes in the cell and in the blood must be equal; the solutions must be [15]_____. You signal the control team to inject a small amount of concentrated salt solution into the blood, making the blood slightly [16]_____ relative to the cell contents. This causes water to flow [17]_____ the cell, until the two solutions are again in equilibrium. This diffusion of water through a [18]_____ permeable membrane is called [19]_____.

Some sugar molecules floating in the blood are simply too large and polar to pass easily through the plasma membrane. The sugar molecules simply bounce off, unless they happen to pass through pores in special [20]_____ proteins. This is a type of passive transport, because the molecules move down a concentration gradient without the expenditure of [21]_____. Because transport proteins help out, it is called [22]_____ diffusion.

Your chemscanner detects a high concentration of potassium ions inside the cell. Transport proteins here and there in the membrane are able to move potassium into the cell against the concentration gradient. This must be [23]_____ transport; the cell expends [24]_____ to provide energy to "pump" the potassium into the cell.

Suddenly there is a tug at your foot. You look down to see your flipper engulfed by a rippling membrane. A white blood cell the size of a house quickly pins you against the vessel wall. The phospholipids of its membrane are pressed against your face mask. The cell is engulfing you, protecting the body from a foreign invader! Taking in a substance in this way is called [25]_____, more specifically [26]_____, if the substance is a solid particle. Suddenly the pressure diminishes, and you are inside the white blood cell, floating free in a membrane-enclosed bag, or [27]_____. Another sac is approaching; it is a [28]_____, full of digestive enzymes. You manage to get your legs outside of the vacuole and move it back toward the inner surface of the cell membrane. As the vacuole fuses with the plasma membrane, you tear your feet free and swim away from the voracious cell, realizing that [29]_____ expelled you almost as fast as endocytosis trapped you!

You swim to the exit point, and the control team removes you by syringe. You are soon back in the lab, restored to normal size, and telling your colleagues about your close call.

After reading Modules 5.10–5.14, review energy, chemical reactions, and the function of enzymes by filling in the blanks in the following story.

If you were to stop eating, you would probably starve to death in weeks or months. If you were unable to breathe, you would die in minutes. Organisms need the energy that is released when food and oxygen combine. This energy is used not only to move the body but also to keep it from falling apart.

Energy is the ability to perform [1]_____. The sun is the source of the energy that sustains living things. Sunlight is pure [2]_____ energy, energy of movement that is actually doing work. In the process of photosynthesis, plants are able to use the energy of sunlight to produce food molecules. This process obeys the laws of [3]_____, the principles that govern energy transformations. Plants do not make the energy in food. According to the [4]_____ law of thermodynamics, energy can be [5]_____ or transferred, but it cannot be created or destroyed. In photosynthesis, no energy is created. Rather, the plant transforms the energy of sunlight into chemical energy, a form of [6]_____ energy, stored in the chemical bonds of molecules of glucose.

No energy change is 100% efficient, and the changes that occur in photosynthesis are no exception to this rule. Some of the energy of sunlight is not stored in glucose, but rather is converted to [7]_____, which is random molecular motion. This energy is "lost" as far as the plant is concerned, and this random motion contributes to the disorder of the plant's surroundings. The [8]_____ law of thermodynamics says that energy changes are always accompanied by an increase in [9]_____, a measure of disorder. One of the reasons living things need a constant supply of energy is to counter this natural tendency toward disorder.

The products of photosynthesis contain [10]_____ potential energy than the reactants. This means that, overall, photosynthesis is an [11]_____ reaction. Such a reaction consumes energy, which in photosynthesis is supplied by the sun.

Photosynthesis produces food molecules, such as glucose, which store energy. An animal might obtain this food by eating a plant or an animal that has eaten a plant. The food molecules enter the animal's cells, where their potential energy is released in the process of cellular respiration. The products of this chemical reaction (actually a series of reactions) contain less potential energy than the reactants. Therefore, cellular respiration is an [12]_____ process; it [13]_____ energy. In fact, this is the same overall change that occurs when glucose in a piece of wood or paper burns in air. When paper burns, the energy escapes as the heat and light of the flames. In a cell, the reaction occurs in a more controlled way, and some of the energy is captured for use by the cell.

Energy released by the exergonic "burning" of glucose in cellular respiration is used to make a substance called [14]_____. A molecule of [15]_____ and a [16]_____ group are joined to form each molecule of ATP. This is an endergonic reaction, because it takes energy to assemble ATP. The covalent bond connecting the phosphate group to the rest of the ATP molecule is unstable and easily broken. This arrangement of atoms stores [17]_____ energy. The [18]_____ of ATP is an exergonic reaction. When ATP undergoes hydrolysis, a [19]_____

is removed, ATP becomes ^{20}_____, and energy is released. Thus, ATP is a kind of energy "currency" that can be used to perform cellular ^{21}_____.
There are three kinds of cellular work: ^{22}_____, ^{23}_____, and
^{24}_____. Most cellular activities depend on ATP energizing other molecules by transferring its phosphate group to them—a process called
^{25}_____. This happens in mechanical work, when ATP causes molecules in muscle cells to move. It should be noted that energy is not destroyed when ATP is used to do work. When an ATP molecule is hydrolyzed to make muscles move, some of its energy moves the body, and some ends up as random molecular motion, or
^{26}_____. Similarly, ATP is used to move substances through
^{27}_____; this is called transport work.

A less obvious but important function of ATP is supplying the energy for fighting the natural tendency for a system to become disordered. A cell constantly needs to manufacture molecules to replace ones that are used up or damaged. This is chemical work. Building a large molecule from smaller parts is an ^{28}_____ reaction. Energy released by the exergonic hydrolysis of ATP is used to drive essential endergonic reactions. The linking of exergonic and endergonic processes is called energy
^{29}_____, and ATP is the critical connection between the processes that release energy and those that consume it.

What prevents a molecule of ATP from breaking down until its energy is needed? Molecules can break down spontaneously; that is why ATP energy is needed to repair them. Fortunately for living things, it takes some additional energy, called energy of ^{30}_____, to get a chemical reaction started. This creates an energy
^{31}_____ that prevents molecules from breaking down spontaneously. Energy barriers exist for both exergonic and endergonic reactions. Most of the time, most molecules in a cell lack the extra energy needed to clear the barrier, so chemical reactions occur slowly, if at all.

So what enables the vital reactions of metabolism to occur when and where they are needed, at a rate sufficient to sustain life? This is where enzymes come in. An enzyme is a special ^{32}_____ molecule that acts as a biological ^{33}_____.
It ^{34}_____ the rate of a chemical reaction without being ^{35}_____
by it. An enzyme holds reactants in such a way as to ^{36}_____ the energy barrier that prevents them from reacting. Even though reactants would not normally possess the activation energy needed to start the reaction, the enzyme creates conditions that make the reaction possible. Enzymes enable the cell to carry out vital chemical changes when and where they are needed, enabling it to control the many chemical reactions that make up cellular ^{37}_____.

Exercise 6   (Modules 5.10 – 5.14)

Briefly summarize the differences between the words or phrases in each of the following sets.

1.   Kinetic energy and potential energy

2. Exergonic reactions and endergonic reactions

3. Reactants and products

4. ATP and ADP

5. A reaction without an enzyme and a reaction with an enzyme

6. Photosynthesis and cellular respiration

7. First and second laws of thermodynamics

8. Mechanical, transport, and chemical work

Exercise 7   (Modules 5.15 – 5.16)

Review enzyme action by completing the activities below.

1. Complete the diagram below so that it shows the cycle of enzyme activity. Imagine that the reaction carried out by this enzyme is splitting a substrate molecule into two parts. Color the diagram as suggested, and label the items in boldface type. Color the **enzyme** purple. Sketch the **substrate** as a dark green shape. Sketch the **products,** and color them light green. Also label the **active site.**

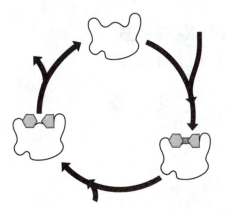

2. Make a sketch showing how heat or change in pH might change the enzyme and alter its ability to catalyze its chemical reaction. Color and label the **enzyme**, its **active site**, and its **substrate**, as above.

3. On the left side of the space below, make a sketch showing how a competitive inhibitor might interfere with activity of the enzyme. Label the **competitive inhibitor**, and color it blue. On the right side, make a sketch showing how a noncompetitive inhibitor might interfere with activity of the enzyme. Label the **noncompetitive inhibitor**, and color it yellow.

Test Your Knowledge

Multiple Choice

1. The movement of molecules from an area of higher concentration to an area of lower concentration is called
   a. diffusion.
   b. endocytosis.
   c. catalysis.
   d. active transport.
   e. osmosis.

2. Which of the following is *not* true of an enzyme? An enzyme
   a. is a protein.
   b. acts as a biological catalyst.
   c. supplies energy to start a chemical reaction.
   d. is specific.
   e. lowers the energy barrier for a chemical reaction.

3. Phospholipid molecules in a membrane are arranged with their____ on the exterior and their____ on the interior.
   a. hydrophobic heads . . . hydrophilic tails
   b. hydrophilic heads . . . hydrophobic tails
   c. nonpolar heads . . . polar tails
   d. hydrophobic tails . . . hydrophilic heads
   e. hydrophilic tails . . . hydrophobic heads

4. In osmosis, water always moves toward the ____ solution, that is, toward the solution with the ____ solute concentration.
   a. isotonic . . . greater
   b. hypertonic . . . greater
   c. hypertonic . . . lesser
   d. hypotonic . . . greater
   e. hypotonic . . . lesser

5. Which of the following enables a cell to pick up and concentrate a specific kind of molecule?
   a. passive transport
   b. diffusion
   c. osmosis
   d. receptor-mediated endocytosis
   e. pinocytosis

6. A cell uses energy released by ____ reactions to drive the ____ reaction that makes ATP. Then it uses the energy released by the hydrolysis of ATP, an ____ reaction, to do various kinds of work in the cell.
   a. exergonic . . . exergonic . . . endergonic
   b. endergonic . . . exergonic . . . endergonic
   c. exergonic . . . endergonic . . . exergonic
   d. endergonic . . . endergonic . . . exergonic
   e. exergonic . . . endergonic . . . endergonic

7. Energy of activation
   a. is released when a large molecule breaks up.
   b. gets a reaction going.
   c. is released by an exergonic reaction.
   d. is stored in an endergonic reaction.
   e. is supplied by an enzyme.

8. The laws of thermodynamics state that whenever energy changes occur, ____ always increases.
   a. disorder
   b. order
   c. kinetic energy
   d. potential energy
   e. energy of activation

9. Living things transform kinetic energy into potential chemical energy in the ____ , when ____ is made.
   a. mitochondrion . . . ADP
   b. chloroplast . . . ADP
   c. chloroplast . . . an enzyme
   d. mitochondrion . . . glucose
   e. chloroplast . . . glucose

10. Why does heating interfere with the activity of an enzyme?
    a. It kills the enzyme.
    b. It changes the enzyme's shape.
    c. It increases the energy of substrate molecules.
    d. It causes the enzyme to break up.
    e. It kills the cell, so enzymes can't work.

11. An enzyme is specific. This means
    a. it has a certain amino acid sequence.
    b. it is found only in a certain place.
    c. it functions only under certain environmental conditions.
    d. it speeds up a particular chemical reaction.
    e. it occurs in only one type of cell.

12. Diffusion of water across a selectively permeable membrane is called
    a. active transport.
    b. osmosis.
    c. exocytosis.
    d. passive transport.
    e. facilitated diffusion.

Essay

1. Describe the kinds of molecules that cannot easily diffuse through cell membranes. How do proteins facilitate diffusion of these substances?

2. Make a sketch showing why an enzyme acts only on a specific substrate.

3. Most enzyme-catalyzed chemical reactions in humans occur most readily around body temperature, 37°C. Why do these reactions slow down at lower temperatures? Why do they slow down at higher temperatures?

4. Which contains more potential energy, a large, complex molecule like a protein, or the smaller amino acid subunits of which it is composed? Is the joining of amino acids to form a protein an exergonic or endergonic reaction? Why must this be the case? Where does the cell obtain energy to carry out such reactions?

5. Describe the circumstances under which plant and animal cells gain and lose water by osmosis. Which of the following is the least serious problem: water gain by a plant cell, water loss by a plant cell, water gain by an animal cell, or water loss by an animal cell? Why?

Apply the Concepts

Multiple Choice

1. If a cell is like a factory, then enzymes are like
   a. the plans for the factory.
   b. the machines in the factory.
   c. the power plant for the factory.
   d. the raw materials used by the factory.
   e. the walls of the factory.

2. A molecule that has the same shape as the substrate of an enzyme would tend to
   a. speed metabolism by guiding the enzyme to its substrate.
   b. speed metabolism by acting as a cofactor for the enzyme.
   c. speed metabolism because it would also be a catalyst.
   d. save the cell energy by substituting for the substrate.
   e. slow metabolism by blocking the enzyme's active site.

3. A plant cell is placed in a solution whose solute concentration is twice as great as the concentration of the cell cytoplasm. The cell membrane is selectively permeable, allowing water but not the solutes to pass through. What will happen to the cell?
   a. No change will occur because it is a plant cell.
   b. The cell will shrivel because of osmosis.
   c. The cell will swell because of osmosis.
   d. The cell will shrivel because of active transport of water.
   e. The cell will swell because of active transport of water.

4. A white blood cell is capable of producing and releasing thousands of antibody molecules every second. Antibodies are large, complex protein molecules. How would you expect them to leave the cell?
   a. active transport
   b. exocytosis
   c. receptor-mediated endocytosis
   d. passive transport
   e. pinocytosis

5. Which of the following would be least likely to diffuse through a cell membrane without the help of a transport protein?
   a. a large polar molecule
   b. a large nonpolar molecule
   c. a small polar molecule
   d. a small nonpolar molecule
   e. Any of the above would easily diffuse through the membrane.

6. Red blood cells shrivel when placed in a 10% sucrose solution. When first placed in the solution, the solute concentration of the cells is ____ the concentration of the sucrose solution. After the cells shrivel, their solute concentration is ____ the concentration of the sucrose solution.
   a. less than . . . greater than
   b. greater than . . . less than
   c. equal to . . . equal to
   d. less than . . . equal to
   e. greater than . . . equal to

7. A nursing infant is able to obtain disease-fighting antibodies, which are large protein molecules, from its mother's milk. These molecules probably enter the cells lining the baby's digestive tract via
   a. osmosis.
   b. passive transport.
   c. exocytosis.
   d. active transport.
   e. endocytosis.

8. Some enzymes involved in the hydrolysis of ATP cannot function without the help of sodium ions. Sodium in this case functions as
   a. a substrate.
   b. a cofactor.
   c. an active site.
   d. a noncompetitive inhibitor.
   e. a vitamin.

9. The relationship between an enzyme's active site and its substrate is most like which of the following?
   a. a battery and a flashlight
   b. a car and a driver
   c. a key and a lock
   d. a glove and a hand
   e. a hammer and a nail

10. In which of the following do both examples illustrate kinetic energy?
   a. positions of electrons in an atom—a ball rolling down a hill
   b. heat—arrangement of atoms in a molecule
   c. a rock resting on the edge of a cliff—heat
   d. a ball rolling down a hill—heat
   e. light—arrangement of atoms in a molecule

11. Which of the following is a difference between active transport (AT) and facilitated diffusion (FD)?
   a. AT involves transport proteins, and FD does not.
   b. FD can move solutes against a concentration gradient, and AT cannot.
   c. FD requires energy from ATP, and AT does not.
   d. FD involves transport proteins, and AT does not.
   e. AT requires energy from ATP, and FD does not.

12. An enzyme and a membrane receptor molecule are similar in that they
    a. are always attached to membranes.
    b. act as catalysts.
    c. require ATP to function.
    d. supply energy for the cell.
    e. bind to molecules of a particular shape.

13. Zoologists discovered that the blood cells of a certain African lungfish were much slower to swell or shrink with water when faced with changes in blood solute concentration, a useful adaptation to drought and dehydration. The researchers suspected that this might have something to do with the number of _____ in the blood cells
    a. phospholipids
    b. aquaporins
    c. ATPs
    d. competitive inhibitors
    e. enzymes

Essay

1. The burning of glucose molecules in paper is an exergonic reaction, which releases heat and light. If this reaction is exergonic, why doesn't the book in your hands spontaneously burst into flame? You could start the reaction if you touched this page with a burning match. What is the role of the energy supplied by the match?

2. Seawater is hypertonic in comparison to body tissues. Explain what would happen to his stomach cells if a shipwrecked sailor drank seawater.

3. The laws of thermodynamics have imaginatively been described as the house rules of a cosmic energy card game: "You can't win, you can't break even, and (if you want to stay alive) you can't get out of the game." State the law that says living things can't win the energy game. State the law that says they can't break even.

4. A farm worker accidentally was splashed with a powerful insecticide. A few minutes later he went into convulsions, stopped breathing, and died. The insecticide acted as a competitive inhibitor of an enzyme important in the function of the nervous system. Describe the structural relationship between the enzyme, its substrate, and the insecticide.

5. Lecithin is a substance used in foods such as mayonnaise as an emulsifier, which means that it helps oil and water mix. Lecithin is a phospholipid; a lecithin molecule has a polar (hydrophilic) "head" and a nonpolar (hydrophobic) "tail." How might the structure of lecithin allow water to surround fat droplets? Sketch a microscopic view of some fat droplets in mayonnaise, and show how you think the fat, surrounding water, and lecithin molecules might be arranged.

Put Words to Work

Correctly use as many of the following words as possible when reading, talking, and writing about biology:

active site, active transport, aquaporin, cellular respiration, chemical energy, competitive inhibitor, concentration gradient, diffusion, endergonic reaction, endocytosis, energy, energy coupling, energy of activation (E_A), entropy, enzyme, exergonic reaction, exocytosis, facilitated diffusion, feedback inhibition, first law of thermodynamics, fluid mosaic, hypertonic solution, hypotonic solution, isotonic solution, kinetic energy, metabolic pathway, metabolism, noncompetitive inhibitor, osmosis, osmoregulation, passive transport, phagocytosis, phosphorylation, pinocytosis, potential energy, second law of thermodynamics, selective permeability, substrate

Use the Web

There is more on membranes, energy, and enzymes at www.mybiology.com.

Exercise 3   (Module 6.6)

Figure 6.6 introduces the three stages of cellular respiration. After studying it, see if you can label the diagram below without referring to the text. Include **oxidative phosphorylation, pyruvate, mitochondrion, CO_2, electrons carried by NADH, citric acid cycle, glycolysis, cytoplasm, ATP, glucose,** and **electrons carried by NADH and $FADH_2$.** (Note: 3, 6, and 7 are processes, 8 and 11 are places, and the rest are inputs and outputs.)

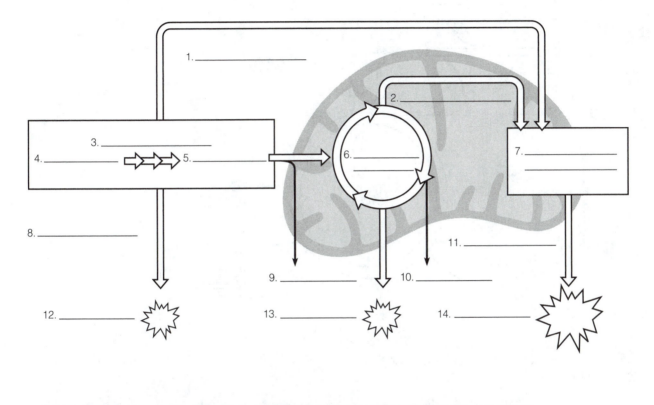

Exercise 4   (Module 6.7)

Glycolysis is the first of three steps in cellular respiration. Review glycolysis by matching each phrase on the left with a term on the right. Some terms are used twice.

_____ 1. Compound formed as glucose is changed to pyruvic acid
_____ 2. Steps in glycolysis that produce ATP and pyruvate
_____ 3. Fuel molecule broken down in glycolysis
_____ 4. Produced by substrate-level phosphorylation
_____ 5. Invested to energize glucose molecule at start of process
_____ 6. Substance that is reduced as glucose is oxidized
_____ 7. Glucose is converted to two molecules of this
_____ 8. Steps in glycolysis that consume energy
_____ 9. "Splitting of sugar"
_____ 10. Carries hydrogen and electrons from oxidation of glucose
_____ 11. When an enzyme transfers a phosphate from a substrate to ADP
_____ 12. Assembled to make ATP
_____ 13. Not involved in glycolysis

A. NADH
B. Pyruvate
C. ATP
D. NAD1
E. Glucose
F. Glycolysis
G. ADP and P
H. Oxygen
I. Intermediate
J. Preparatory phase
K. Energy payoff phase
L. Substrate-level
    phosphorylation

Exercise 5   (Modules 6.8–6.9)

Pyruvate from glycolysis is chemically altered and then enters the citric acid cycle, a series of steps that completes the oxidation of glucose. The energy of pyruvate is stored in NADH and $FADH_2$. To review these processes, fill in the blanks in the diagram below. (Try to do as many as you can without referring to the text.) Include the following: **NAD^+, pyruvate, CO_2, $FADH_2$, NADH, coenzyme A, ATP,** and **acetyl CoA.**

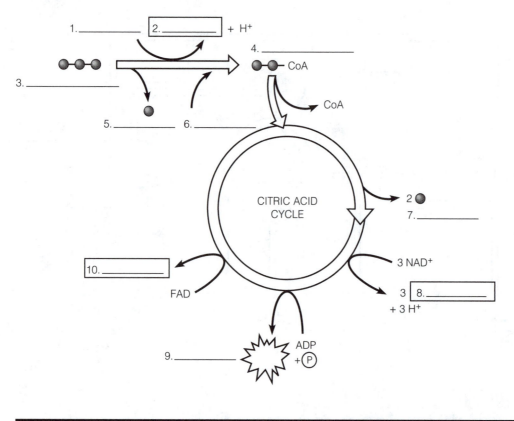

Exercise 6   (Module 6.10)

Circle the correct words or phrases in parentheses to complete each sentence.

The [1](*final, second*) stage of cellular respiration is the electron transport chain and synthesis of [2](*glucose, ATP*) by a process called [3](*oxidative phosphorylation, active transport*). The electron transport chain is a sequence of [4](*electron, proton*) carriers built into the [5](*outer, inner*) membrane of the mitochondrion. Molecules of $FADH_2$ and [6](*ADP, NADH*) bring high-energy electrons to the chain from glycolysis and [7](*the citric acid cycle, oxidative phosphorylation*). The electrons move along the chain from carrier to carrier in a series of redox reactions, finally joining with [8](H_2O, CO_2, O_2) and H1 from the surrounding solution to form [9](H_2O, CO_2, O_2). Energy released by the electrons is used to move protons— [10](*H1 ions, ADP molecules*)—by [11](*active transport, passive transport*) into the space between the inner and outer mitochondrial membranes.

The buildup of protons in the intermembrane space—a proton gradient—constitutes [12](*kinetic, potential*) energy that the cell can tap to make [13](*ATP, glucose*). The concentration of protons tends to drive them back through the membrane into the [14](*inner compartment of the mitochondrion, cytoplasm of the cell*), but protons can cross the membrane only by passing through special protein complexes, called [15](*coenzyme As, ATP synthases*).

As each of these complexes allows protons back through the membrane, a component of the complex rotates like a turbine, causing catalytic sites to phosphorylate [16](*NAD, ADP*) and make [17](*NADH, ATP*). Thus, oxidative phosphorylation transforms energy extracted from glucose into the phosphate bonds of ATP.

Exercise 7   (Modules 6.10–6.11)

These diagrams will help you review oxidative phosphorylation (electron transport and chemiosmosis) and how poisons disrupt them. In the first diagram, show how the processes work normally. Trace movement of an electron with an orange arrow, movement of H1 ions (active transport and chemiosmosis) with green arrows, and formation of ATP with a black arrow.

In the second diagram, draw arrows showing the movement of electrons and H1 and the formation of ATP, as in the first diagram. Then draw a red line where each poison acts to show how each of the poisons short-circuits the normal processes. Label the poisons **rotenone, cyanide, carbon monoxide, DNP,** and **oligomycin.**

1. Normal electron transport and chemiosmosis:

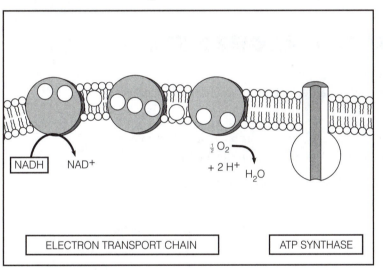

2. Effects of poisons:

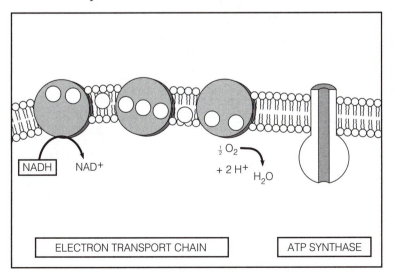

Check your overall understanding of cellular respiration by matching each of the phrases below with one of the three stages of the process. Use G for glycolysis, CA for the citric acid cycle, and OP for oxidative phosphorylation.

_____ 1. Generates most of the ATP formed by cellular respiration
_____ 2. Begins the oxidation of glucose
_____ 3. Occurs outside the mitochondrion
_____ 4. Produces four ATPs per glucose by substrate-level phosphorylation, but two ATPs per glucose are used to get it started
_____ 5. Oxidizes NADH and $FADH_2$, producing NAD1 and FAD
_____ 6. Carried out by enzymes in the matrix (fluid) of the mitochondrion
_____ 7. Here electrons and hydrogen combine with O_2 to form H_2O
_____ 8. Occurs along the inner mitochondrial membrane
_____ 9. Generates most of the CO_2 produced by cellular respiration
_____10. $FADH_2$ and NADH deliver high-energy electrons to this stage
_____11. ATP synthase makes ATP
_____12. Reduces NAD1 and FAD, producing NADH and $FADH_2$

Review fermentation by filling in the blanks below.

1. _____ anaerobes can make their ATP by fermentation or oxidative phosphorylation.

2. _____ are organisms that normally uses aerobic respiration to produce ATP, but can generate ATP without oxygen, via alcoholic fermentation.

3. Fermenters replenish their supply of NAD1 by using NADH to oxidize _____ .

4. When oxygen is scarce, human _____ cells can make ATP by lactic acid fermentation.

5. Fermentation enables cells to make ATP in the absence of _____ .

6. For every molecule of glucose consumed, glycolysis produces two molecules of pyruvic acid, two molecules of ATP, and two molecules of _____ .

7. The waste products of alcoholic fermentation are _____ and carbon dioxide.

8. _____ acid fermentation is used to make cheese and yogurt.

9. Fermentation generates two energy-rich _____ molecules for every molecule of glucose consumed.

10. A cell can use _____ to generate a small amount of ATP, but it must somehow recycle its supply of NAD1.

11. Like aerobic respiration, alcoholic fermentation produces _____ gas as a waste product.

12. Strict _____ require anaerobic conditions and are poisoned by oxygen.

13. Fermentation provides an _____ pathway that allows NADH to get rid of electrons and recycle as NAD1.

14. The ATP yield of fermentation is much _____ than that of aerobic respiration.

15. _____ is the one energy-harvesting process common to all life; early prokaryotes probably used it long before Earth's atmosphere contained oxygen.

16. The buildup of _____ from strenuous exercise can cause muscle fatigue and soreness.

Exercise 10   (Module 6.15)

Review the molecules that can be used as fuel for cellular respiration by writing their names in the blanks in this diagram. Include **glucose, amino acids, fats, fatty acids, proteins, sugars, carbohydrates,** and **glycerol.**

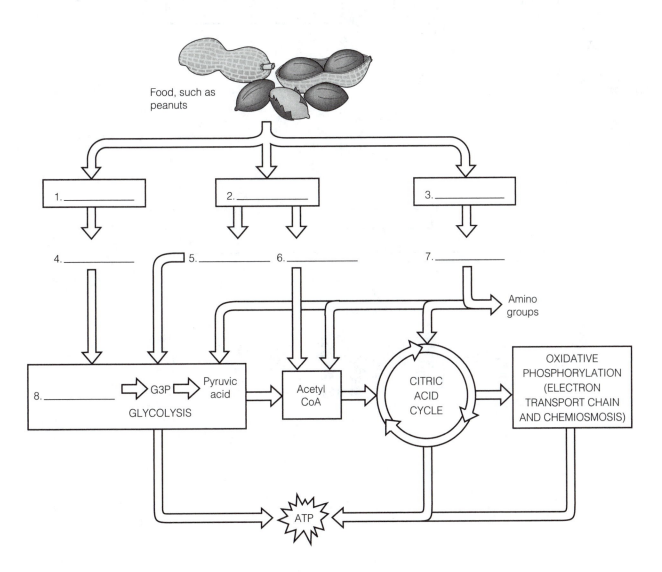

Show how a cell obtains organic molecules for biosynthesis of proteins, polysaccharides, and fats by drawing the missing arrows on this diagram.

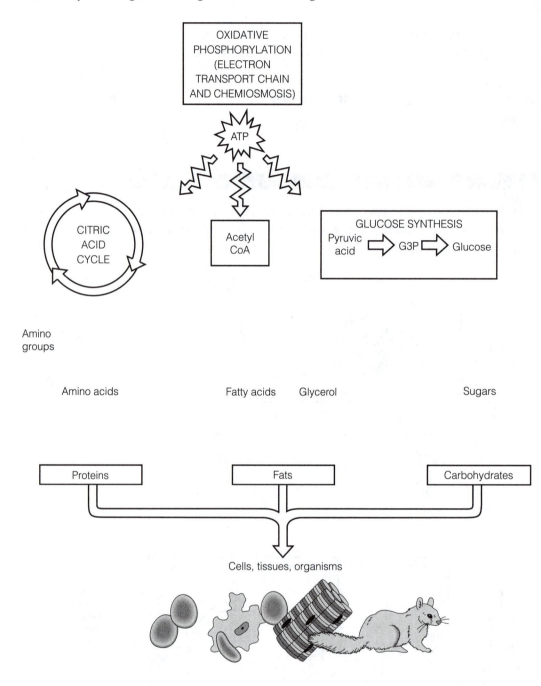

Test Your Knowledge

Multiple Choice

1. The main function of cellular respiration is
   a. breaking down toxic molecules.
   b. making ATP that powers cell activities.
   c. making food.
   d. producing chemical "building blocks" for cell structures.
   e. breaking down ATP, so that ADP and P can be reused.

2. The ultimate source of the energy powers our cells is
   a. glucose
   b. fermentation
   c. oxygen
   d. biosynthesis
   e. the sun

3. In cellular respiration, _____ is oxidized and _____ is reduced.
   a. O_2 . . . ATP
   b. ATP . . . O_2
   c. glucose . . . O_2
   d. CO_2 . . . H_2O
   e. glucose . . . ATP

4. Most of the ATP produced in cellular respiration is produced by
   a. glycolysis.
   b. oxidative phosphorylation.
   c. lactic acid fermentation.
   d. biosynthesis.
   e. the citric acid cycle.

5. _____ is consumed and _____ is produced in the overall process of cellular respiration.
   a. CO_2 . . . H_2O
   b. O_2 . . . glucose
   c. H_2O . . . ATP
   d. glucose . . . CO_2
   e. ATP . . . O_2

6. The energy given up by electrons as they move through the electron transport chain is used to
   a. break down glucose.
   b. make NADH and $FADH_2$.
   c. pump H1 through a membrane.
   d. oxidize water.
   e. manufacture glucose.

7. Fermentation is essentially glycolysis plus an extra step in which pyruvate is reduced to form lactic acid or alcohol and CO_2. This last step
   a. removes poisonous oxygen from the environment.
   b. extracts a bit more energy from glucose.
   c. enables the cell to recycle NAD1.
   d. inactivates toxic pyruvic acid.
   e. enables the cell to make pyruvic acid into substances it can use.

8. A small amount of ATP is made in glycolysis and the citric acid cycle
   a. by transfer of a phosphate group from a fragment of glucose to ADP.
   b. using energy from the sun to perform the process of photosynthesis.
   c. by transport of electrons through a series of carriers.
   d. when electrons and hydrogen atoms are transferred to NAD1.
   e. as a product of chemiosmosis.

9. The ATP synthase in a human cell gets energy for making ATP directly from
   a. sunlight.
   b. flow of H1 down a concentration gradient.
   c. oxidation of glucose.
   d. movement of electrons through a series of carriers.
   e. reduction of oxygen.

10. Which of the following describes glycolysis?
    a. It begins the oxidation of glucose.
    b. It produces a small amount of ATP.
    c. It generates NADH.
    d. It splits glucose to form two molecules of pyruvate.
    e. All of the above.

11. Most of the NADH that delivers high-energy electrons to the electron transport chain comes from
    a. chemiosmosis.
    b. the cytoplasm.
    c. glycolysis.
    d. biosynthesis.
    e. the citric acid cycle.

12. When protein molecules are used as fuel for cellular respiration, _____ are produced as waste.
    a. amino groups
    b. fatty acids
    c. sugar molecules
    d. molecules of lactic acid
    e. ethanol and CO_2

Essay

1. Describe the relationship between breathing and cellular respiration.

2. Compare the advantages and disadvantages of aerobic cellular respiration with the advantages and disadvantages of fermentation as methods of making ATP for cellular activities.

3. Compare the two mechanisms that generate ATP in cellular respiration—oxidative phosphorylation and substrate-level phosphorylation. In what stage(s) of cellular respiration does each occur? Where does each get the energy for making ATP? Which produces the most ATP under aerobic conditions? Under anaerobic conditions?

4. Describe three ways in which poisons can interfere with cellular respiration.

5. Explain the roles of glycolysis and the Krebs cycle in the biosynthesis of organic molecules.

Apply the Concepts

Multiple Choice

1. Which of the following illustrates oxidation?
   a. $CO_2 + H_2O \rightarrow C_6H_{12}O_6$
   b. $C_6H_{12}O_6 \rightarrow CO_2 + H_2O$
   c. $ADP + P \rightarrow ATP$
   d. $ATP \rightarrow ADP + P$
   e. $H_2O \rightarrow O_2 + H$

2. In an experiment, mice were fed glucose ($C_6H_{12}O_6$) containing a small amount of radioactive oxygen. The mice were closely monitored, and in a few minutes radioactive oxygen atoms showed up in
   a. CO_2.
   b. NADH.
   c. H_2O.
   d. ATP.
   e. O_2.

3. In a second experiment, mice were allowed to breathe oxygen gas (O_2) laced with radioactive oxygen. In this experiment, the radioactive oxygen atoms quickly showed up in
   a. CO_2.
   b. NADH.
   c. H_2O.
   d. ATP.
   e. glucose, $C_6H_{12}O_6$.

4. A chemist has discovered a drug that blocks glucose phosphate isomerase, an enzyme that catalyzes the second reaction in glycolysis. He thought he could use the drug to kill bacteria in people with infections. But he can't do this because
   a. bacteria are facultative anaerobes; they usually don't need to do glycolysis.
   b. glycolysis produces so little ATP that the drug will have little effect.
   c. human cells also do glycolysis; the drug might also poison them.
   d. bacteria do not perform glycolysis.
   e. glycolysis can occur without the action of enzymes.

5. A glucose molecule is completely broken down in glycolysis and the citric acid cycle, but these two processes yield only a few ATPs. Where is the rest of the energy the cell obtains from the glucose molecule?
   a. in FAD and NAD1
   b. in the oxygen used in the electron transport chain
   c. lost as heat
   d. in NADH and $FADH_2$
   e. in the CO_2 molecules released by the processes

6. Which of the following contains energy that a cell could use to make ATP?
   a. O_2
   b. CO_2
   c. NAD1
   d. NADH
   e. H_2O

7. NADH is sometimes used by the cell to assist in the biosynthesis of needed organic molecules. Based on what you know about NADH, which of the following might be its function in biosynthesis?
   a. oxidizing organic molecules
   b. aiding in direct phosphorylation
   c. reducing organic molecules
   d. producing NAD1
   e. breaking down ATP

8. Gram for gram, sugars are not as good as fats as a source of energy for cellular respiration, because sugars
   a. produce toxic amino groups when broken down.
   b. contain more hydrogen.
   c. usually bypass glycolysis and the Krebs cycle.
   d. contain fewer hydrogen atoms and electrons.
   e. are not as easily reduced.

9. A microbiologist discovered a new antibiotic that slowed the growth of bacteria by interfering with cellular respiration. She found that bacteria treated with the antibiotic produced about 15 ATP molecules for every glucose molecule they consumed. Which of the following hypotheses could explain the antibiotic's effect? The treated bacteria
   a. cannot perform glycolysis.
   b. have partially crippled electron transport chains.
   c. cannot produce NADH.
   d. have to rely at least partially on biosynthesis for their ATP.
   e. are forced to rely on fermentation for ATP.

10. A drug that blocks dehydrogenase enzymes would cause cells to
    a. run out of NADH.
    b. suffocate.
    c. deplete their supply of ADP.
    d. rely totally on electron transport and chemiosmosis for energy.
    e. be poisoned by lactic acid.

11. Unlike turkey breast, the breast of a duck is "dark meat." Why?
    a. Ducks fly longer distances, so their breast muscles consist of fast fibers.
    b. Ducks fly faster than turkeys, so duck breast muscles consist of fast fibers.
    c. The fibers of duck breast muscles contain less myoglobin.
    d. Ducks fly longer distances, so their breast muscles contain more slow fibers.
    e. Duck muscle fibers are specialized for anaerobic ATP production.

12. A novel scheme stores energy by moving water between two reservoirs. When electricity is abundant, the excess energy is used to pump water into an upper reservoir. During times of high power demand, the water is allowed to flow through a dam back to the lower reservoir, generating electricity. Compare this system to oxidative phosphorylation. The role of the water flowing between the two reservoirs is similar to the role of _____ in oxidative phosphorylation.
    a. hydrogen ions (H+)
    b. oxygen (O_2)
    c. electrons
    d. NAD1 and FAD
    e. glucose

13. A Burger King double Whopper with cheese contains 1,070 Calories (1,070 kilocalories). This is about _____ of an average person's daily energy needs.
    a. 100%
    b. 90%
    c. 50%
    d. 40%
    e. 10%

Essay

1. Fermentation is a much less efficient way to make ATP than aerobic cellular respiration via oxidative phosphorylation. This being the case, why do you think the fermenters have not been driven to extinction by competition with aerobes?

2. Without oxygen, cellular respiration grinds to a standstill, although glycolysis can continue to make some ATP anaerobically for a short time. When oxygen runs out, why does electron transport stop? Why do you think the citric acid cycle stops?

3. FAD and NAD1 are made from the B vitamins riboflavin and niacin. Why do you think these substances are required in such tiny amounts in your diet? How would a deficiency in one of these vitamins interfere with cell function?

4. After a biochemical analysis of a murder victim's tissues, forensics expert Ben Zyme announced his findings: "Contrary to the conclusions of the police at the scene, the victim did not suffocate. The electron carriers in his mitochondria were all in the oxidized state. We will need to perform a second autopsy to determine the actual cause of death." Explain how the data led Zyme to his conclusion.

5. A microbiologist poured a test tube full of yeast into a flask of sugar water and periodically took samples from the flask. At first, the amount of sugar in the flask decreased gradually. Then there was a sharp drop in sugar, accompanied by the appearance of ethanol in the flask. Explain these results.

Put Words to Work

Correctly use as many of the following words as possible when reading, talking, and writing about biology:

acetyl CoA (acetyl coenzyme A), alcohol fermentation, ATP synthase, cellular respiration, chemiosmosis, citric acid cycle, dehydrogenase, electron transport chain, facultative anaerobe, FAD, glycolysis, intermediate, kilocalorie (kcal), lactic acid fermentation, NAD1, obligate anaerobe, oxidation, oxidative phosphorylation, redox reaction, reduction, substrate-level phosphorylation

Use the Web

Words and pictures in a book fail to do justice to a process as elegant (and complicated!) as cellular respiration. To get a better idea of what happens in cellular respiration, see the activities and animations at www.mybiology.com.

Photosynthesis: Using Light to Make Food   7

Focus on the Concepts

Photosynthesis is the process by which living things use light energy to make sugar from carbon dioxide and water. It is arguably the most important chemical process on Earth. As you study this chapter, focus on these major concepts:

- Plants and other photoautotrophs (such as algae and bacteria) carry out photosynthesis and are the source of organic molecules for most life on Earth. Overall, photosynthesis is a redox process; it reduces carbon dioxide to produce sugar, and oxidizes water to release oxygen gas.

- In plant cells, photosynthesis occurs in organelles called chloroplasts. The process occurs in two steps—the light reactions, which occur in the membranous thylakoids, and the Calvin cycle, which takes place in the fluid of the stroma.

- In the light reactions, light excites electrons of chlorophyll molecules. The electrons pass though carriers, and are picked up by $NADP^+$, forming NADPH. Electrons are replaced by splitting water, leaving O_2 gas. As electrons pass though carriers, their energy is used to make ATP. Thus the light reactions consume light and water, plus ADP + P and $NADP^+$, to store energy in NADPH and ATP. An important by-product is O_2.

- In the Calvin cycle, ATP and high-energy electrons from the light reactions drive the synthesis of sugar. CO_2 is linked to a larger molecule in a process called carbon fixation. In a repeating series of steps, this molecule breaks up to form smaller sugars, with ATP providing energy and NADPH supplying reducing power. Altogether, three molecules of CO_2 are consumed to make a molecule of G3P, which is turned into glucose. Overall, the Calvin cycle uses ATP and NADPH from the light reactions plus CO_2 from the air. It stores energy in G3P and regenerates $NADP^+$ and ADP + P.

- Under hot, dry conditions, stomata in leaves close to save water. This prevents CO_2 from entering leaves, and as CO_2 drops, the Calvin cycle can be disrupted. C_4 plants (corn, sugarcane) capture CO_2 in special cells, and CAM plants (pineapples, cacti) store CO_2 at night, so there is sufficient CO_2 available to the Calvin cycle even when it is hot and dry.

- Carbon dioxide is a greenhouse gas, trapping heat in the atmosphere and moderating Earth's climate. But human consumption of fossil fuels has increased atmospheric CO_2 and caused global warming, which may have dramatic effects on climate and life. Because plants consume billions of tons of CO_2 each year, photosynthesis could mitigate the CO_2 increase, but deforestation and rising fossil fuel consumption contribute to warming.

Review the Concepts

Work through the following exercises to review the concepts in this chapter. For additional review, check out the activities at www.mybiology.com. The animations are particularly helpful. The website offers a pre-test that will help you plan your studies.

Exercise 1   (Module 7.1)

Autotrophs are able to produce their own organic molecules. All organisms that use light energy to make food are called photoautotrophs. Circle all the following organisms that are photoautotrophs: mushroom, pine tree, squirrel, green bacterium, rosebush, seaweed, moss, bread mold, parasitic bacterium, alga, sponge, grass

---

Exercise 2   (Modules 7.1–7.7)

Review some of the basic terminology of photosynthesis by completing this crossword puzzle.

Across

5. ____ is oxidized in the process of photosynthesis.
6. The ____ is the cell organelle where photosynthesis takes place.
10. ____ are the light-catching membranes in a chloroplast.
11. Stacks of thylakoids in a chloroplast are called ____.
13. ____ energy travels through space as rhythmic waves.
14. When chlorophyll absorbs a photon, an ____ becomes excited.
16. Sugar is actually made in the ___ cycle.
18. The Calvin cycle occurs in the ____
—the fluid of the chloroplast.
19. ____ is the green pigment in a leaf.
20. Carbon dioxide enters a leaf via pores called____.
22. Shorter wavelengths of light have____ energy than longer wavelengths.

Down

1. A ____ is a cluster of light-harvesting complexes in a thylakoid.
2. A ____ is a fixed quantity of light energy.
3. ____ is the process by which plants make food from carbon dioxide and water.
4. The color of light is related to its ____.
7. ____ is the source of energy for photosynthesis.
8. Carbon ____ is the incorporation of carbon dioxide into organic compounds.
9. The reaction ____ complex is the part of a photosystem that donates excited electrons.
12. ____ are yellow-orange pigments in a chloroplast.
15. ____ is the green tissue in the interior of a leaf.
17. An ____ is an organism that makes its own food from inorganic molecules.
21. Photosynthesis occurs in ____ main stages.

Exercise 3 (Modules 7.3–7.4)

Write the overall equation for photosynthesis in the boxes below. Show the substances used on the left, and those produced on the right. Use different colors for carbon, hydrogen, oxygen in carbon dioxide, and oxygen in water, and then use your color code to show where atoms of C, H, and O on the left end up in the products on the right. On the lines under the substances used, state which is oxidized and which is reduced.

[_____] + [_____] $\xrightarrow{\text{LIGHT}}$ [_____] + [_____] + [_____]

_____   _____

Exercise 4   (Modules 7.3–7.5)

Label this diagram summarizing the two stages of photosynthesis. Include **outer membrane of chloroplast, thylakoids, granum, stroma, light reactions, Calvin cycle, light,** H_2O, O_2, **electrons, NADPH, ATP,** CO_2, **sugar, ADP + P,** and **$NADP^+$.** (Note: 5 and 7 are processes, 3, 9, 10, and 11 are places or structures, and the rest are inputs and outputs.)

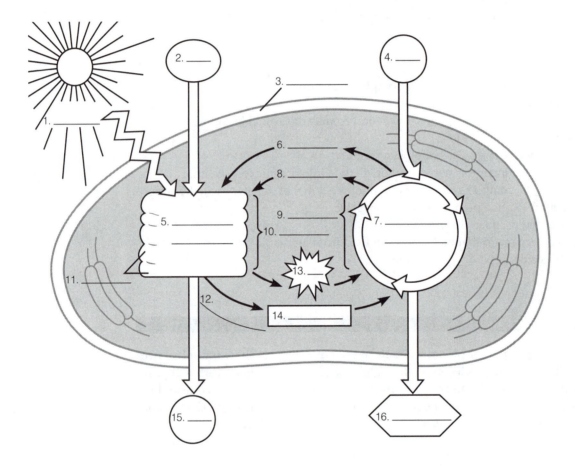

Exercise 5   (Modules 7.3–7.5)

Refer to the equations and diagrams in the modules to match each of the phrases on the right with one of the ingredients or products of photosynthesis listed on the left.

_____ 1. Oxidized in the light reactions
_____ 2. Reduced in the Calvin cycle
_____ 3. Carries H and electrons from the light reactions to the Calvin cycle
_____ 4. Food produced by photosynthesis
_____ 5. Source of H and electrons that end up in glucose
_____ 6. Source of O atoms that end up in glucose
_____ 7. Where O atoms from water end up
_____ 8. Oxidized in the Calvin cycle
_____ 9. Reduced in the light reactions
_____ 10. Supplies energy to the Calvin cycle
_____ 11. Where C and O atoms in carbon dioxide end up
_____ 12. Recycled from the Calvin cycle to make ATP
_____ 13. Supplies energy to the light reactions
_____ 14. Gas produced by reactions in the thylakoids
_____ 15. Gas consumed by reactions in the stroma
_____ 16. Source of carbon for carbon fixation
_____ 17. Source of H for the Calvin cycle
_____ 18. Picks up energized electrons from reactions in the thylakoids

A. Carbon dioxide, CO_2
B. Water, H_2O
C. Glucose, $C_6H_{12}O_6$
D. Oxygen, O_2
E. ADP + P
F. ATP
G. $NADP^+$
H. NADPH
I. Light

Exercise 6   (Modules 7.6–7.7)

Order the following forms of electromagnetic energy from the shortest wavelength (1) to the longest (9). Which photons have the most energy? Which are used by plants in photosynthesis?

_____ a. Green light
_____ b. Radio waves
_____ c. X-rays
_____ d. Red light
_____ e. Ultraviolet light
_____ f. Infrared
_____ g. Microwaves
_____ h. Blue light
_____ i. Gamma rays

Exercise 7   (Modules 7.7–7.11)

To review photosynthesis, fill in the blanks in the following story.

The next time you eat an apple, reflect on the apple tree's ability to make the sugars it contains, using sunlight to assemble simple substances from air and soil. This process is called [1]_____ . It takes place in structures called [2]_____ in cells of tissues called the [3]_____ inside the leaves of the apple tree. Photosynthesis actually consists of two processes: In the

4_____ reactions, 5_____ molecules, in chloroplast membranes called 6_____, capture light energy. In the 7_____ cycle, which takes place in the fluid 8_____ surrounding the thylakoids, this energy is captured in molecules of 9_____.

Chlorophyll molecules absorb 10_____, packets of light energy. Chlorophyll absorbs only certain wavelengths, or colors, of light, mainly in the 11_____ and 12_____ parts of the spectrum. It reflects 13_____ light. Other pigments, such as 14_____, can absorb colors that chlorophyll cannot use directly, and transfer this energy to chlorophyll. Chlorophyll, other pigments, and proteins are clustered on the thylakoid membranes in groups called photosystems. A photosystem consists of several pigment-and-protein light-harvesting complexes, which act as antennas to capture light energy. They pass their energy on to a central 15_____, which consists of two chlorophyll molecules and a protein called the primary electron acceptor. There are two kinds of photosystems, photosystem I and photosystem II, which absorb slightly different colors of light.

When light strikes a leaf, and pigment molecules absorb photons, they pass their energy to a photosystem II reaction center complex, and there the energy excites a chlorophyll 16_____ to a higher energy level. This electron is passed to the primary electron acceptor and on to an electron transport chain. On their way down the electron transport chain, electrons from photosystem II perform important work. One of the electron carriers in the chain uses the energy released by the electrons to transport 17_____ ions from the 18_____ into the space inside the 19_____. This creates a buildup of H^+ ions, a concentration 20_____ of H^+ across the membrane. The H^+ ions then diffuse through the membrane via a protein complex called 21_____, which captures their energy to make 22_____. In photosynthesis, this chemiosmotic production of ATP is called 23_____, because its energy source is light.

How does photosystem II replace its lost electrons? It gets them by splitting 24_____. When the electrons of photosystem II are jarred loose, its reaction center complex develops a strong attraction for electrons. It obtains them by breaking apart a molecule of 25_____. This leaves two H^+ ions (which pass into the thylakoid space) and an 26_____ atom. This atom combines with one from another water molecule to form a molecule of 27_____ gas, which diffuses out of the leaf—a product of photosynthesis important to us and other animals.

Meanwhile, 28_____ energy excites an electron in a 29_____ molecule in the reaction center complex of photosystem I. (This electron is replaced by one from the electron transport chain from photosystem II.) The primary electron 30_____ passes the electron via a short electron transport chain to $NADP^+$, reducing it to a molecule of 31_____.

At this point in the story, the cells of the apple leaf have captured the energy of the sun in molecules of NADPH and ATP, and its leaf has released some O_2 gas, but so far no sugar has been produced. NADPH and ATP are used, and sugar is made, in the [32]_____ cycle, the second portion of [33]_____ that takes place in the [34]_____ of the chloroplast, around the thylakoids. Using carbon from [35]_____ obtained from the air, energy from [36]_____, and hydrogen and high-energy electrons carried by [37]_____, the enzymes of the Calvin cycle construct [38]_____, a high-energy three-carbon sugar molecule. In a series of steps, these molecules are combined to form the important six-carbon sugar [39]_____ and other organic compounds. This incorporation of carbon from CO_2 into organic molecules is called carbon [40]_____.

The cellulose that gives an apple its crunch and the sugar that gives it its sweet taste are made from the glucose made in photosynthesis. In your intestine, the sugars enter your blood and are transported to your body cells. There the chemical pathways of cellular [41]_____ release the energy in the sugar molecules and use it to build [42]_____, which is in turn used to power cellular work. Energy from the sun, captured by the apple and passed on to you, enables you to see, to move, and to contemplate this amazing story.

Exercise 8   (Module 7.12)

Plants employ a variety of ways of fixing CO_2 and saving water. State whether each of the following statements relates to C_3 plants, C_4 plants, or **CAM** plants.

_____ 1. Tend to waste energy on photorespiration on a hot day
_____ 2. Enzyme in mesophyll cells fixes carbon in 4-C compound, for use by bundle-sheath cells
_____ 3. Corn and sugarcane
_____ 4. Open stomata and trap CO_2 in five-carbon compound at night, for later use during daytime
_____ 5. Most plants
_____ 6. Soybeans, oats, wheat, rice
_____ 7. Can grow in hot, dry climates
_____ 8. Also can grow in hot, dry climates
_____ 9. Pineapple and many cacti
_____10. Calvin cycle uses CO_2 directly from the air

Exercise 9   (Module 7.13)

Test your understanding of this module by stating whether you think each of the following would be likely to **warm** or **cool** Earth through alteration of the greenhouse effect.

_____ 1. Increased burning of coal to produce electricity
_____ 2. Increased rate of growth of algae in the oceans
_____ 3. Increased cloud cover, which would reflect more sunlight
_____ 4. Hybrid cars that get better gas mileage
_____ 5. Increased cutting of tropical rain forests
_____ 6. Using nuclear power instead of coal and oil to make electricity
_____ 7. Reforesting deforested and overgrazed land
_____ 8. Increased rate of decomposition of organic matter
_____ 9. Slowing the population growth rate
_____ 10. Using solar cells to generate electricity
_____ 11. Warming of the ocean, which would release carbon dioxide to the air
_____ 12. Melting of Arctic ice, so less sunlight is reflected

Exercise 10   (Module 7.14)

Summarize this module by describing the relationship between CFCs and Earth's protective ozone layer *in exactly 25 words.* (Distilling your answer down like this will make you think and choose your words wisely.)

Test Your Knowledge

Multiple Choice

1. The ultimate source of energy in the sugar molecules produced by photosynthesis is
   a. sugar.
   b. the sun.
   c. oxygen.
   d. ATP.
   e. chlorophyll.

2. Which of the following is produced by the light reactions of photosynthesis and consumed by the Calvin cycle?
   a. NADPH
   b. O_2
   c. H_2O
   d. sugar
   e. ADP + P

3. Which of these wavelengths is *least* useful for photosynthesis?
   a. green
   b. yellow
   c. blue
   d. orange
   e. red

4. When chloroplast pigments absorb light,
   a. they become reduced.
   b. they lose potential energy.
   c. their electrons become excited.
   d. the Calvin cycle is triggered.
   e. their photons become excited.

5. The light reactions of photosynthesis generate high-energy electrons, which end up in _____ . The light reactions also produce _____ and _____ .
   a. ATP . . . NADPH . . . O_2
   b. O_2 . . . sugar . . . ATP
   c. chlorophyll . . . ATP . . . NADPH
   d. water . . . sugar . . . O_2
   e. NADPH . . . ATP . . . O_2

6. The overall function of the Calvin cycle is
   a. capturing sunlight.
   b. making sugar.
   c. producing CO_2.
   d. splitting water.
   e. oxidizing glucose.

7. Which of the following correctly matches each of the inputs of the Calvin cycle with its role in the cycle?
   a. CO_2: high-energy electrons; ATP: energy; NADPH: oxidation
   b. CO_2: carbon; ATP: energy; NADPH: high-energy electrons
   c. CO_2: high-energy electrons; ATP: carbon; NADPH: energy
   d. CO_2: energy; ATP: carbon; NADPH: high-energy electrons
   e. CO_2: hydrogen; ATP: carbon; NADPH: energy

8. The main photoautotrophs in aquatic environments are
   a. plants and animals.
   b. plants and fungi.
   c. animals and algae.
   d. algae and bacteria.
   e. plants and bacteria.

9. Which of the following is *not* a product of the light reactions of photosynthesis?
   a. O_2
   b. sugar
   c. high-energy electrons
   d. ATP
   e. NADPH

10. Which of the following is oxidized in photosynthesis?
    a. O_2
    b. CO_2
    c. $C_6H_{12}O_6$
    d. ATP
    e. H_2O

11. In photosynthesis, plants use carbon from _____ to make sugar and other organic molecules.
    a. water
    b. the air
    c. chlorophyll
    d. the sun
    e. soil

12. The carbon-fixation processes in C_4 and CAM plants differs from that of the majority of plants. C_4 and CAM plants are better-adapted to
    a. hot, dry conditions
    b. polar regions
    c. living in the shade
    d. low levels of carbon dioxide
    e. growing underwater

13. Aside from their importance in passing energy on to chlorophyll, carotenoid pigments also function in
    a. the Calvin cycle
    b. C_4 metabolism
    c. photorespiration
    d. photoprotection
    e. carbon fixation

Essay

1. Photosynthesis uses water and carbon dioxide to produce sugar and oxygen gas. Scientists long wondered whether the oxygen atoms in the oxygen gas produced in photosynthesis were obtained from carbon dioxide or water. Describe the experiments that enabled them to find out and the results of these experiments.

2. Draw two squares, one labeled to represent the light reactions and the other to represent the Calvin cycle. Using arrows, show the inputs and outputs of each process. Include the following: NADPH, ADP + P, O_2, light, CO_2, sugar, H_2O, ATP, $NADP^+$, electrons.

3. Photosynthesis has been called "the most important chemical process on Earth." Explain why.

4. State two activities of humans that tend to intensify the greenhouse effect. Why are people concerned about this? State two actions we could take that would reduce our contribution to the greenhouse effect.

Apply the Concepts

Multiple Choice

1. A photon of which of these colors would carry the most energy?
   a. blue
   b. yellow
   c. green
   d. orange
   e. red

2. A plant is placed in a sealed greenhouse with a fixed supply of water, soil, and air. After a year, the plant weighs 5 kg more than at the start of the experiment, and the ____ weighs almost 5 kg less.
   a. soil in the pot
   b. water left in the room
   c. organic matter in the soil
   d. air in the room
   e. soil in the pot together with the water in the soil

3. In a rosebush, chlorophyll is located in
   a. chloroplasts, which are in mesophyll cells in the thylakoids of a leaf.
   b. mesophyll cells, which are in the thylakoids in chloroplasts in a leaf.
   c. thylakoids, which are in mesophyll cells in the chloroplasts in a leaf.
   d. chloroplasts, which are in thylakoids in the mesophyll cells of a leaf.
   e. thylakoids, which are in chloroplasts in the mesophyll cells of a leaf.

4. Certain bacteria use smelly hydrogen sulfide gas, H_2S, instead of water, H_2O, as a source of electrons and hydrogen for the light reactions of photosynthesis. If their method of photosynthesis is similar to plants in other ways, where do you think the sulfur would end up?
   a. sulfur dioxide, SO_2
   b. chlorophyll
   c. solid sulfur, S_2
   d. a sugar containing sulfur
   e. NADPH

5. The *photo* part of the word *photosynthesis* refers to _____, whereas *synthesis* refers to _____.
   a. the reactions that occur in the thylakoids . . . carbon fixation
   b. the reactions in the stroma . . . the reactions in the thylakoids
   c. the Calvin cycle . . . carbon fixation
   d. the Calvin cycle . . . the reactions in the stroma
   e. the light reactions . . . reactions in the thylakoids

6. The energy used to produce ATP in the light reactions of photosynthesis comes from
   a. the "burning" of sugar molecules.
   b. splitting water.
   c. movement of H^+ through a membrane.
   d. carbon fixation.
   e. fluorescence.

7. The following (*P* through *U*) are the main steps of chemiosmotic ATP synthesis in the light-dependent reactions of photosynthesis. Which answer places them in the correct order?

P. H^+ concentration gradient established.

Q. H^+ diffuses through ATP synthase.

R. Carriers use energy from electrons to move H^+ across membrane.

S. Electrons from photosystem II pass along electron transport chain.

T. Light excites electrons in photosystem II.

U. Energy of H^+ flow used by ATP synthase to make ATP.

   a. PQTSRU
   b. STPQRU
   c. TSRPQU
   d. TSRUQP
   e. PQUSTR

8. The way ATP synthase captures the energy of H^+ ions in the light reactions appears to be similar to the way
   a. a turbine in a dam harnesses running water.
   b. an automobile engine uses hydrocarbons for fuel.
   c. heating of water forms steam in a nuclear reactor.
   d. electricity makes a lightbulb glow.
   e. a rotating wheel stores mechanical energy.

9. Oceanographers have suggested slowing the rate of greenhouse warming by fertilizing the ocean to increase the growth of algae. How would this reduce the greenhouse effect?
   a. It would produce oxygen, which reflects sunlight from the atmosphere.
   b. It would "repair" Earth's ozone layer.
   c. It would use CO_2, which would otherwise trap heat in the atmosphere.
   d. It would change the color of the ocean, reflecting the sun's heat.
   e. It would trap sunlight that would otherwise warm Earth.

10. Which of the following would not be capable of performing photosynthesis?
   a. bacterium
   b. alga
   c. peach tree
   d. seaweed
   e. mushroom

11. Ozone depletion might lead to
   a. warming of the climate.
   b. storms.
   c. cooling of the climate.
   d. increases in skin cancer.
   e. depletion of atmospheric CO_2.

Essay

1. In an experiment, plants were grown under colored filters that allowed equal amounts of light of different colors to strike different plants. Under which filter do you think plants grew the slowest? Why?

2. Carotenoids are yellow and orange pigments involved in photosynthesis. What colors of light must carotenoids absorb? Reflect? How is this useful to the plant?

3. Remember that the greater the concentration of hydrogen ions (H^+), the lower the pH of a solution. How do you suppose the pH of the solution in the thylakoid space compares with the pH of the solution in the stroma? What is responsible for the difference?

4. Compare chemiosmotic ATP production in photosynthesis (photophosphorylation) with ATP production in cellular respiration (oxidative phosphorylation): What is the source of high-energy electrons for each? What harnesses the energy of the electrons, and what is the energy used for? Where does each process take place? Where do the electrons end up? What manufactures the ATP? What is the immediate source of energy for ATP synthesis?

5. ATP is used to power your muscles as you turn the pages of this book. Where did the energy in the ATP come from? Trace the energy in the ATP molecules back to the sun.

6. In the mid-1600s, Belgian physician and chemist Jan Baptista van Helmont grew a small willow tree in a pot, adding only water to the soil. After five years, he found that the soil in the pot had lost only 60 grams, while the tree had grown by nearly 75 kilograms—more than 1,000 times the material lost from the soil. Van Helmont concluded that the tree had gained most if its substance not from the soil, but rather from the water he supplied. Was van Helmont right? Explain.

Put Words to Work

Correctly use as many of the following words as possible when reading, talking, and writing about biology:

autotroph, C$_3$ plant, C$_4$ plant, Calvin cycle, CAM plan, carbon fixation, Chorophyll, electromagnetic spectrum, global warming, granum, greenhouse effect, light reactions, mesophyll, photoautotroph, photon, photophosphorylation, photorespiration, photosynthesis, photosystem, producer, reaction center complex, stoma, stroma, thylakoid, wavelength

Use the Web

It is much easier to understand the chain of events in the light reactions and the molecular square dance of the Calvin cycle if you can see the processes in motion. To do so, see the activities and animations at www.mybiology.com.

The Cellular Basis of Reproduction and Inheritance

8

Focus on the Concepts

Reproduction and inheritance are fundamental characteristics of life, and the process of cell division is essential to reproduction and inheritance. As you study the chapter, focus on the following concepts:

- Single-celled organisms can reproduce asexually, duplicating their chromosomes and then undergoing cell division to produce identical offspring. In sexual reproduction, offspring inherit a mix of genes and chromosomes from two parents, so they are not identical to their parents or each other.

- Cell division takes place in the context of the cell cycle. Between divisions, a cell replicates its DNA. At the start of division, each chromosome consists of a pair of chromatids—the two duplicate DNAs. During mitosis, the chromosomes line up and the chromatids separate. The cell splits in a process called cytokinesis, producing two daughter cells identical to the original. Mitosis provides for growth, cell replacement, and asexual reproduction.

- Cells usually do not divide unless signaled by chemical growth factors that control the cell cycle. Most animal cells must be in contact with a surface to divide, and will stop dividing when crowded. Cancer cells do not respond to these cell cycle controls; they divide excessively and can invade other tissues.

- Cells of each species contain a characteristic number of chromosomes. In humans, somatic (body) cells are diploid, containing two sets of 23 chromosomes—23 from the mother and 23 from the father—for a total of 46. The two chromosomes of a matching pair are called homologous chromosomes. A process called meiosis produces haploid eggs and sperm (gametes)—each containing one set of 23 chromosomes. Egg and sperm join in fertilization to produce a diploid zygote with 46 chromosomes, which divides repeatedly to grow into an adult.

- In diploid organisms, meiosis produces haploid gametes, in a process that involves two consecutive cell divisions. In the first division, unlike mitosis, homologous pairs of chromosomes line up and then the pairs separate, producing two haploid daughter cells. In the second division, the sister chromatids of each chromosome separate in each cell, producing a total of four haploid cells.

- Each chromosome carries genes for certain traits, and homologous chromosomes can carry different versions of genes. The mixing of chromosomes that occurs during meiosis and fertilization can generate a variety of offspring in three ways: First, homologous chromosomes line up independently in meiosis, creating many possible sets of 23 when the pairs separate. Homologous chromosomes cross over and exchange segments while they are paired, generating greater variety. With such a diversity of eggs and sperm, fertilization produces further variation among offspring.

Review the Concepts

Work through the following exercises to review the concepts in this chapter. For additional review, refer to the activities at www.mybiology.com. The website offers a pre-test that will help you plan your studies.

Exercise 1   (Introduction–Module 8.3)

Review the concepts introduced in these modules by filling in the blanks.

"Like begets [1]_____." This old saying means offspring look like their parents. Technically, only offspring produced by [2]_____ reproduction look exactly like their parents, because they inherit all their [3]_____ from a single parent. For example, when an amoeba divides, its [4]_____ is duplicated, and identical sets of [5]_____ (the structures that contain most of the amoeba's DNA) are allocated to opposite sides of the cell. The parent amoeba splits, and the two daughter amoebas that are formed are genetically [6]_____ to each other and to the [7]_____ cell.

Prokaryotes also reproduce asexually, via a type of cell division called [8]_____ . Most genes in a prokaryote are carried on a single [9]_____ DNA molecule, which is much [10]_____ than the multiple chromosomes of eukaryotes. The prokaryote [11]_____ its DNA, and the copies move toward opposite ends of the cell. As new [12]_____ grows between them, the chromosomes become separated. Finally, the plasma membrane and cell [13]_____ split the cell in two. Like the reproduction of an amoeba, binary fission produces daughter cells identical to the parent cell. Parent and offspring share identical [14]_____, or sets of genetic information.

The offspring produced by sexual reproduction resemble their parents, but they are not identical to their parents or to each other. Sexual reproduction begins with the production of an [15]_____ and a [16]_____, specialized cells that join to produce an offspring. The egg and sperm fuse—a process called [17]_____—and the fertilized egg inherits a unique combination of genes from both parents. Through repeated cell divisions, the fertilized egg develops into an organism with a unique combination of traits—for example, a cat with long, gray fur or a human with blue eyes and freckles. Thus, sexual reproduction produces [18]_____ among offspring. Through sexual reproduction "like begets like," but not exactly.

Exercise 2   (Module 8.4)

Test your knowledge of chromosomes by completing this crossword puzzle.

Across

4. ____ cells have more chromosomes and genes than prokaryotes do.

5. A typical ____ has about 3,000 genes.

6. Chromosome duplication produces two sister ____.

8. A prokaryote has a ____ simple chromosome.

9. A chromosome contains one long ____ molecule.

10. Chromosomes are named for their attraction to colored ____ used in microscopy.

11. Eukaryotic chromosomes involve more ____ molecules than those of bacteria.

13. When a chromosome divides, one sister chromatid goes to each ____ cell.

14. Proteins help determine chromosome structure and control ____ activity.

Down

1. Sister chromatids are joined at the ____.

2. The number of chromosomes in a eukaryotic cell depends on the ____.

3. When a cell is not dividing, chromosomes form long, thin fibers called ____.

7. Genes in eukaryotic cells are grouped into ____ chromosomes.

9. Chromosomes are clearly visible only when a cell is ____.

12. ____ cells carry about 25,000 genes in 46 chromosomes.

Exercise 3   (Module 8.5)

Review the cell cycle: First identify the parts of the cycle and place them in order by writing the name of each phase or process on the diagram. Choose from **S, interphase, mitosis, G₁, mitotic phase, cytokinesis,** and **G₂.** Then add a brief description of what is happening during that portion of the cycle. Choose from **growth and DNA synthesis, cell growth following division, division of cytoplasm, activity between divisions, division of nucleus and chromosomes, growth and activity between DNA replication and division,** and **mitosis plus cytokinesis.**

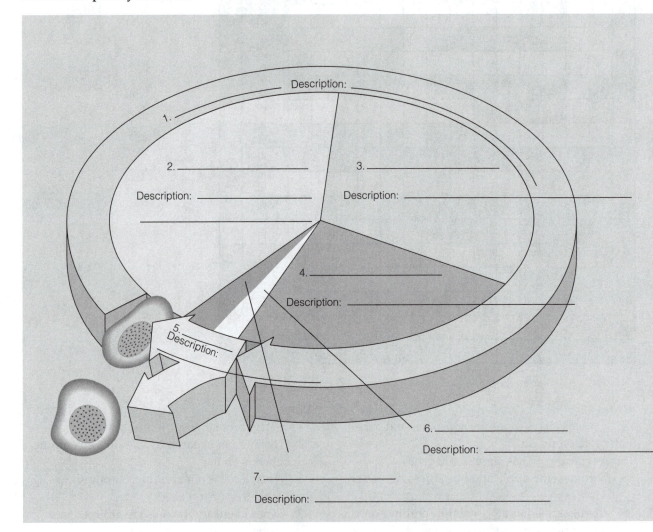

Exercise 4   (Module 8.6)

Summarize mitotic cell division. Briefly describe the appearance and activities of each of
these cell parts during interphase and the five stages of mitosis. Include a simple sketch
for each phase (just cell outlines and chromosomes).

	Interphase	Prophase	Prometaphase	Metaphase	Anaphase	Telophase
Nucleus and nuclear envelope						
Mitotic spindle						
Chromosomes						
Cell size and shape						
Sketch						

Exercise 5   (Module 8.6)

This is a photograph of cells in an onion root tip, an area of rapid cell division. In which
stage of mitosis (or interphase) is each of the numbered cells?

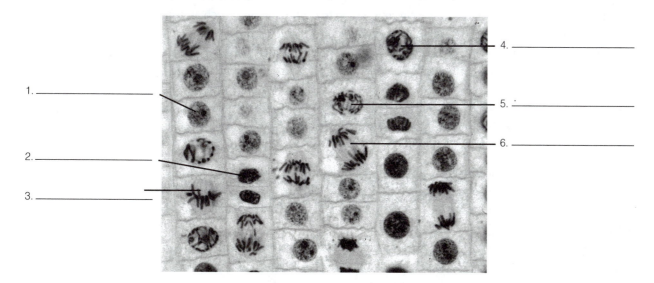

1. _____

2. _____

3. _____

4. _____

5. _____

6. _____

Exercise 6   (Module 8.6)

Match each structure on the left with its correct role in mitosis in an animal cell on the right.

_____ 1. Metaphase plate
_____ 2. Kinetochores
_____ 3. Sister chromatids
_____ 4. Spindle microtubules
_____ 5. Centrosome
_____ 6. Motor proteins

A. Where spindle microtubules attach to chromosomes
B. Move chromosomes
C. Pulled apart by spindle microtubules
D. Material around centrioles from which mitotic spin-
   dle grows
E. "Walk" chromosomes along microtubules toward
   cell poles
F. Chromosomes come to rest here during metaphase

Exercise 7   (Module 8.7)

Read this module and then write a statement containing exactly 30 words (no more, no less!) comparing cytokinesis in plant and animal cells. (Writing *exactly* 30 words will help you to think about the processes and choose your words carefully. It's fun to try it.)

Exercise 8   (Modules 8.8–8.11)

Review the functions of cell division and the factors that control it by filling in the blanks below.

Mitotic cell division has several important functions. Some animals rely on cell division for ^{1}_____ reproduction. *Hydra*, for example, produces buds that detach from the parent and take up life on their own. Cell division is also responsible for ^{2}_____, as seen in human embryos and plant roots. In an adult human, some cells, such as most nerve and muscle cells, cease to divide. Others, such as cells of the ^{3}_____, divide only if the organ is damaged. This process ^{4}_____ wounds. Some cells, such as those on the surface of the ^{5}_____ and the lining of the ^{6}_____, are constantly being abraded and lost. These cells are ^{7}_____ by cell division. In each of these cases, the new cells have exactly the same ^{8}_____ and ^{9}_____ of chromosomes as the parent cells because of the way duplicated chromosomes divide in the process of ^{10}_____.

Growth, cell replacement, and reproduction require control of the rate and tim-ing of cell division. Much has been learned by studying cells grown in laboratory ^{11}_____. Cells growing in a laboratory dish will divide only when in con-tact with a solid ^{12}_____. In the body, this ^{13}_____depen-dence may keep normal cells from dividing if separated from their normal surroundings. Cells will multiply only until they touch one another, a phenomenon known as ^{14}_____. Apparently, cells rely on proteins called ^{15}_____ to tell them when to divide, and will stop when cells are crowded and these substances are depleted.

It appears that growth factors influence cell division by acting on the cell-cycle
16_____ system, a set of molecules that triggers and coordinates events in
the cell cycle. The system automatically 17_____ cell division at several
major checkpoints unless the "brakes" are overridden by go-ahead signals. There are
checkpoints in the G_2 and M phases of the cell cycle, but the most important checkpoint
for many cells is the 18_____ checkpoint. If a cell receives a go-ahead signal
in the form of a growth factor at the G_1 checkpoint, the cell will proceed into the
19_____ phase of the cell cycle, replicate its DNA, and eventually divide. A
growth factor probably acts on a cell by attaching to a 20_____ protein in
the cell membrane. This protein in turn generates a signal that acts on the cell cycle control system within the cell. In the absence of a go-ahead signal, a cell will cease dividing.
Many of our cells that can no longer divide—21_____ cells, for example—
are stopped at the G_1 checkpoint.

Sometimes cells escape these control mechanisms, divide uncontrollably, and invade other body tissues. These 22_____ cells can kill the organism. In cell
culture, they can grow without being attached to a solid surface, are unaffected by density-dependent inhibition, and are less affected than normal cells by growth factors and
23_____ signals. Cancer cells can go on dividing indefinitely (unlike normal cells, which can divide in culture for only 24_____ generations). A
mass of cancer cells is called a 25_____. If the cells of a tumor remain at the
original site, the tumor is called a 26_____ tumor; if the cells spread, it is a
27_____ tumor. The spread of cancer cells is called 28_____.
Carcinomas are cancers that arise in body coverings, such as the skin.
29_____ arise in support tissues, such as muscle or bone.
30_____ are cancers of blood-forming tissues.

Cancer treatments, such as 31_____ and 32_____,
slow cancer by interfering with 33_____. The anticancer drugs vinblastin
and taxol prevent cell division by disrupting the mitotic 34_____.

Exercise 9   (Modules 8.12–8.13)

Describe the relationship between the terms or items in each of the following pairs.

1. Sex chromosomes and autosomes

2. The two chromosomes of a homologous pair

3. The two sister chromatids of a single chromosome

4. A diploid cell and a haploid cell

5. A somatic cell and a gamete

6. An egg and a zygote

7. Fertilization and meiosis

8. Mitosis and meiosis

9. X and Y chromosomes

10. Gene and locus

Exercise 10   (Module 8.14)

Review meiosis by drawing in the chromosomes to complete this sequence of diagrams. Some have been done for you. Label **meiosis I, meiosis II,** the **phases** of meiosis I and II, a pair of **homologous chromosomes,** two **sister chromatids,** and an example of **crossing over.** To make you think carefully about meiosis, the diploid number of chromosomes is 6 in this example. (It is 4 in Module 8.14.)

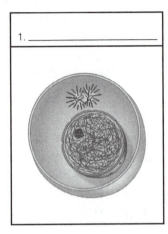

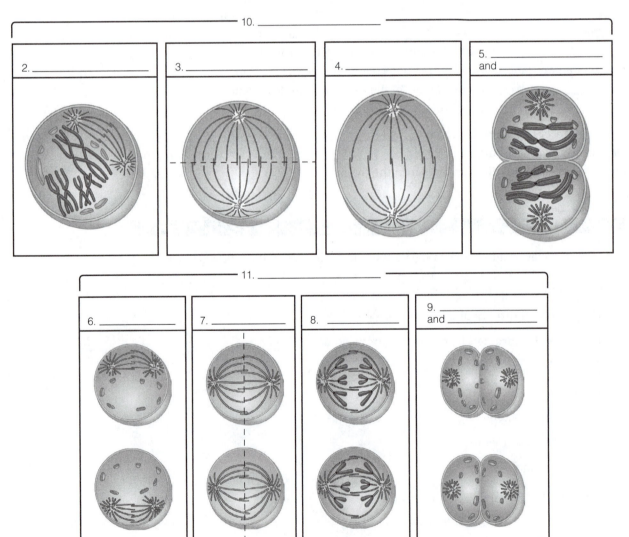

Compare mitosis and meiosis by completing this chart.

Mitosis	Meiosis
1.	Produces haploid daughter cells unlike parent cell
Involves one cell division	2.
Produces two daughter cells	3.
4.	Homologous chromosomes pair and then separate
Individual chromosomes line up at metaphase plate	5.
No crossing over occurs	6.
7.	Needed for sexual reproduction

These modules discuss how independent orientation of chromosomes, random fertilization, and crossing over can lead to varied offspring. The diagram below shows the two homologous pairs of chromosomes in a cell with a diploid number of 4. Three different genes are also shown. On separate paper, complete the sketches described in questions 1 through 3.

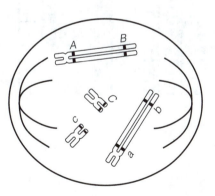

1. Show how two different orientations of the chromosomes during metaphase I of meiosis could lead to four different combinations of genes in gametes (assuming crossing over does not occur). (You don't need to show meiosis step by step—just the outcome.)

2. Show how crossing over could recombine genes on the larger pair of chromosomes, producing different gametes.

3. How many different combinations of genes in gametes are possible if these two processes happen simultaneously? Try to sketch all of them.

Errors in meiosis can produce gametes with extra or missing chromosomes. Fertilization of these abnormal gametes produces offspring with genetic abnormalities. Sometimes new species arise from chromosomal errors. A karyotype is an inventory of chromosomes. Review alterations in chromosome number and karyotyping by completing this crossword puzzle.

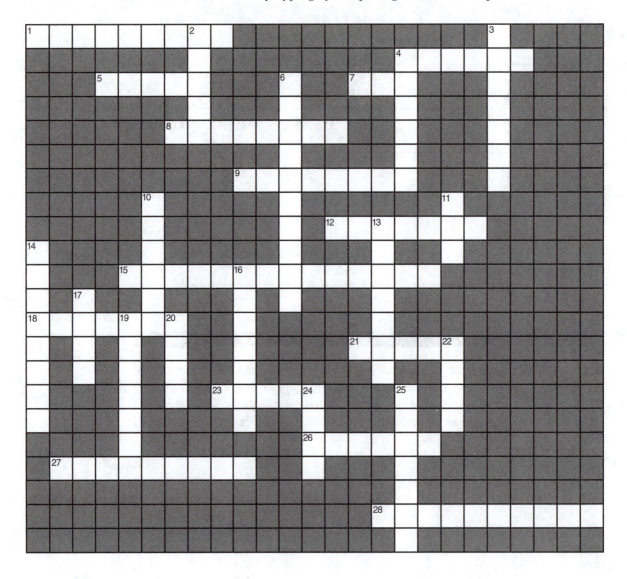

Across

1. An abnormal number of chromosomes often results in ____.
4. A woman with only one X chromosome is said to have ____ syndrome.
5. The Y chromosome carries ____ genes than the X chromosome.
7. The incidence of Down syndrome increases with the ____ of the mother.
8. Down syndrome individuals are prone to ____ and Alzheimer's disease.
9. A single Y chromosome is enough to produce "____."
12. Nondisjunction results in extra or missing chromosomes in ____.
15. ____ is when members of a pair of chromosomes fail to separate during meiosis.
18. Most human offspring with abnormal numbers of chromosomes are spontaneously ____.
21. Down syndrome is the most serious common ____ defect in the United States.
23. Pregnant women over 35 years of age are candidates for ____ testing.
26. In meiosis II, ____ chromatids may fail to separate.
27. A ____ is a photographic inventory of an individual's chromosomes.
28. A plant whose cells have four sets of chromosomes is said to be ____.

Down

2. A person with Down syndrome has ____ number 21 chromosomes.
3. ____ who are lacking an X chromosome are designated XO.
4. XXY or XXYY males have small ____ and feminine body contours.
6. In meiosis I, a pair of ____ chromosomes may fail to separate.
10. If a sperm fertilizes an egg with an extra chromosome, the ____ will have an extra chromosome.
11. Nondisjunction can alter the number of ____ chromosomes, as well as autosomes.
13. Chromosomal defects usually result from errors in ____.
14. Individuals with Down syndrome are usually mentally ____.
16. People with Down syndrome live ____-than-normal life spans.
17. Meiosis begins in a woman's ovaries before she is ____ and is completed years later.
19. An extra chromosome 21 is called ____ 21.
20. A person with an extra number 21 chromosome has ____ syndrome.
22. ____ of the eggs of a Down syndrome woman will have an extra chromosome.
24. An unusual number of sex chromosomes has ____ effect than an unusual number of autosomes.
25. Many new ____ of plants seem to have arisen from errors in cell division.

Chromosomes sometimes break, their parts can become scrambled, and abnormalities can result. Match each of the diagrams of chromosome alterations with its name (A-D) and a description of its effects (W-Z).

Diagram	*Name*	*Effects*

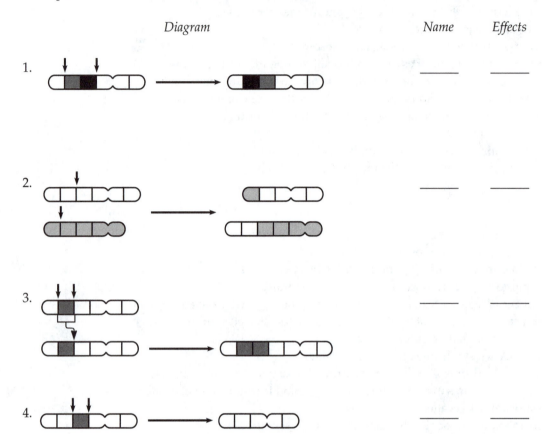

Names: A. Deletion
  B. Duplication
  C. Inversion
  D. Translocation

Effects: W. May cause chronic myelogenous leukemia in somatic cells
  X. Least likely to have serious effects, because genes are still present in normal numbers
  Y. Likely to have the most serious effects, as in *cri du chat* syndrome
  Z. A chromosome fragment breaks off and joins a homologous chromosome

Test Your Knowledge

Multiple Choice

1. There are a number of differences between binary fission of a bacterium and human cell division. Which of the following is *not* one of them?
   a. A bacterium has only one chromosome.
   b. Human cells undergo mitosis and cytokinesis.
   c. Bacteria are smaller and simpler than human cells.
   d. Bacteria have to duplicate their DNA before dividing; human cells do not.
   e. Human chromosomes are larger and more complex.

2. You would be unlikely to see which of the following human cells dividing?
   a. muscle cell
   b. skin cell
   c. cancer cell
   d. cell from an embryo
   e. intestinal lining cell

3. Which of the following correctly matches a phase of the cell cycle with its description?
   a. M—replication of DNA
   b. S—immediately precedes cell division
   c. G_2—cell division
   d. G_1—immediately follows cell division
   e. All of the above are correctly matched.

4. Which of the following is *not* true of human somatic cells?
   a. They arise by mitotic cell division.
   b. They are haploid.
   c. They are body cells other than eggs and sperm.
   d. They are larger and more complex than bacterial cells.
   e. They contain 46 chromosomes.

5. In telophase of mitosis, the mitotic spindle breaks down and nuclear envelopes form. This is essentially the opposite of what happens in
   a. prophase.
   b. interphase.
   c. metaphase.
   d. S phase.
   e. anaphase.

6. Sister chromatids
   a. cross over during prophase I of meiosis.
   b. separate during the first meiotic division.
   c. are produced during S phase between cell divisions.
   d. cross over during prophase II of meiosis.
   e. are also called homologous chromosomes.

7. Which of the following is *not* a function of mitotic cell division in animals?
   a. asexual reproduction
   b. growth
   c. repair of damaged organs
   d. production of gametes
   e. cell replacement

8. Meiosis
   a. is responsible for body growth and repair.
   b. halves the number of chromosomes in cells.
   c. is the process by which the body produces diploid cells.
   d. follows mitosis and splits the cytoplasm in two.
   e. is important in asexual reproduction.

9. Crossing over is
   a. important in genetic recombination.
   b. what makes a cell become cancerous.
   c. a key process that occurs during mitosis.
   d. an important mechanism of chromosome repair.
   e. what prevents cells from multiplying indefinitely in cell culture.

10. Human _____ are diploid, and human _____ are haploid.
    a. sex chromosomes . . . autosomes
    b. autosomes . . . sex chromosomes
    c. somatic cells . . . gametes
    d. gametes . . . somatic cells
    e. chromosomes . . . chromatids

11. Which of the following does *not* lead to genetic variability?
    a. random fertilization
    b. crossing over during meiosis
    c. division of chromosomes during anaphase of mitosis
    d. orientation of chromosomes during metaphase I of meiosis
    e. mutation

12. Most cells will divide if they receive the proper signal at a checkpoint in the ____ phase of the cell cycle.
    a. M
    b. G_1
    c. S
    d. G_2
    e. cytokinesis

13. Geneticists suspect that the extra chromosome seen in Down syndrome usually comes from the egg, rather than the sperm, because
    a. eggs are produced so rapidly that there is more chance for error.
    b. Down syndrome is due to a dominant gene in women, a recessive gene in men.
    c. most women inherit Down syndrome from their mothers.
    d. eggs are produced in much larger numbers than sperm.
    e. meiosis takes much longer in the ovary, increasing the likelihood of error.

14. Which of the following chromosomal alterations would you expect to have the most drastic consequences?
    a. inversion
    b. duplication
    c. translocation
    d. deletion
    e. a and b are equally the most serious

15. Disorders involving unusual numbers of sex chromosomes show that "maleness" is caused by the
    a. presence of an X chromosome.
    b. presence of a Y chromosome.
    c. absence of an X chromosome.
    d. absence of a Y chromosome.
    e. absence of an X chromosome and presence of a Y chromosome.

Essay

1. Explain why, strictly speaking, the phrase "like begets like" applies only to asexual reproduction.

2. Briefly describe mitosis and cytokinesis and state their functions.

3. Describe how cancer cells differ from normal body cells.

4. Compare mitosis and meiosis. What are their functions? Which produces haploid cells? Diploid cells? What kinds of cells undergo mitosis and meiosis? What kinds of cells are produced by each? How many cells are produced?

5. How might the genes on the sister chromatids of a certain chromosome compare with each other and with the genes on the sister chromatids of the homologous chromosome?

6. Describe three aspects of sexual reproduction that lead to the production of varied offspring.

7. Explain how an error in meiosis can cause a baby to be born with an extra or missing chromosome.

8. About one in 700 babies born in the United States possesses an extra chromosome 21, resulting in Down syndrome. Why are few individuals seen who have extra copies of other chromosomes?

Apply the Concepts

Multiple Choice

1. In certain fungi and algae, cells undergo mitosis repeatedly without subsequently undergoing cytokinesis. What would result from this?
    a. a decrease in chromosome number
    b. inability to duplicate DNA
    c. division of the organism into many cells, most lacking nuclei
    d. large cells containing many nuclei
    e. a rapid rate of sexual reproduction

2. A human bone marrow cell, in prophase of mitosis, contains 46 chromosomes. How many chromatids does it contain altogether?
    a. 46
    b. 92
    c. 23
    d. 23 or 46, depending on when during prophase you look
    e. 46 or 92, depending on when during prophase you look

3. Which of the following is the most significant difference between mitosis and meiosis?
    a. Chromosomes are duplicated before mitosis.
    b. Meiosis is not followed by cytokinesis.
    c. Homologous pairs of chromosomes are split up in meiosis.
    d. A spindle formed of microtubules moves the chromosomes in mitosis.
    e. Crossing over occurs in mitosis.

4. If there are 22 chromosomes in the nucleus of a toad skin cell, a toad egg would contain ____ chromosomes.
   a. 22
   b. 44
   c. 11
   d. 33
   e. 88

5. Which of the following carry the same genetic information?
   a. sister chromatids
   b. X and Y chromosomes
   c. all autosomes
   d. homologous chromosomes
   e. all haploid cells

6. A cell biologist carefully measured the quantity of DNA in grasshopper cells growing in cell culture. Cells examined during the G_2 phase of the cell cycle contained 200 units of DNA. What would be the amount of DNA in one of the grasshopper daughter cells seen in telophase of mitosis?
   a. 50 units
   b. 100 units
   c. between 50 and 100 units
   d. 200 units
   e. 400 units

7. What would be the quantity of DNA in one of the grasshopper cells (question 6) produced by telophase II of meiosis?
   a. 50 units
   b. 100 units
   c. between 50 and 100 units
   d. 200 units
   e. 400 units

8. The two chromosomes of a homologous pair
   a. carry identical genetic information at corresponding locations.
   b. carry information for the same characteristics at different locations.
   c. carry identical genetic information at different locations.
   d. carry information for the same characteristics at corresponding locations.
   e. Any of the above is possible.

9. A picture of a dividing pigeon cell taken through a microscope shows that the cell contains seven chromosomes, each consisting of two chromatids. This picture must have been taken during
   a. metaphase of mitosis.
   b. prophase I of meiosis.
   c. telophase II of meiosis.
   d. prophase II of meiosis.
   e. telophase of mitosis.

10. A culture of mouse cells is treated with a chemical that interferes with the activity of microfilaments. Which of the following will probably be affected the most?
    a. mitosis
    b. chromosome duplication
    c. pairing of homologous chromosomes
    d. cytokinesis
    e. joining of sister chromatids at the centromere

11. A zoologist examined an intestine cell from a crayfish and counted 200 chromosomes, each consisting of two chromatids, at prophase I of mitosis. What would he expect to see in each of the four cells at telophase II of meiosis if he looked in the crayfish ovary?
    a. 50 chromosomes, each consisting of two chromatids
    b. 50 chromosomes, each consisting of one chromatid
    c. 100 chromosomes, each consisting of two chromatids
    d. 100 chromosomes, each consisting of one chromatid
    e. 200 chromosomes, each consisting of one chromatid

12. One chromosome of a homologous pair carries the genes *J* and *K*. The other chromosome of the pair carries the genes *j* and *k* at corresponding loci. Crossing over results in exchange of chromosome segments and production of gametes with new combinations of genes. A "recombinant"-type gamete resulting from this crossover might contain
    a. genes *J* and *K*.
    b. genes *j* and *K*.
    c. genes *J* and *j*.
    d. genes *j* and *k*.
    e. genes *K* and *k*.

13. Humans have 23 pairs of chromosomes, while our closest relatives, chimpanzees, have 24. Chromosome studies indicate that at some point early in human evolution, two chromosomes simultaneously broke into a large portion and a small portion. The large parts combined to form a large chromosome, and the small parts combined to form a much smaller chromosome, which was subsequently lost. This important chromosomal change could best be described as
    a. nondisjunction followed by deletion.
    b. translocation followed by deletion.
    c. duplication followed by deletion.
    d. translocation followed by inversion.
    e. nondisjunction followed by inversion.

14. It would be most difficult to spot which of the following by looking at a karyotype?
    a. an extra chromosome
    b. part of a chromosome duplicated
    c. a missing chromosome
    d. part of a chromosome inverted
    e. a translocation

Essay

1. Recall how chemotherapy slows the growth of cancerous tumors. Common side effects of chemotherapy are hair loss, nausea, and loss of appetite. Why do you think that chemotherapy has some of its strongest side effects on the skin and lining of the digestive tract?

2. Most of the grasshopper somatic cells examined by the cell biologist in Multiple Choice questions 6 and 7 above contained either 100 or 200 units of DNA. Some of the interphase cells, however, contained between 100 and 200 units. Explain what was happening in these cells.

3. A slide of dividing cells in an onion root tip (Figure 8.11A in the text) is a "snapshot" in time. Each cell is stopped at the particular point in its cell cycle when the slide was made. A biology student examined such a slide under a microscope. Out of 100 cells she caught in the act of dividing, 38 were in prophase, 15 in prometaphase, 8 in metaphase, 10 in anaphase, and 29 in telophase. Assuming that the cells are growing and dividing independently, what do these data tell you about the phases of mitosis in onion cells?

4. A white blood cell from a female golden retriever was found to contain a total of 78 chromosomes. How many different kinds (sizes and shapes) of chromosomes would you expect to find in the cell? Why?

5. The somatic cells of a mosquito contain three pairs of chromosomes—two large ones, two medium-sized ones, and two small ones. One large chromosome bears the *A* gene; its homologue bears the *a* gene. One medium-sized chromosome bears the *B* gene; its homologue bears the *b* gene. One small chromosome bears the *C* gene; its homologue bears the *c* gene. Sketch cells at metaphase I of meiosis in the ovary of the mosquito, showing the different alignments of chromosomes that are possible. Then show how these lead to different combinations of genes in the gametes. How many different combinations are possible in the eggs? In the sperm of a male mosquito with the same genes, how many different gene combinations are possible? If the male and female mate and her eggs are fertilized with his sperm, how many different combinations are possible in the zygotes?

6. One chromosome of a homologous pair carries genes *Q* and *R*. Its homologue carries genes *q* and *r*. Show how crossing over during meiosis could produce gametes with new combinations of genes. What combinations of genes occur in parental-type gametes? In recombinant-type gametes?

7. Which of the following abnormalities in sex chromosomes would result in a predominantly male phenotype? A predominantly female phenotype? Explain why. XXXX, XXY, XYYY, X, XXYY

Put Words to Work

Correctly use as many of the following words as possible when reading, talking, and writing about biology:

anaphase, anchorage dependence, asexual reproduction, autosome, benign tumor, binary fission, carcinoma, cell cycle, cell cycle control system, cell division, cell plate, centromere, centrosome, chiasma (plural, *chiasmata*)*, chromatin, chromosome, cleavage, cleavage furrow, crossing over, cytokinesis, deletion, density-dependent inhibition, diploid cell, Down syndrome, duplication, fertilization, gamete, genetic recombination, genome, growth factor, haploid cell, homologous chromosomes, interphase, inversion, karyotype, leukemia, life cycle, loci* (plural, *locus*)*, lymphoma, malignant tumor, meiosis, metastasis, mitotic phase (M phase), mitotic spindle, mitosis, nondisjunction, prometaphase, prophase, sarcoma, sex chromosome, sexual reproduction, sister chromatid, somatic cell, telophase, tetrad, translocation, trisomy 21, tumor, zygote*

Use the Web

Cell division is another of those processes best understood if seen in motion. See cell division activities, videos, and animations at www.mybiology.com.

Patterns of Inheritance

9

Focus on the Concepts

The characteristics of organisms are shaped by genes inherited from their parents. A few fundamental rules underlie patterns of inheritance. In studying this chapter, focus on the flowing concepts:

- Discrete factors called genes govern inheritance. An organism's phenotype (characters and traits) is shaped by genotype (genes and alleles). For each character, an organism inherits two alleles, one from each parent. These alleles occur on homologous chromosomes and may be the same or different.

- Homologous chromosomes, and their allele pairs, segregate from one another during the production of gametes. An egg or sperm carries only one allele for each character. Fertilization results in offspring with various combinations of allele pairs. The Punnett square predicts combinations of alleles in gametes and offspring, and ratios of resulting offspring phenotypes: The F_2 of a mono-hybrid cross shows a 3:1 phenotypic ratio.

- Each pair of chromosomes, and each pair of alleles, separate independently of other pairs during meiosis. The Punnett square can also predict combinations of alleles in gametes and offspring, and ratios among offspring phenotypes, when tracking two different characters at once: The F_2 of a dihybrid cross shows a 9:3:3:1 phenotypic ratio.

- Rules of probability can be used to predict patterns of inheritance. The probability of a compound event is the product of separate probabilities of independent events. The probability of an event that can occur two different ways is the sum of the probabilities of the different ways.

- Many inherited human disorders are caused by a single gene. Many of these disorders are recessive and occur in offspring of heterozygous carrier parents.

- There are more complex patterns of inheritance: Sometimes there are more than two alleles for a character, and sometimes alleles are not clearly dominant or recessive. Sometimes one gene shapes many characters, and sometimes one character is shaped by many genes. Linked genes (those on the same chromosome) do not assort independently, confounding Mendel's ratios but allowing us to map chromosomes.

- In many species, chromosomes determine sex. In humans, sex is determined by the X and Y chromosomes. Females are XX, and males are XY. The *SRY* gene on the Y produces a male. Sex-linked genes (almost always on the X) exhibit unique patterns of inheritance; males are more likely to exhibit sex-linked recessive defects.

Review the Concepts

Work through the following exercises to review the concepts in this chapter. For additional review, check out the activities at www.mybiology.com. The website offers a pre-test that will help you plan your studies.

Exercise 1   (Modules 9.1–9.4)

These modules discuss the basic principles of heredity and introduce the vocabulary of genetics. Read the modules carefully, and then practice using the vocabulary by matching each phrase on the left with a word or phrase on the right. (This is just a start; you will remember these terms better as you use them.)

_____ 1. A unit that determines heritable characteristics
_____ 2. Varieties that always produce offspring identical to parents
_____ 3. The offspring of two different varieties
_____ 4. When two alleles of a pair differ, the one that is hidden
_____ 5. A feature that varies among individuals, such as flower color
_____ 6. Parent organisms that are mated
_____ 7. A diagram that shows possible combinations of gametes
_____ 8. A breeding experiment that uses parents different in one character
_____ 9. One of the alternative versions of a gene for a character
_____ 10. Relative numbers of organisms with various traits
_____ 11. An organism that has two different alleles for a gene
_____ 12. Each variant of a character, such as purple and white flower color
_____ 13. An organism's genetic makeup
_____ 14. Separation of allele pairs that occurs during gamete formation
_____ 15. Fertilization of a plant by pollen from a different plant
_____ 16. An organism that has two identical alleles for a gene
_____ 17. The science of heredity
_____ 18. The location of a gene on a chromosome
_____ 19. What an organism looks like; its expressed traits
_____ 20. Offspring of the F_1 generation
_____ 21. When pollen fertilizes eggs from the same flower
_____ 22. Separation of allele pairs during gamete production
_____ 23. When two alleles of a pair differ, the one that determines appearance
_____ 24. Where the two alleles for a certain character are located
_____ 25. Offspring of the P generation
_____ 26. A hybridization

A. Allele
B. Homozygous
C. Hybrid
D. Genotype
E. Segregation
F. F_2 generation
G. True-breeding
H. Heterozygous
I. Self-fertilization
J. Dominant
K. P generation
L. Monohybrid cross
M. Locus
N. Phenotype
O. Cross
P. F_1 generation
Q. Recessive
R. Homologous chromosomes
S. Gene
T. Phenotypic ratio
U. Trait
V. Cross-fertilization
W. Punnett square
X. Character
Y. Genetics
Z. Segregation

Test your knowledge of Mendel's principles by answering the following questions. You may want to test your ideas on scratch paper.

1. A pea plant with green pods is crossed with a plant with yellow pods. (Note: We're talking about the *pods* here, not the peas inside!) All their offspring have green pods.

   a. Which allele is dominant? Which allele is recessive?

   b. Using letters, what is the genotype of the green parent? The yellow parent?

   c. What are the genotypes of the offspring?

2. F_1 pea plants from the above cross are crossed. Use a Punnett square to figure out the genotypic and phenotypic ratios in the F_2 generation.

   a. Genotypic ratios:

   b. Phenotypic ratios:

3. Two black mice mate. Six of their offspring are black and two are white.

   a. What are the genotypes of the parents?

   b. For which offspring are you sure of the genotypes?

Exercise 3   (Module 9.5)

Gregor Mendel studied the inheritance of two characteristics at once and found that each pair of alleles segregates independently during the formation of gametes. In other words, if a tall pea plant with purple flowers is crossed with a short plant with white flowers, some of their descendants can be tall with white flowers. The tall and purple alleles do not have to stick together—they are independent.

So far, the textbook has discussed inheritance in peas and dogs. Just to be different, let's look at a genetic cross involving rabbits. In rabbits, the allele for brown coat is dominant, the allele for white coat recessive. The allele for short fur is dominant, the allele for long fur recessive. Imagine mating a true-breeding brown, short-haired rabbit with a white, long-haired rabbit. Using Module 9.5 as a model, write the genotypes of rabbits and gametes in the P, F_1, and F_2 generations in the blanks in the Punnett square. You may want to modify the drawings (add some hair and color) to show the phenotypes of the rabbits in the F_2 generation. Then use the Punnett square to figure out the phenotypic ratios in the F_2 generation—the proportion of rabbits that you can expect to be brown and short-haired, brown and long-haired, white and short-haired, and white and long-haired. Write their phenotype and their proportions in the blanks at the bottom.

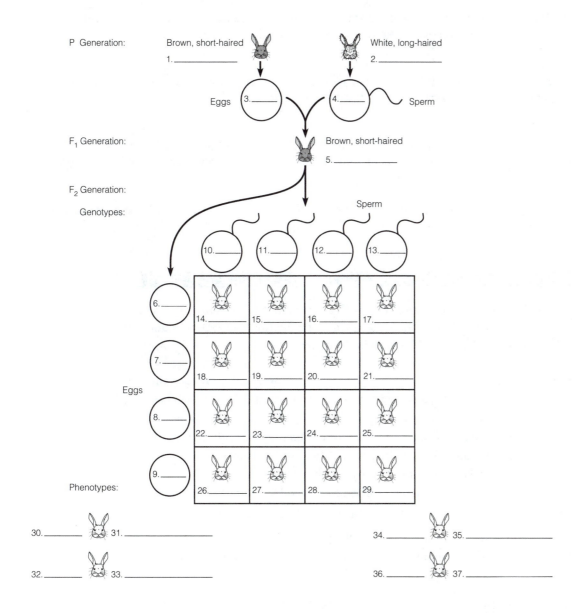

Exercise 4   (Module 9.6)

After reading this module on testcrosses, test your understanding by answering the following questions.

1. Recall that brown coat color in rabbits is dominant and white color is recessive. Suppose you have a group of rabbits—some brown and some white.

    a. For which phenotype(s) do you know the genotype(s)?

    b. For which phenotype(s) are you unsure of the genotype(s)?

2. Using *B* and *b* to symbolize the brown and white alleles:

    a. What are the possible genotypes of a white rabbit in your group?

    b. What are the possible genotypes of a brown rabbit?

3. Suppose you wanted to find out the genotype of a brown rabbit. What color rabbit would you mate it with?

4. A brown buck (male) is mated with a white doe (female). In their litter of 11 young, six are white and five are brown. Using a Punnett square to check your answer, what is the genotype of the buck?

5. Use a Punnett square to figure out the ratio of brown and white offspring that would have been produced by the above mating if the brown buck had been homozygous.

6. If half the offspring from a testcross are of the dominant phenotype and half are of the recessive phenotype, is the parent of the dominant phenotype (but unknown genotype) homozygous or heterozygous?

7. If all the offspring from a testcross are of the dominant phenotype, is the parent with the dominant phenotype (but unknown genotype) homozygous or heterozygous?

Exercise 5   (Module 9.7)

The rules of probability can be used to predict the flip of a coin, the drawing of a card from a deck, or the roll of a pair of dice. They also govern segregation and recombination of genes. Read Module 9.7 carefully, and then fill in the blanks below.

The probability scale ranges from [1]_____ (an event that is certain not to occur) to [2]_____ (an event that is certain to occur). The probabilities of all possible outcomes for an event must add up to [3]_____ . Imagine rolling a pair of dice, one die at a time. Each of the six faces of a die has a different number of dots, from one to six. If you roll a die, the probability of rolling a one is [4]_____ . The probability of rolling any number other than one is [5]_____ . The outcome of a given roll is unaffected by what has happened on previous rolls. In other words, each roll is a(n) [6]_____ event.

If you roll two dice simultaneously, what is the probability of "snake eyes" (both ones)? The roll of each die is an independent event. The probability of such a compound event (both dice coming up ones) is the [7]_____ of the separate probabilities of the independent events. Therefore, the probability of rolling two ones is [8]_____ × [9]_____ = [10]_____ . This is called the rule of [11]_____ .

This rule also governs the combination of genes in genetic crosses. The probability that a heterozygous *(Pp)* individual will produce an egg containing a *p* allele is [12]_____ . The probability of producing a *P* egg is also [13]_____ . If two heterozygous individuals are mated, what is the probability of a particular offspring being [14]_____ recessive *(pp)*? The probability of producing a *p* egg is 1/2. The probability of producing a *p* sperm is also 1/2. The production of egg and sperm are independent events, so to calculate their combined probability we use the rule of [15]_____ . Thus the chance that two *p* alleles will come together at fertilization to produce a *pp* offspring is [16]_____ × [17]_____ = [18]_____ .

Back to the dice for a moment. What is the probability that a roll of two dice will produce a three and a four? There are two different ways this can occur. The first die can come up a three and the second a four, or the first can come up a four and the second a three. The probability of the first combination is $1/6 \times 1/6 = 1/36$. The probability of the second is also $1/6 \times 1/6 = 1/36$. According to the rule of [19]_____ , the probability of an event that can occur in two or more alternative ways is the [20]_____ of the separate probabilities of the different ways. The probability of rolling a three and a four is therefore [21]_____ + [22]_____ = [23]_____ .

Similarly, what is the probability that a particular offspring of two heterozygous parents will itself be heterozygous? The probability of the mother producing a *P* egg is [24]_____ . The probability of the father producing a *p* sperm is also [25]_____ . Therefore, the probability of a *P* egg and a *p* sperm joining at fertilization is [26]_____ × [27]_____ = [28]_____ . Or a *p* egg and a *P* sperm could join. The probability of this occurring is also [29]_____ . According to the rule of addition, the probability of an event that can occur in two alternative ways is the sum of the separate probabilities. Therefore, the probability of heterozygous parents producing a heterozygous offspring is [30]_____ + [31]_____ = [32]_____ .

Exercise 6   (Module 9.8)

After you read this module, use the information in the illustration to solve the following problems. You will probably want to work out Punnett squares on scratch paper.

1. A man and woman, both without freckles, have four children. How many of the children would you expect to have freckles?

2. Both Fred and Wilma have widow's peaks. Their daughter Shirley has a straight hairline. What are Fred and Wilma's genotypes?

3. A man and woman both have free earlobes, but their daughter has attached earlobes. What is the probability that their next child will have attached earlobes?

Exercise 7   (Module 9.8)

Family trees called pedigrees are used to trace the inheritance of human genes. The two pedigrees below show the inheritance of sickle-cell disease (described in Modules 9.9 and 9.13), which is caused by an autosomal recessive allele. In the first pedigree, the square and circle symbols are colored, as far as genotypes are known. Fill in the genotypes—*SS*, *Ss*, or *ss*—below the symbols. Use question marks to denote unknown genotypes. Complete the second pedigree by coloring in the symbols, following the rules described in Module 9.8. Again denote unknowns with question marks.

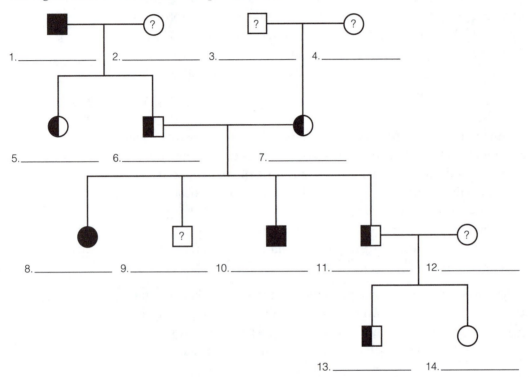

1._____   2._____   3._____   4._____

5._____   6._____   7._____

8._____   9._____   10._____   11._____   12._____

13._____   14._____

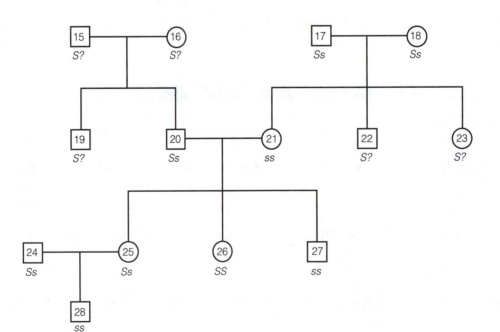

Exercise 8   (Module 9.9)

This module discusses common human genetic diseases and their inheritance. Indicate whether each of the statements below is true or false, and change false statements to make them true.

_____ 1. Cystic fibrosis is the most common fatal genetic disease in the United States.
_____ 2. A genetic disorder is expected in half the children of two carriers of a recessive allele.
_____ 3. About 50 human genetic disorders are known to be inherited as Mendelian traits.
_____ 4. Most people afflicted with PKU are born to afflicted parents.
_____ 5. Most serious human genetic disorders are recessive.
_____ 6. Cystic fibrosis is most common among Asian Americans.
_____ 7. Most genetic diseases are evenly distributed among ethnic groups.
_____ 8. Most societies have taboos and laws against marriage between close relatives.
_____ 9. Half of the offspring of two carriers of a recessive allele are likely to be carriers.
_____ 10. Dominant alleles are always more common in the population than recessive alleles.
_____ 11. Tay-Sachs disease is seen among Jews from Central Europe.
_____ 12. Lethal dominant alleles are much more common than lethal recessive alleles.
_____ 13. Achondroplasia is a form of dwarfism caused by a dominant allele.
_____ 14. Geneticists agree that inbreeding always increases the risk of inherited diseases.
_____ 15. Huntington's disease is lethal, but it does not strike until middle age.
_____ 16. Symptoms of PKU include excess mucus in the lungs and digestive tract.
_____ 17. Extra fingers and toes is a dominant trait.
_____ 18. Sickle-cell disease is common among African-Americans.
_____ 19. There is no way for individuals to know whether they may develop Huntington's.
_____ 20. Children with galactosemia are usually offspring of heterozygous parents.

Exercise 9   (Modules 9.9–9.10 and 9.15)

Choose the correct words to complete the following story.

Greg and Amy were excited and happy that she was pregnant, but their joy was mixed with anxiety. The couple had just received some bad news: Greg's sister had just given birth to a baby boy diagnosed with cystic fibrosis. Greg and Amy were at the clinic for genetic [1]_____ to discuss the possibility that Greg was a [2]_____ of cystic fibrosis and to determine their unborn child's chances of inheriting the disease.

Sharon, the genetic counselor, reviewed Amy's and Greg's family histories. She said, "Our first priority is to figure out whether the two of you are carriers. We knew that Amy could be, because her brother died of cystic fibrosis, but until Greg's nephew was diagnosed, we didn't know that there was CF in his family, too. Greg, if your sister is a carrier, you could be as well."

Amy interjected, "What does this mean for us and for our baby?"

"It means that you and Greg need to be tested for the cystic fibrosis allele. If you both are carriers, then we can talk about [3]_____ testing to determine whether your baby might have it."

Later that day, a technician withdrew blood for the tests, and the following week, Amy and Greg were back in the genetic counselor's office. Sharon breezed through the door. "You're in the clear for cystic fibrosis," she said matter-of-factly.

"What?"

"The CF tests were both negative: Neither of you are carriers. Plus, the
[4]_____ of Amy's blood was fine—there doesn't appear to be much chance of a
neural tube defect."

Greg and Amy both sighed with relief.

Then Sharon's expression became a bit more serious. "Unfortunately, we did find
something else that concerns me. Besides testing for the CF allele, we did a routine screen
for several other disorders, including [5]_____ disease—even though you are not
Jewish—and PKU. Turns out you both are carriers for PKU."

Amy groaned, "Oh no."

Sharon quickly added, "Don't worry yet. Even though you are both carriers, the
probability that the baby will have the disease is only [6]_____ ."

Amy asked, "What exactly is PKU? Is it a serious problem?"

Sharon explained that PKU, short for [7]_____ , is an inherited inability to break down an [8]_____ called phenylalanine. "The phenylalanine can
build up in the blood and cause mental retardation. As I said, don't start worrying yet.
We can test your fetus. If PKU is detected early, retardation can be prevented by putting
the child on a special [9]_____ , low in phenylalanine. Like most genetic characteristics, even genetic diseases can be modified by the [10]_____ ."

Amy asked, "How will you test the baby?"

"We'll have to perform [11]_____—taking a sample of the
[12]_____ fluid. We can check for PKU by testing for certain chemicals in the fluid
itself. While we're at it we will culture some of the fetal [13]_____ from the fluid
and do a [14]_____—take a picture of the chromosomes—to check for
[15]_____ syndrome. It will take a couple of weeks to culture the cells. Or we
could get the karyotype results right away by using a newer technique called
[16]_____ sampling. Then—"

Greg interrupted. "Wait a minute. Do you have to get samples? Can't you just do
[17]_____ imaging to look at the baby? Doctor Portillo did that before Kelly was
born."

"We really can't check chromosomes or PKU by just looking at the fetus with ultrasound. Amy is over [18]_____ years old, so I think it is important to get a
sample of amniotic fluid so we can check for Down syndrome. I'm sure everything will
be okay, but it's best to be prepared. Plus, the karyotype will answer another question
I'm sure you are eager to know the answer to—whether you are going to have a boy or a
girl."

Exercise 10   (Modules 9.9–9.16)

These modules discuss examples of inheritance that are a bit more complex—and common—than the simple patterns of heredity observed by Mendel. After reading the modules, see if you can match each description with a pattern of inheritance. Choose from:

A. incomplete dominance
B. multiple alleles
C. codominance
D. pleiotropy
E. polygenic inheritance

_____ 1. There are three different alleles for a blood group—I^A, I^B, and i—but an individual has only two at a time.

_____ 2. Crosses between two cremello (off-white) horses always produce cremello offspring. Crosses between chestnut (brown) horses always result in chestnut offspring. A cross between chestnut and cremello horses produces palomino (a golden-yellow color somewhat intermediate between chestnut and cremello) offspring. If two palominos are mated, their offspring are produced in the ratio of 1 chestnut:2 palominos:1 cremello.

_____ 3. The sickle-cell allele, s, is responsible for a variety of phenotypic effects, from pain and fever to damage to the heart, lungs, joints, brain, or kidneys.

_____ 4. In rabbits, an allele for full color *(C)* is dominant over an allele for chinchilla *(c′)* color. Both full color and chinchilla are dominant over the white allele *(c)*. A rabbit can be *CC, Cc′, Cc, c′c′, c′c,* or *cc*.

_____ 5. In addition to the A and B molecules found on the surface of red blood cells, humans also have M and N molecules. The genotype $L^M L^M$ produces the M phenotype. The genotype $L^N L^N$ gives the N phenotype. Individuals of genotype $L^M L^N$ have both kinds of molecules on their red blood cells, and their phenotype is MN.

_____ 6. If a red shorthorn cow is mated with a white bull, all their offspring are roan, a phenotype that has a mixture of red and white hairs.

_____ 7. Independent genes at four different loci are responsible for determining an individual's HLA tissue type, important in organ transplants and certain diseases.

_____ 8. A recessive allele causes a human genetic disorder called phenylketonuria. Homozygous recessive individuals are unable to break down the amino acid phenylalanine. As a consequence, they have high levels of this substance in their blood and urine, reduced skin pigmentation, lighter hair than their normal brothers and sisters, and often some degree of mental impairment.

_____ 9. When graphed, the number of individuals of various heights forms a bell-shaped curve.

_____10. Chickens homozygous for the black allele are black, and chickens homozygous for the white allele are white. Heterozygous chickens are gray.

Exercise 11 (Module 9.16)

Genes are located on chromosomes. Genes undergo segregation and independent assortment because the chromosomes that carry them undergo segregation and independent assortment during meiosis. The illustration below is similar to that in Module 9.16. It shows how alleles and chromosomes are arranged in an F_1 rabbit and how meiosis sorts the alleles into their gametes. The diagram below shows only the chromosomes. Put a letter (*B*, *b*, *S*, or *s*) in each of the numbered boxes to show how segregation and independent assortment of chromosomes cause segregation and independent assortment of alleles.

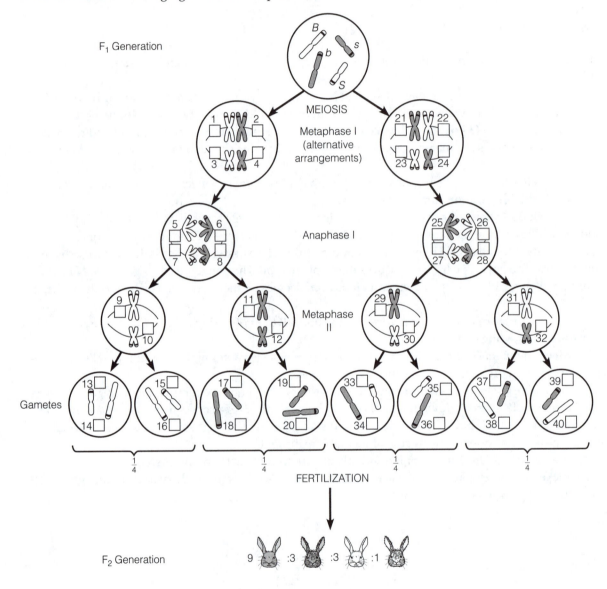

Exercise 12   (Modules 9.17–9.19)

These three modules discuss the inheritance of linked genes—genes that are on the same chromosome and therefore tend to be inherited together. Their pattern of inheritance is inconsistent with Mendel's "rules," but they illustrate important principles of chromosome structure and behavior. (In fact, linked genes were one of the original clues that genes are on chromosomes!) After reading the modules, match each of the observations below with the statement that explains the observation. Take your time and be careful; this exercise is not easy.

Observations

_____ 1. When two heterozygous round yellow peas are crossed, their offspring are produced in a 9:3:3:1 ratio (9 round yellow:3 round green:3 wrinkled yellow:1 wrinkled green).

_____ 2. When two peas heterozygous for purple flowers and long pollen are crossed, the expected 9:3:3:1 ratio is not seen. The ratio is close to 3 purple long:1 red round. Similarly, when a fruit fly with red eyes and long wings (*SsCc*) is crossed with a fly with scarlet eyes and curled wings (*sscc*), offspring are not produced in the expected 1:1:1:1 ratio. Most offspring are red long and scarlet curled.

_____ 3. When two heterozygous purple long peas are crossed, most of their offspring are purple and long or red and round. But a very small number of offspring are purple and round or red and long. Similarly, when the *SsCc* and *sscc* fruit flies are mated, nearly all their offspring are *SsCc* and *sscc*. However, a small number of offspring (about 6% of the total) are *Sscc* and *ssCc*.

_____ 4. When a fruit fly with red eyes and long wings (*SsCc*) is crossed with a fly with scarlet eyes and curled wings (*sscc*), 94% of their offspring are *SsCc* and *sscc*, and 6% are *Sscc* and *ssCc*. In other words, the recombination frequency between the *s* and *c* alleles is 6%. When a fly with red eyes and pale body (*SsEe*) is crossed with a fly with scarlet eyes and ebony body, 27% of their offspring are *Ssee* and *ssEe*. The recombination frequency between alleles *s* and *e* is 27%.

_____ 5. When a fly with long wings and pale body (*CcEe*) is crossed with a fly with curled wings and ebony body (*ccee*), 21% of their offspring are *Ccee* and *ccEe*. The recombination frequency between alleles *c* and *e* is 21%.

Explanations

A. The greater the distance between two genes, the greater the opportunity for crossing over to occur between them. If crossing over is more likely, more recombinant offspring will result. If two genes are farther apart, the recombination frequency will be greater between them.

B. Pairs of alleles on different chromosomes segregate independently during gamete formation. They follow Mendel's principle of independent assortment. In other words, genes for different traits on different chromosomes do not tend to "stick together" when passed on to offspring.

C. If two genes are on the same chromosome, or linked, they tend to be inherited together. Alleles on the same chromosome do not segregate independently. They tend to "stick together," violating Mendel's principle of independent assortment.

D. Recombination frequencies can tell you how far apart genes are on a chromosome. If you know the distance from *a* to *b*, the distance from *a* to *c*, and the distance from *b* to *c*, you can map the sequence of genes on the chromosome.

E. Homologous chromosomes cross over during meiosis and exchange segments. This recombines linked genes into assortments not seen in the parents.

Exercise 13   (Module 9.20)

What determines an individual's sex? Sex is generally determined by genes and chromosomes, but the process of sex determination works differently in different species. Match each group of organisms below with their system of sex determination.

_____ 1. Most plants, including peas, and corn (earthworms and land snails too)

_____ 2. Humans and fruit flies

_____ 3. Date palms and marijuana

_____ 4. Some butterflies, birds, and fishes

_____ 5. Grasshoppers and roaches

_____ 6. Ants, bees

_____ 7. Wild strawberries

A. Females are ZW, males ZZ.

B. Females are diploid, males haploid.

C. Females are XX, males XO (one X).

D. Females are XX, males XY.

E. Sexes not separate; all individuals produce both eggs and sperm.

Exercise 12   (Modules 9.20–9.23)

Genes located on the sex chromosomes—called sex-linked genes—determine many traits unrelated to maleness or femaleness. Due to the X-Y sex-determination system of humans, sex-linked traits show a peculiar pattern of inheritance. Red-green color blindness is a recessive sex-linked trait in humans. After reading these modules, see if you can describe the inheritance of color blindness by filling in the blanks below.

The genes for normal color vision and red-green color blindness, like most human sex-linked traits, are carried on the [1]_____ chromosome. A capital letter C represents the [2]_____ allele for normal vision; a small *c* represents the color-blindness allele. A male with normal color vision has the genotype [3]_____. (Because these genes are carried on the X chromosome, their symbols are shown as superscripts on the letter X.) A color-blind male has the genotype [4]_____.

A color-blind male will transmit the allele for color blindness to all his [5]_____ but none of his [6]_____. This is because only his daughters inherit his [7]_____ chromosome, and only his [8]_____ chromosome is passed to all his sons. All the children of a color-blind male and a homozygous dominant female will have normal color vision. Their sons will inherit only the normal vision allele, but their daughters will be [9]_____ of the color-blindness allele, thus possessing the genotype [10]_____.

A heterozygous female carrier transmits the color-blindness allele to [11]_____ of her offspring. If she and a male with normal vision have children, [12]_____ of their sons will be normal and [13]_____ will be color blind. [14]_____ of their daughters will be normal, because they inherit at least one dominant allele from their [15]_____. But half these daughters will be [16]_____ of the color-blindness trait, because they inherit the color-blindness allele from their mother.

Color blindness is much more common in men than in women. If a man inherits a single color-blindness allele from his [17]_____, the gene will be expressed and he will be color blind. Because a man has only one [18]_____chromosome, whatever genes it carries are seen in the man's phenotype. If a woman inherits just one color-blindness allele, she has relatively normal vision, because the dominant normal allele on her other X chromosome masks most of the effects of the color-blindness allele. For a woman to be color blind, she would have to inherit [19]_____ alleles from both her mother and her father, which is much less likely.

Because the Y chromosome is so much smaller than the X, it carries many fewer genes. One important gene on the Y chromosome is called the [20]_____ gene. It seems to code for proteins that control other genes on the Y related to "maleness." The Y chromosome is special—it is the only chromosome unique to one sex. (Both men and women, after all, possess [21]_____ chromosomes.) Because the Y is passed essentially unchanged from father to son, it is useful for tracing human [22]_____. But lest men become smug about this unique genetic endowment, it is perhaps relevant to note that the Y chromosome is the only chromosome that an individual can live without. (Think about it!)

Test Your Knowledge

Multiple Choice

1. How did Mendel's studies in genetics differ from earlier studies of breeding and inheritance?
   a. Mendel worked with plants; earlier studies used animals.
   b. Mendel was able to explain the "blending" hypothesis.
   c. Mendel's work was more quantitative.
   d. Mendel worked with wild species, not domesticated ones.
   e. Mendel found that offspring inherit characteristics from both parents.

2. A true-breeding fruit fly would be _____ for a certain characteristic.
   a. homozygous dominant
   b. homozygous recessive
   c. heterozygous
   d. Any of the above can be true-breeding.
   e. a or b

3. When looking at the inheritance of a single characteristic, Mendel found that a cross between two true-breeding peas (between purple and white, for example) always yielded a _____ in the F_2 generation.
   a. 1:1 phenotypic ratio
   b. 3:1 genotypic ratio
   c. 1:2:1 phenotypic ratio
   d. 3:1 phenotypic ratio
   e. 1:1 genotypic ratio

4. Alternative forms of genes for a particular character are called
   a. traits.
   b. alleles.
   c. linked genes.
   d. genotypes.
   e. phenotypes.

5. A fruit fly has two genes for eye color, but each of its sperm cells has only one. This illustrates
   a. independent assortment.
   b. linked genes.
   c. pleiotropy.
   d. polygenic inheritance.
   e. segregation.

6. Mendel made some crosses where he looked at two characteristics at once—round yellow peas crossed with wrinkled green peas, for example. He did this because he wanted to find out
   a. how new characteristics originated.
   b. whether different characteristics were inherited together or separately.
   c. how plants and animals adapt to their environments.
   d. whether the characteristics influence each other—whether color affects degree of roundness, for example.
   e. Actually, Mendel never had a clear purpose in mind.

7. A pea plant with purple flowers is heterozygous for flower color. Its genotype is *Pp*. The *P* and *p* alleles in the pea plant's cells are located
   a. next to each other on the same chromosome.
   b. at corresponding locations on homologous chromosomes.
   c. on the X and Y chromosomes.
   d. at different locations on the same chromosome.
   e. at different locations on homologous chromosomes.

8. When an individual has both I^A and I^B blood group alleles, both genes are expressed and the individual has group AB blood. This is an example of
   a. codominance.
   b. a dihybrid.
   c. pleiotropy.
   d. incomplete dominance.
   e. linked genes.

9. How many genes are there on one chromosome?
   a. one
   b. two
   c. hundreds
   d. thousands
   e. millions

10. Which of the following is *not* true of linked genes?
   a. They tend to be inherited together.
   b. They violate Mendel's principle of independent assortment.
   c. They are on the same chromosome.
   d. They can form new combinations via crossing over.
   e. They are relatively rare; most genes are unlinked.

11. T. H. Morgan and his students were able to map the relative positions of genes on fruit fly chromosomes by
   a. coloring chromosomes with dyes and observing them under a microscope.
   b. scrambling the chromosomes and observing how the flies changed.
   c. crossing various flies and looking at the proportions of offspring.
   d. transplanting chromosomes from one fly to another.
   e. looking at crosses that showed independent assortment.

12. The sex chromosomes of a human female are ____. The sex chromosomes of a human male are ____.
   a. XX . . . XY
   b. YY . . . XX
   c. XX . . . YY
   d. XY . . . XX
   e. YY . . . XY

13. Most sex-linked traits in humans are carried on the ____ chromosome, and the recessive phenotypes are seen most often in ____.
   a. X . . . women
   b. X . . . men
   c. Y . . . women
   d. Y . . . men

14. The most common fatal genetic disease in the United States is
   a. sickle-cell disease.
   b. cystic fibrosis.
   c. Huntington's disease.
   d. hemophilia.
   e. PKU.

15. The Y chromosome is useful for tracing ancestry because
   a. every human has one.
   b. it carries a large number of genes.
   c. it carries a small number of genes.
   d. it is the location of most sex-linked genes.
   e. it is passed on virtually unchanged.

16. Which of the following human genetic disorders is sex linked?
   a. hemophilia
   b. PKU
   c. cystic fibrosis
   d. sickle-cell disease
   e. all of the above

17. There are various procedures that can be used to detect genetic disorders before birth. Among the tests discussed in this chapter, ____ is the least invasive, while ____ carries the highest risk.
   a. chorionic villus sampling . . . amniocentesis
   b. ultrasound imaging . . . genetic screening
   c. ultrasound imaging . . . chorionic villus sampling
   d. chorionic villus sampling . . . ultrasound imaging
   e. amniocentesis . . . ultrasound imaging

Essay

1. Explain why Gregor Mendel was able to figure out the principles of heredity, while many other investigators before (and some after) Mendel failed to do so.

2. If you flip two coins, the probability that you will get two heads is 1/4, but the probability that you will get one head and one tail is 1/2. Explain why.

3. Why are organisms such as peas and fruit flies better subjects for genetics studies than human beings?

4. What determines a human's sex? Describe two other systems of sex determination in different organisms.

5. What are some of the ethical questions posed by genetic testing?

Apply the Concepts

Multiple Choice

1. A brown mouse is mated with a white mouse. All of their offspring are brown. If two of these brown offspring are mated, what fraction of their offspring will be white?
   a. all
   b. none
   c. 1/4
   d. 1/2
   e. 3/4

2. Suppose you wanted to know the genotype of one of the brown F_2 mice in question 1. The easiest way to do it would be to
   a. keep careful records of the parent mice.
   b. mate it with a brown mouse.
   c. mate it with a mouse of its own genotype.
   d. mate it with a white mouse.
   e. It can't be done.

3. Some dogs bark while trailing; others are silent. The barker gene is dominant, the silent gene recessive. The gene for normal tail is dominant over the gene for screw (curly) tail. A barker dog with a normal tail who is heterozygous for both characteristics is mated to another dog of the same genotype. What fraction of their offspring will be barkers with screw tails?
   a. 3/4
   b. 9/16
   c. 3/16
   d. 1/4
   e. 1/16

4. Two heterozygous tall pea plants with purple flowers are crossed. The probability that one of their offspring will have white flowers is 1/4. The probability that one of their offspring will be short is 1/4. What is the probability that one of their offspring will be short with white flowers?
   a. 0
   b. 1/16
   c. 1/8
   d. 1/4
   e. 1/2

5. A young woman had a baby. Her blood group is AB. The baby's blood group is A. There are two possible fathers: Jim is group A, and Michael is group O. Which man could be the father?
   a. either
   b. Jim
   c. Michael
   d. neither
   e. impossible to tell given this evidence

6. Which of the following illustrates pleiotropy?
   a. In fruit flies, the genes for scarlet eyes and hairy body are located on the same chromosome.
   b. Matings between earless sheep and long-eared sheep always result in short-eared offspring.
   c. Wheat kernels can range from white to red in color, a trait controlled by several genes.
   d. The human cystic fibrosis gene causes many symptoms, from respiratory distress to digestive problems.
   e. An individual with both I^A and I^B alleles has blood group AB.

7. When two gray-bodied fruit flies are mated, their offspring total 86 gray-bodied males, 81 yellow-bodied males, and 165 gray-bodied females. The allele for yellow body is
   a. sex-linked and dominant.
   b. not sex-linked and dominant.
   c. sex-linked and recessive.
   d. not sex-linked and recessive.
   e. impossible to say on the basis of this information.

8. In fruit flies, the allele for red eyes is dominant, and the allele for purple eyes is recessive. Normal gray body is dominant, and black body is recessive. A geneticist mated a heterozygous red-gray male with a purple-black female. She predicted that there would be four phenotypes of offspring in equal numbers, but she was wrong. Instead, 48% of the offspring were red-gray, 46% were purple-black, 3% were red-black, and 3% were purple-gray. She concluded that in the male, the red and gray genes were linked, and in the female the purple and black genes were linked. If this is the case, how would you account for the red-black and purple-gray offspring?
   a. This is an example of pleiotropy.
   b. Body color and eye color are quantitative characteristics.
   c. Crossing over during meiosis recombined the genes.
   d. This cross shows incomplete dominance at work.
   e. Segregation of alleles occurred during meiosis.

9. Red-green color blindness is a human recessive sex-linked trait. A man and a woman with normal vision have a color-blind son. What is the probability that their next child will also be a color-blind son?
   a. 0
   b. 1/8
   c. 1/4
   d. 1/2
   e. 3/4

10. On a pedigree tracing the inheritance of PKU, a horizontal line joins a black square and a half-black circle. What fraction of this couple's children would you expect to suffer from PKU?
   a. none
   b. 1/4
   c. 1/2
   d. 3/4
   e. all

11. Duchenne muscular dystrophy is caused by a sex-linked recessive allele. Its victims are almost invariably boys, who usually die before the age of 20. Why do you suppose this disorder is almost never seen in girls?
   a. Sex-linked traits are never seen in girls.
   b. The allele is carried on the Y chromosome.
   c. Nondisjunction occurs in males but not in females.

   d. Males carrying the allele don't live long enough to be fathers.
   e. A sex-linked allele cannot be passed on from mother to daughter.

12. Which of the following would be most useful for preventing a particular genetic disorder?
   a. knowing how the allele causes its phenotypic effects
   b. being able to identify carriers
   c. a test that can determine whether a fetus suffers from the disorder
   d. knowing which chromosome bears the allele that causes the disorder
   e. tracing the trait back through parents and grandparents

Essay

1. Two apparently normal parents have a daughter who suffers from agammaglobulinemia, an inherited defect of the immune system. Use a Punnett square to show how two normal parents could have a child afflicted with an inherited disease. What are the parents' genotypes? The daughter's genotype? What is the probability that their second child will also have agammaglobulinemia?

2. A pea plant with purple flowers and green pods is crossed with a plant that has white flowers and yellow pods. All the offspring have purple flowers and green pods. If two of these F_1 peas are crossed, what phenotypes will be seen in the F_2 generation, and in what proportions?

3. Freckles is dominant, no freckles recessive. A man with freckles and a woman with no freckles have three children with freckles and one with no freckles. What are the genotypes of the parents and children?

4. The inheritance of flower color in snapdragons illustrates incomplete dominance: When a red snapdragon is crossed with a white one, all their offspring are pink. What offspring would be produced, in what proportions, if two of these pink snapdragons were crossed? What offspring would be produced, in what proportions, if a pink snapdragon was crossed with a white one?

5. A man whose blood group is A and a woman whose blood group is B have a son whose blood group is O. What are their genotypes? What is the probability of the couple's next child having blood group B?

6. Recall that some characteristics, such as skin color, appear to be controlled by several genes. This creates a continuum of variation. If this polygenic explanation for the inheritance of human skin pigmentation is correct, how do the skin colors of the following four individuals compare? Which of the couples could have children with the widest range of skin colors? Why? Couple 1: *aaBbCC* and *aaBbCC*. Couple 2: *AaBbCc* and *AaBbCc*.

7. In fruit flies, the allele for red eyes is dominant over the allele for pink eyes. Straight wings are dominant over curled wings. Imagine that a red-eyed, straight-winged fly that is heterozygous for both characteristics is mated with a fly with pink eyes and curled wings. Predict the offspring that would be produced by this cross (genotypes, phenotypes, and fraction of each) if these two genes were on different chromosomes. When a geneticist actually carried out this mating, the offspring were as follows: 49% red eyes and straight wings, 49% pink eyes and curled wings, 1% red eyes and curled wings, and 1% pink eyes and straight wings. Does this agree with your prediction? How would you explain these results?

8. Numerous fruit fly matings show that the *h* allele for hairy body, the *b* allele for spineless bristles, and the *s* allele for striped body are all located on the same chromosome. The recombination frequency between alleles *h* and *b* is 4%. The recombination frequency between alleles *s* and *b* is 15%. Why are the recombination frequencies between *h* and *b* and between *s* and *b* different? The recombination frequency between alleles *h* and *s* is 10%. What is the order of these three genes on the chromosome?

9. In humans, the presence of a fissure (gap) in the iris of the eye (called "coloboma iridis") is due to a sex-linked recessive gene. Show how an apparently normal couple could have a child with this condition. Is the affected child more likely to be a boy or a girl?

10. Imagine that you are a genetic counselor, and a couple that is planning to have children comes to you for advice. Diane's brother has hemophilia. There is no history of hemophilia in Craig's family. What is the probability that their child will have hemophilia? (Recall that hemophilia is caused by a sex-linked recessive allele.)

Put Words to Work

Correctly use as many of the following words as possible when reading, talking, and writing about biology:

ABO blood group, allele, amniocentesis, carrier, character, chromosome theory of inheritance, codominant, complete dominance, cross, cross-fertilization, cystic fibrosis, dihybrid cross, dominant allele, F_1 generation, F_2 generation, genotype, hemophilia, heterozygous, homozygous, Huntington's disease, hybrid, inbreeding, incomplete dominance, law of independent assortment, law of segregation, linked genes, monohybrid cross, P generation, pedigree, phenotype, pleiotropy, polygenic inheritance, Punnett square, recessive allele, recombination frequency, red-green color blindness, rule of addition, rule of multiplication, self-fertilize, sex chromosome, sex-linked gene, testcross, trait, true-breeding

Use the Web

For more practice using the concepts of genetics, see the activities at www.mybiology.com.

Molecular Biology of the Gene

10

Focus on the Concepts

Genes are made of nucleic acids. This chapter covers the structure and function of DNA, the role of DNA and RNA in the manufacture of proteins, and the molecular biology of bacteria and viruses. Focus on these concepts:

- Experiments with viruses and bacteria demonstrated that DNA is the genetic material. DNA and RNA are polymers of nucleotides. DNA consists of two long polynucleotide strands, twisted in a "double helix" and held together by pairs of complementary nucleotide bases. It is the sequence of bases that carries genetic information.

- The structure of DNA—base pairing—relates to one of DNA's important functions—replication. Helped by enzymes and other cell machinery, the two strands of a DNA separate, and each strand acts as a template for the nucleotide-by-nucleotide assembly of a complementary strand. The end result is two identical double-stranded DNAs.

- A gene consists of hundreds or thousands of DNA nucleotides. The genetic code consists of three-base codons. Each codon represents an amino acid; a sequence of codons spells out an amino acid sequence. Thus the information in the DNA base sequence is expressed in proteins, which shape phenotype.

- The genetic code is "read" to build proteins in a two-step process, consisting of transcription and translation. First, a complimentary messenger RNA molecule is transcribed from a portion of the DNA. In eukaryotes, this RNA may then be "edited," or processed. At a ribosome, mRNA code is then translated into protein. Transfer RNAs match each mRNA codon with a specific amino acid, and the amino acids are linked to form a polypeptide with a specific amino acid sequence.

- Errors in DNA base sequence, called mutations, may change the meaning of genes. Because there are multiple codons representing each amino acid, base substitutions may or may not change the protein. Base deletions or additions alter the three-base reading frame of every codon "downstream" and usually have more drastic effects. Most mutations are harmful, but they do spawn the genetic diversity that makes natural selection and adaptation possible.

- Viruses are simple packages of nucleic acid—DNA or RNA—wrapped in protein, sometimes covered by a membranous envelope. Bacteriophages are viruses that attack bacteria. Other viruses infect animals and plants. A virus infects a host by slipping its nucleic acid into a host cell and taking over. The host cell helps make viral nucleic acid and protein, which then assemble into more viruses, which infect new cells.

- Bacteria reproduce asexually by replicating their DNA and undergoing binary fission. They can produce new combinations of genes by taking up foreign DNA from their surroundings (transformation), acquiring DNA from another bacterium via a virus (transduction), or "mating" and sharing DNA with another bacterium (conjugation).

Review the Concepts

Work through the following exercises to review the concepts in this chapter. For additional review, check out the activities at www.mybiology.com. The website offers a pre-test that will help you plan your studies.

Exercise 1   (Introduction and Modules 10.1–10.3)

Review the discovery that DNA is the genetic material and the structures of DNA and RNA. Then match each phrase on the left with the correct term(s) on the right. Note that some answers are used more than once, and some questions have multiple answers.

_____ 1. The basic chemical unit of a nucleic acid	A. Adenine (A)
_____ 2. The "transforming factor" that alters pneumonia bacteria	B. Base
_____ 3. The two kinds of nucleic acids	C. Cytosine (C)
_____ 4. The three parts of every nucleotide	D. DNA
_____ 5. A pair of these forms a "rung" in the DNA ladder	E. *E. coli*
_____ 6. Used to "label" DNA and protein in experiments	F. Double helix
_____ 7. The component of a bacteriophage that enters the host cell	G. Guanine (G)
_____ 8. Two alternating parts that form the nucleic acid "backbone"	H. Hydrogen bond
_____ 9. The four bases in DNA	I. Radioactive isotope
_____ 10. The DNA base complementary to T	J. Covalent bond
_____ 11. A virus that attacks bacteria	K. Bacteriophage
_____ 12. The substance a phage leaves outside its host cell	L. Protein
_____ 13. Ribose in RNA and deoxyribose in DNA	M. Nucleic acid
_____ 14. Watson and Crick deduced the structure of this molecule	N. Nucleotide
_____ 15. Attacked by herpes virus	O. Centrifuge
_____ 16. The DNA base complementary to G	P. Phosphate
_____ 17. A bacterium attacked by T2 phages	Q. Polynucleotide
_____ 18. The sequence of these encodes DNA information	R. RNA
_____ 19. Eukaryotic chromosomes consist of this and DNA	S. Sugar
_____ 20. The overall shape of a DNA molecule	T. Thymine (T)
_____ 21. Links adjacent nucleotides in a polynucleotide chain	U. Uracil (U)
_____ 22. Machine used to separate particles of different weights	V. Herpesvirus
_____ 23. Links a complementary pair of bases together	W. Nerve cell
_____ 24. The four bases in RNA	
_____ 25. RNA base that is not in DNA	
_____ 26. A polymer of nucleotides	
_____ 27. Causes cold sores, chickenpox, and other diseases	

Review the structure of DNA by labeling these diagrams. Include **nucleotide, polynucleotide, sugar (deoxyribose), phosphate group, sugar-phosphate backbone, pyrimidine bases, purine bases, thymine (T), adenine (A), guanine (G), cytosine (C), hydrogen bond, complementary base pair,** and **double helix.**

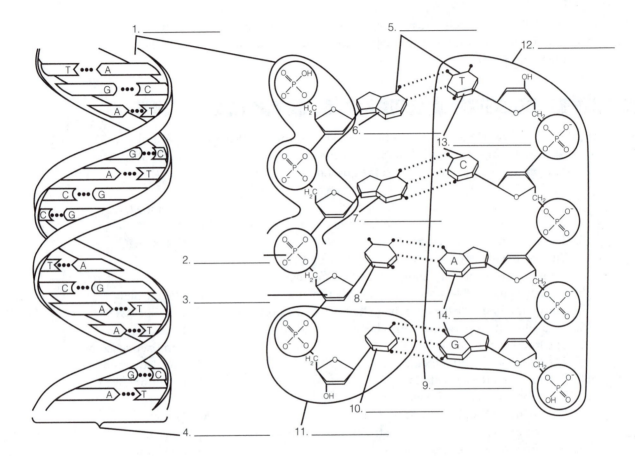

Exercise 3   (Module 10.4)

Reproduction and inheritance involve copying DNA instructions so that they can be passed to the next generation. This process is carried out by DNA polymerases, enzymes that use each strand of the DNA helix as a template on which to build a complementary strand. Review DNA replication by completing the simplified diagrams below. The first diagram shows the parent DNA molecule; label the nucleotides in the right-hand strand. Add five or six nucleotides to the second diagram so that it shows the parent strands separating and being used as templates. (Make sure you match complementary nucleotides correctly!) Complete the third diagram so that it shows two completed daughter molecules of DNA. Color the original DNA strands blue and the new strands gray.

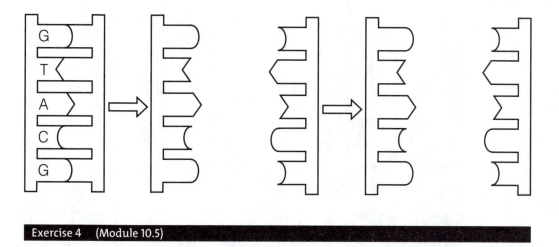

Exercise 4   (Module 10.5)

This module describes some of the ins and outs of DNA replication. Look at the diagrams carefully, and then see if you can match each of the numbers in the boxes on the diagram below with one of the lettered choices. Choices may be used more than once.

A. 5′ end of daughter strand
B. 3′ end of daughter strand
C. 5′ end of parental strand
D. 3′ end of parental strand
E. DNA polymerase
F. where DNA ligase will
    unite pieces

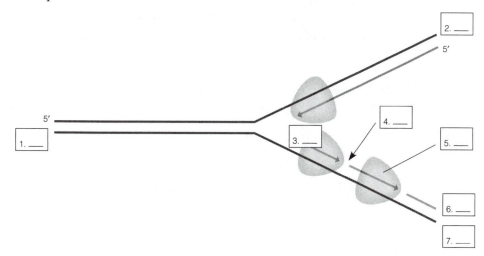

In a cell, the genotype (genetic information in DNA) is expressed as phenotype in the form of proteins—structural proteins that shape the organism and enzymes that carry out metabolism. Review the relationship between genotype and phenotype by completing this crossword puzzle.

Across

3. A gene consists of hundreds or ____ of nucleotide bases.

6. The information in DNA specifies the synthesis of ___.

8. Archibald Garrod noted the gene-protein link in "inborn errors of ____."

9. Genetic instructions are written in three-base "words" called ____.

10. An organism's expressed traits (what it looks like) make up its ____.

12. To make a protein, DNA information is first transcribed into ____.

13. The DNA language consists of a linear sequence of nucleotide ____.

14. An organism's genetic makeup is called its ____.

15. Phenotype is expressed via structural proteins, ____, and other proteins.

17. Making a polypeptide according to an RNA message is called ____.

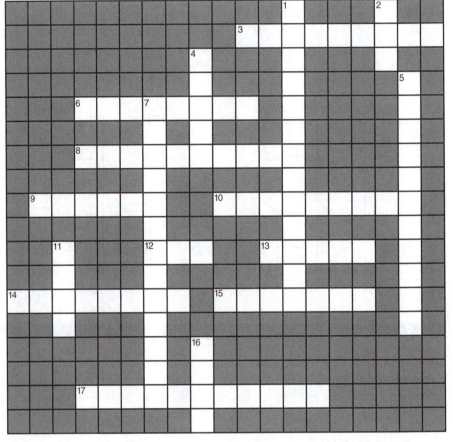

Down

1. The base sequence of RNA is ____ to the DNA from which it is transcribed.

2. Genotype is the inheritable information encoded in ____.

4. Each codon in DNA and RNA specifies a certain ____ acid in a polypeptide.

5. Translation is conversion of an RNA message into a ____.

7. Transfer of information from DNA into an RNA molecule is called ____.

11. One ____ specifies how to build one polypeptide.

16. Using bread ____, Beadle and Tatum showed that a gene codes for an enzyme.

These modules explain how the information in a gene is used to build a protein. Review the processes of transcription and translation by filling in the blanks below.

The first step in making a protein is transcription of a gene. This occurs in the [1]_____ of a eukaryotic cell. An enzyme called [2]_____ carries out the process of transcribing RNA from the DNA. It starts at a specific nucleotide sequence called a [3]_____, next to the gene. RNA polymerase attaches, and the two DNA strands separate. RNA polymerase moves along one strand, and as it does, RNA [4]_____ take their places one at a time along the DNA template. They hydrogen-bond with complementary bases, following the same pairing rules as in DNA—C with G, and U (replacing T in RNA) with A. As the RNA molecule elongates, it peels away from the DNA. Finally, RNA polymerase reaches the [5]_____, a base sequence that signals the end of the gene, and the enzyme lets go of the gene and the RNA molecule. In a prokaryote, the RNA transcribed from a gene, called [6]_____ (mRNA), can be used immediately in polypeptide synthesis. In a eukaryotic cell, the RNA is further modified, or [7]_____, before leaving the nucleus as mRNA. Extra nucleotides are added to the ends of the transcript, and noncoding regions called [8]_____ are removed. The remaining [9]_____ are spliced together to from a continuous coding sequence. The finished mRNA leaves the nucleus and enters the [10]_____, where translation takes place.

Translation of the "words" of the mRNA message into the [11]_____ sequence of a protein requires an interpreter—[12]_____ (tRNA)—which links the appropriate [13]_____ with each [14]_____ in the mRNA message. A tRNA molecule is a folded strand of RNA. At one end, a special [15]_____ links the tRNA to a specific amino acid. The other end of the tRNA molecule bears three bases called the [16]_____, which is complementary to a particular mRNA codon. During the translation process, the tRNA matches its amino acid with an mRNA codon.

[17]_____ are the "factories" where the information in mRNA is translated and polypeptide chains are constructed. A ribosome consists of protein and [18]_____ (rRNA). Each ribosome has a groove that serves as a binding site for mRNA. There are two binding sites for tRNA: The P site holds the tRNA carrying the growing [19]_____, while the A site holds a tRNA bearing the next amino acid.

Translation begins with initiation. An mRNA and a special [20]_____ tRNA bind to the ribosome and a specific mRNA codon, the [21]_____, where translation begins. The initiator tRNA generally carries the amino acid methionine (Met). Its anticodon UAC binds to the start codon, AUG. The initiator tRNA fits into the P site on the ribosome.

The next step in [22]_____ synthesis is elongation—adding amino acids to the growing chain. The anticodon of an incoming tRNA, carrying its amino acid, pairs with the mRNA codon at the open A site. With help from the ribosome, the polypeptide attached to the tRNA in the P site separates from its tRNA and forms a peptide bond with the [23]_____ attached to the tRNA in the A site. Then the "empty" tRNA in the P site leaves the ribosome, and the tRNA in the A site, with the polypeptide chain, is shifted to the P site. The mRNA and tRNA move as a unit, allowing the next codon to enter the A site. Another tRNA, with a complementary anticodon, brings its amino acid to the A site. Its amino acid is added to the chain, the tRNA leaves, and the complex shifts again. In this way, 24_____ are added to the chain, one at a time.

Finally, a 25_____ reaches the A site of the 26_____, terminating the polypeptide. A stop codon causes the polypeptide to separate from the last tRNA and the 27_____. The polypeptide folds up, and it may join with other polypeptides to form a larger 28_____ molecule.

Exercise 7   (Module 10.15)

This module summarizes the key steps in the flow of genetic information from DNA to protein. Study the diagrams carefully, then label the numbered parts and processes.

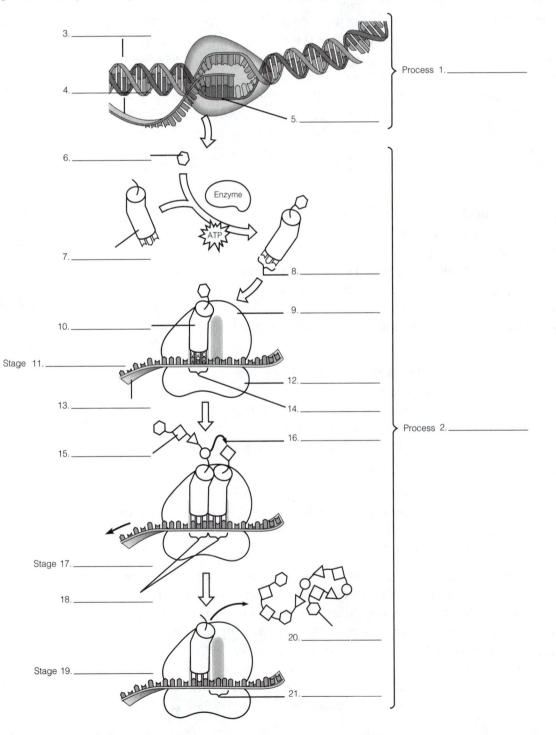

Exercise 8   (Modules 10.8 and 10.16)

These modules describe the genetic code, how biologists cracked the code, and how mutations change the meaning of the coded genetic message. Use the genetic code chart (Figure 10.8A in the textbook) to translate the following mRNAs into amino acid sequences and answer the questions.

mRNA nucleotide sequence:
(mRNA 1)

A U G C C A G A C A A U A U U A A G U G A

1. Amino acid sequence:

Mutation in mRNA:
(mRNA 2)

A U G C C A G A C C A U A U U A A G U G A

2. Amino acid sequence:
3. Number of bases changed in mRNA:
4. Type of mutation:
5. Number of amino acids changed:

Mutation in mRNA:
(mRNA 3; compare to 1)

A U G C C A G A C G A A U A U U A A G U G A

6. Amino acid sequence:
7. Number of bases changed in mRNA (look carefully!):
8. Type of mutation:
9. Number of amino acids changed (compared to mRNA 1):
10. Which mutation had the greatest effect and why?

Exercise 9   (Modules 10.17–10.21)

These modules describe the structures and life cycles of viruses, viroids, and prions. Match each phrase on the left with a term from the right. Some answers are used more than once.

_____ 1. Consists of nucleic acid packaged in protein

_____ 2. When infection leads quickly to breaking open of host cell

_____ 3. Phage DNA inserted into bacterial chromosome

_____ 4. When virus "hides" for a while as part of bacterial chromosome

_____ 5. Responsible for toxins of diphtheria, botulism

_____ 6. Infectious protein that causes mad cow disease

_____ 7. This or DNA may be virus genetic material

_____ 8. Cause of flu, colds, polio, mumps, AIDS

_____ 9. How plant viruses pass from cell to cell

_____10. Used by mumps virus or HIV to attach to host receptors

_____11. Mumps virus reproduces here

_____12. Mumps virus makes this and protein from RNA template

_____13. Mumps virus gets envelope from this part of host cell

_____14. Herpesvirus reproduces here

_____15. Genetic material of herpesvirus

_____16. DNA of herpesvirus "hiding" in host cell DNA

_____17. Small, circular RNA molecule that infects a plant

_____18. Virus protein coat

_____19. Ineffective against viral diseases

_____20. RNA virus that reproduces by first making DNA

_____21. Enzyme that can make DNA from RNA template

_____22. Form in which HIV "hides" in host cell

_____23. Viral disease that may spread from birds to people

_____24. Kind of cell infected by HIV

_____25. Causes an African hemorrhagic fever

_____26. Virus like T2 that infects bacteria

_____27. A deadly respiratory disease caused by an emerging coronavirus

_____28. Virus that causes acquired immunodeficiency syndrome

_____29. Can be used to prevent a viral disease

_____30. Genetic material of HIV

A. RNA viruses
B. Prophage
C. Avian flu
D. Glycoprotein spikes
E. Virus
F. DNA
G. Lytic cycle
H. Vaccine
I. Nucleus
J. Viroid
K. Bacteriophage
L. Provirus
M. HIV
N. Reverse transcrip-
tase
O. Lysogenic cycle
P. Retrovirus
Q. White blood cell
R. Prion
S. Plasmodesmata
T. Plasma membrane
U. RNA
V. Cytoplasm
W. SARS
X. Ebola virus
Y. Antibiotic
Z. Capsid

Exercise 10   (Modules 10.22–10.23)

Bacteria have their own unique genetic characteristics and processes. They can transfer DNA via conjugation, transformation, and transduction. Match the following statements with one of the methods of bacterial DNA transfer. (Some statements are true of all methods of DNA transfer.)

_____ 1. What happened in Griffith's experiment with pneumonia bacteria

_____ 2. DNA may be integrated into chromosome of recipient cell

_____ 3. Taking up of DNA from the surrounding environment

_____ 4. Alters genetic makeup of recipient cell

_____ 5. Figure 5 on the next page

_____ 6. Male and female cells joined by sex pili

A. Conjugation
B. Transformation
C. Transduction
D. All three of the
above

_____ 7. Figure 7 below
_____ 8. Bacterial "mating"
_____ 9. Figure 9 below
_____10. Creates a recombinant cell
_____11. Transfer of genes by a bacteriophage
_____12. May involve transfer of genes by a plasmid
_____13. Usually controlled by a piece of DNA called an F factor
_____14. May transfer R plasmids which enable bacteria to resist antibiotics

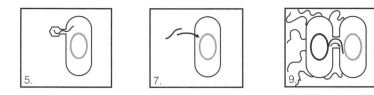

Test Your Knowledge

Multiple Choice

1. In an important experiment, bacteriophages were allowed to infect bacteria. In the first trial, the phages contained radioactive DNA, and radioactivity was detected in the bacteria. Next, other phages containing radioactive protein were allowed to infect bacteria, and no radioactivity was detected in the bacteria. When the experimenters compared the results of these two trials, they concluded that
   a. genes are made of DNA.
   b. bacteriophages can infect bacteria.
   c. DNA is made of nucleotides.
   d. genes carry information for making proteins.
   e. genes are on chromosomes.

2. An RNA or DNA molecule is a polymer made of subunits called
   a. bases.
   b. amino acids.
   c. nucleotides.
   d. nucleic acids.
   e. pyrimidines.

3. The information carried by a DNA molecule is in
   a. its amino acid sequence.
   b. the sugars and phosphates forming its backbone.
   c. the order of the bases in the molecule.
   d. the total number of nucleotides it contains.
   e. the RNA units that make up the molecule.

4. A gene is
   a. the same thing as a chromosome.
   b. the information for making a polypeptide.
   c. made of RNA.
   d. made by a ribosome.
   e. made of protein.

5. DNA replication occurs
   a. whenever a cell makes protein.
   b. to repair gene damage caused by mutation.
   c. before a cell divides.
   d. whenever a cell needs RNA.
   e. in the cytoplasm of a eukaryotic cell.

6. The flow of information in a cell proceeds
   a. from RNA to DNA to protein.
   b. from protein to RNA to DNA.
   c. from DNA to protein to RNA.
   d. from RNA to protein to DNA.
   e. from DNA to RNA to protein.

7. Which of the following is *not* needed for DNA replication?
   a. ribosomes
   b. DNA
   c. nucleotides
   d. enzymes
   e. All of the above are needed.

8. Which of the following processes occur(s) in the cytoplasm of a eukaryotic cell?
   a. DNA replication
   b. translation
   c. transcription
   d. DNA replication and translation
   e. translation and transcription

9. Beadle and Tatum showed that each kind of mutant bread mold lacked a specific enzyme. This experiment demonstrated that
   a. genes carry information for making proteins.
   b. mutations are changes in genetic information.
   c. genes are made of DNA.
   d. enzymes are required to repair damaged DNA information.
   e. cells need specific enzymes in order to function.

10. During the process of translation (polypeptide synthesis), ____ matches a nucleic acid codon with the proper amino acid.
    a. a ribosome
    b. DNA polymerase
    c. ATP
    d. transfer RNA
    e. messenger RNA

11. How does RNA polymerase "know" where to start transcribing a gene into mRNA?
    a. It starts at one end of the chromosome.
    b. Transfer RNA acts to translate the message to RNA polymerase.
    c. It starts at a certain nucleotide sequence called a promoter.
    d. The ribosome directs it to the correct portion of the DNA molecule.
    e. It looks for the AUG start codon.

12. When RNA is being made, the RNA base ____ always pairs with the base ____ in DNA.
    a. U . . . T
    b. T . . . G
    c. U . . . A
    d. A . . . U
    e. T . . . A

13. A mutagen is
    a. a gene that has been altered by a mutation.
    b. something that causes a mutation.
    c. an organism that has been changed by a mutation.
    d. the portion of a chromosome altered by a mutation.
    e. any change in the nucleotide sequence of DNA.

14. How do retroviruses, such as HIV, differ from other viruses?
    a. They are much simpler than other viruses.
    b. They contain DNA that is used as a template to make RNA.
    c. They can reproduce only inside of living cells.
    d. They contain nucleic acids that code for the making of proteins.
    e. They contain RNA that is used as a template to make DNA.

15. The primary difference between bacterial sex and sexual reproduction in plants and animals is that
    a. bacterial sex involves more than two individuals.
    b. bacterial sex does not involve genetic recombination.
    c. bacteria exchange RNA, not DNA.
    d. bacterial sex does not produce offspring.
    e. eggs and sperm are different, but bacterial gametes are all alike.

16. Sometimes a bacteriophage transfers a gene from one bacterium to another. This process is called
    a. transduction.
    b. conjugation.
    c. cloning.
    d. DNA splicing.
    e. transformation.

Essay

1. Sketch a short piece of a DNA molecule, five base pairs long. Use simple shapes to represent bases, sugars, and phosphates. Show proper base pairing, and label a nucleotide, a base, a phosphate group, a sugar, A, C, T, G, the double helix, and hydrogen bonds.

2. Explain why in DNA T pairs only with A and not with C or G.

3. Why does it take a group of three DNA nucleotides to specify one amino acid in a protein? Wouldn't it be simpler to have a one-to-one code, where one nucleotide specified one amino acid?

4. What is a mutation? What causes mutations? Why are most mutations harmful? Why aren't all mutations harmful?

5. Which type of mutation—a base substitution or a base deletion—is likely to have the greatest effect on the organism? Why?

6. Describe step by step, but in simple terms, the roles of mRNA, tRNA, ribosomes, and amino acids in making a polypeptide.

Apply the Concepts

Multiple Choice

1. Which of the following are arranged in the correct order by size, from largest to smallest?
   a. chromosome-gene-codon-nucleotide
   b. nucleotide-chromosome-gene-codon
   c. codon-chromosome-gene-nucleotide
   d. gene-chromosome-codon-nucleotide
   e. chromosome-gene-nucleotide-codon

2. A geneticist raised a crop of T2 bacteriophages in a medium containing radioactive phosphorus so that the DNA of the bacteriophages was labeled with radioactivity. The labeled phages were then allowed to infect nonradioactive bacteria. In a few hours, these bacteria burst open, releasing many bacteriophages. Some of these phages contained labeled
   a. DNA.
   b. RNA.
   c. protein.
   d. all of the above.
   e. DNA and protein only.

3. A messenger RNA molecule for making a protein is made in the nucleus and sent out to a ribosome. The ribosome reads the mRNA message and makes a protein containing 120 amino acids. The mRNA consisted of at least how many codons?
   a. 30
   b. 40
   c. 120
   d. 360
   e. 480

4. The nucleotide sequence of a DNA codon is ACT. A messenger RNA molecule with a complementary codon is transcribed from the DNA. In the process of protein synthesis, a transfer RNA pairs with the mRNA codon. What is the nucleotide sequence of the tRNA anticodon? (Careful—this one is harder than it appears.)
   a. TGA
   b. UGA
   c. ACT
   d. TGU
   e. ACU

5. Imagine an error occurring during DNA replication in a cell so that where there is supposed to be a T in one of the genes there is instead a G. What effect will this probably have on the cell?
   a. Each of its kinds of proteins will contain an incorrect amino acid.
   b. An amino acid will be missing from each of its kinds of proteins.
   c. One of its kinds of proteins might contain an incorrect amino acid.
   d. An amino acid will be missing from one of its kinds of proteins.
   e. The amino acid sequence of one of its kinds of proteins will be completely changed.

6. A cell is grown in a solution containing radioactive nucleotides, so that its DNA is labeled with radioactivity. It is removed from the radioactive solution and grown in a normal medium so that any new DNA strands it makes will not be radioactive. In the normal medium, the cell replicates its DNA and divides. The two daughter cells also replicate their DNA and divide, producing a total of four cells. If a dotted line represents a radioactive DNA strand and a solid line represents a nonradioactive DNA strand, which of the following depicts the DNA of the four cells?

   a.

   b.

   c.

   d.

   e.

7. A particular ____ carry the information for making a particular polypeptide, but ____ can be used to make any polypeptide.
   a. gene and ribosome . . . a tRNA and an mRNA
   b. gene and mRNA . . . a ribosome and a tRNA
   c. ribosome and mRNA . . . a gene and a tRNA
   d. gene and tRNA . . . a ribosome and an mRNA
   e. tRNA and ribosome . . . a gene and an mRNA

8. A sequence of pictures of polypeptide synthesis shows a ribosome holding two transfer RNAs. One tRNA has a polypeptide chain attached to it; the other tRNA has a single amino acid attached to it. What does the next picture show?
   a. The polypeptide chain moves over and bonds to the single amino acid.
   b. The tRNA with the amino acid leaves the ribosome.
   c. The amino acid moves over and bonds to the polypeptide chain.
   d. The tRNA with the polypeptide chain leaves the ribosome.
   e. A third tRNA with an amino acid joins the pair on the ribosome.

9. A microbiologist analyzed chemicals obtained from an enveloped RNA virus (similar to a mumps virus) that infects monkeys. He found that the virus envelope contained a protein characteristic of monkey cells. Which of the following is the most likely explanation for this?
   a. The virus gets its envelope when it leaves its host cell.
   b. The virus forced the monkey cell to make proteins for its envelope.
   c. The virus has a lysogenic life cycle.
   d. The virus gets its envelope when it enters its host cell.
   e. The virus fools its host cell by mimicking its proteins.

10. At one point as a cell carried out its day-to-day activities, the nucleotides G A T were paired with the nucleotides C U A. This pairing occurred
    a. in a double-stranded DNA molecule.
    b. during translation.
    c. during transcription.
    d. when an RNA codon paired with a tRNA anticodon.
    e. It is impossible to say, given this information.

11. Which of the following does not take part in polypeptide synthesis?
    a. an exon
    b. mRNA
    c. an intron
    d. tRNA
    e. a ribosome

12. A microbiologist analyzed the DNA of *E. coli* before and after conjugation. She found that
    a. both cells lost some genes and gained others.
    b. both cells gained genes but lost none of their original genes.
    c. one cell lost genes, and the other gained genes.
    d. one cell gained genes, and the genes of the other were unchanged.
    e. the genes of both cells remained unchanged.

Essay

1. *E. coli* bacteria are used in many genetic studies. Type A *E. coli* can live on a simple nutrient medium because they have all the genes necessary to produce the chemicals they need. Type V *E. coli* can live only on a nutrient medium to which a certain vitamin has been added because they lack a gene that enables them to make this vitamin for themselves. It has been found that bacteria can absorb genes from other dead, ground-up bacteria. Describe an experiment using type A and type V *E. coli* to demonstrate that genes are made of DNA and not protein.

2. It is possible to extract DNA from cells and analyze it to determine the relative amounts of the four DNA bases. The DNA of a goldfish contains more T and less G than human DNA, but in both goldfish and human DNA the amount of T is equal to the amount of A. Explain why.

3. Eric said to Renee, "The amino acid sequence of the proteins in your hair determines how curly or straight your hair will be." Renee replied, "I don't think that's right. Your genes determine whether your hair is curly or straight. That's why it's inherited." Who is right? Explain.

4. The DNA base sequence for a short gene is:

   TATGATACCTTGATAGCTATCTGATTG.

   What is the amino acid sequence of the polypeptide produced according to this DNA information? Use the genetic code chart (Figure 10.8A in the text) and your knowledge of transcription and translation to figure out the message.

5. A biochemist found that a bacterium produced an mRNA molecule consisting of 852 nucleotides and translated this mRNA into a polypeptide containing 233 amino acids. How many nucleotides in the mRNA message would actually be needed to carry the message for the polypeptide, and how many were "extras"? How would the bacterium know *which* nucleotides made up the message?

6. The virus that causes chickenpox can disappear for years and then reappear in a line of painful sores ("shingles") where a nerve cell passes through the skin. How can viruses go away and then reappear like this? Where are the viruses during the intervening period of time?

7. A gene can be removed from a eukaryotic cell and spliced into the DNA of a prokaryotic cell. The prokaryotic can transcribe the gene into mRNA and translate this mRNA into a polypeptide, but the polypeptide has an incorrect amino acid sequence, very different from the polypeptide normally produced by the eukaryotic cell. Why?

8. A mutant strain of *E. coli* bacteria will not grow unless they are supplied with the amino acid lysine. Another strain will not grow without a different amino acid, proline. When *E. coli* of the two strains are mixed, a few bacteria appear in the culture that are able to grow without either of the amino acids. Name and briefly describe three possible mechanisms that might account for this change.

9. Flu viruses and polio viruses are both RNA viruses. Normally, flu viruses attack lung cells and polio viruses attack nerve cells. In the lab, virologists assemble a hybrid RNA virus that has the membranous envelope of a flu virus and the RNA of a polio virus. The researchers then try to infect lung cells and nerve cells with the hybrid. They find that the hybrid can only infect one of the kinds of cells. Which kind of cell would you expect that to be, a lung cell or a nerve cell? Why? After the viruses replicate and fill the infected cell, what kind of RNA would you expect them to contain, flu or polio virus RNA? Why? Which kind of proteins would you expect them to have, flu or polio proteins? Why?

Put Words to Work

Correctly use as many of the following words as possible when reading, talking, and writing about biology:

adenine (A), AIDS, anticodon, bacteriophage, capsid, codon, conjugation, cytosine (C), DNA polymerase, double helix, emerging virus, exon, F factor, genetic code, guanine (G), HIV, intron, lysogenic cycle, lytic cycle, messenger RNA (mRNA), molecular biology, mRNA, mutagen, mutation, nucleotide, phage, plasmid, polynucleotide, prion, promoter, prophage, R plasmid, reading frame, retrovirus, reverse transcriptase, ribosomal RNA (rRNA), RNA polymerase, RNA splicing, rRNA, start codon, stop codon, sugar-phosphate backbone, terminator, thymine (T), transcription, transduction, transfer RNA (tRNA), transformation, translation, triplet code, tRNA, uracil (U), viriod, virus

Use the Web

There is more on molecular biology at www.mybiology.com. The "models" and animations of DNA replication, transcription, translation, and viral and bacterial life cycles are very informative.

The Control of Gene Expression 11

Focus on the Concepts

This chapter describes how genes are controlled and how this relates to differentiation, development, cloning, signal transduction, and what can happen when cells escape their normal controls—cancer. Focus on the following concepts:

- The flow of information from genes to proteins is called gene expression. Regulation of gene expression helps organisms respond to environmental changes. In bacteria such as *E coli*, groups of genes called operons respond to changes. In the absence of the sugar lactose, a repressor protein keeps the operator sequence controlling a group of genes turned off. When lactose is present, it pulls the repressor from the DNA, RNA polymerase attaches to the promoter, the genes are transcribed, and proteins are made that enable the bacterium to use lactose.

- In eukaryotes, differentiation of cells results from differences in gene expression. Regulation can occur at many points from gene to protein: Coiling or "packing" can control access to genes, even inactivating an entire X chromosome. The most important level of control is carried out by transcription factors (and other proteins) that enhance or inhibit attachment of RNA polymerase and thus regulate gene transcription. mRNA can be spliced, blocked, or broken down before translation. Finally, polypeptides and proteins may be modified, activated, or destroyed.

- Cascades of gene expression direct animal development. For example, communication between a fruit fly egg and surrounding cells determines where in the egg genes are transcribed, where certain RNAs and proteins accumulate, and as a result, which end of the fly will be the head and which end the tail. Similar changes determine left and right and division of the fly into segments. Master control genes called homeotic genes regulate batteries of other genes that determine where and when organs develop.

- Signal transduction pathways are important in gene control. Typically, the pathway is activated by a signal molecule that binds to a receptor in the cell membrane. The receptor activates a chain of relay proteins, which actives a transcription factor. The transcription factor triggers a specific gene, which codes for a protein that carries out some function in the cell.

- Cloning demonstrates that differentiated cells may retain all of their genetic potential. Individual plant cells can be made to dedifferentiate and develop into whole plants. Animals are typically cloned via nuclear transplantation— transferring the nucleus of a somatic cell into an egg or zygote. Reproductive cloning can produce genetically identical individuals. Therapeutic cloning generates stem cells, which retain the ability to give rise to all or most kinds of cells in the organism.

- Cancer results from mutations in genes that control the cell cycle, causing uncontrolled multiplication of cells. The affected genes usually code for proteins in signal transduction pathways. Mutation of a proto-oncogene into an oncogene results in manufacture of an overactive protein or excess of a protein that stimulates cell division. Mutation of a tumor-suppressor gene alters a protein that normally activates the production of a cell-division inhibitor.

Review the Concepts

Work through the following exercises to review the concepts in this chapter. For additional review, refer to the activities at www.mybiology.com. The website offers a pre-test that will help you plan your studies.

Exercise 1   (Module 11.1)

Natural selection has favored bacteria that express only those genes whose functions are needed by the cell—in other words, bacteria that can turn genes on and off in response to changes in their environment. In bacteria, genes are grouped, with control sequences called operators and promoters, into clusters called operons. The *lac* and *trp* operons are two such gene clusters that enable the bacterium *E. coli* to respond to its environment. Study the diagrams in Module 11.1 and then match each of the components of the *lac* and *trp* operon systems with its function.

lac operon:

_____ 1. Regulatory gene
_____ 2. Repressor protein + lactose
_____ 3. Repressor protein without lactose
_____ 4. RNA polymerase
_____ 5. Promoter
_____ 6. Operator
_____ 7. Operon genes
_____ 8. Enzymes

A. Keeps RNA polymerase from attaching to promoter and transcribing genes
B. Transcribes genes into mRNA for protein synthesis
C. Repressor protein attaches here
D. Use lactose
E. Information for making repressor protein
F. Where RNA polymerase starts transcribing genes
G. Allows RNA polymerase to transcribe genes
H. Information for making enzymes that use lactose

trp operon:

_____ 1. Regulatory gene
_____ 2. Repressor protein + tryptophan
_____ 3. Repressor protein without tryptophan
_____ 4. RNA polymerase
_____ 5. Promoter
_____ 6. Operator
_____ 7. Operon genes
_____ 8. Enzymes

A. Keeps RNA polymerase from attaching to promoter and transcribing genes
B. Transcribes genes into mRNA for protein synthesis
C. Repressor protein attaches here
D. Make tryptophan
E. Information for making repressor protein
F. Where RNA polymerase starts transcribing genes
G. Allows RNA polymerase to transcribe genes
H. Information for making enzymes that make tryptophan

In eukaryotic organisms, cells become specialized—differentiated—for different functions. Cloning demonstrates that differentiated cells retain a complete set of genes for making all the specialized cells in the whole organism; one specialized cell can be cloned into a whole sheep or cat. Apparently, each differentiated cell *has* all the genes, but *different* genes are *active* in different kinds of cells. That's what makes them different! Check your understanding of this concept by coloring in the boxes below to show which genes you think would be active in each of the cell types.

	Stomach Gland Cell	*Hair Follicle Cell*	*Stem Cell* *(Becomes Red Blood Cell)*
Hemoglobin gene	☐	☐	☐
Keratin (hair protein) gene	☐	☐	☐
Citric acid cycle enzyme gene	☐	☐	☐
Digestive enzyme gene	☐	☐	☐
Insulin gene	☐	☐	☐

Exercise 3   (Modules 11.3 – 11.5)

The first level of genetic control is a variety of mechanisms that can "lock" and "unlock" the DNA itself by controlling access to genes and allowing or disallowing RNA transcription. Complete this crossword puzzle to review the roles of DNA packing and protein activators in gene expression.

Across

2. The color pattern of a ____ cat reflects the influence of chromosome inactivation.

4. Scientists think most eukaryotic regulatory proteins act as ____.

7. One X chromosome in each of a woman's cells is ____.

10. The DNA-histone beaded fiber is further wrapped into a tight ___ fiber.

12. Nucleosomes may control gene ____ by limiting access to DNA.

13. The DNA supercoil is further wrapped and folded into a ____.

15. In eukaryotes, many ____ proteins interact with DNA and one another to turn genes on and off.

17. DNA fits into a cell because of a system of folding, or ____.

18. In eukaryotes, structural and regulatory genes are ____ around the genome.

19. The twisted DNA further coils into a ____ with a diameter of 200 nm.

Down

1. The DNA-histone complex looks like "____ on a string."

3. DNA is wound around small proteins called ____.

5. The first step in initiating gene transcription is binding of activators to sites called ____.

6. A transcription ____ is a protein involved in turning on a eukaryotic gene.

8. A ____ is a complex of DNA wrapped around eight histone molecules.

9. DNA packing seems to control gene expression at the ____ stage.

11. Activators and other proteins help trigger RNA ____ to begin transcription.

14. Activators and enhancers may help position RNA polymerase on a gene's ____.

16. In most eukaryotic cells, most ____ are not expressed.

Exercise 4   (Modules 11.6 – 11.8)

Often gene expression is controlled at the transcription step, but in eukaryotes, gene expression is also regulated after transcription of genes into mRNA and during and after translation of mRNA into protein. Review these processes by matching each of the processes on the left (listed in order of occurrence) with a description on the right.

_____ 1. First step in RNA splicing
_____ 2. Second step in RNA splicing
_____ 3. Alternative RNA splicing
_____ 4. RNA interference
_____ 5. Selective breakdown of mRNA
_____ 6. Inhibition of translation
_____ 7. Activation of finished protein
_____ 8. Selective breakdown of proteins

A. Altering a protein to form an active final product
B. Retaining or destroying mRNA molecules, controlling how much they are translated
C. Action of inhibitors that may block synthesis of protein from mRNA message
D. Joining exons in different ways to produce more than one kind of mRNA (and polypeptide) from a single gene
E. Removal of noncoding introns from RNA
F. Joining of exons to produce mRNA
G. Retaining or destroying proteins, depending on cell's needs
H. Binding of microRNA (miRNA) to mRNA, blocking translation

Exercise 5   (Module 11.9)

This module summarizes the major steps in gene expression in eukaryotes and the key mechanisms that regulate gene expression. After reviewing the module, match each of the mechanisms of regulation with the stage of gene expression at which it acts. Choose from **mRNA breakdown, DNA unpacking and changes, cleavage/modification/activation, protein breakdown, addition of cap and tail, TRANSCRIPTION, splicing, TRANSLATION,** and **flow through nuclear envelope.**

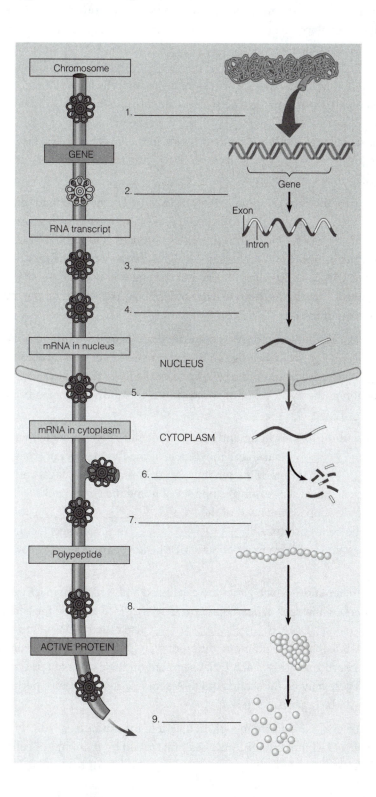

As an animal develops, master genes activate other genes, and these genes signal still others. A chain reaction of gene expression shapes the body from head to tail. Review this cascade of gene expression, how cells in a developing embryo signal each other, and how researchers are unraveling this complex process, by filling in the blanks in the story below.

Powerful new techniques of molecular biology have enabled scientists to explore how gene regulation controls animal development. Researchers have found that one of the first events in fruit fly development is a sequence of changes that determine which end of an egg will develop into the fly's [1]_____ and which will develop into the [2]_____. One of the first [3]_____ that "turns on" in the egg cell codes for a protein that leaves the egg and signals nearby cells in its follicle, or egg chamber. In a follicle cell, the egg protein activates a signal-transduction pathway. The signal protein binds to a specific [4]_____ in the membrane of a follicle target cell, which in turn activates a series of relay proteins in the target cell. The last relay protein activates a [5]_____ factor that triggers transcription of a specific target cell gene. The mRNA produced is then [6]_____ into a protein.

Via this mechanism, the egg cell signals the follicle cells. The new protein formed in the follicle cells then [7]_____ follicle cell genes, and they produce proteins that signal back to the egg cell. One of the egg cell's responses is to localize "head" [8]_____ at the opposite end of the egg cell. This marks where the fly's [9]_____ end will develop. The other end of the egg will become the [10]_____. Similar processes establish the other body axes and thus the layout of the overall body plan of the fly.

After the egg is fertilized, the zygote is transformed into a multicellular embryo by repeated [11]_____. Further signaling and translation creates a cascade of [12]_____ that diffuse through the cell layers of the fly embryo, activating further signal transduction pathways. One of the outcomes is to subdivide the body into a series of sections, or [13]_____.

Protein products of the axis-forming and segment-forming genes now activate another set of [14]_____ that shape the details of the fly. Master control genes called [15]_____ genes determine what body parts—antennae, legs, and so on—will develop in each segment. For example, one set of homeotic genes causes [16]_____ to develop on the head of the fly and [17]_____ on the thorax. Errors in homeotic genes produce [18]_____ flies with spectacular changes in body structure, such as extra pairs of wings, or heads bearing legs instead of antennae.

How can a researcher know which genes are expressed and which are inactive in particular cells during particular stages of development? One way is to analyze the activity of genes in different cells using a DNA [19]_____, also called a DNA chip or gene chip. This is a glass slide that can hold many different [20]_____ stranded DNA fragments in a checkerboard array. Each DNA segment on the chip is obtained from a particular gene. There may be thousands of DNA sequences on the chip, perhaps representing every gene in the organism's genome!

To find out which genes are active in a particular cell, for example, a fruit fly follicle cell, a researcher collects all the different mRNAs transcribed by that kind of cell. This collection of mRNAs is mixed with [21]_____ transcriptase enzyme,

which produces a mix of DNA fragments called copy DNAs or [22]_____.
These cDNAs are modified so they will fluoresce, or glow. Then a small amount of the
cDNA mixture from the cell in question (the follicle cell in this example) is added to each
of the DNA fragments on the chip. If a molecule in the cDNA mixture is
[23]_____ to a DNA fragment at a particular spot on the chip, it will bind to
it. After nonbinding DNA is rinsed away, the pattern of glowing spots enables the researcher to identify which genes are being [24]_____ in the cells from which
the mRNA was obtained. In this way, one can find out which genes are active in different
parts of the organism during different stages in development.

DNA microarrays are valuable tools in other areas of research and medicine. For
example, a DNA chip can test for many different kinds of infectious bacteria at one time,
or distinguish among different subtypes of [25]_____ based on the activity of
17 genes.

Exercise 7   (Modules 11.12 – 11.13)

Cell signaling systems evolved early in the history of life, and the signal transduction pathways of many organisms, from microorganisms to multicellular eukaryotes, show many
common features. Review signal transduction by matching each of the parts and processes
in the diagram below with its name (A–H) and description (P–W).

Names
  A. Transcription
  B. Relay protein
  C. Gene
  D. Receptor protein
  E. Transcription factor
  F. Translation
  G. New protein
  H. Signal molecule

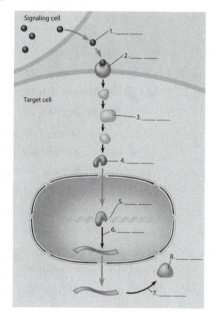

Descriptions
  P. Making a new protein
  Q. Information for making target cell protein
  R. Binds to receptor
  S. Conducts message within target cell
  T. Carries out new function in target cell
  U. Synthesis of specific mRNA coding for
      new protein
  V. When activated, it binds to DNA and
      "turns on" gene
  W. Receives signal molecule

Exercise 8   (Modules 11.14 – 11.17)

Cloning experiments show that differentiated cells retain all of their genetic potential. Stem cells of embryos and adults are able to differentiate into many kinds of cells—useful for reproduction and treating disease. Review cloning and stem cells by matching each phrase with a term from the list on the right.

_____ 1. Partially differentiated cells present in mature animals	A. reproductive cloning
_____ 2. Producing genetically identical organisms for agriculture, research, or saving endangered species	B. nuclear transplantation
	C. differentiation
_____ 3. Naturally occurring animal or human clones	D. embryonic stem cells
_____ 4. Cells that give rise to all specialized cells in the body	E. dedifferentiation
_____ 5. Regrowth of lost body parts	F. adult stem cells
_____ 6. The process of cell specialization	G. regeneration
_____ 7. Ability of adult cell to reverse specialization	H. clones
_____ 8. Growing cells for replacement or repair of damaged or diseased organs	I. therapeutic cloning
	J. identical twins
_____ 9. Genetically identical organisms	
_____10. Replacing the nucleus of an egg or zygote with a nucleus from a differentiated cell	

Exercise 9   (Modules 11.18 – 11.21)

Review the causes and mechanisms of cancer by filling in the blanks in the following story.

In the United States, lung cancer kills about 160,000 people per year. Long one of the most common kinds of cancer in men, lung cancer recently passed breast cancer to become the most frequent cancer in women. There has been a 136% increase in lung cancer deaths among women over a 20-year period.

Cancer is uncontrolled multiplication of cells. Cancer cells have escaped from the normal [1]_____ systems responsible for regulating cell [2]_____. Cancer cells form abnormal masses called tumors, which displace nearby normal tissues and can spread through the body. The cell changes that lead to cancer are caused by accumulated [3]_____ in genes that [4]_____ cell division and still other genes that [5]_____ cell division.

Because the growing tumors block breathing passages, the first symptoms of lung cancer are usually coughing and difficulty breathing. The tumorous masses show up on chest X-rays, and usually a small sample of lung tissue is taken to examine the tumor cells.

What causes lung cancer? Cancer-causing agents are called [6]_____. Radiation, such as X-rays and UV light, are known to cause some cancers, but most are caused by chemicals. Carcinogens in tobacco smoke appear to be the major cause of lung cancer. An increase in cigarette smoking over the last century has been paralleled by a rise in lung cancer rates. Tobacco has also been linked to other forms of cancer, such as

cancer of the mouth and throat. Chemicals in tobacco smoke act as [7]_____,
triggering mutations in lung cells exposed to the smoke.

Scientists have learned a lot about the cellular mechanisms of cancer by studying
cancers caused by viruses in humans and other animals. Researchers were surprised to
find that cancer-causing viruses carry cancer-causing genes, called [8]_____,
as part of their genome. When the viruses insert their genes into the chromosomes of a
host cell, the cancer-causing genes are inserted as well. Even more surprising, researchers
found that oncogenes are simply altered versions of genes normally found in all cells.
These normal genes, called [9]_____, usually code for proteins called
[10]_____ factors—which normally stimulate cell [11]_____, or for
other proteins that affect the cell cycle. Most cancers seem to begin with a
[12]_____ in a proto-oncogene in a body (somatic) cell, turning it into an
oncogene. The altered gene may produce a hyperactive growth stimulating protein, or an
[13]_____ of the normal protein.

Changes in genes whose products normally *inhibit* cell division—so-called
[14]_____ genes—also contribute to the development of cancer. One kind of
tumor-suppressor gene acts to [15]_____ damaged DNA. Others work
through signal-transduction pathways to create inhibitory proteins. It appears that it
takes at least [16]_____ active oncogene and the mutation or loss of
[17]_____ tumor-suppressor genes for a cell to become fully cancerous. This
may explain why the likelihood of cancer increases with [18]_____.

Generally, the normal products of proto-oncogenes and tumor-suppressor genes
are involved in [19]_____-transduction pathways. Normally, the
[20]_____ product of a proto-oncogene (such as one called *ras*) might act to
conduct a signal from a [21]_____ factor to the interior of a cell, activating a
gene that makes a protein that causes [22]_____ to occur. When the proto-
oncogene mutates into an oncogene, it might produce a hyperactive protein that signals
an increase in cell division even in the [23]_____ of the growth factor.

Cell division can also be affected by a mutant tumor-suppressor gene. The tumor
suppressor gene *p53* produces a protein that acts as a [24]_____ factor, which
normally acts at the end of a pathway that promotes production of a protein that blocks
cell division. A mutation in *p53* could produce a defective transcription factor, which cannot trigger transcription. In this case, the inhibitory protein is not transcribed, allowing
an [25]_____ rate of cell division.

We are beginning to understand the genetic and cellular changes that cause cancer to develop. Lifestyle choices can decrease the chances of developing cancer. Regular
[26]_____ can detect tumors early and increase the chances for successful
treatment. Not [27]_____, avoiding overexposure to the [28]_____,
and a [29]_____-fiber, low-[30]_____ diet all reduce cancer risk.

Exercise 10   (Module 11.21)

After reading this module, match each of the following human cancers with the correct associated risk factor(s). Some answers are used more than once.

_____ 1. Lung
_____ 2. Colon and rectum
_____ 3. Breast
_____ 4. Prostate
_____ 5. Urinary bladder
_____ 6. Kidney
_____ 7. Lymphomas
_____ 8. Melanoma (skin)
_____ 9. Liver

A. Ultraviolet light
B. African heritage, possibly dietary fat
C. Tobacco smoke
D. Estrogen
E. Alcohol; hepatitis viruses
F. High dietary fat, smoking, alcohol
G. Viruses (some types)

Test Your Knowledge

Multiple Choice

1. Your muscle and bone cells are different because
   a. they contain different sets of genes.
   b. they are differentiated.
   c. they contain different operons.
   d. different genes are switched on and off in each.
   e. they contain different histones.

2. Operons enable bacteria to
   a. function in frequently changing environments.
   b. increase their genetic diversity
   c. correct mutations that might interfere with their genetic instructions.
   d. differentiate.
   e. mutate and evolve more rapidly.

3. If the nucleus of a frog egg is destroyed and replaced with the nucleus of an intestine cell from a tadpole, the egg can develop into a normal tadpole. This demonstrates that
   a. intestine cells are fully differentiated.
   b. there is little difference between an egg cell and an intestine cell.
   c. an intestine cell possesses a full set of genes.
   d. intestine cells are not differentiated.
   e. frogs can regenerate lost parts.

4. DNA packing—the way DNA is folded into chromosomes—affects gene expression by
   a. controlling access to DNA.
   b. positioning related genes near each other.

   c. protecting DNA from mutations.
   d. enhancing recombination of genes.
   e. allowing "unpacked" genes to be eliminated from the genome.

5. The genes that malfunction in cancer normally
   a. function in DNA replication
   b. are responsible for organizing DNA packing.
   c. code for enzymes that transcribe DNA.
   d. are not present in most body cells unless inserted by a virus.
   e. regulate cell division.

6. In most eukaryotic cells, most genes are not expressed. This suggests that most eukaryotic regulatory proteins act as
   a. exons.
   b. repressors.
   c. introns.
   d. enhancers.
   e. activators.

7. After an mRNA molecule is transcribed from a eukaryotic gene, portions called ____ are removed and the remaining ____ are spliced together to produce an mRNA molecule with a continuous coding sequence.
   a. operators . . . promoters
   b. exons . . . introns
   c. silencers . . . enhancers
   d. introns . . . exons
   e. promoters . . . operators

8. Which of the following mechanisms of gene regulation operates after mRNA transcription but before translation of mRNA into protein?
   a. mRNA splicing
   b. DNA packing
   c. repressors and activators
   d. protein degradation
   e. all of the above

9. Homeotic genes
   a. are responsible for the cellular changes that occur in cancer.
   b. coordinate development by controlling other genes.
   c. are found only in certain cells.
   d. are any genes transferred from cell to cell by viruses.
   e. repair DNA and defend against mutations.

10. In a eukaryote, which of the following may block gene expression by binding to DNA?
    a. an operon.
    b. a histone.
    c. an enhancer.
    d. a promoter.
    e. a silencer.

11. Which of the following appear to have the most potential for use in therapeutic cloning?
    a. adult stem cells
    b. somatic cells
    c. gametes
    d. cancer cells
    e. embryonic stem cells

12. Gene expression in animal development seems to be regulated largely by
    a. controlling gene packing and unpacking.
    b. controlling the transcription of genes into mRNAs.
    c. controlling the translation of mRNAs into protein.
    d. selectively eliminating certain genes from the genome.
    e. selectively breaking down certain proteins so they cannot function.

13. Which of the following is a known or likely carcinogen?
    a. ultraviolet light
    b. cigarette smoke
    c. alcohol
    d. X-rays
    e. all of the above

14. Which of the following is the first thing that happens when a signal molecule acts on a target cell?
    a. A transcription factor acts on the DNA.
    b. The signal molecule binds to the DNA.
    c. A new protein is made in the target cell.
    d. A specific gene is transcribed.
    e. The signal molecule binds to a receptor.

15. Researchers want to test whether a particular combination of 56 suspected genes are active in cancer cells. They might find out by using
    a. reverse transcriptase
    b. a signal transduction pathway
    c. a DNA microarray
    d. therapeutic cloning
    e. nuclear transplantation

Essay

1. In the proper growth medium, a single cell from a Boston fern can be stimulated to grow into an entire plant. (This is how nurseries propagate many houseplants.) What does this signify with regard to cellular differentiation in plants?

2. What are introns and exons? Discuss three possible biological functions of introns.

3. What is a homeotic gene? Why does a mutation in a homeotic gene have a much more drastic effect on the organism than a mutation in other genes?

4. Briefly explain how genes control development of the head-to-tail axis of a fruit-fly embryo.

5. Describe the changes in a cell that can make the cell become cancerous.

6. What is the difference between reproductive and therapeutic cloning?

7. Compare the relative advantages and disadvantages of using embryonic stem cells versus adult stem cells for therapeutic cloning.

8. If a person wishes to avoid cancer, what factors in the environment should he or she try to avoid? What dietary and health habits would you recommend?

Apply the Concepts

Multiple Choice

1. When a certain bacterium encounters the antibiotic tetracycline, the antibiotic molecule enters the cell and attaches to a repressor protein. This keeps the repressor from binding to the bacterial chromosome, allowing a set of genes to be transcribed. These genes code for enzymes that break down the antibiotic. This set of genes is best described as
   a. an exon.
   b. a signal transduction pathway.
   c. an operon.
   d. a homeotic gene.
   e. a nucleosome.

2. A genetic defect in humans results in the absence of sweat glands in the skin. Some men have this defect all over their bodies, but in women it is usually expressed in a peculiar way. A woman with the defect typically has small patches of skin with sweat glands and other patches where sweat glands are lacking. This pattern suggests the phenotypic effect of
   a. a mutation.
   b. chromosome inactivation.
   c. RNA splicing.
   d. an operon.
   e. miRNAs.

3. A bacterium makes the amino acid glycine or absorbs it from its surroundings. A biochemist found that glycine binds to a repressor protein and causes the repressor to bind to the bacterial chromosome, "turning off" an operon. If this is like other operons, the genes of this operon probably code for enzymes that
   a. control bacterial cell division.
   b. break down glycine.
   c. produce glycine.
   d. cause the bacterium to differentiate.
   e. manufacture the repressor protein.

4. In humans, the hormone testosterone enters cells and binds to specific proteins, which in turn bind to specific sites on the cells' DNA. These proteins probably act to
   a. help RNA polymerase transcribe certain genes.
   b. alter the pattern of DNA splicing.
   c. stimulate protein synthesis.
   d. unwind the DNA so that its genes can be transcribed.
   e. cause mutations in the DNA.

5. It is possible for a cell to make proteins that last for months; hemoglobin in red blood cells is a good example. However, many proteins are not this long-lasting. They may be degraded in days or even hours. Why do you think cells make proteins with such short lifetimes if it is possible to make them last longer?
   a. Most proteins are used only once.
   b. Most cells in the body live only a few days.
   c. Cells lack the raw materials to make most of the proteins they need.
   d. Only cancer cells, which can keep dividing, contain long-lasting proteins.
   e. This enables cells to control the amount of protein present.

6. Dioxin, produced as a by-product of various industrial chemical processes, is suspected of causing cancer and birth defects in animals and humans. It apparently acts by entering cells and binding to proteins, altering the pattern of gene expression. The proteins affected by dioxin are probably
   a. enzymes.
   b. DNA polymerases.
   c. transcription factors.
   d. enhancers.
   e. nucleosomes.

7. Which of the following would be most likely to lead to cancer?
   a. multiplication of a proto-oncogene and inactivation of a tumor-suppressor gene
   b. hyperactivity of a proto-oncogene and activation of a tumor-suppressor gene
   c. inactivation of a proto-oncogene and multiplication of a tumor-suppressor gene
   d. inactivation of both a proto-oncogene and a tumor-suppressor gene
   e. hyperactivity of both a proto-oncogene and a tumor-suppressor gene

8. A cell biologist found that two different proteins with largely different structures were translated from two different mRNAs. These mRNAs, however, were transcribed from the same gene in the cell nucleus. What mechanism below could best account for this?
   a. Different systems of DNA unpacking could result in two different mRNAs.
   b. A mutation might have altered the gene.
   c. Exons from the same gene could be spliced in different ways to make different mRNAs.
   d. The two mRNAs could be transcribed from different chromosomes.
   e. Different chemicals activated different operons.

9. Researchers have found homeotic genes in humans, but they are not yet certain how these genes shape the human phenotype. Considering the functions of homeotic genes in other animals, which of the following is most likely to be their function in humans?
   a. determining skin and hair color
   b. regulating cellular metabolic rate
   c. determining head and tail, back and front
   d. determining whether an individual is male or female
   e. regulating the rate and timing of cell division

10. Look at the signal transduction pathway in Figure 11.20A. Which of the following changes might cause a speed-up in cell division? A mutation that
   a. alters the middle relay protein so that it does not respond to the first relay protein.
   b. alters the receptor protein so that it overreacts to growth factor.
   c. alters the receptor protein so that growth factor does not fit.
   d. alters the transcription factor so that it does not attach to the DNA.
   e. alters growth factor molecules so they do not fit the receptor.

Essay

1. Mutations sometimes affect operons. Imagine a mutation in the regulatory gene that produces the repressor of the *lac* operon in *E. coli*. The altered repressor is no longer able to bind to the operator. What effect will this have on the bacterium?

2. Describe how three different types of cells in your body are specialized for different functions. How do their differences reflect differences in gene expression? Suggest a gene that might be active in each of the cells but none of the others. Suggest a gene that might be active in all the cells. Suggest a gene that is probably not active in any of the cells.

3. A biochemist was studying a membrane-transport protein consisting of 258 amino acids. She found that the gene coding for the transport protein consisted of 3,561 nucleotides. The mRNA molecule from which the transport protein was transcribed contained 1,455 nucleotides. What is the minimum number of nucleotides needed to code for the protein? How can the protein be transcribed from an mRNA that is larger than necessary? How can this mRNA be made from a gene that is so much larger?

4. Explain how, in a eukaryotic cell, a gene on one chromosome might affect the expression of a gene on a different chromosome. How might a gene in a certain cell affect expression of a gene in a different cell?

5. A certain kind of leukemia can be caused by a virus, a chemical, or radiation. Explain how these different factors can all trigger identical forms of cancer in the same kind of tissue.

6. Researchers have suggested that it might be possible to clone an extinct woolly mammoth (an ice age elephant) from a tissue sample obtained from a mammoth frozen in a glacier. Describe how this might be done.

Put Words to Work

Correctly use as many of the following words as possible when reading, talking, and writing about biology:

activator, adult stem cell, alternative RNA splicing, carcinogen, clone, differentiation, embryonic stem cell (ES cell), enhancer, gene expression, histone, homeotic gene, nuclear transplantation, oncogene, operator, operon, promoter, proto-oncogene, regeneration, regulatory gene, repressor, reproductive cloning, signal transduction pathway, silencer, therapeutic cloning, transcription factor, tumor-suppressor gene, X chromosome inactivation

Use the Web

For further review of gene control, be sure to access the exercises and questions at www.mybiology.com.

DNA Technology and Genomics

12

Focus on the Concepts

This chapter discusses the methods that allow us to analyze, combine, compare, and study genes—methods that have led us to a new understanding of genes, genomes, and evolution. In studying the chapter, focus on the following concepts:

- Recombinant DNA is formed when nucleotide sequences from different sources are combined. Restriction enzymes are used to cut DNA, and DNA ligase is used to "paste" DNA fragments together. In a vector—a bacterial plasmid, phage, or virus—genes can then be transferred to a different organism. Different DNA fragments in different colonies of cells constitute a genomic library.

- Reverse transcriptase is used to transcribe DNA from mRNA, producing DNA free of introns and identifying the genes expressed in a cell. Specific nucleic acid sequences can be tagged by nucleic acid probes with complementary sequences.

- Recombinant cells and organisms—those whose genes have been augmented by recombinant DNA—can mass-produce genes and gene products. Genetically modified (GM) bacteria, yeasts, and mammalian cells produce hormones, enzymes, and vaccines. GM chickens can produce gene products in their eggs, and sheep and cattle secrete them in their milk. Transgenic farm animals can be modified to grow faster or resist disease. GM crops resist herbicides and pests and manufacture desired nutrients, such as vitamin A. GM organisms raise some health and environmental concerns.

- Genetic diseases may someday be treated with gene therapy—altering an afflicted individual's genes. This has been done by using a retrovirus to insert the normal allele into a patient's cells. Unfortunately, this can damage normal genes in the recipient, and it is difficult to get the transferred gene to function at the right time in the right part of the body. There are also ethical concerns about altering the genome of the recipient and future generations.

- Genetic profiling allows us to identify the DNA of a particular individual or to locate particular nucleotide sequences. The polymerase chain reaction (PCR) is used to amplify small samples of DNA, and gel electrophoresis sorts DNA fragments by size. STP analysis sorts out short random repeats, small segments of repetitive DNA that are unique to an individual. RFLP analysis sorts out nucleotide sequences cut from the DNA at particular restriction sites. DNA profiling is useful in forensics, tracing ancestry, finding disease markers, and determining evolutionary relationships.

- Researchers have so far sequenced the genomes of a variety of animals, plants, fungi, and a large number of microorganisms. The Human Genome Project revealed some surprises: Humans only have about 20,000 genes (about the same as a mouse or a fish), and 98.5% of our DNA is "noncoding" DNA. Cross-species comparison of genomes (and proteomes) offers clues to gene function, disease, and evolution.

Review the Concepts

Work through the following exercises to review the concepts in this chapter. For additional review, check out the activities at www.mybiology.com. The website offers a pre-test that will help you plan your studies.

Exercise 1   (Modules 12.1–12.5)

"Plasmid," "sticky ends," "vector"—DNA technology has its own lingo. Start building your genetic engineering vocabulary by matching each of the terms on the right with a description on the left. (Each answer is used only once.)

_____ 1. Used to cut DNA at a specific location for splicing

_____ 2. Using organisms or their components to make useful products

_____ 3. Direct manipulation of genes for practical purposes

_____ 4. Making multiple copies of gene-sized pieces of DNA

_____ 5. A small piece of bacterial DNA used for gene transfer

_____ 6. DNA transcribed from RNA

_____ 7. Used to "splice" pieces of DNA

_____ 8. A set of techniques for combining genes from different sources

_____ 9. Large "plasmid" that can carry more foreign DNA

_____10. An organism used to clone genes

_____11. A collection of cloned DNA fragments

_____12. Specific location where an enzyme cuts DNA

_____13. An enzyme used to make DNA from an RNA master

_____14. DNA in which nucleotide sequences from different sources are combined

_____15. A gene carrier, such as a plasmid or virus

_____16. A segment of DNA produced by a restriction enzyme

_____17. A virus that attacks bacteria; used to clone genes

A. *E. coli*
B. Genetic engineering
C. Reverse transcriptase
D. Recombinant DNA technology
E. Vector
F. Restriction enzyme
G. Complementary DNA
H. Biotechnology
I. Restriction fragment
J. DNA ligase
K. Recombinant DNA
L. Gene cloning
M. Bacteriophage
N. Restriction site
O. Genomic library
P. Plasmid
Q. Bacterial artificial chromosome

We can engineer bacteria, yeasts, mammal cells, and even entire animals to produce desired genes or proteins. Continue your review of techniques used to cut, splice, clone, and identify genes by filling in the blanks.

Gene engineers use plasmids as [1]_____ to insert genes into bacteria or eukaryotic cells. Imagine that you wanted to build a bacterium capable of making large quantities of human growth hormone (HGH), which is a protein. Your first step would be to obtain the [2]_____ that codes for HGH. One way to do this is to use a [3]_____ enzyme to cut up all the DNA in a human cell. The enzyme recognizes short nucleotide [4]_____ within DNA molecules and cuts the DNA at these [5]_____ sites. Restriction enzymes cut the two DNA strands unevenly, leaving single-stranded ends that can hydrogen-bond with complementary single-stranded [6]"_____ ends." A restriction enzyme can chop up a cell's DNA into thousands of restriction fragments, each consisting of a few genes or parts of genes.

The next step in making human growth hormone is isolating a supply of [7]_____ to use as vectors, for carrying the DNA fragments into bacteria. These are treated with the same restriction enzyme that was used to cut up the human DNA, producing plasmids with sticky ends that are [8]_____ to sticky ends of the human DNA fragments.

Now the human DNA fragments are mixed with plasmids. The sticky ends on the fragments base-pair with the sticky ends on the plasmids, but these connections are weak and temporary. An enzyme called DNA [9]_____, which naturally functions in DNA [10]_____, is used to catalyze the formation of covalent bonds between adjacent nucleotides in the DNA fragments and plasmids. This forms [11]_____ DNA, a DNA molecule with a new, human-made combination of genes. DNA fragments can also be spliced into bacterial [12]_____ chromosomes, or BACs, which are larger than plasmids and can carry more spliced DNA. Or DNA fragments can be spliced into phages, [13]_____ that infect bacteria. The phages reproduce in bacteria to produce libraries of cloned DNA pieces.

In the next step, each recombinant plasmid is added to a bacterium. Under specific conditions, a bacterium will take up the plasmid DNA from solution by the process of [14]_____. The bacterium, with its recombinant plasmid, is allowed to grow and reproduce on a nutrient medium. Each bacterium replicates its own DNA and the plasmid DNA and then divides repeatedly. Each bacterium grows into a colony of identical cells, all containing the recombinant DNA. This production of multiple copies of the genes is called gene [15]_____. Cloning all the different DNA fragments obtained from the human cell produces a genomic [16]_____ of DNA segments. But here are some problems: There are a lot of genes to sort through in a library produced from an entire eukaryotic genome. Plus, eukaryotic genes contain noncoding

17 _____, which must be removed before bacteria can read them. Why not simplify things by looking only at genes expressed in the particular cell you are interested in? You can identify the genes expressed in a particular kind of cell by using the enzyme [18]_____ transcriptase to produce intron-free genes from the mRNA produced in the cell. If you wanted to obtain a human growth hormone gene, the place to start would be a cell from the pituitary gland, where HGH is made. In the cell, the HGH gene (and others) is transcribed into RNA. Enzymes then remove the introns from the RNA and splice the remaining [19]_____ together to make mRNA. If you extract the mRNA from a cell, and add reverse transcriptase (obtained from a [20]_____), the enzyme transcribes a strand of DNA along the mRNA. The RNA is then broken down, and a second DNA strand is synthesized, producing double-stranded DNA. The artificial genes produced this way lack introns, so they are more manageable than the original genes. They also can be transcribed and translated by [21]_____, which lack the ability to deal with introns. Complementary DNA produced in this fashion is cut and pasted into plasmids, using restriction enzymes and ligase, and then cloned in bacteria. There are many mRNA molecules in a pituitary gland cell, so this method also produces a [22]_____ library; however, this library is smaller than a library produced by cutting up the entire genome, because it is limited to the genes actually [23]_____ in a pituitary gland cell.

At this point you have isolated and cloned the HGH gene, but where is it? A genomic library can consist of thousands of bacterial colonies. The bacteria of one of the colonies contain the HGH gene, but which one? One way to look for the gene is to look for HGH, its [24]_____ product. But usually you look for the gene itself, a search that is made easier by using a nucleic acid [25]_____. If you know part of the amino acid sequence of HGH, you can work backward to figure out the probable nucleotide sequence of part of the HGH gene—AAGTGTAG, for example. Now you can produce an artificial RNA (or DNA) molecule with a complementary base sequence—[26]_____, in this case. This complementary molecule is labeled with a [27]_____ isotope or a fluorescent dye, and is called a probe because it can be used to find the gene. To find the bacterial clone that holds the gene, DNA is obtained from each colony of bacteria and treated to separate the DNA strands. The probe is then mixed with the DNA strands, and it hydrogen-bonds only with the recombinant DNA with a complementary base sequence—the HGH gene. Once you have figured out which bacterial colony in the library contains the HGH gene, you can grow these bacteria in larger amounts.

The final step in engineering bacteria to produce human growth hormone is to grow the bacteria in large quantities (usually done in large vats) and extract the protein. The bacteria will manufacture the protein on command if you have spliced the proper control sequences into your recombinant plasmids. Now it is only necessary to collect and purify the protein (and get approval from the Food and Drug Administration!) to start treating patients with recombinant DNA HGH.

Gene-cloning procedures like these have been used to modify organisms to produce a variety of protein products. For example, *E. coli* can be grown in large quantities and induced to [28]_____ protein products into the medium where they live.

They have been engineered to make [29]_____ such as insulin; taxol, a substance used to treat ovarian [30]_____; and proteins from pathogens that are put into [31]_____ that protect us from infectious diseases. Other vaccines are made with viruses and bacteria that have been genetically modified to do no harm. Recombinant DNA techniques are also used to track down and identify [32]_____, such as HIV, and to diagnose and target treatment for cancers.

Sometimes it works better to use eukaryotic cells rather than bacteria to produce a particular protein. [33]_____ cells are the easiest eukaryotes to grow in large quantities, and they also have the ability to take up DNA from the environment and [34]_____ it into their genomes. Yeast cells are currently being used to produce interferons (used to treat cancer and infections) and hepatitis B vaccine. Finally, some proteins can only be produced by modifying cells from mammals. Many proteins secreted by mammalian cells are glycoproteins—proteins with chains of [35]_____ attached to them. Only mammalian cells can make them properly. And why not take this one step farther and harvest the gene products of whole animals or plants rather than cells grown in cell cultures? For example, adding a human blood protein gene to the genome of a sheep enables us to obtain the gene's protein from the animal's [36]_____!

Exercise 3   (Modules 12.1–12.5)

Can you visualize how recombinant DNA techniques are used? The concepts and methods of DNA technology are much easier to understand if you can visualize (in simple terms) what is going on. Test your visual memory by matching each of the diagrams below with one of the following: **isolating a plasmid from *E. coli*; extracting DNA from a eukaryotic cell; obtaining copies of a gene and protein from cloned bacteria; cutting DNA with a restriction enzyme; joining a plasmid and DNA fragment using DNA ligase; cloning recombinant DNA; using reverse transcriptase to make cDNA; using a nucleic acid probe to find a nucleic acid sequence; inserting a plasmid into a bacterium via transformation; and mixing plasmids and DNA fragments with sticky ends**

1. _____

_____

2. _____

_____

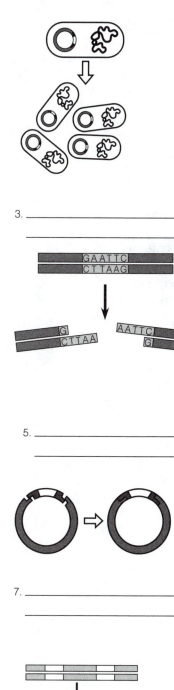

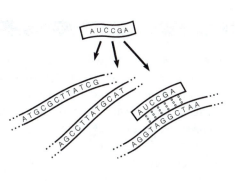

3. _____

4. _____

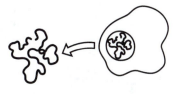

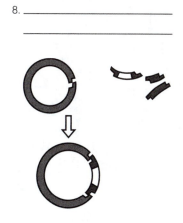

5. _____

6. _____

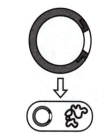

7. _____

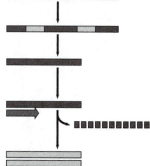

9. _____

8. _____

10. _____

Genetically modified organisms (GM organisms) not only produce pharmaceuticals and medicines, but they have also become important in agriculture. Genes can be introduced into plant cells via Ti plasmids or injected into animal zygotes to produce transgenic crops, farm animals, and "pharm" animals. Match each of the GM organisms below with one of the ways in which it has been genetically modified.

_____ 1. Soybeans, corn, cotton
_____ 2. Rice
_____ 3. Rice
_____ 4. Sheep
_____ 5. Pig
_____ 6. *S. cerevisiae* (yeast)
_____ 7. *E. coli*

A. Gene for beta-carotene production
B. Production of bovine growth hormone (BGH)
C. Hepatitis B gene; produces protein used in vaccine
D. Fat-metabolism gene; produces omega-3 fatty acids
E. Genes for herbicide and pest resistance
F. Gene for blood protein used to treat cystic fibrosis
G. Salinity-resistance gene

Exercise 5 (Modules 12.8–12.9)

Genetically modified organisms promise great benefits, but also present some hazards and ethical questions. Review some of the techniques, uses, and potential problems and safeguards associated with GM organisms by completing this crossword puzzle.

Across

2. Wild and genetically modified plants might exchange genes via their _____.

4. In matters regarding genetically modified organisms, _____ risk is probably unattainable.

5. Transgenic chickens can make foreign proteins in their _____.

11. Genetically modified crops are monitored by the U.S. Department of _____.

13. Genetically modified crop plants can resist pests or _____.

15. _____ rice makes a vitamin A precursor not normally found in rice.

18. Genetically modified microorganisms are _____ so they cannot survive outside the laboratory.

19. To prevent gene transfer between GM and wild plants, GM plants can be engineered so they cannot _____.

20. Researchers use a _____ from a soil bacterium to transfer genes into plant cells.

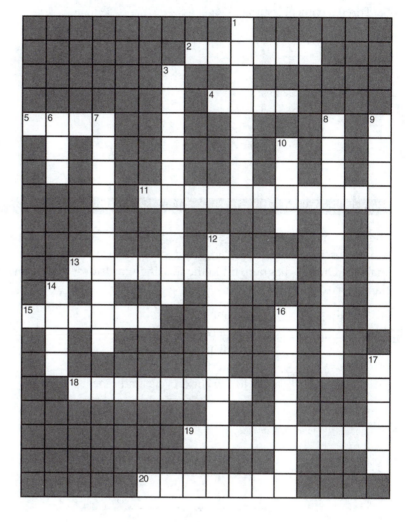

Down

1. There is concern that some people might be _____ to proteins in GM food plants.
3. An organism with a gene from another species is called a _____ organism.
6. _____ is the abbreviation for "genetically modified organism."
7. There is concern that genetically modified plants might give rise to _____ that are difficult to control.
8. Strict government _____ are designed to protect us from genetically modified microorganisms.
9. A transgenic animal can be created by _____ DNA into a fertilized egg.
10. Transgenic mammals can secrete useful proteins in their _____.
12. In the year 2000, 130 countries agreed on a _____ Protocol that requires identification of GM organisms.
14. The U.S. _____ and Drug Administration regulates biotechnology in medicine.
16. A genetically _____ organism has acquired one or more genes by artificial means.
17. A transgenic animal has genes from at least _____ parents.

Exercise 6   (Module 12.10)

Why not simply correct a genetic disease or defect by supplying an affected individual with the correct allele? Gene therapy has great promise, but it has proved difficult to apply. List three technical and three ethical issues raised by gene therapy.

1.   Three technical issues:

2.   Three ethical issues:

Exercise 7   (Modules 12.11–12.12)

The PCR method is used to amplify small DNA samples, from such sources as crime scenes, fossils, or infected cells, in order to produce larger amounts of DNA for study or analysis. On this diagram, fill in the missing words describing how scientists and technicians carry out PCR.

1.   Obtain _____ sample.
2.   _____ to separate strands.
3.   _____ to allow primers to bond.
4.   Allow _____ to add nucleotides to primers.
5.   _____ again to start next cycle, and repeat.

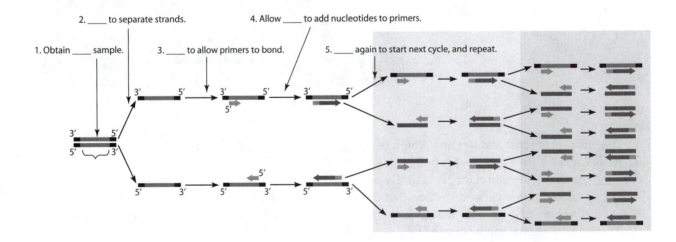

Each person has a unique DNA profile. DNA profiling has been used to identify criminals, exonerate the falsely convicted, trace paternity, identify human remains, reconstruct human ancestry, and even trace the origin of contraband animal products, such as ivory and skins. STR analysis is a widely-used method of profiling that uses gel electrophoresis to separate segments of DNA containing short tandem repeats (STRs) from specific sites in the genome. Imagine that you have been called upon to investigate a rape/homicide. Samples of the rapist's DNA have been recovered from the victim. Meanwhile, the police have taken two suspects into custody—Sam and Joe. Both were seen with the victim the night of the crime, but both say they are innocent. Your job is to compare the perpetrator's DNA (obtained from the crime scene) with samples of Sam's and Joe's DNA, to determine whether either of them should be tried for the crime. Study the diagrams in Module 12.14, and compare them to the diagrams below. First look at the DNA profiles in the electrophoresis gel:

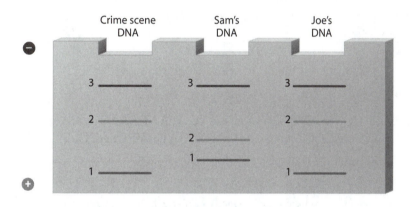

1. Which of the suspects should be held and which should be released?
2. Why did DNA segments from different STR sites move different distances through the gel?
3. Which of the diagrams below depicts Sam's DNA? Which depicts Joe's? How do you know?

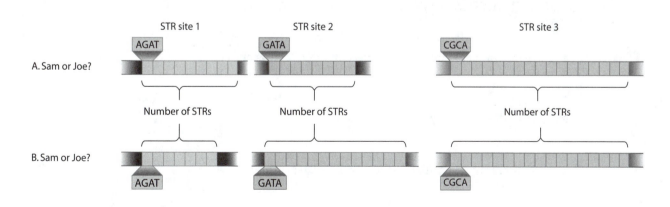

Exercise 9   (Module 12.16)

A genetic marker is a DNA sequence that varies in a population. Single base-pair variations that occur in at least 1% of the population are called single nucleotide polymorphisms, or SNPs. SNPs can occur in coding or noncoding portions of the DNA and can be used to profile DNA or detect particular genes. SNPs can alter a restriction site—where a restriction enzyme cuts DNA—and therefore can change the lengths of restriction fragments produced when the DNA is cut. The DNA fragments are sorted by electrophoresis. Different possible patterns of cutting, sorted by electrophoresis, are called restriction fragment length polymorphisms, or RFLPs. RFLPs can serve as genetic markers for disease-causing alleles.

In this highly simplified example, you will test a brother and sister, Jake and Jennifer, for the allele for Huntington's disease, and compare their DNA to the DNA of their father, John, who has Huntington's. First you will have to cut up DNA samples with a restriction enzyme that cuts at the CCGG restriction site. Start by drawing a line showing where each DNA molecule is cut. (Hint: Remember, DNA strands have 3' and 5' ends. Read the top DNA strand left to right and the bottom strand right to left. Note that CCGG is *not* the same as GGCC!)

John's DNA:
(Huntington's)

    AT GCCGGT ACAT T AGT AGCC GGCAT T TGAACGAT CGT AAT AAAT GGCA
    T ACGGCCAT GT AAT CAT CGGCCGT AAACT T GCT AGCAT T AT T T ACCGT

Jake's DNA:

    AT GCCGGT ACAT T AGT AGCC GGCAT T TGAACGAT CGT AAT AAAT GGCA
    T ACGGCCAT GT AAT CAT CGGCCGT AAACT T GCT AGCAT T AT T T ACCGT

Jennifer's DNA:

    AT GCCGGT ACAT T AGT AGCC GGCAT T TGAACGAT CGT AAT AAAT GGCA
    T ACGGCCAT GT AAT CAT CGGCCGT AAACT T GCT AGCAT T AT T T ACCGT

1. Where does the restriction enzyme cut each piece of DNA? Show above.

2. In how many places does the enzyme cut each sample?

3. How many restriction fragments are produced from each sample?

4. Now show how each collection of restriction fragments will move through the electrophoresis gel. (Do the larger or smaller fragments move faster and farther?)

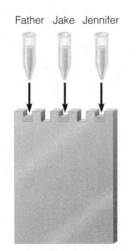

Father   Jake   Jennifer

In actual practice, only a few selected portions of the DNA from each individual would be examined using radioactive probes to identify specific nucleotide sequences, or markers, among the bands on the gel. But in our example, we will just look at the overall pattern of bands. This is the moment of truth.

5. How would you describe the three DNA fingerprints?

6. Did Jake or Jennifer's DNA match the RFLP pattern of Huntington's from their father?

The science of genomics is giving us new information about the genomes of humans and other species. This understanding will help us answer questions about genome organization, regulation of genes, growth and development, and evolution. Mapping the human genome and comparing it to the genomes of other species is yielding new insights into genetic defects and disease. Review genomics and the Human Genome Project by matching each phrase on the left with a term from the list on the right. (Each answer is used only once.)

_____ 1. A complete set of genes
_____ 2. The first organism to have its genome sequenced
_____ 3. Number of genes in a bacterial genome
_____ 4. Number of genomes sequenced and in progress
_____ 5. Most genomes known for these organisms
_____ 6. The first eukaryote whose entire genome was known
_____ 7. The first animal to have its genome sequenced
_____ 8. A plant whose genome has been sequenced
_____ 9. Animals that are like "little people with wings"
_____10. Another animal whose genome is known
_____11. Percentage of our genes that we share with the chimpanzee
_____12. Identified the location and sequence of every human gene
_____13. Human genome database on the Internet
_____14. Number of nucleotide pairs in the human genome
_____15. Estimated number of human genes
_____16. Percentage of human DNA that does not code for protein, tRNAs, or rRNAs
_____17. DNA that has a few nucleotides repeated many times
_____18. Structure at chromosome end with a lot of repetitive DNA
_____19. "Jumping genes" that can disrupt other genes
_____20. Gene for this, related to Alzheimer's, found in rats and finches too
_____21. Number of markers mapped in the first stage of the HGP
_____22. Chopping up the whole genome and mapping it by computer
_____23. The full set of proteins encoded by a genome
_____24. Number of proteins in the human proteome
_____25. Comparison of genomes confirms that these, Bacteria, and Eukarya are the three domains of life
_____26. A speech gene that has evolved rapidly in humans

A. Shotgun method
B. Repetitive DNA
C. 100,000
D. Transposable elements
E. Proteome
F. 5,000
G. Prokaryotes
H. Pneumonia bacterium
I. *FOXP2*
J. Fruit flies
K. *Arabidopsis*
L. Yeast
M. Archaea
N. 96%
O. Telomere
P. 20,000
Q. Genome
R. 1,500
S. Human Genome Project
T. 1,709
U. Mouse
V. Genbank
W. 98.5%
X. Roundworm
Y. Parkinson's disease
Z. 2.9 billion

Test Your Knowledge

Multiple Choice

1. Comparing whole sets of genes, especially among different organisms, is called
   a. transgenics.
   b. genomics.
   c. recombinant DNA technology.
   d. molecular biology.
   e. genetics.

2. Which of the following has been produced by genetically modified microorganisms?
   a. human insulin
   b. human growth hormone
   c. cancer drugs
   d. growth factor for burn treatment
   e. all of the above

3. There are thought to be about ____ genes in a human cell.
   a. 23
   b. 46
   c. 5,000
   d. 20,000
   e. 2.9 billion

4. Comparing the genome of humans with fruit flies, yeasts, and mice
   a. has revealed clues to diseases and evolutionary relationships.
   b. clarifies evolution, but there are few similarities that might be useful in medicine.
   c. has been disappointing to genetic scientists.
   d. shows that these organisms are almost genetically identical.
   e. has revealed clues to diseases, but evolutionary relationships are murky.

5. A genetic marker is
   a. a place where a restriction enzyme cuts DNA.
   b. a chart that traces the family history of a genetic trait.
   c. a nucleotide sequence near a particular gene.
   d. a radioactive probe used to find a gene.
   e. an enzyme used to cut DNA.

6. In recombinant DNA experiments, ____ is used to cut pieces of DNA, and ____ joins these segments to form recombinant DNA.
   a. a restriction enzyme . . . DNA ligase
   b. a transposon . . . a restriction enzyme
   c. a plasmid . . . DNA ligase
   d. DNA ligase . . . a restriction enzyme
   e. a transposon . . . a plasmid

7. A genomic library is
   a. where you look to find out how to make recombinant DNA.
   b. a listing of the known nucleotide sequences for a particular species.
   c. all the genes contained in one kind of cell.
   d. a collection of cloned DNA pieces from an organism's genome.
   e. a place where one can obtain DNA samples from various species.

8. A nucleic acid probe might be used to
   a. insert genes into a host cell.
   b. make DNA for gene cloning.
   c. splice pieces of DNA.
   d. cut pieces of DNA down to manageable size.
   e. find a nucleotide sequence.

9. It is sometimes necessary to genetically engineer mammalian cells to produce proteins because they
   a. can produce larger quantities of protein than bacteria.
   b. can read eukaryotic genes, and bacteria cannot.
   c. can add sugars to make glycoproteins, and bacteria cannot.
   d. are easier to grow than bacteria.
   e. can be induced to secrete proteins into their environment.

10. Which of the following is cited as a possible risk of genetically modified crop plants?
    a. allergic reactions
    b. hybridization with wild relatives
    c. creation of new pests that might be hard to control
    d. all of the above
    e. GM crops actually present no risks

11. Gel electrophoresis is used to
    a. separate fragments of DNA.
    b. clone genes.
    c. cut DNA into fragments.
    d. match a gene with its function.
    e. amplify small DNA samples to obtain enough for analysis.

Essay

1. Explain how DNA segments can be cut and spliced together to produce recombinant DNA. How do the segments "find" each other and stick together? How is recombinant DNA then cloned to produce multiple copies of the gene?

2. Explain why bacteria and yeast are often used as "factories" for gene products.

3. Describe some uses of recombinant DNA technology in agriculture.

4. Describe some uses of recombinant DNA technology in medicine.

5. What are some potential risks of genetically engineered organisms being accidentally or purposefully released into the environment? What kinds of safety measures guard against accidental release? Are you concerned about these risks? Why or why not?

6. Discuss some of the ethical questions raised by recombinant DNA technology. What is the most difficult ethical question concerning human gene therapy?

7. Describe how restriction fragment analysis is done: How is DNA cut into fragments? How are the fragments separated? What does the DNA fingerprint look like, and why? Why do different people have different DNA profiles?

8. Briefly explain what we can learn by comparing the human genome with the genomes of other organisms.

Apply the Concepts

Multiple Choice

1. Which of the following DNA technology procedures does *not* take advantage of the hydrogen bonding of complementary nucleotide bases?
   a. joining a gene with a plasmid vector
   b. Separating nucleic acid fragments using gel electrophoresis
   c. PCR amplification of small amounts of DNA
   d. finding a gene using a nucleic acid probe
   e. using reverse transcriptase to make DNA

2. Connection of "sticky ends" of restriction fragments is
   a. the opposite of separation of fragments via gel electrophoresis.
   b. similar to a nucleic acid probe locating a base sequence.
   c. carried out by DNA ligase.
   d. used to make complementary DNA (cDNA).
   e. allows the polymerase chain reaction (PCR) to copy DNA.

3. A mule is a cross been a horse and a donkey. Why do you think a mule is not considered a "genetically modified organism"?
   a. because horses and donkeys are so much alike
   b. because a mule only inherits horse genes
   c. because a mule is produced naturally, not by artificial means
   d. because half the genes come from each parent
   e. Actually, a mule is considered a genetically modified organism.

4. Because eukaryotic genes contain introns, they cannot be translated by bacteria, which lack RNA-splicing machinery. If you want to engineer a bacterium to produce a eukaryotic protein, you can synthesize a artificial gene without introns (if you know the nucleotide sequence) or
   a. alter the bacteria used so that they can splice RNA.
   b. use a molecular probe to find a gene without introns.
   c. work backward from mRNA to a piece of DNA without introns.
   d. use a phage to insert the desired gene into a bacterium.
   e. use a restriction enzyme to remove introns from the gene.

5. Scientists wished to create an organism capable of breaking down several kinds of toxic wastes, so they combined the genes of several bacteria to create a single "superbacterium." They probably did not need to use which of the following in creating the superbacterium?
   a. nucleic acid probes
   b. gel electrophoresis
   c. plasmids
   d. restriction enzymes
   e. DNA ligase

6. A crop scientist spliced genes for disease resistance into Ti plasmids and then treated tomato plants with the plasmids. Some parts of some plants resisted the disease, but most of the plants eventually died. The researcher could increase his chances for success by
   a. treating single cells and cloning whole plants from the cells.
   b. using molecular probes to figure out where to put the genes.
   c. using bacteriophages rather than Ti plasmids to introduce the genes.
   d. inserting the genes into the cells of the tomato plants with a needle.
   e. employing reverse transcriptase to get the genes into the plants.

7. A molecular biologist used a virus to introduce a gene coding for a certain enzyme into mouse cells. Most of the mouse cells were able to make the enzyme, but most of them lost the ability to make some other protein (different ones in different cells), and many died. Which of the following best explains these results?
   a. The viruses caused the mouse cells to become diseased.
   b. The viruses transferred genes from one mouse cell to another.
   c. The viruses inserted the enzyme gene into mouse cell genes.
   d. The viruses activated transposable elements, which disrupted other genes.
   e. The enzyme acted as a restriction enzyme, cutting up mouse DNA.

8. DNA fingerprints were used to determine whether Evan could be the father of Becky's baby. Which of the following would show that Evan is not the father? If ____ genetic fingerprint showed some bands not in ____ genetic fingerprint.
   a. Evan's . . . the baby's
   b. Becky's . . . the baby's
   c. the baby's . . . Evan's
   d. the baby's . . . Becky's
   e. the baby's . . . Evan's or Becky's

9. Archaeologists unearthed a human skull with a small dried fragment of the scalp still attached. They extracted a tiny amount of DNA from the scalp tissue. How could they obtain sufficient DNA for an analysis of the ancient man's genes?
   a. subject the DNA to electrophoresis
   b. use a nucleic acid probe
   c. use reverse transcriptase
   d. use the polymerase chain reaction
   e. subject the DNA to restriction enzymes

10. There is more than 6,000 times as much DNA in a human cell as in an *E. coli* cell but only about 5 times as many genes. Why?
    a. A human cell has much more noncoding DNA.
    b. The DNA is much more tightly coiled in a human cell.
    c. Most of the genes in a human cell are turned off.
    d. *E. coli* genes are less efficient than human genes.
    e. Human genes are much smaller than *E. coli* genes.

11. Researchers wanted to find out which genes were expressed in muscle tissue, so they ground up several pounds of hamburger (muscle!) and extracted the DNA and RNA from the cells. Their next step should be to
    a. amplify the DNA sample via the polymerase chain reaction (PCR).
    b. cut up the DNA with a restriction enzyme.
    c. subject the DNA to RFLP analysis.
    d. locate the desired genes with a nucleic acid probe.
    e. use the RNA and reverse transcriptase to make complementary DNA (cDNA)

12. Different methods of DNA profiling are useful in different ways. _____ is used to match DNA with that of a particular person, while _____ is used to determine whether particular nucleotide sequences are present.
    a. STR analysis . . . . RFLP analysis
    b. PCR . . . . RFLP analysis
    c. proteomics . . . . genomics
    d. RFLP analysis . . . . STR analysis
    e. RFLP analysis . . . . PCR

Essay

1. Describe how you would go about genetically engineering a bacterium to produce human epidermal growth factor (EGF), a protein used in treating burns. Use the following: DNA ligase, *E. coli*, plasmids, genetic code chart, restriction enzyme, machine for synthesizing a gene, description of the amino acid sequence of EGF, and glassware and equipment for growing and handling bacteria and extracting protein.

2. What are short tandem repeats (STRs)? Briefly describe how STR analysis is used to identify crime suspects.

3. A researcher is searching for the bacterial clone containing a particular cloned gene. She knows that part of the nucleotide sequence of the gene is ATGGCTATC. Explain how she might find the bacteria that contain the gene.

4. A microbiologist developed a strain of *E. coli* that was easily killed by sunlight and whose diet required two unusual amino acids not normally found outside the laboratory. Why would such a bacterium be useful in recombinant DNA work?

5. Manuel, a political activist, "disappeared" during the reign of a dictator. After the dictator was overthrown, human remains thought to be those of Manuel were found buried in a prison compound. A sample of DNA was extracted from the remains and subjected to RFLP analysis, along with DNA samples from Manuel's parents. The DNA profiles are shown below. Could the remains be those of Manuel, the missing activist? Explain your answer.

6. A certain genetic disorder results from the lack of a blood enzyme that is secreted by bone marrow cells. A second disease occurs when nerve cells are unable to produce a particular enzyme. Which of these disorders do you think would be a better candidate for gene therapy and why?

Put Words to Work

Correctly use as many of the following words as possible when reading, talking, and writing about biology:

biotechnology, clone, complementary DNA (cDNA), DNA fingerprinting, DNA ligase, DNA profiling, DNA technology, forensics, gel electrophoresis, gene cloning, gene therapy, genetic engineering, genetically modified (GM) organism, genomic library, genomics, Human Genome Project (HGP), nucleic acid probe, plasmid, polymerase chain reaction (PCR), primers, proteomics, recombinant DNA, repetitive DNA, restriction enzyme, restriction fragments, restriction fragment length polymorphism (RFLP), restriction site, reverse transcriptase, short tandem repeat (STR) analysis, single nucleotide polymorphism (SNP), STRs (short tandem repeats), telomere, Ti plasmid, transgenic, transposable element, vaccine, vector, whole-genome shotgun method

Use the Web

Learn more about DNA technology and genomics, and carry out your own genetic engineering experiments, by trying the activities at www.mybiology.com.

How Populations Evolve

13

Focus on the Concepts

Evolution is the core theme of biology. It explains the diversity of life, the relatedness of all living things, and adaptation of organisms to their environments. As you study this chapter, focus on the following concepts:

- Charles Darwin first proposed natural selection as the mechanism of evolution. He noted that members of a population vary, and most traits are inherited. He further observed that organisms can produce more offspring than the environment can support. Those individuals whose inherited traits give them a greater chance of surviving and reproducing leave more offspring than others. Unequal production of offspring causes favorable traits to accumulate in the population over the generations.

- There is ample evidence for evolution: Scientists have observed many examples of natural selection. Selective breeding illustrates artificial selection. The fossil record documents changes in organisms over time, and biogeography shows how living things adapt to new environments. Comparative anatomy and molecular biology reveal homologies that record the descent and relationships of living things.

- The population is the unit of evolution. The gene pool is all the alleles in the population. Microevolution is change in relative frequencies of alleles in the gene pool.

- Mutation and sexual reproduction (crossing over, independent assortment, and random fertilization) produce genetic variation. Diploidy and balancing selection preserve variation.

- If a population is large, there is no gene flow in or out, there are no mutations, mating is random, and there is no natural selection, the allele and genotype frequencies in the population remain stable from generation to generation. This hypothetical situation is called Hardy-Weinberg equilibrium.

- If the H-W conditions are not met, microevolution occurs: In small populations, sampling error leads to genetic drift. Migration and mutations change allele frequencies. Sexual selection leads to nonrandom mating, which alters the gene pool. The "sorting" effect of natural selection causes change.

- Natural selection—differential reproduction—favors certain phenotypes, genotypes, and alleles. It is measured in terms of fitness, the contribution an individual makes to the next generation. Natural selection is the only process that leads to adaptive evolution. Stabilizing selection favors intermediate phenotypes, directional selection shifts the population in response to a changing environment, and disruptive selection favors extreme types and can lead to two or more contrasting phenotypes.

Review the Concepts

Work through the following exercises to review the concepts in this chapter. For additional review, refer to the activities at www.mybiology.com. The website offers a pre-test that will help you plan your studies.

Exercise 1   (Modules 13.1–13.3)

Charles Darwin was not the first person to ponder the origin of species, and biologists continue to study and document evolutionary change. Match each of the following with their place in unraveling the history of life. Don't focus on names and dates, but rather on how ideas about the origin and history of life have continued to . . . evolve.

_____ 1. Greek philosopher who believed species to be perfect and permanent
_____ 2. Stated that all species were designed by a supreme creator
_____ 3. Research on Galapagos finches documents natural selection
_____ 4. Proposed that acquired characteristics may be inherited
_____ 5. His book about gradual geological change influenced Darwin.
_____ 6. Asserted that populations grow faster than resources
_____ 7. Wrote *The Origin of Species,* explaining "descent with modification"
_____ 8. Conceived a theory of evolution almost identical to Darwin's

A. Charles Darwin
B. Charles Lyell
C. Alfred Russell Wallace
D. Book of Genesis
E. Aristotle
F. Jean Baptiste Lamarck
G. Peter and Rosemary Grant
H. Thomas Malthus

Exercise 2   (Modules 13.1–13.3)

Darwin's key insight was the mechanism of evolution—natural selection. Review Darwin's theory by filling in the blanks in this story.

If you think that the more you mow your lawn, the meaner the weeds get, you may be right. Researchers have found that in lawns that are mown regularly, the dandelions fight back! Dandelions in a regularly mown lawn are shorter and reproduce faster than their ancestors in more "natural" environments. This allows them to produce seeds in between mowings. But how does it happen? What causes the dandelions to change?

[1]_____, the English scientist who first devised the theory of [2]_____, would have explained it this way: Not all dandelions are alike; they [3]_____ in color, size, and rate of maturation. Many of these characteristics are [4]_____, or passed on to offspring. Some dandelions grow slowly and are taller when they mature and produce seeds. Others are faster-growing and shorter when they reach maturity. Apparently height and growth rate are inherited.

Every dandelion flower is capable of producing hundreds, perhaps thousands, of white-tufted seeds in a season. This constitutes an overproduction of offspring, because the [5]_____ can only support so many; only a fraction of offspring will mature and leave offspring of their own. But which offspring will make it? Darwin speculated that those individuals whose inherited characteristics suited them best to their environment would be more likely to [6]_____ and [7]_____ than other, less-suited individuals. This would cause [8]_____ traits to accumulate over generations. When a lawn is mown often, the taller slower-growing dandelions get lopped off. The shorter, faster-growing ones survive to reproduce. Over time, the dandelions get shorter, and they grow faster. Darwin called this unequal reproduction [9]_____. He likened it to the process by which humans pick out and breed individuals with [10]_____ traits, a process called artificial selection.

Many examples of natural selection are known. In a long-term study of Galapagos ground finches, Peter and Rosemary Grant found that individuals with bigger, stronger [11]_____ become more numerous during dry years, when small seeds are in short supply and the birds eat more large seeds. Another example, similar to the dandelion scenario, is the evolution of [12]_____ resistance in insects. The poison might kill most members of an insect [13]_____, but individuals with alleles that protect them best from being poisoned survive and reproduce, passing those alleles onto their descendents. Resistant insects are seen in greater and greater numbers in succeeding generations. Less resistant individuals die out.

Natural selection does not *create* the [14]_____ for dandelion height or pesticide resistance or beak size. It just selects for individuals in the population that already have those alleles. Also, natural selection favors those characteristics that fit the local [15]_____. In a crowded, unmowed field, the situation might be reversed, and taller, slower-growing dandelions might have the advantage. Organisms do not *try* to evolve; evolution is not [16]_____ -directed. It does not lead to the perfect dandelion, insect or finch.

Finally, note that natural selection involves differences among individuals, but *individual* dandelions do not evolve. An individual does not change its growth rate. But because there is heritable variation in the survival and reproduction of individuals with different alleles for different traits, the [17]_____ of dandelions changes. Over time, the dandelions in your yard evolve; they [18]_____ to their environment.

Review fossils by completing this crossword puzzle.

Across

1. Most fossils are the hard parts of animals, such as bones, teeth, and ____.
5. A ____ is a scientist who studies fossils.
10. Fossils trace the evolution of mammals from a ____ ancestor.
11. Rock layers are called ____.
12. Biologists rely mainly on fossils preserved in ____ rock.
16. An entire organism may be preserved if bacteria and fungi do not ____ the corpse.
17. The oldest fossils are ____ from 3.5 billion years ago.
18. The ____ ages of fossils can be deduced from the layers in which they are found.

Down

2. Many fossil species no longer exist; they are ____.
3. Scientists have studied whole ____ frozen in ice.
4. A fossil seashell might be an empty ____ filled with sediments.
6. Fossils show that ____ evolved from four-legged land mammals.
7. The fossil ____ is the sequence of fossils within layers of rock.
8. Ankle bones show that whales are related to pigs, ____, cows, camels, and deer.
9. Sedimentary rocks form from layers of mud or ____.
13. Footprints, burrows, and tracks are called ____ fossils.
14. Fossils found in the deepest strata are generally the ____.
15. Entire insects can be perfectly preserved in ____.
16. Fossils demonstrate that past organisms ____ from existing ones.

Exercise 4 (Modules 13.3–13.4)

In addition to fossils, examples of natural selection, and examples of artificial selection, there are other kinds of evidence for evolution: biogeography, comparative anatomy, and molecular biology. Each of the examples below illustrates which of these kinds of evidence?

Category   *Example*

_____ 1. Fertilized eggs of earthworms, insects, and snails all go through the same pattern of cell division.

_____ 2. The same gene seems to be involved in the formation of eyes in many animals.

_____ 3. The genes of humans and chimpanzees are about 96% identical.

_____ 4. Remains of upright-walking but small-brained apes have been found in Africa.

_____ 5. All animals with backbones have 12 pairs of nerves extending from the brain.

_____ 6. Bacteria quickly become resistant to antibiotics.

_____ 7. The American Kennel Club recognizes more than 150 dog breeds.

_____ 8. Many unique marsupial mammals live in Australia.

_____ 9. A protein called albumin is very similar in dogs and wolves, less similar in dogs and cats.

_____ 10. The farther an island is from the mainland, the more unique its plants and animals.

_____ 11. The limbs of all land vertebrates have one upper "leg" bone and two lower "leg" bones.

_____ 12. Humans have small tail bones, and occasionally a baby is born with a short tail.

_____ 13. Animals called trilobites were common in the oceans 300 million years ago, but they have been extinct for millions of years.

_____ 14. In all living things, AUG is the mRNA "start" codon.

_____ 15. Some whales possess vestigial hind leg and foot bones.

_____ 16. Cabbage, cauliflower, and broccoli were all bred from a wild mustard.

Exercise 5 (Modules 13.5–13.6)

A human arm and a bat wing are homologous structures—features that may have different functions but are structurally similar because of common ancestry. Such homologies, whether anatomical or molecular, are important clues to evolutionary relationships. Check off each of the examples in Exercise 4 (above) that illustrate homologous structures.

Exercise 6   (Modules 13.7–13.8)

These modules introduce evolution of populations. After reading the modules, circle the word that best matches each statement

1. A group of individuals of the same species: *gene pool, population*

2. What natural selection acts on: *individual, population*

3. What actually evolves: *individual, population*

4. All the alleles in all the individuals in the population: *genome, gene pool*

5. Change in relative frequencies of alleles in the gene pool: *microevolution, macroevolution*

6. Causes of variation (circle all that apply): *mutation, natural selection, sexual reproduction*

7. Portion of variation relevant to natural selection: *acquired, genetic*

8. Produces new alleles: *mutation, sexual reproduction*

9. Where most mutations occur: *body (somatic) cells, gametes*

10. Mutations that affect population's variability: *somatic mutations, mutations in gametes*

11. Most mutations: *harmful, helpful*

12. Important source of helpful variation: *deletion mutation, duplication mutation*

13. Organisms in which mutation alone generate most variation: *eukaryotes, prokaryotes*

14. Mutation rate in most plants and animals: *1/100,000, 1/1000*

15. Generates most new gene combinations in animals and plants: *mutation, sexual reproduction*

16. Sexual processes that generate variation: (circle all that apply) *crossing over, independent assortment, random fertilization*

Exercise 7   (Modules 13.9–13.10)

Microevolution is the change of frequencies of alleles in the gene pool. To see what happens when microevolution occurs, it is helpful to first look at a hypothetical population that is *not* evolving—a population at "Hardy-Weinberg equilibrium." Imagine a population of 100 annual wildflowers, some red and some yellow. The red allele, R, is dominant; the yellow allele, r, is recessive. There are 36 RR plants in the population, 48 Rr plants, and 16 rr (yellow) plants. If the population is at Hardy-Weinberg equilibrium, what will be the frequencies of the various genotypes and the frequencies of the two alleles, R and r, in the next generation? Follow the example in Module 13.9 as a guide, and fill in the blanks below to figure out the frequencies for this example.

First, figure out genotype frequencies for the current generation:

	Red	Red	Yellow
A. Phenotypes			
B. Genotypes	____	____	____
C. Number of plants (total 100)	____	____	____
D. Genotype frequencies (number of genotypes/100)	____	____	____

Next, figure out the frequencies of *R* and *r* alleles in the gene pool:

E. Number of *R* alleles in gene pool ____ ____ ____

F. Number of *r* alleles in gene pool ____ ____ ____

G. Allele frequencies
   (number of *R* alleles/200
   or number of *r* alleles/200)   Frequency of R: p = ____   Frequency of r: q = ____

Now you know the frequency of *R* and *r* gametes these plants will produce:

H. Gamete frequencies
   (= allele frequencies)   Frequency of R: ____   Frequency of r: ____

Now you can use the rule of multiplication to calculate the frequencies of the three possible genotypes of plants in the second generation:

I. Phenotype   Red   Red   Yellow

J. Genotype   ____   ____   ____

K. Genotype frequencies ____   ____   ____

Now you can figure out the frequencies of *R* and *r* alleles in the gene pool for the second generation (assuming the population stays at 100 individuals):

L. Number of *R* alleles in gene pool ____ ____ ____

M. Number of *r* alleles in gene pool ____ ____ ____

N. Allele frequencies
   (number of *R* alleles/200
   or number of *r* alleles/200)   Frequency of R: p = ___ Frequency of r: q = ____

O. What happened to the genotype and allele frequencies in the second generation?

What would you predict for the third generation? Why?

P. Did the number of red plants or red alleles increase? Isn't the red allele dominant? Explain.

Hardy-Weinberg equilibrium is an idealized model. Equilibrium is maintained only if five conditions are met. This happens only in the fertile imaginations of biologists, not in real populations. Real populations always deviate from one or more of the conditions, and their gene pools change over time. The Hardy-Weinberg equilibrium is nevertheless a useful standard with which to compare real populations whose gene pools are changing slowly. (This makes it useful in figuring out the frequency of alleles and carriers of genetic diseases, for example.)

Let's continue to look at the wildflower population introduced in Exercise 7. If it is like other real populations, its gene pool is changing. For each of the scenarios below, state which of the Hardy-Weinberg conditions the population deviates from, and explain what agent of microevolution causes the gene pool to change. Also state which of these deviations would cause the flowers to adapt to their environment.

1. A windstorm blows in hundreds of seeds from a nearby meadow, where nearly all the flowers are yellow.

2. A pollutant molecule from the soil gets in the way just as a developing egg cell in one of the red flowers is replicating its DNA. Quite by chance, a red allele is transformed into a yellow allele.

3. The flowers tend to grow in red or yellow patches. A landslide buries and kills most of the red flowers.

4. The red pigment in the petals of the red flowers is poisonous and protects them from beetles that eat the developing seeds. The yellow flowers are not protected in this way.

5. The bees that pollinate the flowers tend to develop a "search image." Once they start visiting flowers of a certain color, they stick to that color. So pollen from red flowers is more likely to be delivered to other red flowers, and pollen from yellow flowers is more likely to fertilize other yellow flowers.

Returning one last time to the wildflowers discussed in Exercises 7 and 8, complete the following scenario regarding fitness and natural selection.

An early writer on evolution described natural selection as "nature red in tooth and claw." This may be true for cheetahs and antelope on the Serengeti, but natural selection is usually more subtle. All living things are engaged in what Charles Darwin called a 1"_____ for existence." Natural selection may be "survival of the 2_____," but fitness is more than simply brute strength. Biologists define 3_____ as the relative contribution that an individual makes to the gene pool of the next generation. It has more to do with reproduction than strength or cunning.

For the wildflowers in our previous example, the struggle for existence involves physical traits such as color and shape of leaves and metabolic characteristics such as efficiency in capturing sunlight and resistance to cold. The [4]_____ of the plant is a composite of all its characteristics. Only the [5]_____, not the [6]_____, is exposed to the environment. A red flower may be protected from predation, while a yellow flower is eaten. Genes for red color increase fitness, and these genes are more likely to be passed on to the next generation. The frequency of red alleles in the [7]_____ increases, as does the frequency of red individuals in the [8]_____.

If the red flowers have an advantage, how is it that we still see yellow individuals? Why has the yellow [9]_____ not been eliminated from the gene pool? One reason is that the flowers are [10]_____—each individual has two sets of chromosomes and two genes for each character. In heterozygous red flowers, yellow alleles are hidden, or protected, from natural selection. These hidden alleles constitute a "reserve" of genetic [11]_____, keeping the gene pool from becoming too uniform. The yellow alleles might become useful if the [12]_____ changes.

It is also possible for natural selection to maintain both alleles in the gene pool—a phenomenon called [13]_____ selection. For example, the heterozygous red flowers could have some advantage over both the homozygous red or yellow forms—improved disease resistance, for example. This is called [14]_____ advantage, and would maintain the yellow allele in the gene pool. Another type of balancing selection is [15]_____-dependent selection, which maintains two different phenotypic forms in a population by selecting against either form if it becomes too common. An herbivore might develop a "search image" and eat the common red flowers, while ignoring the yellow ones. As the red flowers disappear, the herbivore then might notice the yellow ones. Finally, much of the genetic variation in population probably has little or no impact reproductive success and is considered [16]_____ variation.

Most flowers contain both male and female parts, but because males and females are distinct in most animal species, natural selection often shapes differences in appearance, such as antlers or bright plumage. Such difference is called sexual [17]_____. Among some species, these [18]_____ sexual characteristics might be used in fighting over females. This is called [19]_____ selection. A more common form of sexual selection, called [20]_____ selection, involves [21]_____ choice. Usually males display their bright "plumage" and females do the choosing. According to the "good [22]_____" hypothesis, bright plumage, a long tail, or a loud mating call might reflect overall male health.

There are three different ways in which natural selection can act on the variation in a population. Many characteristics are not simple "either/or" alternatives like red and yellow flower color. Characteristics such as height vary continuously and can be described by a [23]_____-shaped curve. There may be a few very short plants in the population, a majority of plants of medium height, and a few very tall plants. Imagine our wildflowers growing in a cold, windy environment. Very tall plants might freeze before their seeds mature. Shorter plants would stay warmer, but very short plants might have trouble dispersing their seeds to favorable environments. In this kind of situation, [24]_____ selection favors the intermediate variants, not too tall and not too short. Next imagine a situation where the environment is gradually becoming drier. In this case, [25]_____ selection might favor those individuals genetically programmed to grow the deepest roots. This kind of natural selection is most common during periods of environmental [26]_____. Finally, [27]_____ selection occurs when two different sets of environmental conditions favor the extreme phenotypes and act against [28]_____ types. For example, plants with shallow, spreading roots might be at an advantage in dry rocky soil, where water tends to penetrate quickly. At the same time, a deep taproot system might be favored in richer soil that holds water longer. Intermediate root systems would be at a disadvantage in both environments.

Will natural selection ever shape a perfect flower? Probably not. [29]_____ can only act on existing variations, which depend on history and chance. And because a plant must do many different things, adaptations are often [30]_____.

Test Your Knowledge

Multiple Choice

1. During his voyage around the world, Charles Darwin was inspired to think about evolution by
   a. books that he read.
   b. fossils he collected.
   c. studying adaptations of organisms to their environments.
   d. unique organisms he saw in the Galápagos Islands.
   e. all of the above.

2. _____ and _____ generate variation, while _____ results in adaptation to the environment.
   a. genetic drift . . . natural selection . . . mutation
   b. mutation . . . sexual reproduction . . . natural selection
   c. overproduction of offspring . . . mutation . . . sexual reproduction
   d. natural selection . . . mutation . . . sexual reproduction
   e. sexual reproduction . . . natural selection . . . mutation

3. Breeding of plants and animals by humans is called
   a. natural selection.
   b. balancing selection.
   c. founder effect.
   d. artificial selection.
   e. intersexual selection.

4. Microorganisms can adapt to changes in the environment by means of mutation alone because
   a. they are so small in size.
   b. their populations are very isolated from one another.
   c. most of their mutations are helpful, rather than harmful.
   d. they multiply so rapidly.
   e. their populations are so large.

5. The smallest unit that can evolve is a
   a. species.
   b. genotype.
   c. population.
   d. gene.
   e. individual.

6. "Differential reproduction" is just another way of saying
   a. natural selection.
   b. mutation.
   c. variation.
   d. recombination.
   e. genetic drift.

7. Which of the following changes in the gene pool results in adaptive evolution?
   a. nonrandom mating
   b. genetic drift
   c. natural selection
   d. gene flow
   e. mutation

8. The ultimate source of all genetic variation is
   a. natural selection.
   b. genetic drift.
   c. sexual recombination.
   d. the environment.
   e. mutation.

9. In evolutionary terms, an organism's fitness is measured by its
   a. health.
   b. contribution to the gene pool of the next generation.
   c. mutation rate.
   d. genetic variability.
   e. stability in the face of environmental change.

10. Organisms that possess homologous structures probably
    a. are headed for extinction.
    b. evolved from the same ancestor.
    c. have increased genetic diversity.
    d. by chance had similar mutations in the past.
    e. are not related.

11. Recombination of alleles occurs when chromosomes are shuffled in _____ and fertilization.
    a. mitosis
    b. genetic drift
    c. natural selection
    d. mutation
    e. meiosis

12. Darwin
    a. was the first person to conceive that organisms could change over time.
    b. believed that organisms could pass on acquired changes to offspring.
    c. was the first biologist to win the Nobel Prize.
    d. worked out the mechanism of evolution—natural selection.
    e. was the first to realize that fossils are remains of ancient organisms.

13. Endangered species are often subject to _____ which _____ genetic diversity.
    a. balancing selection . . . reduces
    b. the bottleneck effect . . . reduces
    c. disruptive selection . . . increases
    d. neutral variation . . . reduces
    e. genetic drift . . . increases

Essay

1. Explain how heritable variations, overproduction of offspring, and limited natural resources cause a species to adapt to its environment.

2. Briefly describe five categories of evidence for evolution.

3. What is the difference between a population and a species?

4. What are homologous structures? Give some examples. Why are homologous structures considered important evidence for evolution?

5. Horses look a lot like zebras. How would Darwin have explained this?

6. Sometimes a harmful allele may be present in the gene pool at a relatively high frequency. How might this be explained in terms of genetic drift? How might it relate to the fact that most organisms are diploid? What might it have to do with heterozygote advantage?

7. Describe how each of the following might alter the gene pool: genetic drift, nonrandom mating, gene flow, mutation, and natural selection.

Apply the Concepts

Multiple Choice

1. In a population of black bears, which would be considered the fittest?
   a. the biggest bear
   b. the bear having the most mutations
   c. the healthiest bear
   d. the strongest, fiercest bear
   e. the bear that leaves the most descendants

2. A geneticist mixed together many different kinds of fruit flies—some with long wings, some with short wings, some with red eyes, some with brown eyes, and so on. He allowed the flies to feed, mate randomly, and reproduce by the thousands. After many generations, most of the flies in the population had medium wings and red eyes, and most of the extreme types had disappeared. This experiment appears to demonstrate
   a. stabilizing selection.
   b. neutral variation.
   c. diversifying selection.
   d. genetic drift.
   e. fitness.

3. Because of global climate change, researchers have discovered that Arctic wildflowers are blooming weeks earlier than they did a few decades ago. The earlier blooming time seems to be genetic. This appears to be an example of
   a. directional selection.
   b. diversifying selection.
   c. bottleneck effect.
   d. sexual dimorphism.
   e. genetic drift.

4. From its first settlement in the 1700s and well into the 1900s, the island of Martha's Vineyard, Massachusetts, had a particularly high incidence of hereditary deafness. Off the island, about one person in 6,000 is deaf. In Chilmark, one of the villages on Martha's Vineyard, the incidence of deafness was one person out of 25! Which of the following is the most likely explanation for this high incidence of deafness?
   a. natural selection
   b. mutation
   c. founder effect
   d. heterozygote advantage
   e. sexual selection

5. Biologists have noticed that most human beings enjoy sex. How would they explain this, in evolutionary terms?
   a. If sex were not enjoyable, the human species would have died out.
   b. Early humans who enjoyed sex most had the most babies.
   c. Only body structures evolve, not behavior, so enjoyment cannot evolve.
   d. This was due to a random mutation, so it did not affect evolution.
   e. Like most people, biologists are baffled by the phenomenon of sex.

6. Which of the following would result in evolutionary adaptation of a mouse population to its environment?
   a. Half the mice are killed by an avalanche.
   b. A mutation for spotted fur occurs.
   c. Several mice leave the area and mate with individuals elsewhere.
   d. Mice with thicker fur best survive a cold winter.
   e. Mice are most likely to mate with close neighbors.

7. The relationship of genome to organism is the same as that of _____ to population.
   a. species
   b. gene
   c. gene pool
   d. mutation
   e. variation

8. Some critics of evolution believe that the theory of evolution is flawed because it is based on random changes—mutations. They say that a random change in an organism (or a car or a TV set) is likely to harm it, not make it function better. What logical statement could a defender of evolution make in reply to this criticism?
   a. The fossil record proves without a doubt that mutations drive evolution.
   b. Mutation is random, but natural selection has a nonrandom "sorting" effect.
   c. Mutation has little to do with evolution.
   d. This is a weak spot in the theory that remains to be worked out.
   e. Mutations are not actually random.

9. A zoologist found that in the population of frogs in MacGregor's pond, half the genes for skin color in the gene pool were alleles for green spots and half the genes were alleles for brown spots. Which of the following could cause these proportions to change?
   a. A drought shrinks the pond so that only five frogs remain.
   b. Females prefer to mate with brown-spotted males.
   c. Green-spotted frogs can hide more easily among the pond weeds.
   d. Filling in a nearby pond causes those frogs to move to MacGregor's pond.
   e. Any of the above could cause the proportions to change.

10. Each of us is part of the ongoing evolution of the human species. Which of the following occurrences would have the greatest impact on the future biological evolution of the human population?
   a. You work out every day so that you stay physically fit and healthy.
   b. A mutation occurs in one of your skin cells.
   c. You move to Hawaii, the state with the longest life expectancy.
   d. A mutation occurs in one of your sperm or egg cells.
   e. You encourage your children to develop their intellectual abilities.

11. The frequencies of dominant and recessive alleles in a gene pool are 0.2 and 0.8. If this population is at Hardy-Weinberg equilibrium, what is the frequency of heterozygotes in the population?
   a. 0.04
   b. 0.16
   c. 0.2
   d. 0.32
   e. 0.64

12. We know a lot about fossil crabs, snails, and corals, but not much about ancient seaweeds. Why do you suppose this is the case?
    a. There were no seaweeds in ancient oceans.
    b. Seaweeds were too soft to fossilize well.
    c. Animal life was much more abundant than seaweeds in ancient times.
    d. Plants moved onto land, leaving only animals in the sea.
    e. A mass extinction wiped out the seaweeds, but animals survived.

Essay

1. A butterfly has a long tubelike proboscis that it uses to suck the nectar from a certain kind of deep, tubular flower. Its close relatives have much shorter proboscises. How might Lamarck have explained the existence of this long proboscis? How would Darwin have explained it?

2. A long section of *The Origin of Species* describes the breeding of pigeons and how pigeon breeders have produced many shapes, sizes, and colors of pigeons, all starting from a common wild pigeon. Why do you suppose Darwin thought this was important?

3. Jason and Lindsay collected some wildflower seeds from a meadow and scattered them along the roadside near their home. As the weather got hotter, some of the seedlings began to wither. Jason said, "Hey, those scrawny plants are going to have to adapt or they are going to die." Lindsay replied, "*They* may not adapt, but the wildflower *population* might." What did she mean?

4. A population of 100 fish in an aquarium consists of 49 homozygous dominant green individuals, 42 heterozygous green individuals, and 9 white individuals. What are the frequencies of the three genotypes in the population? What are the frequencies of the green and white alleles in the gene pool? If the population is at Hardy-Weinberg equilibrium, what will be the genotype and gene frequencies in the next generation of fish?

5. Describe five ways in which the fish population in Question 4 could be caused to deviate from Hardy-Weinberg equilibrium. Is this likely? Which of these deviations might cause the fish to become better adapted to their environment?

6. An ecologist studying predators and their prey in White Sand Dunes State Park found that nocturnal pocket mice exist in dark and light morphs. In this area, 42% of the mice are dark and 58% are light. Because owls swallow their prey whole and then cough up "pellets" consisting of bones and fur, the ecologist was able to discover that 61% of the mice caught by owls were dark, and 39% were light. What is likely to happen to the pocket mouse population in the future, and why?

Put Words to Work

Correctly use as many of the following words as possible when reading, talking, and writing about biology:

adaptation, artificial selection, balancing selection, biogeography, bottleneck effect, directional selection, disruptive selection, evolution, evolutionary tree, extinction, fitness, fossil record, fossil, founder effect, frequency-dependent selection, gene flow, gene pool, genetic drift, Hardy-Weinberg equilibrium, heterozygote advantage, homologous structures, homology, microevolution, molecular biology, mutation, natural selection, neutral variation, paleontologist, population, sexual dimorphism, sexual selection, stabilizing selection, strata, vestigial organ

Use the Web

For further review of the concepts of evolution, see the activities at www.mybiology.com.

The Origin of Species

14

Focus on the Concepts

Charles Darwin titled his famous book *The Origin of Species*. Speciation—the evolution of new species—is the bridge between microevolutionary change and broader macroevolutionary themes. As you study this chapter, focus on these major concepts:

- There are several ways to define a species: The biological species concept defines a species as a group of populations whose members have the potential to interbreed in nature and produce fertile offspring. The morphological species concept is based on size, shape, and other body features. The ecological species concept identifies a species in terms of its roles in biological community, and the phylogenetic species concept defines a species as the smallest group with a common ancestor.

- Reproductive barriers keep species separate by restricting gene flow between populations. Prezygotic barriers, such as differences in structure or behavior, prevent mating or fertilization. Postzygotic barriers, such as reduced hybrid viability or fertility, prevent the development of fertile adults.

- In allopatric speciation, geographic isolation can lead to speciation. Populations split off by a canyon or mountain range can be changed by genetic drift and natural selection. Speciation occurs if the gene pool of an isolated population undergoes changes that establish reproductive barriers between it and the parent population.

- Sympatric speciation occurs without geographic isolation. Sometimes this is a result of habitat preference or sexual selection. A common form of sympatric speciation in plants involves polyploidy, a change in the number of chromosomes. Errors in cell division may produce fertile tetraploid plants that are reproductively isolated from the diploid parent population. More often, hybridization of two species followed by chromosome duplication can result in a self-fertile plant isolated from both parent species.

- Reproductive barriers may result from genetic changes as populations adapt to different environments. Natural selection can reinforce a genetic barrier, reducing the formation of unfit hybrids. Or, if reproductive barriers are not strong, two hybridizing species may fuse into one. Sometimes, stability develops in the hybrid zone, with gene exchange between two species via hybrids.

- Formation of many diverse species from a common ancestor is called adaptive radiation. This occurs when a few organisms enter a new, unexploited area, or when environmental changes cause numerous extinctions, and new species can fill new habitats or roles. The fossil record suggests that much speciation occurs relatively rapidly, with periods of stability in between. Other examples are more gradual. In either case, timelines are long—tens of thousands to millions of years.

Review the Concepts

Work through the following exercises to review the concepts in this chapter. For additional review, check out the activities at www.mybiology.com. The website offers a pre-test that will help you plan your studies.

Exercise 1   (Modules 14.1–14.2)

There are several ways that biologists define a species. Several are listed below. Match each of the descriptions on the left with a species concept on the right.

_____ 1. The smallest group that shares a common ancestor; one branch on the tree of life

_____ 2. Body form; shape, size and other features; what the organism looks like

_____ 3. Populations whose members can interbreed in nature and produce fertile offspring.

_____ 4 Organisms with a particular niches or roles in the biological community.

A. Biological species concept

B. Ecological species concept

C. Phylogenetic species concept

D. Morphological species concept

Exercise 2   (Modules 14.2–14.3)

The biological species concept is one of the most useful ways to define a species, but it is not foolproof. Briefly explain why it might be difficult to apply the biological species concept in each of the following situations.

1. Fossils of "Java Man" and "Peking Man" are both thought to represent a single species—*Homo erectus*.

2. A tiger and a lion can interbreed in a zoo and produce a hybrid offspring called a tiglon.

3. Dogs come in many shapes and sizes, from Chihuahuas to Saint Bernards.

4. There are many strains and species of *Streptococcus* bacteria, which reproduce asexually.

5. Among *Clarkia* wildflowers in California, flowers of population A can interbreed and produce fertile offspring when crossed with flowers from population B. Similarly, B can interbreed with C. But A and C cannot successfully interbreed.

6. One bird guide calls flycatchers of the genus *Empidonax* "the bane of bird-watchers." Several species look so much alike that birders can distinguish them only by their songs.

7. A song sparrow population in Baja, California, is separated from other song sparrows by over a hundred miles of desert.

Review the reproductive barriers that separate species by categorizing the following examples. State whether each barrier is prezygotic (pre) or postzygotic (post), and then name the specific kind of barrier (such as temporal isolation or reduced hybrid viability) it exemplifies. Table 14.3 is a helpful summary.

Pre or Post	*Kind of Barrier*	*Example*
1. _____	_____	The salamanders *Ambystoma tigrinum* and *A. maculatum* breed in the same areas. *A. tigrinum* mates from late February through March. *A. maculatum* does not start mating until late March or early April.
2. _____	_____	Two species of mice are mated in the lab and produce fertile hybrid offspring, but offspring of the hybrids are sterile.
3. _____	_____	When fruit flies of two particular species are crossed in the lab, their offspring are unable to produce eggs and sperm.
4. _____	_____	A zoologist observed two land snails of different species that were trying to mate with little success because they apparently did not "fit" each other.
5. _____	_____	Male fiddler crabs (genus *Uca*) wave their large claws to attract the attention of females. Each species has a slightly different wave.
6. _____	_____	When different species of tobacco plants are crossed in a greenhouse, the pollen tube usually bursts before the eggs are fertilized.
7. _____	_____	Blackjack oak (*Quercus marilandica*) grows in dry woodlands, and scrub oak (*Q. ilicifolia*) grows in dry, rocky, open areas. Pollen of one species seldom pollinates the other.
8. _____	_____	The tiglon offspring of a lion and a tiger are often weak and unhealthy.

Allopatric speciation often begins with geographical isolation. For each of the organisms listed below, name two geographical barriers that might block gene flow between populations and possibly lead to allopatric speciation.

1. Daisy

2. Mouse

3. Trout

4. Oak tree

5. Sparrow

6. Sea star

Exercise 5   (Modules 14.5–14.6)

Many species of plants have arisen from accidents in cell division that result in extra sets of chromosomes. A change in chromosome number produces a reproductive barrier that isolates organisms from their parent populations. This is one way that sympatric speciation—speciation in the same geographic area as the parent species—can occur. Study the examples given in Modules 14.5 and 14.6. Then state whether you think each of the organisms *in italics* below would be able to reproduce (yes or no), whether it could represent a new species (yes or no), and how many chromosomes would be present in its cells.

Repro-duction	New Species	No. of Chrom.	Example
1. ____	____	____	A flower has 18 chromosomes ($2n = 18$). There is an error in meiosis, and diploid gametes are formed. The *zygotes formed* are tetraploids, capable of self-fertilization when mature.
2. ____	____	____	A tree has 22 chromosomes ($2n = 22$). An error in meiosis produces a diploid pollen grain, which fertilizes a normal haploid egg. The resulting polyploid zygote develops into a *full-grown tree.*
3. ____	____	____	Antelope of two different species mate in a zoo. Species A has 36 chromosomes; species B has 34 chromosomes. They produce a *hybrid offspring.*
4. ____	____	____	A lily of species A has 16 chromosomes. Species B has 20 chromosomes. A hybrid with 18 chromosomes undergoes chromosome duplication, producing a *lily* that is capable of self-pollination.
5. ____	____	____	A botanist treats some tissue from a strawberry plant ($2n = 14$) with a chemical that disrupts the mitotic spindle. This doubles the number of chromosomes in a cell. This cell is then cultured on a special medium, until it eventually develops into a *strawberry plant* that can self-pollinate.
6. ____	____	____	Pollen from the cell-cultured strawberry in Question 5 is placed on the flower of a plant of the parent species, producing a triploid *hybrid zygote.*

Exercise 6   (Modules 14.4–14.11)

Speciation can occur when populations are geographically isolated (allopatric speciation) or when polyploidy or other factors restrict gene flow within an area (sympatric speciation). Reproductive barriers may then be reinforced or may break down. Review speciation and its aftermath by using the concepts and terms from this chapter to complete the following story about (imaginary) butterflies and asters.

The yellowspot butterfly is found over hundreds of square miles of land in the delta of an African river. Its primary food source is a species of purple aster—a flower related to daisies and dandelions. Patches of asters are scattered in sunny meadows in the delta, some several miles apart. The butterflies do not usually venture far for food. Each patch of asters supports a separate [1]_____ of yellowspots, but because the butterflies sometimes wander and mate with butterflies in other areas, until recently all the yellowspot butterflies have been classified as members of the same [2]_____.

Insect taxonomists noted that one population of yellowspots was centered across the main river channel from the other populations. They suspected that the river might act as a [3]_____ to the butterflies, since they do not usually fly far over water. The researchers examined the butterflies from the isolated population and found that the butterflies from across the river were a bit smaller than most yellowspots and more orange in color, so the researchers nicknamed them "orangespots." The biologists found that the differences in appearance were inherited. They suspected these could reflect [4]_____ due to chance differences in the butterflies that founded the orangespot population. The researchers also noted that the environment was slightly warmer and drier on the far side of the river, so [5]_____ may have caused the orangespots to adapt to conditions there.

The scientists started to suspect that the two populations could represent different species. To test this hunch, they had to find out whether butterflies from the far side of the river could [6]_____ with individuals from the main population. The biologists captured some butterflies from both areas and placed them in a cage. Surprisingly, the orangespots and yellowspots largely ignored each other. The researchers found that the female butterflies rest on leaves and flash their wing spots to attract the males. Yellowspot females flash their wings at a much faster rate than orangespot females. Apparently the wing-flashing display acts as a [7]_____zygotic reproductive barrier. Apparently, [8]_____ isolation keeps the two populations of butterflies from interbreeding. In a few instances, the eggs of an orangespot female were fertilized by a yellowspot male, and vice-versa, but the embryos soon died. Apparently, there is also a [9]_____zygotic reproductive barrier between the butterflies. This particular type of barrier is called [10]_____. When hybrids are less fit than parent species, natural selection can [11]_____ reproductive barriers, reducing the production of unfit hybrids, and preventing [12]_____ of the two populations in the [13]_____ zone. (Studies of other species, such as snails and monkeyflowers, have shown that a change in a single [14]_____ can create a such a barrier.) The researchers concluded that the reproductive barriers were effectively preventing interbreeding of the butterflies, making the yellowspots and orangespots separate [15]_____, according to the [16]_____ species concept.

A study of river sediments showed that the channel separating the two butterfly populations formed about a thousand years ago, when the river shifted course. It has indeed acted as a geographical barrier, [17]_____ the two populations from one another and eventually leading to [18]_____ speciation. A thousand years seems like a long time to us, but this is a relatively rapid rate of speciation. It would seem to support the [19]_____ equilibrium model, where bursts of speciation alternate with long periods of stability.

While studying the habits of the butterflies, the biologists turned their attention to the flowers, and found a group of asters with oval-shaped leaves and slightly larger flowers than all the others. A study of their chromosomes showed that the unusual plants were [20]_____, having more than the usual two sets of chromosomes. All the other purple asters in the area had two sets of chromosomes and a $2n$ chromosome number of 26. The unusual plants were [21]_____, having four sets of chromosomes, for a total of 52. An error must have occurred during [22]_____ in one of the diploid asters, creating diploid gametes. Because the plant could self-pollinate, zygotes were formed with four sets of chromosomes. The plants that developed from these zygotes could interbreed with one another, but they were reproductively 23_____ from the parent species. They represented a new 24_____ of aster, produced in one generation through the process of 25_____ speciation.

Exercise 7   (Modules 14.9–14.10)

Speciation and adaptive radiation often occur on islands. The Galápagos and Hawaiian Islands provide many examples. But an island does not have to be a bit of land in the midst of an ocean for geographic isolation to occur. As the cichlids of Lake Victoria illustrate, an evolutionary island could be, for example, a lake surrounded by land. Try to come up with four other examples and the kinds of species that could be isolated and evolve in each.

1.

2.

3.

4.

Exercise 8   (Modules 14.1–14.11)

This chapter describes many examples of speciation. For one last review, try to match each description below with the correct species.

_____ 1. Mammal species overlap on Great Plains, but mate at different times.

_____ 2. Eighty percent of these are descended from ancestors formed by polyploid speciation.

_____ 3. Different colors of two species attract different pollinators.

_____ 4. Different diets led to reproductive barriers in a laboratory experiment.

_____ 5. These animals are thought to have originated via polyploid speciation.

_____ 6. Hundreds of species evolved in Lake Victoria.

_____ 7. Different species have different beaks and diets; sometimes hybridize.

_____ 8. Diverged after a population was split by the Grand Canyon.

_____ 9. Change in one gene made it physically impossible for them to mate.

A. Gray tree frogs
B. Cichlids
C. Antelope squirrels
D. Darwin's finches
E. Monkeyflowers
F. Fruit flies
G. Japanese land snails
H. Spotted skunks
I. Plants

Test Your Knowledge

Multiple Choice

1. According to the biological species concept, two animals are considered different species if they
   a. look different.
   b. cannot interbreed.
   c. live in different habitats.
   d. are members of different populations.
   e. are geographically isolated.

2. Which of the following is the first step in allopatric speciation?
   a. genetic drift
   b. geographic isolation
   c. polyploidy
   d. hybridization
   e. formation of a reproductive barrier

3. Bacteriologists usually use the morphological species concept to define species because
   a. bacteria often hybridize.
   b. bacterial reproduce sexually.
   c. bacteria are so small.
   d. bacteria reproduce very rapidly.
   e. bacteria reproduce asexually.

4. A new species can arise in a single generation
   a. through geographic isolation.
   b. in a very large population that is spread over a large area.
   c. if a change in chromosome number creates a reproductive barrier.
   d. if allopatric speciation occurs.
   e. because of hybrid breakdown.

5. The evolution of numerous species, such as Darwin's finches, from a single ancestor is called
   a. adaptive radiation.
   b. sympatric speciation.
   c. punctuated equilibrium.
   d. polyploidy.
   e. geographic isolation.

6. According to the _____ model, evolution occurs in spurts; species evolve relatively rapidly, then remain unchanged for long periods.
   a. allopatric
   b. gradual
   c. adaptive radiation
   d. punctuated equilibrium
   e. geographic isolation

7. Most of the time, species are identified by their appearance. Why?
   a. If two organisms look alike, they must be the same species.
   b. This is the criterion used to define a biological species.
   c. If two organisms look different, they must be different species.
   d. This is the most convenient way of identifying species.
   e. Most organisms reproduce asexually.

8. Most species of _____ are descended from ancestors that underwent sympatric speciation by polyploidy.
   a. plants
   b. bacteria
   c. land mammals
   d. insects
   e. fish

9. Individuals of different species living in the same area may be prevented from interbreeding by responding to different mating dances. This is called
   a. ecological isolation.
   b. hybrid breakdown.
   c. mechanical isolation.
   d. temporal isolation.
   e. behavioral isolation.

10. It is unlikely that the human population will give rise to a new species because
    a. the human population is too large.
    b. geographic isolation is unlikely to occur.
    c. a change in chromosome number would be fatal.
    d. the human population is too diverse.
    e. natural selection cannot affect humans.

Essay

1. Give three situations in which it might be difficult to use the biological species concept to decide whether two organisms are of the same or different species.

2. Describe step-by-step how geographic isolation could lead to speciation.

3. Describe what has to happen for two species with different numbers of chromosomes to interbreed and produce a fertile hybrid. Will this hybrid be able to interbreed with either of its parents? Why? How common is this in nature?

4. State whether a large, widely distributed population or a small, isolated population is more likely to undergo speciation, and explain why.

5. Compare the gradual and the punctuated equilibrium models of species evolution. What would the fossil record of speciation look like if it supported the gradual model? What would it look like if punctuated equilibrium were the case? Which model does the fossil record seem to support most often? Why?

Apply the Concepts

Multiple Choice

1. Three species of frogs—*Rana pipiens, Rana clamitans,* and *Rana sylvatica*—all mate in the same ponds, but they pair off correctly because they have different calls. This illustrates a _____ and _____ .
   a. prezygotic barrier . . . behavioral isolation.
   b. postzygotic barrier . . . hybrid breakdown.
   c. prezygotic barrier . . . temporal isolation.
   d. postzygotic barrier . . . behavioral isolation.
   e. prezygotic barrier . . . gametic isolation.

2. Two closely related species of rabbits may occasionally mate, but their embryos are never viable. Which of the following is a possible consequence?
   a. hybridization
   b. reinforcement of reproductive barriers
   c. fusion of gene pools
   d. polyploidy
   e. a stable situation with gene flow between populations

3. Bullock's oriole and the Baltimore oriole are closely related, but are they the same species? To find out, you could see whether they
   a. sing similar songs.
   b. look alike.
   c. live in the same areas.
   d. have the same number of chromosomes.
   e. successfully interbreed.

4. Sometimes two quite different populations interbreed to a limited extent, so that it is difficult to say whether they are clearly separate species. This does not worry biologists much because it
   a. is quite rare.
   b. is true for almost every species.
   c. supports the theory of punctuated equilibrium.
   d. may illustrate the formation of new species in progress.
   e. happens only among plants, not among animals.

5. Two species of water lilies in the same pond do not interbreed because one blooms at night and the other during the day. The reproductive barrier between them is an example of
   a. temporal isolation.
   b. gametic isolation.
   c. mechanical isolation.
   d. hybrid breakdown.
   e. ecological isolation.

6. Comparison of fossils with living humans seems to show that there have been no significant physical changes in *Homo sapiens* in 30,000 to 50,000 years. What might an advocate of punctuated equilibrium say about this?
   a. It is about time for humans to undergo a burst of change.
   b. That is about how long we have been reproductively isolated.
   c. It is impossible to see major internal changes by looking at fossils.
   d. You would expect lots of changes in the skeleton in that time period.
   e. Lack of change is consistent with the punctuated equilibrium model.

7. Which of the following is an example of a postzygotic reproductive barrier?
   a. One species of frog mates in April; another mates in May.
   b. Two fruit flies of different species produce offspring that are sterile.
   c. The sperm of a marine worm only penetrate eggs of the same species.
   d. One species of flower grows in forested areas, another in meadows.
   e. Two pheasant species perform different courtship dances.

8. The Hawaiian Islands are home to more than twenty species of birds called honeycreepers, each with a slightly different diet and habits. All these birds probably evolved from one ancestor, an example of
   a. sympatric speciation.
   b. hybrid breakdown.
   c. adaptive radiation.
   d. gradualism.
   e. punctuated equilibrium.

9. A botanist found that a kind of white daisy had a diploid chromosome number of 16. In the same area, he found a yellowish daisy. Its cells contained 24 chromosomes. He found that the yellowish daisy was a polyploid descendant of the white daisies. Which of the following would

describe this unusual plant? P: tetraploid, Q: triploid, R: probably sterile, S: a new species.

a. P R S

b. Q R

c. Q R S

d. P S

e. Q S

10. A fossil expert finds an impression of an ancient marine creature called a trilobite in a layer of rock. In the adjacent layer is another species of trilobite, clearly related to the first but quite different in form. If the expert is a gradualist, how might he or she interpret this?

a. This kind of change is exactly what gradualism would predict.

b. Sympatric speciation must have occurred.

c. Intermediate forms could have existed but were not fossilized.

d. Internally, the creatures were identical; only the outer shell changed.

e. This kind of abrupt transition is rare in the fossil record.

Essay

1. There are dozens of species of small rodents, such as rats and mice, in western North America, but relatively few species, ranging over much larger areas, in the east. Suggest a hypothesis related to speciation that might explain this.

2. The mating of a horse and a donkey produces a mule, which is strong and hard working, but sterile. A horse cell contains 64 chromosomes, a donkey cell 62 chromosomes. How many chromosomes would you expect to find in a cell from a mule? Why? Explain why a mule is sterile.

3. A dog (*Canis familiaris*) and a coyote (*Canis latrans*) will readily mate in captivity, and their offspring are healthy and fully fertile. Are we justified in saying they are distinct species? What might this tell us about the history of dogs and coyotes?

4. In terms of speciation, how might freshwater streams in a desert be like islands in the ocean?

5. Various sections of the Hawaiian Islands were formed by lava flows at different times. Kenneth Kaneshiro compared older populations of fruit flies living in older habitats with presumably younger populations living in more recently formed habitats. He found that male fruit flies in both areas are generally eager to mate, but females of older populations are much more selective about their partners than females of newer populations. Suggest a hypothesis to explain why this might be the case.

6. In the Hawaiian Islands, there are several native species of moths in the genus *Hedylepta* that feed exclusively on bananas. Other related species specialize in palms, grasses, or legumes. Bananas were brought to Hawaii by the native Polynesians about a thousand years ago. Why might the banana-eating species be of particular interest to biologists working out the details of the punctuated equilibrium theory?

7. Critics of evolution often say things like, "Sure, insects and bacteria can adapt to a changing environment. But fruit flies are still fruit flies, and *Streptococcus* is still *Streptococcus*. Nobody has ever seen a new *species* evolve." Is this criticism valid? How would you respond to this comment?

Put Words to Work

Correctly use as many of the following words as possible when reading, talking, and writing about biology:

adaptive radiation, allopatric speciation, biological species concept, ecological species concept, hybrid zone, morphological species concept, phylogenetic species concept, polyploidy, postzygotic barrier, prezygotic barrier, punctuated equilibrium (plural, *equilibria*), *reproductive barrier, reproductive isolation, speciation, species, sympatric speciation, taxonomy*

Use the Web

For further review, see the activities and questions on speciation at www.mybiology.com.

Tracing Evolutionary History

15

Focus on the Concepts

This chapter considers the subject of macroevolution—large-scale changes in living things over vast spans of time. Focus on the following concepts:

- Earth formed about 4.6 billion years ago, and the first fossils of microorganisms date from about 3.5 billion years ago. Biologists believe that physical and chemical processes on the early Earth gave rise to simple cells perhaps 3.9 billion years ago. Experiments suggest that this happened in four stages: (1) abiotic synthesis of organic monomers such as amino acids, (2) joining of monomers to form simple polymers, such as polypeptides and nucleic acids, (3) packaging of these molecules into "protobionts," droplets with simple biological activity, and (4) origin of self-replicating (probably RNA) molecules that made inheritance possible.

- The geologic and fossil records, along with radiometric dating, outline the history of life. Prokaryotes possessed Earth for a billion years. As photosynthesis added O_2 to the atmosphere, cellular respiration and more complex eukaryotic cells appeared. Multicellular eukaryotes first evolved about 1.5 billion years ago, but more complex forms don't appear in the fossil record until about 600 million years ago. Fungi, plants, and animals first colonized land about 500 million years ago.

- Geologic changes and catastrophes have shaped the history of life. Earth's crust is divided into huge plates that "float" on the hot mantle. Circulation of the mantle causes continental drift, resulting in geologic shifts along plate boundaries. About 250 million years ago, all the continents joined to form Pangaea, which later broke up. These changes altered climate, habitats, and distribution of organisms and may have caused mass extinctions. The Cretaceous Mass Extinction, the end of the line for the dinosaurs, may have been caused by an asteroid impact.

- Various biological mechanisms contribute to macroevolution. Major adaptive radiations have followed mass extinctions. Changes in genes (duplication, timing, regulation) that control growth and development, such as homeotic genes, may drastically reshape body parts. Complex structures, such as eyes, may develop step-by-step. Features such as feathers may take on new uses in a new context. Natural selection may "prune" the multibranched tree of life, shaping evolutionary trends.

- Phylogeny is the evolutionary history of a group of organisms. Biologists reconstruct phylogeny by studying fossils and anatomical and molecular homologies. Cladistics is a method that groups organisms into branches called clades. A clade includes an ancestral species and all its descendants that possess shared derived characters.

- Molecules reflect evolutionary relationships. The more recently two species have branched form a common ancestor, the more similar their DNA base sequences. Some regions of the genome mutate rapidly and track rapid, recent change, while other parts change more slowly and reveal more ancient relationships. Some change at known rates and can be used as a "clock." Whole genome comparisons demonstrate the importance of gene duplications, the large number of homologies among very different organisms, and the unity of all life.

- Taxonomists give each organism a binomial (two-part) name and group organisms into genera, families, and larger categories such as phyla and kingdoms, all the way up to one of three domains: Bacteria, Archaea, and Eukarya. Ideally, classification reflects evolutionary relationships, and each grouping is a clade—a branch or twig on the tree of life. Horizontal gene transfer early in the history of life makes the branches of the tree more twisted and intertwined than once thought.

Review the Concepts

Work through the following exercises to review the concepts in this chapter. For additional review, refer to the activities at www.mybiology.com. The website offers a pre-test that will help you plan your studies.

Exercise 1 (Modules 15.1 and 15.4)

Summarize the early history of Earth and some major milestones in the history of life by finding each of the following events on the evolutionary "clock." Some of the events are described in the module and some in Figure 15.4. (Note that it is more important to know the *order* of events than exactly *when* they occurred.)

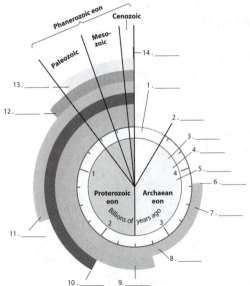

A. Age of oldest known fossils—stromatolites
B. Molten Earth, followed by formation of core and crust
C. Origin of the solar system and Earth
D. First single-celled eukaryotes
E. Formation of the oceans
F. Appearance of humans
G. Buildup of oxygen in atmosphere
H. First land plants
I. Prokaryotes dominant organisms
J. Appearance of animals
K. Appearance of the first living things
L. Volcanoes belch out atmosphere of H_2O, H_2, NH_3, CO_2, etc.
M. The "Big Bang"—formation of the universe
N. Multicellular eukaryotes

Exercise 2   (Modules 15.2–15.3)

The geological record and laboratory experiments suggest how life may have arisen from inorganic chemicals on the early Earth. This flowchart summarizes experiment and theory concerning the origin of life. Fill in the boxes by choosing from the list of components. Fill in the ovals by choosing from the list of processes. Some are done for you.

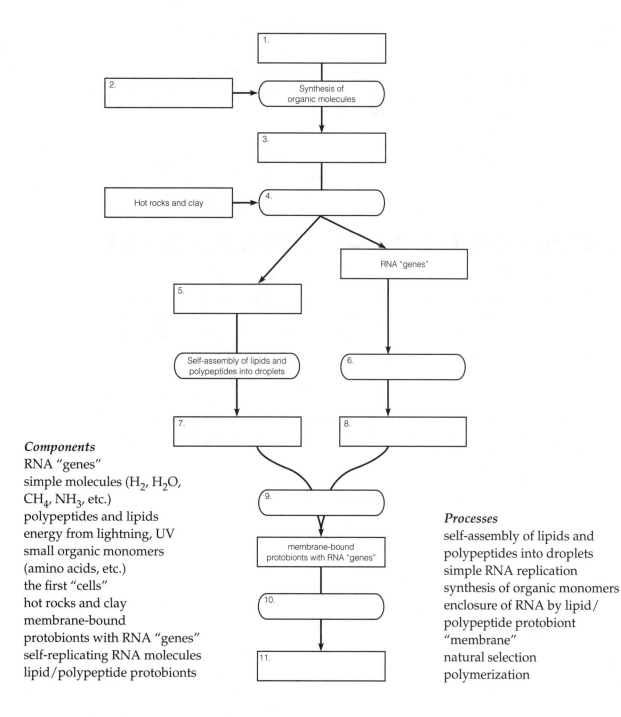

Components
RNA "genes"
simple molecules (H$_2$, H$_2$O, CH$_4$, NH$_3$, etc.)
polypeptides and lipids
energy from lightning, UV
small organic monomers (amino acids, etc.)
the first "cells"
hot rocks and clay
membrane-bound protobionts with RNA "genes"
self-replicating RNA molecules
lipid/polypeptide protobionts

Processes
self-assembly of lipids and polypeptides into droplets
simple RNA replication
synthesis of organic monomers
enclosure of RNA by lipid/polypeptide protobiont "membrane"
natural selection
polymerization

Exercise 3   (Modules 15.4–15.6)

Review the major events in the history of life by numbering each of the following events in sequence (1, 2, 3 . . .) and naming the geologic eon and era (if applicable) when each occurred. (Note once again that it is more important to know the order in which events occurred than the dates or names of the era.)

Sequence	*Eon*	*Era*	*Event*
_____	_____	_____	A. Cone-bearing plants and dinosaurs dominant
_____	_____	_____	B. Humans appear
_____	_____	_____	C. Diverse soft-bodied animals
_____	_____	_____	D. Prokaryotes dominant
_____	_____	_____	E. Diversification of fishes
_____	_____	_____	F. First eukaryotes
_____	_____	_____	G. Radiation of mammals
_____	_____	_____	H. Invasion of land by plants and arthropods
_____	_____	_____	I. Origin of reptiles
_____	_____	_____	J. First animals
_____	_____	_____	K. First living things
_____	_____	_____	L. Permian mass extinction of marine and terrestrial life
_____	_____	_____	M. Flowing plants appear, dinosaurs go extinct
_____	_____	_____	N. "Explosion" of animal phyla

Exercise 4   (Module 15.5)

Use the concept of half-life to answer these questions about the ages of fossils.

1. The half-life of carbon-14 is 5,730 years. If a mammoth has 1/8 the ^{14}C-to-^{12}C ratio that it was thought to have when it was frozen in a Siberian glacier, how old is the mammoth? With a margin of error of plus or minus 10%, what are the maximum and minimum ages of the fossil?

2. The half-life of potassium-40 is 1.3 billion years. If a rock specimen contained 1 g of potassium-40 when it was formed, how much potassium-40 would be left if the rock is 2.6 billion years old?

Exercise 5   (Modules 15.7–15.10)

Review some of the geologic processes that reshape the surface of Earth—and their macroevolutionary consequences—by matching each of the phrases on the left with a word on the right. The illustrations in the textbook will help.

_____	1. The surface of Earth, broken into plates	A. Pangaea
_____	2. Animals that evolved when Pangea was still intact	B. Tsunamis
_____	3. Southern land mass formed when Pangaea broke up	C. Cretaceous
_____	4. A very small plate	D. Earthquake
_____	5. This ocean grows as North America and Eurasia split apart.	E. Himalayas
_____	6. New adaptations that led to adaptive radiations	F. Continental drift
_____	7. Supercontinent formed 250 million years ago	G. Gondwana
_____	8. May have caused Permian mass extinction	H. Mantle
_____	9. Any large, moving segment of Earth's crust	I. Australia
_____	10. "Fallout" from asteroid impact	J. Nazca
_____	11. Formed by collision of Indo-Australian and Eurasian plates	K. Laurasia
_____	12. Movement of continents over Earth's surface	L. Siberian volcanoes
_____	13. Mass extinction probably caused by an asteroid impact	M. Fault
_____	14. Northern land mass formed when Pangaea broke up	N. Atlantic
_____	15. Many marsupials evolved here, isolated from other continents	O. Lungfishes
_____	16. Caused by undersea earthquakes	P. Eurasia
_____	17. Where most important geological processes occur	Q. Permian
_____	18. Predicted to form 250 million years in the future	R. Plate
_____	19. Site of crater that perhaps is related to Cretaceous mass extinction	S. Crust
_____	20. Mass extinctions that had a significant impact on the history of life	T. Mammals
_____	21. Underwent dramatic adaptive radiation after the Cretaceous extinction	U. Iridium
_____	22. Movement resulting from forces at plate edges	V. Plate edges
_____	23. Place where plates slide along one another	W. A new supercontinent
_____	24. Continent formed from the eastern part of Laurasia	X. Yucatan
_____	25. Mass extinction that occurred 250 million years ago	Y. "Big Five"
_____	26. Hot layer that lies beneath the crust	Z. Wings, rigid stems, pollination

Exercise 6   (Modules 15.11–15.13)

Scientists are beginning to understand the biological mechanisms responsible for large-scale evolutionary changes. Review the biological mechanisms underlying macroevolution by matching each of the descriptions on the left with the best evolutionary example on the right.

_____ 1. Changes in body form often result from changes in gene regulation, not in the genes themselves.

_____ 2. Complex structures often evolve step-by-step.

_____ 3. A structure that evolves in one context and takes on a new role is called an exaptation.

_____ 4. Slight changes in the relative rates of growth in different body parts can make big changes in the appearance of adults.

_____ 5. Changes in homeotic genes can radically alter the timing of development and the shape of body parts.

_____ 6. Selection among different species may result in large-scale evolutionary trends or apparent "trends."

_____ 7. Paedomorphosis is a change in timing of development, causing juvenile features to be retained by adults.

A. The horse family
B. Eyes
C. Human and chimp skulls
D. Feathers
E. Fins to legs
F. Salamanders with gills
G. Stickleback spines

Exercise 7   (Modules 15.14–15.19)

Review the principles, methods, and vocabulary of phylogeny and systematics by inserting the correct terms into the following essay.

The evolutionary history of a species or group of species is termed ^1_____. Scientists look to the ^2_____ record to reconstruct phylogeny. Evolutionary history can also be reconstructed by comparing morphological (structural) and molecular features among living species. The teeth and skeletons of lions and bobcats show many ^3_____ that indicate that these animals share a common ancestry. But anatomical comparisons can sometimes be misleading. A process called ^4_____ evolution sometimes causes unrelated organisms to look alike because they have adapted independently to similar environments. The extinct Tasmanian "tiger" (see Chapter 19, Introduction) looks like a cat, but it is actually a marsupial, more closely related to a kangaroo! Such similarity due to convergence is called ^5_____; it can be misleading in reconstructing phylogenies. Often, molecular comparisons allow us to see beyond outward appearance; the DNA of the Tasmanian tiger is very different from the DNA of the two big cats.

^6_____ is the field of biology that focuses on classifying organisms and finding their evolutionary relationships. An important goal of systematics is to name and classify organisms. Biologists called ^7_____ use morphological and molecular comparison to name and group species. Each species is given a two-part name, called a ^8_____. For example, the African lion is *Panthera leo*. The first part of the name is the ^9_____ to which the lion belongs. The second part identifies a particular ^10_____ within that genus.

Naming is only a starting point. The ultimate goal of taxonomy is to place each organism into a hierarchy of taxonomic categories from ^11_____ (the smallest) to ^12_____ (the largest and most inclusive). Ideally, these categories reflect evolutionary history. Biologists depict these relationships in the form of

[13]_____ trees. Species are the twigs of such a tree. The limbs of the tree are larger groupings such as orders, classes, and phyla.

The most widely used method in systematics is called [14]_____. It is a method that that seeks to identify [15]_____—branches that include an ancestral species and all its descendants. Such an inclusive group—a genus, family, or kingdom—is said to be [16]_____. Cladistics makes it possible to construct a classification scheme that reflects the branching of the tree of life.

Cladistics is based on the idea that the evolutionary tree forks when a new heritable trait develops and is passed on to descendants. Groups of organisms that share the new trait are more [17]_____ related than those that have only ancestral traits. For example, the Tasmanian tiger, the lion, and the bobcat all have hair and mammary glands. These are shared [18]_____ characters. But the Tasmanian tiger gives birth to its young very early and nurses them within a pouch. The lion and bobcat retain their young for a much longer period of gestation, nourishing them via a structure called the placenta; this is an added trait, a shared [19]_____ character, that sets the lion and bobcat apart from the Tasmanian tiger. The bobcat and lion are placed in a separate clade, a separate subclass of Class Mammalia, reflecting this evolutionary history. If we are comparing the two cats with the Tasmanian tiger, the cats constitute an ingroup, and the Tasmanian tiger represents an [20]_____, a group known to have diverged before the cat lineage.

The principle of [21]_____, the quest for the simplest explanation, guides cladistics, but this has shaken some branches of the "traditional" evolutionary tree. For example, this approach places [22]_____ within the reptile clade. Several shared [23]_____ characters, such as a four-chambered heart, show that birds and crocodiles are more closely related to each other than crocs are to other reptiles. Similarly, cladistics separates humans and chimps from other apes.

Molecular comparisons can clarify evolutionary relationships. Comparing nucleic acids, proteins, or other molecules to determine relatedness is called molecular [24]_____. Researchers use computers to search through and compare nucleic acid [25]_____ sequences from different species. In general, the more similar the base sequences, the more [26]_____ related the organisms in question. Some nucleic acids, such as the DNA in [27]_____, evolve rather rapidly. Thus, mDNA can be used to trace recent evolutionary events, such as the divergence of various human groups. Other nucleic acids, such as the DNA coding for [28]_____ RNA, change more slowly, so they can track changes occurring over hundreds of millions of years. Because some genes appear to change at a known rate, they allow us to calibrate a molecular [29]_____ that can be used to date evolutionary branch points. Because a good fossil record goes back only about [30]_____ million years, we can use molecular clocks to date divergences thought to have occurred before that time. Returning to the more recent divergence of our cats and the Tasmanian tiger, we would expect that homologous genes of a lion and a bobcat would be more alike than homologous genes of a lion and a Tasmanian tiger. Thus, cladistics and molecular systematics enable us to form testable [31]_____.

Comparing whole genomes has given us some surprising insights into evolutionary relationships. On a molecular level, the genes of humans and chimps are 96% identical. Amazingly [32]_____ of human genes are homologous with genes in mice, and about half our genes are homologous with genes in [33]_____—single-celled eukaryotes! Genomics has revealed that gene [34]_____ has had an important role in evolution, because it increases the number of genes in the genome and provides "raw material" for evolutionary change.

Looking at the larger picture, molecular systematics has catalyzed rethinking the entire tree of life. Back in the day, all life was divided into two kingdoms— [35]_____ and [36]_____. But where did this leave bacteria or photosynthetic organisms that swim? By the 1960s, it looked like the tree of life had five main branches, but soon enough, molecular comparisons showed that that scheme was flawed too. More recently, biologists have adopted a three-[37]_____ system, with two groups of prokaryotes, called [38]_____ and [39]_____, and one group of eukaryotes, the [40]_____. Plants, [41]_____, fungi, and protists (like those swimming green guys) are [42]_____ within Domain Eukarya.

The most recent discoveries suggest that the tree of life might not be a tree at all! During the early history of life, there appear to have been substantial exchanges of genes among the different domains. This took place via [43]_____ gene transfer, a process carried out by exchange of plasmids, [44]_____ infection, or even fusion of whole organisms. Your mitochondria, for example, were once free-living [45]_____. The tree of life thus becomes a tangled thicket of intertwining vines, or even a ring.

Exercise 8   (Module 15.15)

The system of taxonomic categories used by biologists is like a set of boxes into which organisms are sorted. A cocker spaniel—*Canis familiaris*—for example, is first placed in a small box, the specific name *familiaris* that separates it from all other species. This is placed in a slightly larger box, the genus *Canis*, which also holds *Canis lupus* (the wolf) and *Canis latrans* (the coyote). This genus box is placed in a larger box, along with other genera of doglike animals, and so on, all the way up to the last box that separates eukaryotes from prokaryotes. Imagine that the nested boxes below represent the taxonomic categories, starting with species (omitting subphylum). Label the boxes to show the relationships among the categories.

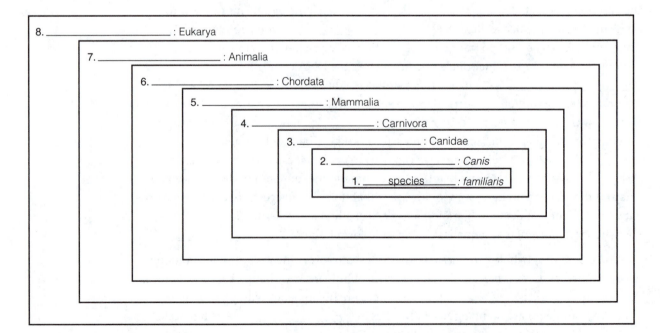

8. _____ : Eukarya

7. _____ : Animalia

6. _____ : Chordata

5. _____ : Mammalia

4. _____ : Carnivora

3. _____ : Canidae

2. _____ : *Canis*

1. ___species___ : *familiaris*

Exercise 9   (Module 15.16)

Cladistics, the most widely used method of systematics, seeks to clarify evolutionary and taxonomic relationships by grouping organisms into clades. A clade is a group of organisms made up of an ancestor and all its descendants. This simplified phylogenetic tree uses cladistics (based on anatomy, but backed up by molecular data) to reconstruct the relationships among four groups of plants and their closest relatives, the green algae. Read Module 15.16, examine the trees in the module and below, and then answer the following questions. This exercise is rather difficult, so take your time.

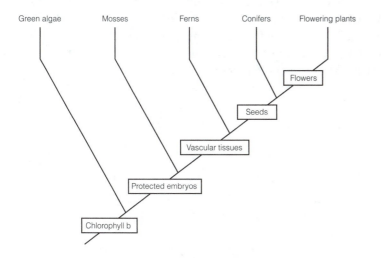

1. Which four groups of organisms above make up the in-group?

2. Which organisms constitute the out-group?

3. Which is more useful in cladistics, analogies or homologies?

4. Which characters are unique to a lineage of organisms, shared derived characters or shared ancestral characters? Which are more useful in differentiating among (separating out) distinct lineages?

5. What is a shared ancestral character common to all plants?

6. What is a shared derived character common to all plants?

7. What is a shared ancestral character common to all plants with seeds?

8. What is a shared derived character common to all plants with seeds?

9. Which characters are most useful in deciding whether an organism is in the out-group or the in-group, shared ancestral characters or shared derived characters?

10. If we are interested in focusing on all plants that have vascular tissues, which groups on the phylogenetic tree constitute the out-group? The in-group?

11. What is the name of a taxonomic group consisting of an ancestor and all its descendants?

12. What other organisms are in the clade that includes the first plants with seeds?

13. Name or describe nine different clades shown on the phylogenetic tree above.

Exercise 10   (Module 15.17)

Homologous structures—similar structures derived from the same structure in a common ancestor—tell us about phylogenetic relationships among organisms. But convergent evolution can make unrelated organisms look alike; their similarities may be analogous, not homologous. Fortunately, we can dig beneath surface similarities and compare biological molecules to measure relatedness between species. For example we can compare protein amino acid sequences, RNA base sequences, or DNA base sequences—even whole genomes. Imagine that you have sequenced mitochondrial DNA (mDNA) for six species of rodents, A through F. All the rodents are thought to have evolved from a common ancestor, X. The number of differences in mDNA sequence are compiled in the table below, which reads like a road map mileage chart. For example, there are four differences between A and C, and nine differences between A and D. Use the differences in sequence to place species A through F on the phylogenetic tree. (Hint: Don't get too mathematical; just "eyeball" overall numbers.)

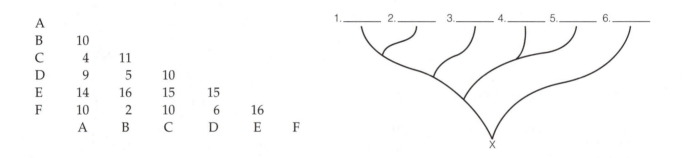

A					
B	10				
C	4	11			
D	9	5	10		
E	14	16	15	15	
F	10	2	10	6	16
	A	B	C	D	E	F

Test Your Knowledge

Multiple Choice

1. The science of naming and classifying organisms is called
   a. biology.
   b. polyploidy.
   c. genetics.
   d. taxonomy.
   e. parsimony.

2. Large-scale changes in the history of life, such as mass extinctions, the development of walking legs from fins, and the appearance of new groups of organisms such as birds, are termed:
   a. macroevolution.
   b. adaptation.
   c. exaptation.
   d. microevolution.
   e. paedomorphosis.

3. Which of the following is thought to have been the first step in the origin of life?
   a. cooperation among molecules
   b. formation of protobionts
   c. formation of organic monomers
   d. replication of primitive "genes"
   e. formation of organic polymers

4. Systematics is concerned with
   a. naming organisms.
   b. studying biological diversity.
   c. taxonomy.
   d. tracing phylogeny.
   e. all of the above.

5. If you want to see a dinosaur, it would be best to set the controls of your time machine for the
   a. Mesozoic era.
   b. Paleozoic era.
   c. Cenozoic era.
   d. Paleozoic era.
   e. Precambrian.

6. Which of the following taxonomic categories contains all the others?
   a. genus
   b. class
   c. family
   d. subclass
   e. order

7. The three domains of life are
   a. bacteria, plants, and animals
   b. prokaryotes, eukaryotes, and plants
   c. plants, animals, and fungi
   d. eukaryotes and two kinds of prokaryotes
   e. bacteria, protists, and eukaryotes

8. Organisms that survive mass extinctions
   a. often diversify, taking advantage of new opportunities.
   b. usually are so reduced in numbers that they soon go extinct.
   c. are often "living fossils" that have existed unchanged for long periods.
   d. usually cannot cope with the new conditions that follow.
   e. usually lack exaptations for their new environment.

9. The ideal taxonomic grouping includes
   a. all known species with shared ancestral characters.
   b. an out-group and an in-group
   c. a group of species that all evolved at about the same time.
   d. species that do not share any derived characters.
   e. an ancestor species and all its descendants.

10. Pangaea
   a. was a land mass that broke up to form the present-day continents.
   b. is the idea that all life on Earth is related.
   c. was an animal common in ancient seas but now extinct.
   d. is the evolutionary history of a species, family, or phylum.
   e. is the theory that crustal plates can move relative to one another.

11. Continental drift is responsible for all of the following except
   a. volcanoes.
   b. formation of river systems.
   c. distribution of animals and plants.
   d. mountain ranges such as the Himalayas.
   e. earthquakes.

12. Which of the following most strongly suggests that an impact by an asteroid may have caused the extinction of the dinosaurs?
   a. Fossils show that dinosaurs suffered from cold and starvation.
   b. There have been several near misses in recent years.
   c. Sedimentary rocks contain a layer of mineral uncommon on Earth.
   d. The dinosaurs disappeared rather abruptly—virtually overnight.
   e. Fossils indicate that most dinosaurs were looking up when they died.

13. Which of the following is least useful in tracing phylogeny?
   a. mitochondrial DNA
   b. convergent evolution
   c. fossils
   d. homologous genes
   e. gene duplications

Essay

1. Describe Stanley Miller's experiment that simulated conditions on the ancient Earth. How was the experiment carried out? What was its purpose? Its result? Looking back, what was the experiment's biggest flaw?

2. Explain how the formation of Pangaea may have led to mass extinctions at the end of the Permian period, about 250 million years ago. How did this lead to new evolutionary developments?

3. Place the following events in the history of life in the proper order: Appearance of humans, origin of eukaryotes, dominance of dinosaurs and cone-bearing plants, origin of animals, appearance of first vertebrates, diversification of mammals, origin of flowering plants, first prokaryotes, movement of plants and animals onto land.

4. Describe two different ways in which continental drift causes mountains to form.

5. The following list shows the levels of classification of a human being, *Homo sapiens.* Name the category that corresponds to each of the taxa listed. (Hint: Vertebrata is a subphylum.)

   Eukarya Animalia Chordata Vertebrata Mammalia Primates Hominidae *Homo sapiens*

6. What is a clade? Explain how cladistics connects phylogeny and taxonomy. How has cladistics shaken up traditional taxonomy?

Apply the Concepts

Multiple Choice

1. Drastic reductions in the number of body segments and pairs of legs may have been responsible for the evolution of the first insects from millipede-like ancestors. This example might illustrate
   a. parsimony.
   b. paedomorphsis.
   c. species selection.
   d. horizontal gene transfer.
   e. changes in homeotic genes.

2. Which of the following discoveries would force scientists to revise their present theories regarding the origin of life on Earth?
   a. Earth is found to be 6 billion years old, rather than 4.6 billion.
   b. Polypeptides can catalyze replication of small RNA molecules.
   c. There was a lot of oxygen gas in the atmosphere 4 billion years ago.
   d. Minerals in lava catalyze formation of polypeptides from amino acids.
   e. Lipids spontaneously form selectively permeable membranes.

3. This phylogenetic tree represents the relationships among several species—A, B, C, D, and E. Which of the following groups of species is a clade?
   a. A, B, C
   b. A, B, C, D
   c. B, C, D
   d. C, D, E
   e. B, C, D, E

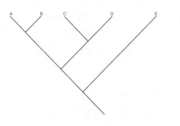

4. The following are some major events in the early history of life.
   P. first cells
   Q. first genes
   R. first eukaryotes
   S. first photosynthetic prokaryotes
   T. first animals

   Which answer below places these events in the correct order?
   a. PQSRT
   b. QSPTR
   c. QPSRT
   d. QSPRT
   e. SPQRT

5. Which of the following would cast doubt on the asteroid-impact hypothesis for the Cretaceous extinction?
   a. finding a crater 200 million years old
   b. finding fossil dinosaur bones beneath a layer of iridium
   c. determining that birds are closely related to dinosaurs
   d. finding fossil dinosaur bones above a layer of iridium
   e. finding that many forms of marine life disappeared at the same time as the dinosaurs

6. The wings of birds and insects have similar functions, but they do not have the same evolutionary origin. Bird and insect wings are
   a. homologous.
   b. phylogenetic.
   c. analogous.
   d. binomial.
   e. monophyletic.

7. Which of the following might be a shared derived character that could be used to classify different kinds of mammals (mice, monkeys, deer, bears, cats, etc.)?
   a. whether or not they have hair
   b. kinds of teeth they possess
   c. presence or absence of a backbone
   d. whether or not they produce milk
   e. analogous skeletal structures

8. Two animals in the same family would not have to be in the same
   a. genus.
   b. domain.
   c. order.
   d. phylum.
   e. class.

9. A phylogenetic tree of bird families would most clearly show which of the following?
   a. characteristics shared by all bird families
   b. evolutionary relationships among families
   c. families that look most alike
   d. analogous structures shared by various species
   e. relative ages of living species of birds

10. Fossils of an ancient reptile called *Lystrosaurus* have been found in Africa, India, and Antarctica. Which of the following best explains this distribution?
    a. They were able to move between continents before the oceans filled.
    b. Movement of India due to continental drift carried them from place to place.
    c. These areas were once next to each other and have since drifted apart.
    d. They were able to migrate over frozen seas during the Ice Ages.
    e. Changes in climate forced them to migrate from place to place.

11. A species of fruit fly is thought to have diverged from another species less than a thousand years ago. Study of which of the following would best clarify these fruit fly relationships?
    a. mitochondrial DNA
    b. the fossil record
    c. ribosomal RNA
    d. comparison of analogous structures
    e. radiometric dating

12. Which of the following taxa is *least* closely related to the others?
    a. Archaea
    b. Plantae
    c. Bacteria
    d. Eukarya
    e. Animalia

13. According to Figure 15.17, which of the following are most closely related?
    a. raccoon and lesser panda
    b. giant panda and raccoon
    c. giant panda and lesser panda
    d. sloth bear and spectacled bear
    e. brown bear and sloth bear

14. In a detailed phylogenetic tree showing the evolutionary relationships within a class, the "trunk" of the tree might represent the _____, while the ends of individual "twigs" might represent _____.
    a. phylum . . . classes
    b. class . . . species
    c. genus . . . species
    d. species . . . families
    e. class . . . phyla

Essay

1. When a tree was buried in a swamp, it contained 400 mg of carbon-14. The half-life of carbon-14 is 5,730 years. How long ago was the tree buried if it now contains 100 mg of carbon-14?

2. Ultraviolet light breaks chemical bonds in large organic molecules. (That is why it is dangerous to unprotected skin.) Does this fact lend support to hypotheses about the possible role of UV radiation in the origin of life, or does it present problems for these hypotheses? Explain.

3. The dinosaurs perished in a mass extinction at the end of the Cretaceous period, 65 million years ago. Briefly describe alternative hypotheses that have been suggested to explain the extinction of the dinosaurs. Did other forms of life disappear during the same period? Does this cast doubt on any of the hypotheses? Why?

4. Biologists like to look at fossils to trace evolutionary relationships. Birds are very delicate and unlikely to be fossilized; only a handful of old bird fossils are known. How might the molecules in the cells of living birds be compared to determine how ducks, sparrows, and hawks are related? Briefly name the molecules, what characteristic of the molecules would be compared, and how this would indicate how closely related the birds are.

5. Molecular comparisons indicate that the vultures of Africa and America are more closely related to separate nonvulture ancestors than they are to each other. If the two groups of vultures are not closely related, why do you think they look so much alike?

6. Are "vultures" a clade? Explain.

7. Advocates of "intelligent design" say that a structure as complex as the eye could not have evolved, but must have been designed by an "intelligent agent." How do biologists counter this assertion?

Put Words to Work

Correctly use as many of the following words as possible when reading, talking, and writing about biology:

analogy, binomial, clade, cladistics, class, continental drift, convergent evolution, domain, "evo-devo," family, genus (plural, *genera*), *geologic record, horizontal gene transfer, in-group, kingdom, macroevolution, molecular clock, molecular systematics, monophyletic, order, out-group, Pangaea, parsimony, phyla, phylogenetic tree, phylogeny, radiometric dating, shared ancestral characters, shared derived characters, species, systematics, taxon, three-domain system*

Use the Web

For further review on the history of life, mechanisms of macroevolution, and the principles of systematics, see the activities and questions at www.mybiology.com.

The Origin and Evolution of Microbial Life: Prokaryotes and Protists

16

Focus on the Concepts

This chapter examines prokaryotes, the first living things on Earth and still the most abundant organisms, and protists, a diverse assortment of eukaryotes. While studying prokaryotes and protists, focus on the following concepts:

- Prokaryotes are single-celled organisms whose cells lack a nucleus or other organelles. They are the most abundant organisms in terms of number and mass. The two kinds of prokaryotes are classified in the domains Bacteria and Archaea. Structural, biochemical, and genetic similarities indicate that bacteria and archaea diverged early in the history of life, and eukaryotes are more closely related to archaea.

- Prokaryotes vary in shape and structure. There are cocci, bacilli, and spirilla. All prokaryotes have cell walls, but archaea lack peptidoglycan. Gram-positive bacteria have thicker cell walls with more peptidoglycan, while gram-negative bacteria have more complex walls with less peptidoglycan. Some bacteria cling to surfaces via pili. Some swim by means of propeller-like flagella. Most reproduce rapidly via binary fission, and some can withstand harsh conditions by forming endospores.

- Prokaryotes exhibit a diversity of nutritional modes. Photoautotrophs get energy from light, and carbon from CO_2 (like plants). Photohetrotrophs get energy from light and carbon from organic molecules. Chemoautotrophs get energy from inorganic chemicals and carbon from CO_2. Chemoheterotrophs get both energy and carbon from organic compounds (like animals).

- Archaea thrive in extreme environments. Extreme halophiles live in salt lakes, while extreme thermophiles exist in hot springs. Methanogens live in anaerobic mud and animal digestive tracts and are important in producing methane. There are nine groups of bacteria. Many kinds recycle nutrients. Some fix nitrogen. Representatives of several groups are pathogenic. Some carry out photosynthesis. Many prokaryotes are useful in sewage treatment, bioremediation, and extracting minerals.

- Most pathogenic bacteria cause illness by producing a toxin. Exotoxins, like those secreted by *Staphlococcus,* are secreted. Endotoxins are membrane components of gram-negative bacteria like *Salmonella,* which are released when the cells die. Some pathogens and toxins, such as anthrax and bubonic plague bacteria and botulinum toxin from *Clostridium botulinum,* are used as biological weapons.

- Protists are a diverse group of mostly unicellular eukaryotes. They obtain nutrition in a variety of ways and live in a variety of moist habitats. Some are simple in structure, and others are among the most complex cells known. Eukaryotic cells evolved via primary and secondary endosymbiosis: First prokaryotic cells took up residence inside larger cells, becoming mitochondria and chloroplasts. Later, certain eukaryotes themselves became endosymbiotic, giving rise to diverse eukaryotic lineages.

- Protists classification is a work in progress. Protists are grouped on the basis of specialized organelles, feeding structures, or method of locomotion. One group can include photosynthetic "algae," animal-like "protozoa," and others that resemble fungi. Many protists are parasitic and pathogenic. Some have complex life cycles. Three major lineages led to multicellular organisms—one to brown algae; one to red algae, green algae, and land plants; and a third to fungi and animals.

Review the Concepts

Work through the following exercises to review the concepts in this chapter. For additional review, access the activities at www.mybiology.com. The website offers a pre-test that will help you plan your studies.

> **Exercise 1** (Modules 16.1–16.2)

Prokaryotes are everywhere! There are two fundamentally different kinds of prokaryotes. State whether each of the following describes bacteria (B) or archaea (A).

_____ 1. Genes rarely contain introns

_____ 2. Some rRNAs match eukaryote rRNAs

_____ 3. Cell walls contain peptidoglycan

_____ 4. Uninhibited by streptomycin

_____ 5. rRNAs different from eukaryotic rRNAs

_____ 6. Some genes contain introns

_____ 7. Several kinds of complex RNA polymerase

_____ 8. Cell walls lack peptidoglycan

_____ 9. One kind of simple RNA polymerase

_____10. Inhibited by streptomycin

_____11. More like eukaryotes

Exercise 2   (Modules 16.2–16.8)

These modules review some of the shapes, characteristics, and types of prokaryotes. Review them by completing this crossword puzzle.

Across

4. ____ are blue-green photosynthetic bacteria.

5. Prokaryotes ____ quickly in a favorable environment.

7. Archaea called ____ inhabit your digestive tract.

9. Prokaryotic ____ are smaller than those of eukaryotes, and differ in RNA and protein content.

10. Certain ____ block prokaryotic ribosomes, but not those of eukaryotes.

11. Archaea produce and consume vast amounts of ____, a greenhouse gas.

12. The ____ diverged from archaea early in the history of life.

14. Gram- ____ bacteria often cause disease.

16. A ____ of cyanobacteria may indicate polluted water.

17. Spherical prokaryotes are called ____.

18. ____ pili link prokaryotes during conjugation.

20. Certain proteobacteria ____ nitrogen.

21. Many archaea live in extreme environments such as hot ____.

22. Many bacteria produce dormant, resistant cells called ____.

23. Photoautotrophs can get energy from ____.

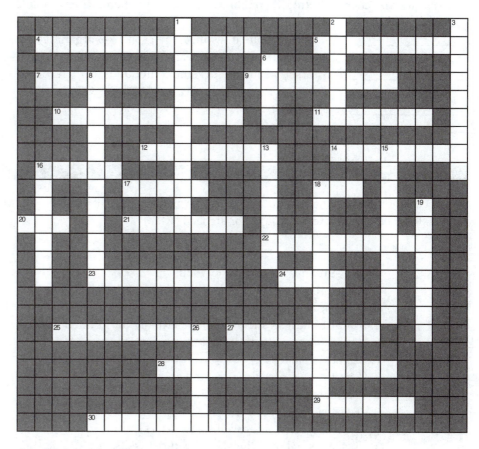

24. We can identify bacteria by using a ____ stain.

25. You might find an extreme ____ in very salty water.

27. A prokaryote may be able to swim using a propeller-like prokaryotic ____.

28. A ____ gets carbon from carbon dioxide and energy from inorganic chemicals.

29. Vibrios, spirilla, and spirochetes are curved or ____-shaped.

30. Archaea and bacteria are both ____.

Down

1. Cocci that occur in chains are called ____.

2. Comparison of ____ sequences supports a three-domain view of life.

3. Some prokaryotes have folded ____ that perform metabolic functions.

6. Hairlike appendages called ____ enable bacteria to stick to surfaces and each other.

8. ____ obtain carbon atoms from organic compounds.

13. The major kinds of prokaryotes are bacteria and ____.

15. ____ make their own organic compounds from inorganic sources.

16. Rod-shaped prokaryotes are called ____.

18. Cyanobacteria formed layered ____ in the ancient seas.

19. Gram- ____ bacteria include actinomycetes and mycoplasmas.

26. Chemoheterotrophs obtain ____ and carbon from organic compounds.

Exercise 3   (Module 16.5)

Prokaryotes (and other organisms, by the way) can be categorized according to their mode of nutrition. Identify whether each of the following prokaryotes is a photoautotroph (as are cyanobacteria), a chemoautotroph, a photoheterotroph, or a chemoheterotroph (as are most prokaryotes).

_____   1. Energy source—sun; carbon source—organic compounds
_____   2. Energy source—inorganic chemicals; carbon source—CO_2
_____   3. Energy source—organic compounds; carbon source—organic compounds
_____   4. Energy source—sun; carbon source—CO_2
_____   5. A plant
_____   6. An animal

Exercise 4   (Modules 16.6–16.7)

Match each of the following descriptions with a particular group of prokaryotes (M-T), and then state whether each are bacteria (B) or archaea (A).

Group B or A

_____ _____ 1. Live in salt lakes and brine ponds
_____ _____ 2. Only prokaryotes with plantlike photosynthesis
_____ _____ 3. *Salmonella, E. coli,* and cholera
_____ _____ 4. Cause blindness and a common sexually transmitted disease
_____ _____ 5. *Rhizobium* and other nitrogen-fixers
_____ _____ 6. Create "marsh gas"
_____ _____ 7. *Streptococcus, Staphlococcus,* and anthrax
_____ _____ 8. Generated ancient stromatolites and oxygenated the atmosphere
_____ _____ 9. Twisted prokaryotes that cause syphilis and Lyme disease
_____ _____ 10. Actinomycetes, soil-dwelling recyclers
_____ _____ 11. Residents of hot springs and deep-sea vents
_____ _____ 12. Enable cattle and deer to digest cellulose

M. Methanogens
N. Gram-positive
   bacteria
O. Proteobacteria
P. Extreme
   thermophiles
Q. Cyanobacteria
R. Spirochetes
S. Extreme halophiles
T. Chlamydias

Exercise 5   (Modules 16.8–16.10)

These modules outline some of the important roles of prokaryotes. To review them, fill in the blanks in this story.

"I wish I could get rid of this cold," Matt said, as he wiped his red and swollen nose. "Bacteria are pests that evolved purely to make my life miserable. I wish they'd all disappear."

Alex looked up from his psychology textbook. "Colds are caused by viruses, not bacteria," he said. "And if bacteria disappeared, we'd be in deep trouble. We couldn't live on this planet if it weren't for bacteria."

"What? Well, I'm not going to live much longer if I don't get rid of this cold."

"You're not gonna die from a cold. Seriously, the first living things on Earth were
[1]_____, the simple kinds of cells that we call bacteria and archaea. Prokaryotes eventually gave rise to creatures with more complex cells—
[2]_____. That includes everything from algae to us."

Matt sniffled. "Yeah, yeah, yeah. Well, maybe they don't cause colds, but they do make people sick."

"Yes, there are many [3]_____, or disease-causing, bacteria," Alex replied. "But bacteria and archaea are important and helpful in many ways—"

"So how do they make you sick?"

"Most bacteria cause illness by producing [4]_____. *Staphylococcus aureus,* for example, is a common skin bacterium, but it secretes substances called [5]_____ that can cause food poisoning or toxic shock in the body. A strain of *E. coli* bacteria from cattle produces an exotoxin that can cause bloody [6]_____."

"Is that the one people get from bad [7]_____ and spinach? Is that like botulism or *Salmonella*?

Both can poison you. The toxin produced by *Clostridium botulinum,* called [8]_____, is an exotoxin, and it is the most potent poison on Earth. *Salmonella* causes food poisoning, too, but its toxin is actually part of its cell wall. It's called an [9]_____. Another kind of *Salmonella* causes [10]_____ fever, and—"

"Achoo!" Matt sneezed again, and gasped "Maybe it's typhoid fever. Or chalmydia!"

Alex groaned. "I don't think so—besides, chalmydia is [11]_____ transmitted. If that's what you've got, I think it is going to affect a rather different part of your body! Trust me, it's a cold."

"I tell you, this doesn't feel like a regular cold. It could be pneumonia. Or maybe I inhaled anthrax! Let's see. Have I been exposed to a white powder?"

"What a hypochondriac! How long have you been sick? A week? If it were anthrax, you'd probably be dead by now. By the way, the powder form of anthrax isn't a toxin. It consists of anthrax [12]_____, dried dormant cells that are easily dispersed in air. The dried fine powder is the [13]_____ form of anthrax used by terrorists . Once they are inhaled and germinate, anthrax starts growing in your body and making its exotoxin."

Matt looked defeated. "Guess it's not anthrax. Could it be pneumonia?"

Alex brightened. "Ahh, now we're getting somewhere. If it is pneumonia, *then* we're talking *bacteria*. But healthy adults in developed countries don't get many bacterial infections these days. Sanitation measures, like [14]_____ and [15]_____ systems, keep many bacterial diseases from spreading, and—"

"You don't get pneumonia from dirty water, Mr. Know-It-All," Matt interjected.

"Right, but most forms of pneumonia, like most bacterial diseases, can be treated with [16]_____. So if it is pneumonia, and not a cold, you might be able to do something. And then you might stop whining!"

Matt sneezed yet again. "I need to get to the emergency room. I think all bacteria should be wiped off the face of the earth."

Alex interrupted, "Have you been listening? We couldn't live without them. Many prokaryotes are involved in the cycling and recycling of various 17_____ between living things and the nonliving environment. They decompose 18_____ matter, like 19_____ and 20_____, and return chemicals to the environment. For example, prokaryotes are important in a chemical cycle that makes 21_____ available to plants—"

"O.K.—I get it. If it weren't for prokaryotes, in no time we'd be up to our necks in . . ."

"Sewage. Right. That's why prokaryotes are important in sewage-22_____ plants. 23_____ prokaryotes work on solid matter, called 24_____, and 25_____ bacteria in a layer called a 26_____ break down liquid wastes. Some bacteria can even break down oil spills, and some are used to extract metals, like 27_____ and 28_____, from low-grade ores and might help us clean up toxic wastes from old mining sites. Using bacteria to clean up the environment is called 29_____".

Matt sneezed. "I could use some to clean up my toxic wastes. This is no ordinary cold. Maybe it's the flu. Maybe it's pneumonia. Maybe it's Lemon disease!"

Alex sighed. "You mean Lyme disease? Been bitten by a 30_____ lately?"

Exercise 6   (Modules 16.11–16.12)

Complex eukaryotic cells evolved when prokaryotic cells took up residence inside other, larger prokaryotic cells. Mitochondria probably got their start this way. The diagram in Module 16.12 takes up the story a bit later, when a cyanobacterium established residence in a heterotrophic eukaryote, eventually becoming a chloroplast. This is called primary endosymbiosis. Still later certain photosynthetic *eukaryotes* became endosymbiotic in heterotrophic eukaryotes, eventually becoming different kinds of chloroplasts in these host cells. This is called secondary endosymbiosis. Make sense? See if you understand the process by identifying each of the following on the diagram. (Some questions have more than one answer.)

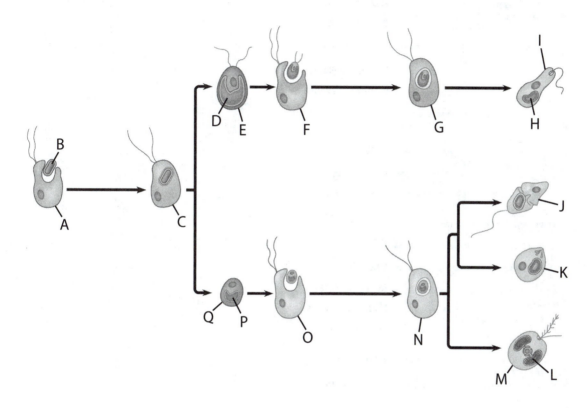

_____ 1. Heterotrophic eukaryotes
_____ 2. Photosynthetic prokaryote
_____ 3. Chloroplasts derived from prokaryote
_____ 4. Green alga
_____ 5. Red alga
_____ 6. Chloroplast derived from red alga
_____ 7. Chloroplast derived from green alga
_____ 8. New cell produced by primary endosymbiosis
_____ 9. New cells produced by secondary endosymbiosis
_____ 10. Diverse forms of existing photosynthetic eukaryotes
       (be careful–don't leave any out!)

This phylogenetic tree tentatively outlines the evolution of various groups of eukaryotes. Most are informally called protists. Protists are a diverse collection of eukaryotes, simple to complex, with varying lifestyles. Read the modules about protists, and match as many statements as you can with the correct branch(es) of the phylogenetic tree. Then look up the ones you don't know. (Names are often clues—ciliates, green algae, etc.) Some answers are used more than once, and some questions have multiple answers.

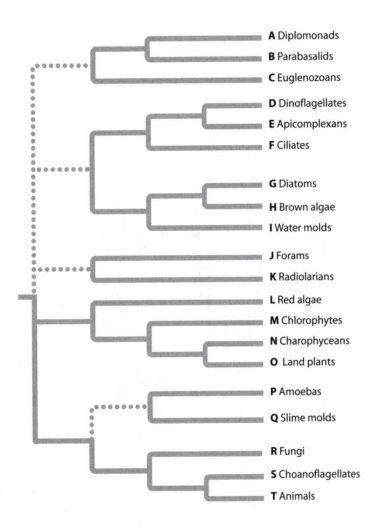

A Diplomonads
B Parabasalids
C Euglenozoans
D Dinoflagellates
E Apicomplexans
F Ciliates
G Diatoms
H Brown algae
I Water molds
J Forams
K Radiolarians
L Red algae
M Chlorophytes
N Charophyceans
O Land plants
P Amoebas
Q Slime molds
R Fungi
S Choanoflagellates
T Animals

_____ 1. Three groups of organisms that are *not* protists
_____ 2. This group gets its name from the green flagellate *Euglena*.
_____ 3. These three groups of organisms make up the stramenopiles, named for their "hairy" flagella.
_____ 4. Some reside within coral animals and are important reef food producers.
_____ 5. Two groups of green algae
_____ 6. The names of these two groups suggest that they are fungi, but they are not.
_____ 7. Probably the most ancient surviving lineage of eukaryotes
_____ 8. Fungus-like protists that decompose dead plants and animals in the water
_____ 9. Mostly free-living protists that move by means of cilia
_____ 10. Kelps
_____ 11. These two groups move via lobe-shaped extensions called pseudopodia.

_____ 12. Algae characteristic of warm coastal tropical waters

_____ 13. You may see *Paramecium,* a representative of this group, in lab.

_____ 14. Two groups that feed with threadlike pseudopodia

_____ 15. Closest relatives of plants

_____ 16. These three groups are called alveolates, named for sacs under their membranes.

_____ 17. Algae with complex, glassy, silica cell walls

_____ 18. Seaweeds are in these three groups

_____ 19. A flagellate in this group causes sleeping sickness.

_____ 20. Fossils of their calcium carbonate shells help identify rock strata

_____ 21. Population explosions of these organisms, called "red tides," can poison fish and people.

_____ 22. Protists that exist part-time as individual cells and part-time as large "blobs"

_____ 23. *Giardia,* an intestinal parasite, is in this group.

_____ 24. Animal parasites that cause diseases such as malaria

_____ 25. Some are serious plant parasites, such as the organism that causes potato blight.

Exercise 8   (Module 16.21)

Multicellular life—seaweeds, plants, animals, and fungi—probably evolved from colonial protists. Test your understanding of this process by completing each sentence on the left with a word on the right. Each answer is used only once.

1. Cell specialization involved separation of somatic cells and _____.
2. The links between unicellular and multicellular life were _____.
3. Organisms moved onto land when green algae gave rise to _____.
4. At least three lineages may have given rise to _____.
5. Multicellularity probably evolved several times among _____.
6. In all _____, all life functions happen in one cell.
7. In _____, different cells do different jobs.
8. The oldest known multicellular organisms include algae and _____, such as corals and worms.

A. Seaweeds
B. Animals
C. Protists
D. Unicellular organisms
E. Plants
F. Colonies
G. Multicellular organisms
H. Gametes

Test Your Knowledge

Multiple Choice

1. There are many similarities between prokaryotes called ____ and eukaryotic organisms.
   a. archaea
   b. bacteria
   c. protozoa
   d. cyanobacteria
   e. dinoflagellates

2. *E. coli* bacteria, which live in human intestines, are shaped like tiny, straight sausages. They are
   a. bacilli.
   b. vibrios.
   c. spirochetes.
   d. cocci.
   e. spirilla.

3. Which of the following is a difference between bacteria and archaea?
   a. Archaea are unicellular and bacteria are colonial.
   b. The genes of bacteria have introns, while archaea lack introns.
   c. They have different chemicals in their cell membranes and cell walls.
   d. Bacteria are autotrophic and archaea are heterotrophic.
   e. They look very different under a microscope.

4. Most prokaryotes
   a. obtain energy from sunlight and carbon from organic compounds.
   b. obtain both energy and carbon from inorganic compounds.
   c. obtain energy from inorganic compounds and carbon from CO_2.
   d. obtain energy from sunlight and carbon from CO_2.
   e. obtain both energy and carbon from organic compounds.

5. Which of the following is true of cyanobacteria?
   a. They are pathogenic.
   b. They are chemoheterotrophs.
   c. They are archaea.
   d. They are protists.
   e. They are photoautotrophs.

6. Archea called _____ live in salty environments, such as salt lakes.
   a. methanogens
   b. actinomycetes
   c. extreme halophiles
   d. apicomplexans
   e. extreme thermophiles

7. All organisms called "algae" are
   a. unicellular.
   b. green.
   c. autotrophic.
   d. prokaryotic.
   e. multicellular.

8. Different groups of seaweeds can generally be distinguished on the basis of
   a. color.
   b. size.
   c. whether they are multicellular or unicellular.
   d. whether or not they have true leaves, stems, and roots.
   e. whether they are autotrophic or heterotrophic.

9. Which of the following groups of bacteria is not paired with the correct description?
   a. gram-positive bacteria—*Staphlococcus* and *Streptococcus*
   b. chlamydias—soil nutrient recyclers
   c. proteobacteria—several kinds fix nitrogen
   d. spiochetes—cause syphilis and Lyme disease
   e. cyanobacteria—carry out plantlike photosynthesis

10. You would *not* expect which of the following to cause disease?
    a. apicomplexan
    b. gram-negative bacterium
    c. euglenozoan
    d. diatom
    e. sprochete

11. Which of the following is *incorrectly* paired with its mode of nutrition?
    a. *Streptococcus*—chemoheterotroph
    b. red alga—photoautotroph
    c. animal—chemoheterotroph
    d. cyanobacterium—chemoheterotroph
    e. plant—photoautotroph

Essay

1. Name four diseases caused by bacteria. How do bacteria cause disease?

2. In what ways are prokaryotes useful and even vital to our well-being?

3. Primitive cyanobacteria radically changed Earth for all life that followed. Their effect on the environment is described as a "revolution." What did these prokaryotes do?

4. Explain the endosymbiosis model for the origin of eukaryotes and protist diversity.

5. How did multicellular eukaryotes—plants, animals, and fungi—probably arise?

Apply the Concepts

Multiple Choice

1. Bacteria that live around deep-sea hot-water vents obtain energy by oxidizing inorganic hydrogen sulfide belched out by the vents. They use this energy to build organic molecules from carbon obtained from the carbon dioxide in the seawater. These bacteria might be described as
   a. photoheterotrophs.
   b. chemoautotrophs.
   c. photoautotrophs.
   d. chemoheterotrophs.
   e. none of the above.

2. The bacterium *Bacillus thuringensis* can withstand heating, dryness, and toxic chemicals that would kill most other bacteria. This indicates that it is probably able to form
   a. pseudopodia.
   b. endotoxins.
   c. endospores.
   d. pili.
   e. peptidoglycans.

3. Certain dinoflagellates enter the cells of coral animals and carry out photosynthesis, producing food for themselves and the coral. This is a lot like
   a. formation of a biofilm.
   b. primary endosymbiosis.
   c. secondary endosymbiosis.
   d. how ancient stromatolites were formed
   e. endospore formation

4. In an experiment, a microbiologist put equal numbers of each of the following organisms into a flask of sterile broth consisting mostly of sugar and a few amino acids. She then placed the flask in the dark. Which of the organisms would be best able to survive and reproduce in this environment?
   a. chemoheterotrophic bacteria
   b. cyanobacteria
   c. diatoms
   d. extreme halophiles
   e. green algae

5. If you collected a sample of plankton—floating organisms—from near the ocean surface, you would be unlikely to find which of the following?
   a. diatoms
   b. cyanobacteria
   c. radiolarians
   d. forams
   e. methanogens

6. Normally, a certain intestinal bacterium is relatively harmless, but if the body's defenses attack it and cells are broken open, it becomes deadly. It sounds like this bacterium
   a. produces an endotoxin
   b. is a methanogen
   c. produces an exotoxin
   d. is a spirochete
   e. has peptidoglycan in its cell wall

7. Which of the following is *not* the product of endosymbiosis?
   a. rabbit
   b. green alga
   c. euglenozoan
   d. spirochete
   e. pine tree

8. Which of the following is *not* evidence for the role of endosymbiosis in the origin of eukaryotes?
   a. Chloroplasts have their own DNA.
   b. The inner membrane of a chloroplast is similar to prokaryotic membranes.
   c. Mitochondria and plant chloroplasts are surrounded by two membranes.
   d. Mitochondria reproduce by binary fission.
   e. The DNA in the eukaryotic nucleus codes for some enzymes in mitochondria.

9. As she peered through the microscope, Paige said, "I know that this thing is supposed to be either a ciliate, a flagellate, or an amoeba, but I can't figure out which." Michelle replied, "That's easy . . . "
   a. "Watch how it moves."
   b. "How big is it?"
   c. "All you have to do is see whether it has a nucleus or not."
   d. "Watch and see what it eats."
   e. "Look at its chloroplasts."

10. Which of the following was probably *not* a direct evolutionary ancestor of a maple tree?
   a. a heterotrophic prokaryote
   b. a green alga
   c. an amoeba
   d. a colonial protist
   e. *All* of the above probably *were* ancestors of a maple tree.

11. Some biologists regard seaweeds as protists, even though most other protists are microscopic unicellular organisms. Other biologists think that at least some seaweeds should be considered plants, not protists. Which of the following would support the latter position?
    a. Certain seaweeds have been found to be heterotrophic.
    b. Certain seaweeds contain several kinds of specialized cells.
    c. Certain seaweeds undergo sexual and asexual reproduction.
    d. Certain seaweeds are found to be prokaryotic.
    e. Certain seaweeds have very complex cells.

12. Protozoans called choanoflagellates live in small clusters. They look very much like choanocytes, special feeding cells found in sponges, which are simple animals. Why might biologists find choanoflagellates of great evolutionary interest?
    a. They show how the very first organisms might have lived.
    b. They might show how the first heterotrophs lived.
    c. They might offer clues about the origin of multicellular organisms.
    d. They suggest what the first eukaryotes might have been like.
    e. They might offer clues about the first organisms to live in the sea.

Essay

1. The diphtheria bacterium *Corynebacterium diphtheriae* grows into a mass at the back of the throat and can kill its victim by suffocation. But diphtheria victims also suffer from nervous tremors, paralysis, and heart failure. How might a bacterium that grows in the throat cause symptoms in other parts of the body?

2. Some bacteria can reproduce as often as every 20 minutes. If a hundred of these bacteria were placed in a large flask of culture medium, how many would there be after 6 hours?

3. Chloroplasts in green algae are surrounded by two membranes, while the chloroplasts of euglenoflagellates have four membranes. Explain this in terms of endosymbiosis. (This difference was, in fact, one of the clues that *led* to the idea of endosymbiosis.)

4. Biologists used to classify all protists into three groups— "protozoa," "algae," and "fungal protists." This was a lot less confusing than the current classification into more than a dozen groups. Why was the old scheme dropped?

Put Words to Work

Correctly use as many of the following words as possible when reading, talking, and writing about biology:

alga (plural, algae), alternation of generations, amoebas, Archaea, autotroph, bacillus (plural, bacilli), Bacteria, biofilm, bioremediation, brown algae, cellular slime molds, chemoautotroph, chemoheterotroph, chlamydia, ciliates, coccus (plural, cocci), cyanobacteria, diatoms, dinoflagellates, endospore, endotoxin, exotoxin, extreme halophiles, extreme thermophiles, gram-positive bacteria, gram stain, green algae, heterotroph, methanogens, parasite, pathogen, peptidoglycan, photoautotroph, photoheterotroph, pilus (plural, pili), protist, protozoan (plural, protozoa), pseudopodium (plural, pseudopodia), red algae, secondary endosymbiosis, slime mold, spirochetes, symbiosis, water molds

Use the Web

Don't forget to check out the material on prokaryotes and protists at www.mybiology.com.

Plants, Fungi, and the Colonization of Land 17

Focus on the Concepts

The colonization of land by plants was a major event in the history of life. Even the earliest plants were accompanied by fungi, and the destinies of plants and fungi continue to be intertwined. In studying this chapter, focus on these concepts:

- Plants are eukaryotic multicellular autotrophs that evolved from green algae. A set of derived characters sets plants apart from algae: Alternation of haploid and diploid generations, walled spores produced in sporangia, male and female gametangia, and multicellular, dependent embryos. Plants produce nearly all food for life on land and many important products for humans.

- Many plant characteristics are adaptations to life on land. Most plants have a waxy cuticle that prevents water loss. Gas exchange occurs through pores called stomata. Aboveground, leaves capture gases and sunlight. Belowground, roots absorb water and minerals. These structures are connected by rigid stems with vascular tissues that support the plant. Leaves and roots develop from apical meristems. Plant gametes and embryos are protected in gametangia. All plants produce spores, which some plants use for dispersal. Other plants disperse seeds.

- Plant evolution occurred in four major stages: (1) After plants split from algae, early diversification led to seedless nonvascular plants, such as mosses, lacking vascular tissues and true roots, stems, and leaves. (2) Vascular plants such as ferns came next. They have well-developed roots, rigid stems, and leaves, with vascular tissues. (3) Vascular plants with seeds were the next big step. In gymnosperms—conifers and their relatives—male gametophytes are protected in pollen grains and embryos are protected and dispersed by seeds. (4) Most modern plants are angiosperms—flowering plants. Flowers facilitate pollination, and seeds are protected and dispersed by fruits.

- Haploid and diploid generations alternate in plant life cycles. Haploid gametophytes produce eggs and sperm. Fertilized eggs develop into diploid sporophytes, which produce spores, which develop into gametophytes. In mosses the gametophyte is dominant. The sporophyte is dominant in ferns. Both mosses and ferns have swimming sperm and are dispersed by spores. Gymnosperms and angiosperms are seed plants. In cones or flowers on the sporophyte, spores develop into gametophytes called ovules and pollen grains. Pollen delivers sperm. Embryos develop in cones or flowers and are dispersed in seeds. Angiosperm seeds are enclosed in fruits.

- Fungi are heterotrophic eukaryotes that digest food outside their bodies and absorb the nutrient molecules. A fungus consists of threadlike hyphae with chitin cell walls that form a feeding network called a mycelium. Fungi spread as spores and grow rapidly through their food. They are found nearly everywhere and are important decomposers. Parasitic fungi cause most plant diseases, and a few are human parasites. Some fungi produce food and antibiotics. Most plants have beneficial mycorrhizal fungi associated with their roots.

- Most fungi reproduce sexually via a three-phase life cycle: Spores germinate and haploid hyphae develop. When mycelia of different mating types meet, hyphae fuse but their nuclei do not, producing heterokaryotic cells with two haploid nuclei. Later, the hertokaryotic mycelium forms a reproductive structure, such as a mushroom. In certain cells, nuclei fuse, producing a diploid cells that immediately undergo meiosis, producing spores. There are five groups of fungi, classified on the basis of sexual reproductive structures.

- Many fungi typically reproduce asexually. A mold is any rapidly growing fungus that produces spores asexually. A yeast is any single-celled fungus that reproduces by fission or budding.

Review the Concepts

Work through the following exercises to review the concepts in this chapter. For additional review, check out the activities at www.mybiology.com. The website offers a pre-test that will help you plan your studies.

Exercise 1   (Module 17.1)

Structural, biochemical, and genetic evidence suggests that plants and green algae called charophytes probably evolved from a common algal ancestor. Land plants make up a clade with a set of derived characters that set them apart from algae. Most of these characters are adaptations to life on land. Complete the chart below by listing the major problems plants have to contend with and the adaptations that set plants apart from green algae. (Use a pine tree or rose bush as your plant example; in some ways simple plants such as mosses are "in between" algae and plants.) Also, note (*) which plant characteristics are the shared derived characters of the plant clade.

Differences Between Green Algae and Plants

Green algae	*Pine tree or rose bush*
1. Surrounded by water, unlikely to dry out	1.
2. Exchange gases through body surface	2.
3. Obtain water, gases, nutrients all from one location	3.
4. Floats in water	4.
5. Gamete-producing cells exposed to water	5.
6. Sperm swim and eggs fertilized in water	6.
7. Embryos develop in water	7.
8. No protected cells or embryos for dispersal	8.

Exercise 2   (Module 17.2)

This module summarizes the four major steps in plant evolution that gave rise to four major modern groups of plants. Each of the statements below describes one of those steps. Number the steps in order, and fill in the names of the plant groups.

____ A. The first plants that produced pollen and seeds evolved. The modern seed plants are _____ and _____.

____ B. Plants evolved from ancestral green algae; one lineage gave rise to the _____, a group that included the mosses, liverworts, and hornworts.

____ C. The first flowering plants, or _____, appeared.

____ D. Vascular plants evolved that had roots and strong stems supported by lignin-hardened vascular tissues, unlike the _____. Their modern-day representatives are the _____ and the seed plants.

Exercise 3   (Modules 17.3 – 17.4)

Haploid and diploid generations alternate in all plant life cycles. This is one of the defining characteristics of plants. Study the diagrams in these modules to review alternation of generations. Then fill in the blanks below to complete the description of the moss life cycle.

The moss life cycle, like that of all plants, is characterized by alternation of generations. Diploid individuals called [1]_____ produce [2]_____ plants called gametophytes, which in turn produce [3]_____ sporophytes. Since it's a cycle, we could start at any point, but let's start with a spore. A haploid moss spore grows into the haploid [4]_____ plant, the green, cushiony growth we see on rocks or logs in a forest or bog. [5]_____ (eggs and sperm) develop in the protection of special organs called [6]_____ that are part of the gametophytes. Moss [7]_____ have [8]_____ that enable them to swim to the eggs, given a film of moisture produced by dew or raindrops. [9]_____, the fusion of egg and sperm, produces a diploid [10]_____, which remains protected in the female gametophyte. The zygote divides by mitosis and develops into the sporophyte, which consists of a [11]_____ attached to the gametophyte by a slender stalk. Within the sporangium, haploid [12]_____ are produced by the process of [13]_____. When these spores are mature, the sporangium opens and they scatter in the wind, beginning the cycle anew.

Ferns and lycophytes have vascular tissues, but do not produce seeds; they are often called "seedless vascular plants."

Identify the stages of the fern life cycle by labeling this diagram. Include the following: **haploid phase, diploid phase, sporophyte, gametophyte, zygote, sporangia, spores, sperm, egg, meiosis, fertilization,** and **mitosis and development.** One answer is used twice. Color the haploid part of the life cycle yellow and the diploid part of the life cycle blue.

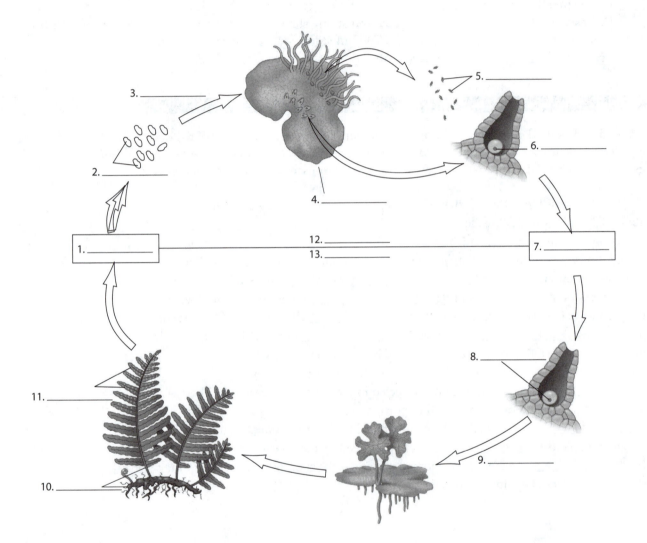

Exercise 5   (Module 17.7)

In seed plants, a specialized structure within the sporophyte houses all reproductive stages—spores, eggs, sperm, zygotes, and embryos. In gymnosperms such as pine trees, this special structure is the cone. Two significant plant adaptations are seen in gymnosperms—pollen and seeds. Complete the following sentences with a structure or stage in the pine life cycle. Select your answers from this list: **seed, pine tree, ovule(s), sperm, egg(s), pollen grain(s), pollen (male) cone(s), ovulate (female) cone(s), embryo, seed coat,** and **zygote.** Some answers are used more than once.

1. A _____ is the diploid sporophyte generation of the pine life cycle.
2. The haploid gametophyte generation develops within the _____ and _____.
3. The small, soft _____ contain many sporangia. Meiosis occurs in the sporangia, producing many spores that develop into _____. These are male gametophytes.
4. Each scale in the larger, woody _____ bears two _____. Each of these develops as a sporangium covered by a tough integument.
5. The wind carries pollen grains to the ovulate cones. Pollination occurs when a pollen grain lands on and enters an _____.
6. After pollination, meiosis occurs in the _____, producing a haploid spore that develops into the female gametophyte.
7. Over a period of months _____ are produced by the female gametophyte in the ovule. At the same time, the male gametophyte (the pollen grain) produces _____.
8. A tiny tube grows out of the _____, releasing a _____ to fertilize an _____.
9. A diploid fertilized egg, or _____, develops into a sporophyte _____. The whole ovule becomes a _____.
10. The _____ consists of the _____ and a food supply made from the remains of the female gametophyte, covered by a seed coat made from the ovule's integument.
11. The seed falls on the ground. When conditions are right, the seed germinates, and the embryo, over decades, grows into a _____, the adult sporophyte. It then produces cones, and the cycle begins again.

Exercise 6   (Module 17.8)

Flowers are responsible for the diversity and success of angiosperms—the flowering plants. Review the flower by matching each flower part with its function.

_____ 1. Ovules contained in this chamber   A. Petal
_____ 2. Consists of filament and anther   B. Style
_____ 3. Produces pollen   C. Sepal
_____ 4. Attracts pollinators   D. Ovary
_____ 5. Female structure with ovary at its base   E. Stamen
_____ 6. Protects the flower before it opens   F. Carpel
_____ 7. Sticky tip that traps pollen   G. Stigma
_____ 8. Stalk that supports anther   H. Filament
_____ 9. Between stigma and ovary   I. Anther

Identify the stages in the life cycle of an angiosperm—a flowering plant—by labeling the diagram below. Include **haploid phase, diploid phase, sporophyte, anther, meiosis, ovary, ovule, female gametophyte, pollen grain (male gametophyte), egg, stigma, pollen tube, sperm, zygote, seed, seed coat, fertilization, embryo, fruit, food supply,** and **pollination.** Then color the haploid part of the life cycle yellow and the diploid part blue.

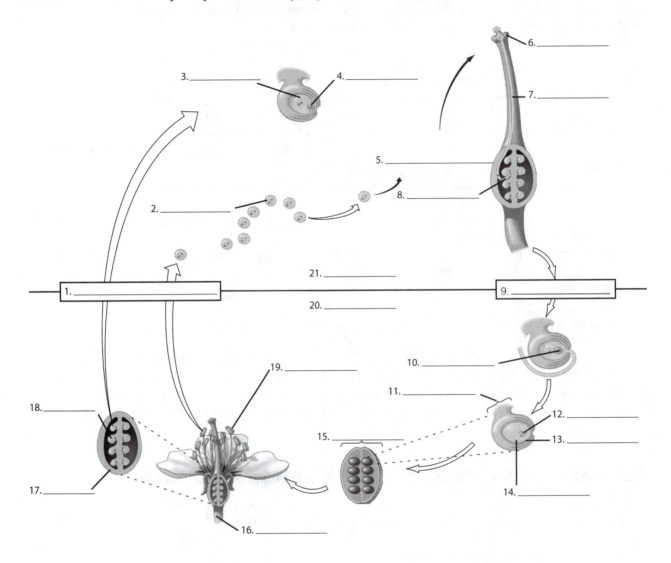

Exercise 8   (Modules 17.7–17.9)

The life cycles of gymnosperms and angiosperms are similar, but angiosperms have added their own tricks of pollination and seed dispersal. Compare gymnosperms (conifers) and angiosperms in the chart below.

Characteristic	Gymnosperms (Conifers)	Angiosperms
Characteristic reproductive structures	1.	2.
Mode of pollination (wind, insects, etc.)	3.	4.
Interval between pollination and fertilization	5.	6.
Time required to produce seeds (pollination to seed dispersal)	7.	8.
Seed protection and dispersal	9.	10.

Exercise 9   (Modules 17.2–17.12)

Review your knowledge of the structure, life cycles, evolution, and uses of the major plant groups. Match each statement with a group (or groups) of plants. Some statements require more than one answer.

_____ 1. Flowering plants
_____ 2. Two types of plants that produce seeds
_____ 3. These plants and their relatives formed coal deposits.
_____ 4. Represent simple vascular plants
_____ 5. Pines, firs, spruces, and cedars
_____ 6. A type of plant in which the gametophyte stage is dominant
_____ 7. Plants that produce fruits
_____ 8. Nonvascular plants
_____ 9. Two types of plants with flagellated swimming sperm
_____ 10. Conifers
_____ 11. Roses, apples, maples, and daisies
_____ 12. Plants with horizontal stems and leaves bearing sporangia
_____ 13. Plants with the shortest gametophyte and longest sporophyte stages
_____ 14. Two types of plants whose spores develop into pollen and ovules
_____ 15. The group that first developed good roots and rigid stems
_____ 16. Source of most lumber and paper
_____ 17. Plants with protected gametes and embryos
_____ 18. Plants that produce seeds but not fruits
_____ 19. Most species of modern plants
_____ 20. Two types of plants without seeds
_____ 21. Source of most of our food
_____ 22. Many of these plants depend on animals for pollination and seed dispersal.
_____ 23. Have walled spores protected in sporangia
_____ 24. The simplest plants

M. Mosses
F. Ferns
G. Gymnosperms
A. Angiosperms

Exercise 10   (Module 17.13)

This module presents some sobering statistics concerning the human threat to world plant diversity. Fill in the number that completes each statement.

1. More than _____ square miles of forest land are cleared every year.
2. About ____ % of the world's forests are found in the tropics.
3. About ____ % of tropical forests were destroyed in the last third of the twentieth century.
4. The forests of North America have shrunk by almost ____ % over the last two centuries.
5. More than ____ % of prescription drugs are extracted from plants.
6. Fewer than _____ of the worlds' _____ known plant species have been investigated as sources of medicines.
7. The All Species Foundation wants to catalogue every species within the next ____ years.

Exercise 11   (Modules 17.14–17.20)

Fungi are strange beasts, but as you get to know them, they will grow on you! These modules introduce some of the structures, roles, and types of fungi. We will review the fungus life cycle in the next exercise, but for now, review fungus vocabulary by matching each fungal phrase with the correct term.

_____ 1. Disperse fungi over great distances	A. Chytrids
_____ 2. Fungal infection	B. Fruiting body
_____ 3. Hard-to-classify fungi with no known sexual stages	C. Mycelium
_____ 4. Threadlike fungal filament	D. Cellulose
_____ 5. Fungi named for their clublike spore-producing structures	E. Mycosis
_____ 6. A branching feeding network of fungal hyphae	F. Lichen
_____ 7. The material of fungal cell walls	G. Absorption
_____ 8. A mutually helpful association between plant roots and fungus	H. Zygomycetes
_____ 9. Systemic mycosis that affects the lungs	I. Mating types
_____ 10. Single-celled fungus that usually lives in liquid or moist habitats	J. Spores
_____ 11. Phase of fungus life cycle where cells have two nuclei	K. Ergot
_____ 12. Mutualistic relationship between fungus and photosynthetic organisms	L. Mycology
_____ 13. Ants (and most other animals) can't break this down, but fungi can.	M. Basidiomycetes
_____ 14. Chemicals from this fungus used to make medicines and LSD	N. Rust
_____ 15. About 90% of plants have partnerships with these fungi.	O. Mold
_____ 16. The closest relatives of fungi	P. Imperfect fungi
_____ 17. A fast-growing furry mycelium on bread or fruit	Q. Hypha
_____ 18. Fungal pathogen of corn	R. Ascomycetes
_____ 19. Their hyphae fuse to start heterokaryotic phase.	S. Heterokaryotic
_____ 20. Aboveground reproductive structure of a sac or club fungus	T. Chitin
_____ 21. A group of fungi that produce spores in saclike structures	U. Yeast
_____ 22. The earliest line of fungi; they have flagellated spores	V. Glomeromycetes
_____ 23. Fungus that attacks wheat crops	W. Mycorrhiza
_____ 24. The scientific study of fungi	X. Smut
_____ 25. How fungi get their food	Y. Animals
_____ 26. Black bread mold is one of these zygote fungi.	Z. Coccidiomycosis

Whether a fungus is a club fungus (basidiomycete), sac fungus (ascomycete), or zygote fungus (zygomycete), its sexual life cycle includes haploid, diploid, and heterokaryotic phases. Fill in the blanks to complete this paragraph about the life cycle of a club fungus.

The life cycle of a club fungus—the familiar supermarket "mushroom"—consists of [1]_____ distinct phases. Under suitable conditions, haploid [2]_____ germinate and produce long filaments called [3]_____. These strands make up a network called a [4]_____. The haploid mycelium grows through the substrate, secreting enzymes, digesting organic matter, and absorbing nutrient molecules. There are different kinds of mycelia, called [5]_____. If compatible mating types come into contact, their hyphae fuse, and the hybrid hyphae form their own mycelium. The [6]_____ within these hybrids do not fuse, however. This begins the [7]_____ phase of the fungus life cycle, which may last for weeks—or years! Each cell in this mycelium contains two genetically distinct nuclei. Eventually, the heterokaryotic mycelium forms a [8]_____—the mushroom that pops up out of the ground. Under the mushroom's cap, haploid nuclei finally fuse, forming [9]_____ cells. Without undergoing mitosis, each of theses cells undergoes [10]_____, forming haploid spores. These spores, produced by the billions, are dispersed by wind, water, and animals to places where they can germinate and grow into new hyphae.

The last few modules in this chapter discuss types of fungi, their ecological roles, and their importance to humans. After reading the modules, review your knowledge of fungi by completing this crossword puzzle.

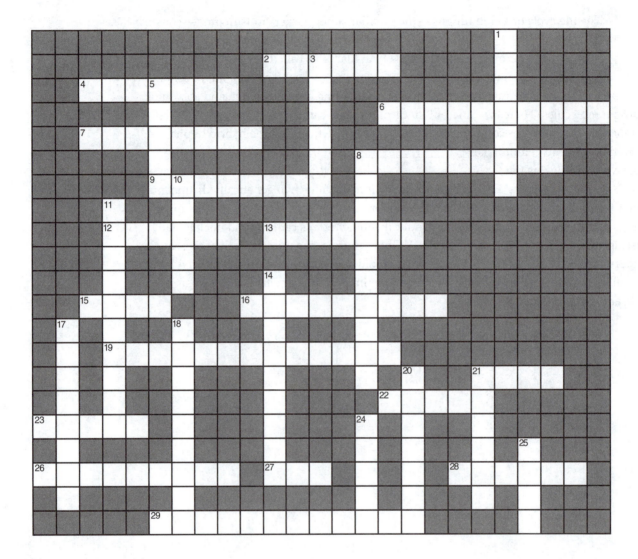

Across

2. A fungus consists of filaments called ____.
4. Fungi are responsible for the flavors of many ____.
6. A ____ is a biologist who studies fungi.
7. ____ is caused by a human fungal parasite.
8. Lichens are very sensitive to air ____.
9. ____ are the most highly prized edible fungi.
12. ____ are unicellular fungi.
13. Yeasts are used in baking and ____.
15. Fungi cannot move, but they ____ rapidly.
16. The word ____ describes the relationship between fungus and algae in a lichen.
19. Unlike plants, which are autotrophic, fungi are ____ eukaryotes.
21. Common mushrooms are ____ fungi, or basidiomycetes.
22. A parasitic fungus causes ____ elm disease.
23. A lichen consists of a fungus and green ____ or cyanobacteria.
26. Molds and ____ are typical fungi.
27. Yeasts, cup fungi, and molds are ____ fungi, or ascomycetes.
28. Club fungi can break down tough ____ in wood.
29. A mushroom is a fungal ____ structure.

Down

1. Athlete's foot is a human fungal disease, or ____.
3. Mycorrhizal fungi help many ____ absorb nutrients.
5. People have been poisoned by ____-infested grain.
8. Fungi produce ____ and other antibiotics.
10. Parasitic fungi called smuts and ____ attack crops.
11. A root and fungus can form a ____, a mutually beneficial partnership.
14. A mycorrhizal fungus helps a plant obtain ____.
17. A network of fungal hyphae make up a ____.
18. Some fungi are parasites, some are mutualists, and some ____ organic matter.
20. Fungi digest their food ____ their bodies.
21. Fungal cell walls are made of ____.
24. Fungi are classified in their own kingdom—Kingdom ____.
25. Leaf-cutting ____ grow fungi for food.

Test Your Knowledge

Multiple Choice

1. Which of the following is *not* a difference between algae and plants?
   a. Plant cells have rigid cellulose walls, and algae cells do not.
   b. Plant zygotes and embryos are protected in moist chambers, and those of algae are not.
   c. Algae lack discrete organs—leaves, stems, roots—characteristic of plants.
   d. Plants have xylem and phloem, and algae do not.
   e. Plants have a waxy, waterproof cuticle, and algae do not.

2. Bryophytes (such as mosses) differ from all other plants in that most bryophytes
   a. do not produce flowers.
   b. have cones but no seeds.
   c. have flagellated sperm.
   d. lack vascular tissues.
   e. produce spores.

3. The gametophyte stage of the plant life cycle is most conspicuous in
   a. ferns.
   b. mosses.
   c. angiosperms.
   d. gymnosperms.
   e. seed plants.

4. Ferns and mosses are mostly limited to moist environments because
   a. their pollen is carried by water.
   b. they lack a cuticle and stomata.
   c. they lack vascular tissues.
   d. they have swimming sperm.
   e. their seeds do not store much water.

5. The diploid generation of the plant life cycle always
   a. produces spores.
   b. is called the gametophyte.
   c. is larger and more conspicuous than the haploid stage.
   d. develops from a spore.
   e. produces eggs and sperm.

6. During the Carboniferous period, forests consisting mainly of ____ produced vast quantities of organic matter, which was buried and later turned into coal.
   a. early angiosperms
   b. seedless vascular plants
   c. giant mosses
   d. gymnosperms
   e. gymnosperms and early angiosperms

7. Which of the following best describes how fertilization occurs in a conifer?
   a. A sperm cell swims through a film of moisture to fertilize the egg.
   b. A pollen grain carried by wind fertilizes the egg.
   c. A pollen grain carried by wind produces a sperm that fertilizes the egg.
   d. A sperm cell carried by wind fertilizes the egg.
   e. A pollen grain swims through a film of moisture to fertilize the egg.

8. Most species of plants are
   a. non-seed-bearing plants.
   b. nonvascular plants.
   c. gymnosperms.
   d. plants other than angiosperms.
   e. angiosperms

9. When you look at a pine or maple tree, the plant you see is
   a. a haploid sporophyte.
   b. a diploid sporophyte.
   c. a haploid gametophyte.
   d. a diploid gametophyte.
   e. none of the above.

10. In a flowering plant, meiosis occurs in the _____, producing a spore that develops into a female gametophyte.
    a. fruit
    b. seed
    c. stamen
    d. anther
    e. ovary

11. A fruit is a ripened
    a. seed.
    b. pollen grain.
    c. bud.
    d. ovary.
    e. anther.

12. How do fungi "find" things to eat?
    a. They produce huge numbers of tiny spores.
    b. They grow rapidly.
    c. They make their own food.
    d. They do all of the above.
    e. They do a and b only.

13. The body of a fungus consists of threadlike ____, which form a network called a ____ .
    a. mycelia . . . sporangium
    b. hyphae . . . gametophyte
    c. mycelia . . . hypha
    d. hyphae . . . mycelium
    e. sporangia . . . fruiting body

14. Where and when does fertilization occur in the mushroom life cycle?
    a. underground, as a mycelium begins to spread
    b. on the surface of the ground, when a spore germinates
    c. in a mushroom, when nuclei of a heterokaryotic cell fuse
    d. underground, when hyphae of different mating types fuse
    e. in a mushroom, when eggs and sperm meet

15. Which of the following kinds of fungi would be considered the least useful or beneficial?
    a. mycorrhizal fungus
    b. yeast
    c. rust
    d. truffle
    e. decomposer

Essay

1. Compare a plant with a multicellular green alga, paying particular attention to plant adaptations to life on land.

2. Most nonbiologists consider seaweeds and fungi to be plants. Why? Why are seaweeds, fungi, and plants placed in separate kingdoms?

3. What are the seed plants? What adaptations have made them so successful?

4. What kinds of products do we obtain from forests? What kinds of trees supply most of our forest product needs? How can we obtain the forest products we need and still preserve our forests?

5. Sketch a flower, name its major parts, and describe their functions.

6. How do animals assist in angiosperm reproduction? How have the structures of angiosperms adapted to reflect this relationship with animals?

7. How do fungi obtain their food?

8. What two components make up a lichen? What are their roles?

Apply the Concepts

Multiple Choice

1. An explorer found a plant that had roots, stems, and leaves. It had no flowers but produced seeds. This plant sounds like a(n)
   a. fern.
   b. bryophyte.
   c. angiosperm.
   d. moss.
   e. gymnosperm.

2. Which of the following stages in the life cycle of a maple tree corresponds to the leafy, spongy plant in the moss life cycle?
   a. egg and sperm
   b. adult tree
   c. flower
   d. pollen grain and ovule
   e. zygote

3. Deep in the tropical rain forest, a botanist discovered an unusual plant with vascular tissues, stomata, a cuticle, flagellated sperm, conelike reproductive structures bearing seeds, and an alternation-of-generations life cycle. He was very excited about this discovery because it is unheard of for a plant to have both
   a. a cuticle and flagellated sperm.
   b. vascular tissues and alternation of generations.
   c. seeds and flagellated sperm.
   d. alternation of generations and seeds.
   e. cones and vascular tissues.

4. The pinyon pine lives in near-desert areas in western North America. This habitat is a bit unusual for gymnosperms because they
   a. have a long life cycle for such harsh growing conditions.
   b. possess flagellated sperm that must swim to the egg.
   c. produce extremely small quantities of pollen.
   d. lack vascular tissues and are unable to transport much water.
   e. produce cones rather than drought-resistant seeds.

5. Unlike most angiosperms, grasses are pollinated by wind. As a consequence, some "unnecessary" parts of grass flowers have almost disappeared. Which of the following parts would you expect to be most reduced in a grass flower?
   a. ovaries
   b. petals
   c. anthers
   d. carpels
   e. stamens

6. Fuchsia flowers are generally reddish, they hang downward, and their nectar is located deep in floral tubes. It sounds like fuchsias are typically pollinated by
   a. bees.
   b. flies.
   c. bats.
   d. butterflies.
   e. birds.

7. Some scum was found growing near the edge of a pond. Under a microscope, each of its cells was found to contain two nuclei. This means the scum must be
   a. some kind of alga.
   b. a fungus.
   c. a plant gametophyte.
   d. a liverwort.
   e. a plant sporophyte.

8. Strolling through the woods, you would be least likely to notice which of the following?
   a. a moss gametophyte
   b. a fern gametophyte
   c. an angiosperm sporophyte
   d. a fern sporophyte
   e. the heterokaryotic stage of a fungus

9. The diploid phase of the life cycle is shortest and smallest in which of the following?
   a. moss
   b. angiosperm
   c. fungus
   d. fern
   e. gymnosperm

10. Which of the following is a difference between plants and fungi?
    a. Plants have diploid and haploid phases.
    b. Fungi have cell walls.
    c. Fungi are autotrophic.
    d. In fungi, zygotes undergo meiosis to produce spores.
    e. Plants undergo sexual reproduction.

Essay

1. What characteristics does a fern share with the seed plants that evolved later? In what ways are ferns similar to mosses?

2. What kinds of plants are dispersed to new habitats by spores? By seeds? What are the advantages and disadvantages of each means of dispersal?

3. When his orchard was attacked by parasitic fungi, a farmer sprayed the trees with a powerful fungicide. The next season, most of the trees were free of the parasite, but they grew poorly and produced even less fruit than they had when they were infected. What might account for this change?

4. If you have ever had athlete's foot, you are probably aware that fungal infections are rather difficult to get rid of. What do you know about the structure and lifestyle of fungi that might make them particularly persistent pests?

5. Biologists give each species a Latin name, such as *Homo sapiens* for humans and *Acer macrophyllum* for the big-leaf maple. Why do you suppose this might be difficult for lichens?

Put Words to Work

Correctly use as many of the following words as possible when reading, talking, and writing about biology:

absorption, alternation of generations, angiosperm, anther, apical meristem, ascomycete, basidiomycete, bryophyte, carpel, club fungus, fossil fuel, fruit, Fungi, fungus (plural, fungi), gametangium (plural, gametangia), gametophyte, gymnosperm, heterokaryotic stage, hypha (plural, hyphae), imperfect fungus, lichen, lignin, mold, mycelium (plural, mycelia), mycorrhiza (plural, mycorrhizae), ovary, parasite, petal, phloem, pollen, pollen grain, pollination, sac fungus, seed, seed coat, seedless vascular plants, sepal, sporangium (plural, sporangia), spore, sporophyte, stamen, vascular plant, vascular tissue, xylem , yeast, zygomycete, zygote fungus

Use the Web

There is more on plants and fungi at www.mybiology.com.

The Evolution of Invertebrate Diversity

Focus on the Concepts

This chapter explores the diversity of the animal kingdom. It examines animal body plans, from simple to complex, and how these plans relate (or don't relate) to the animal phylogenetic tree. In studying the chapter, focus on these concepts:

- Animals are multicellular, heterotrophic eukaryotes that obtain nutrients by ingestion. They probably evolved from a colonial flagellated protist. Animal cells lack walls and are held together by unique proteins and junctions. Animals are diploid, except for eggs and sperm. Their embryos go through a blastula and gastrula stage, and some pass through a larval stage before becoming adults. Development is controlled by *Hox* genes. Most animals have three tissue layers, muscle cells, and nerve cells.

- Animal body plans are used to build phylogenetic trees. Sponges are the lowest branch; they have no true tissues. Among animals with tissues, cnidara are unique in having radial symmetry and only two tissue layers; all other animals have bilateral symmetry and three layers. Bilateral animals split into deuterostomes (echinoderms and chordates) and protostomes (flatworms, mollusks, annelids, arthropods, etc.) on the basis of embryology. Molecular data detect two clades of protostomes, which can be grouped according to their larvae, feeding structures, and exoskeletons.

- Sponges are sessile, mostly marine suspension feeders, with saclike bodies consisting of two layers of cells. They are the simplest animals, with no nerves or muscles.

- Cnidarians are radial animals such as sea anemones, corals, and jellies, with tentacles bearing stinging cells. Their prey are digested in a gastrovascular cavity.

- Flatworms are the simplest bilateral animals, with three tissue layers, organs, and systems. Free-living flatworms suck food into a gastrovascular cavity. Flukes and tapeworms have special adaptations for parasitism.

- Nematodes are tubular worms with a pseudocoelom and complete digestive tract. Many live in soil and water, and some forms are parasitic on plants and animals.

- Clams, snails, and squids are a few of the variety of mollusks. A mollusk has a muscular foot, with a visceral mass, a coelom, and a membranous mantle that usually secretes a shell. Mollusks have circulatory systems, and most have gills.

- Annelids are segmented worms—earthworms, leeches, and marine polychaetes. Their segmented bodies and coeloms allow for flexibility and mobility. Annelids have closed circulatory systems, and exchange gases via gills or the body surface.

- Arthropods are segmented animals with jointed appendages and exoskeletons. Segments can be fused and appendages adapted for various functions. All have open circulatory systems. Most crustaceans are aquatic. Chelicerates, millipedes, centipedes, and insects live on land. Insects are a diverse and successful group of animals.

- Echinoderms, such as sea stars and sea urchins, have spiny skin, an endoskeleton, and a water vascular system with tube feet for movement. These unique marine creatures are the closest relatives of chordates.

- Most chordates have backbones, but a few marine chordates—tunicates and lancelets—are invertebrates. Every chordate has a dorsal hollow nerve cord, a longitudinal notochord, pharyngeal slits, and a muscular post-anal tail.

Review the Concepts

Get to know the invertebrate animals by working though the following exercises. For additional review, explore the activities at www.mybiology.com. The website offers a pre-test that will help you plan your studies.

Exercise 1   (Module 18.1)

This module describes how we "draw the line" between animals and other organisms. Five major groups of living things are listed below. Use colored pens or pencils to draw lines separating the groups on the basis of the following characteristics.

1. Draw a black line between organisms that have simple cells and those that have complex cells (with organelles). Write "simple cells" on one side of the line and "complex cells" on the other.

2. Draw a blue line between organisms that are primarily unicellular and those that are multicellular. Label these groups.

3. Separate heterotrophic and autotrophic multicellular organisms with a green line, and label each group.

4. Draw a red line between those heterotrophs that ingest their food and then digest it and those that digest food outside their bodies and then absorb the nutrients. Label each. On the animal side of the line, also note that animals have unique extracellular proteins and intercellular junctions, lack cell walls, are diploid (except for eggs and sperm), have unique developmental stages, and most have muscle and nerve cells.

Plants   Fungi   Animals

Protists

Prokaryotes

Exercise 2 (Module 18.2)

Using Module 18.2 as a guide, label this series of sketches showing how animals may have evolved from colonial protists. Identify: **colonial protist, somatic cells, reproductive cells, digestive cavity, "proto-animal"**

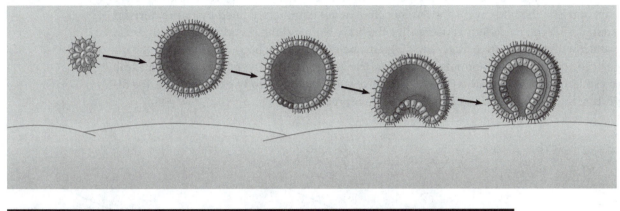

Exercise 3 (Module 18.3)

Symmetry—radial or bilateral—is an important feature of an animal's body plan. Compare animals having bilateral symmetry with those having radial symmetry by completing the following table.

Animals with Radial Symmetry	Animals with Bilateral Symmetry
Examples: sea anemone, jelly	1.
2.	Right and left sides
3.	Dorsal and ventral surfaces
No anterior or posterior ends	4.
No distinct head	5.
6.	Move actively through environment
Encounter environment equally from all directions	7.

Exercise 4   (Module 18.3)

Most animals have a body cavity, a space between the digestive tract and the body wall. (There are many advantages to having a body cavity; these are reviewed in the "Test Your Knowledge" section.) There are two kinds of body cavities: A coelom is a space completely lined by tissue derived from mesoderm, so mesodermal tissues cover the digestive tract and line the body wall. A pseudocoelom is not completely lined by tissue derived from mesoderm; mesodermal tissues line the body wall, but the pseudocoelom is in direct contact with the digestive tract. The diagrams below show the body covering and the digestive tract of three animals. Complete the diagrams by sketching in the middle (mesoderm-derived) tissue layer. Label the layers and color the **body covering** blue, the **digestive tract** yellow, and the **middle tissue layer** red.

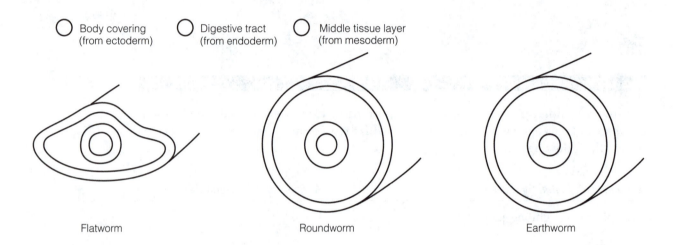

○ Body covering   ○ Digestive tract   ○ Middle tissue layer
(from ectoderm)   (from endoderm)   (from mesoderm)

Flatworm   Roundworm   Earthworm

Exercise 5   (Module 18.4)

Module 18.4 introduces an animal phylogenetic tree based on body plan (tissues, symmetry, body cavity, and so on.) Suppose you found an animal and wanted to know to which phylum it belonged. You could use what biologists call a "key," a series of questions that leads you to the animal's identity. Such a key is given below, in the form of a flowchart. Before you can use this key, you will need to complete it. Some questions and phylum names are given; others are missing. Using information from this module, fill in the blanks to complete the key. You can get some answers by looking at the pictures in Module 18.4. (You might want to try out the key by "keying out" a real animal. Try it on yourself!)

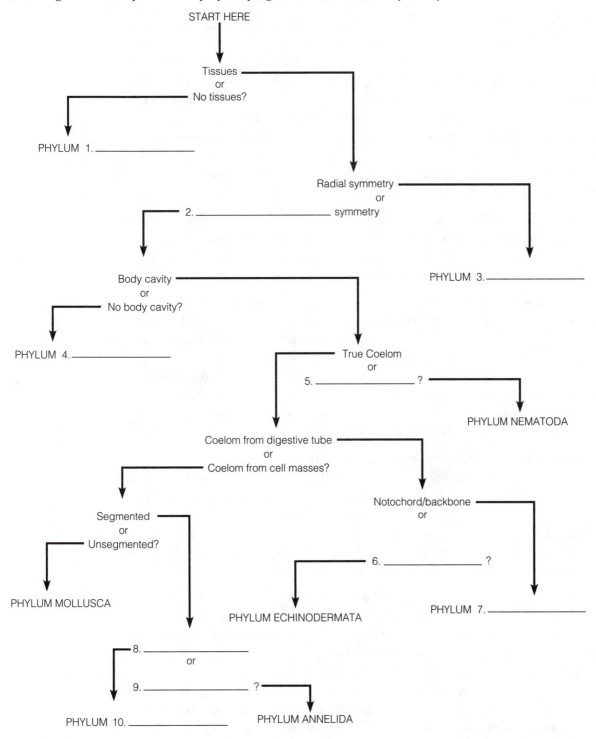

START HERE

Tissues or No tissues?

PHYLUM 1. _____

Radial symmetry or 2. _____ symmetry

PHYLUM 3. _____

Body cavity or No body cavity?

PHYLUM 4. _____

True Coelom or 5. _____ ?

PHYLUM NEMATODA

Coelom from digestive tube or Coelom from cell masses?

Segmented or Unsegmented?

Notochord/backbone or 6. _____ ?

PHYLUM MOLLUSCA

PHYLUM ECHINODERMATA

PHYLUM 7. _____

8. _____ or 9. _____ ?

PHYLUM 10. _____

PHYLUM ANNELIDA

Sponges and cnidarians are the simplest animals. Complete the following description of these two phyla by filling in the blanks.

Sponges, Phylum [1]_____, and cnidarians, Phylum [2]_____, are both simple animals. Most sponges and cnidarians live in the [3]_____, but some are found in fresh water. Sponges are usually irregular in shape, but all cnidarians are characterized by [4]_____ symmetry. This means their body parts are arranged in a circle around a central axis.

Sponges are by far the simpler of the two animals. A sponge is a simple tube perforated by tiny [5]_____. The body wall consists of [6]_____ layers of cells. The outer layer functions to protect the sponge. A gelatinous middle layer contains wandering amoebocytes and a skeleton made of flexible spongin or more rigid mineral-containing particles. The sponge's inner layer consists of cells called choanocytes bearing [7]_____, which move to create a current of water that [8]_____ the sponge through the small pores and [9]_____ through a larger central opening. The choanocytes trap [10]_____ from the water and then engulf them by phagocytosis. The amoebocytes pick up food from the choanocytes and distribute it to other cells. They also make the [11]_____ fibers.

Unlike other animals, sponges lack both [12]_____ and muscles. In fact, their cells are relatively unspecialized, so the cell layers are not considered true [13]_____. It is likely that sponges are early offshoots of ancient colonial [14]_____ called choanoflagellates.

Cnidarians—animals such as [15]_____, sea anemones, and corals—are a bit more complicated. They have a [16]_____ cavity, muscles, and a [17]_____ system that enables them to respond to stimuli and coordinate muscle action. Unlike sponges, their cells are organized into [18]_____, groups of cells adapted to perform specific functions. But unlike more complex animals, they have only [19]_____ tissue layers, and most of their activities are carried out at the tissue level, not by the organs and [20]_____ of more complex creatures.

Cnidarians are radially symmetrical and come in two shapes. A [21]_____ is a tube with tentacles radiating from one end. It is usually fixed in place. A [22]_____ is a disk with a fringe of tentacles on the edge. [23]_____ are medusas and are able to move about in the water. Some cnidarians, such as the freshwater form called a [24]_____, illustrated in the text, exist only in the polyp form; some cnidarians exist only as medusae. Others have both medusa and polyp stages in their life cycles.

A cnidarian captures small prey and pushes it into its mouth with its [25]_____. Special cells called [26]_____ on the tentacles (characteristic only of cnidarians) sting and entangle the prey. The mouth of a polyp faces upward, in the center of the tentacles. A jelly's mouth is [27]_____, in the center of the umbrella. The mouth leads to a digestive sac called the [28]_____ cavity. Food is digested here, and fluid in the cavity circulates food particles around the body. The fluid in the cavity also keeps the flimsy body "inflated" and gives the cnidarian its shape. Because the gastrovascular cavity has only one opening, [29]_____ are expelled through the mouth.

Exercise 7   (Modules 18.6–18.8)

Review and compare the structures and lifestyles of cnidarians, flatworms, and round-worms by completing this chart.

	Cnidarians	*Flatworms*	*Roundworms*
1. Phylum name			
2. Examples			
3. Type of body symmetry			
4. Number of tissue layers			
5. Body shape(s)			
6. Body cavity			
7. Digestive tract			
8. Where they live			
9. Importance to humans			

Exercise 8   (Module 18.9)

This module discusses several of the structural and functional characteristics of mollusks, a varied and successful group of animals. Match each of the statements on the left with a mollusk body structure on the right.

_____ 1. Modified to form a lung in land snails	A. Coelom
_____ 2. Lacking in slugs	B. Radula
_____ 3. Used by a clam to capture food	C. Gill
_____ 4. Divided into hinged halves in bivalves	D. Foot
_____ 5. Functions in locomotion in most mollusks	E. Mantle
_____ 6. Extracts oxygen from the water	F. Circulatory system
_____ 7. Rasping organ used to scrape up food	G. Shell
_____ 8. Distributes nutrients, water, and oxygen around the body	

_____ 1. Modified to form a lung in land snails
_____ 2. Lacking in slugs
_____ 3. Used by a clam to capture food
_____ 4. Divided into hinged halves in bivalves
_____ 5. Functions in locomotion in most mollusks
_____ 6. Extracts oxygen from the water
_____ 7. Rasping organ used to scrape up food
_____ 8. Distributes nutrients, water, and oxygen around the body
_____ 9. Missing or internal in squids and octopuses
_____ 10. Outgrowth of the body surface that drapes over the animal
_____ 11. Modified to form tentacles in cephalopods
_____ 12. Body cavity around heart, kidney, and reproductive organs
_____ 13. Long projections on the back of a sea slug
_____ 14. "Crawling" movements of this structure propel gastropods
_____ 15. Used by a clam for digging and anchoring in mud or sand
_____ 16. Shoots out a jet of water to propel a squid
_____ 17. Eyes of a scallop are along the edge of this structure
_____ 18. A one-piece coiled structure in snails
_____ 19. Lacking in terrestrial snails and slugs
_____ 20. Secretes the shell

A. Coelom
B. Radula
C. Gill
D. Foot
E. Mantle
F. Circulatory system
G. Shell

Exercise 9 (Module 18.10)

This module discusses annelids, the segmented worms, and the importance of segmentation. Review by filling in the blanks.

The next time you dig up an earthworm, or see one wriggling on the sidewalk, pause to appreciate its beauty and complexity. Earthworms are segmented worms of the phylum [1]_____. The name, which means "ringed," refers to the repeating ringlike [2]_____ that make up the worm's body. There are three main groups of annelids. Most live in the [3]_____, but many species live in [4]_____ and moist soil.

The most distinctive external characteristic of annelids is segmentation. Internally, each segment is separated from adjacent ones by membranous [5]_____. The [6]_____ system consists of a simple brain, a ventral nerve cord, and clusters of nerve cells in each segment. There are blood vessels serving each segment, and [7]_____ structures, which dispose of fluid wastes, are also repeated. A dorsal heart (actually an enlarged blood vessel) pumps blood via a [8]_____ circulatory system. The main blood vessels and the [9]_____ system are unsegmented.

What are the advantages of a segmented body? It probably is an adaptation to facilitate [10]_____. It gives the body greater [11]_____ and [12]_____. Longitudinal and [13]_____ muscles in each segment enable an earthworm to burrow, obtaining nutrients from the soil that passes through its digestive tract. Earthworms stir up the soil, and their [14]_____ improve its texture.

The largest group of annelids are the [15]_____. Most of these worms live in the [16]_____, where they wriggle along the bottom, burrow in the mud, or construct protective [17]_____. The mobile polychaetes move by means of segmental [18]_____. In tube-dwellers, these appendages are modified for [19]_____ and respiration.

The third group of annelids are the [20]_____. Some suck [21]_____, but most are free-living [22]_____ that eat small animals. Most leeches live in [23]_____, but some are found in the sea or on land. Leeches have sharp [24]_____, and they secrete an anesthetic that enables them to slice painlessly through the skin and an anticoagulant that keeps blood flowing freely. The latter substance is useful in drug form for dissolving [25]_____.

Annelids are not the only segmented animals. [26]_____ are segmented; this is seen clearly in the abdomen and in the thorax of an insect, where wings and legs are repeated. Animals with backbones are also segmented. In humans, segmentation is most clearly seen in the backbone and in the abdominal muscles.

Exercise 10   (Modules 18.11–18.12)

These modules discuss Phylum Arthropoda, and Class Insecta, the arthropods that are the largest and most diverse group of animals. Review your knowledge of arthropods and insects by completing the crossword puzzle.

Across

1. Crabs and lobsters are ____.
6. The arthropod exoskeleton is made of ____ and protein.
9. ____ make up the largest order of animals.
10. An insect has ____ pairs of legs.
12. The ____ crab is a "living fossil" related to spiders.
13. ____ are sensory appendages on the head.
14. ____ are marine filter-feeding crustaceans.
16. Insects are the only invertebrates with ____.
18. Arthropods have ____ appendages.
20. A spider might hunt insects or catch them in a ____.
22. The ____ is an arachnid with pincers and a sting at the end of its tail.
23. ____ is shedding the old exoskeleton and growing a larger one.
24. Much of insect success can be attributed to their ability to ____.
25. A lobster uses its ____ for defense.
26. Arthropods have an ____ circulatory system.
27. The ____ are the most diverse group of arthropods.
28. ____ are multi-legged carnivores.
29. Many insects undergo ____ during their development.

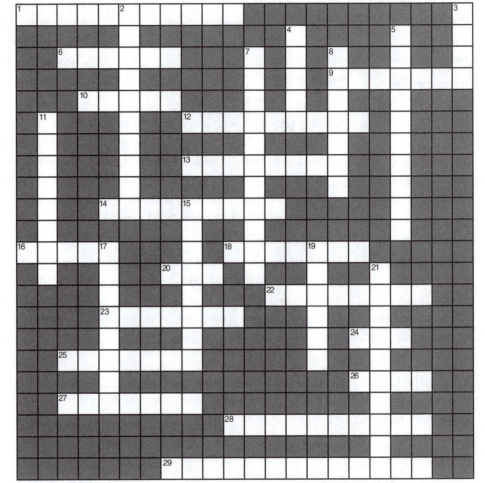

Down

2. Crabs, grasshoppers, and tarantulas are all representatives of phylum ____.
3. ____ are wingless social insects related to bees and wasps.
4. Scorpions, spiders, ____, and mites are all arachnids.
5. The study of insects is called ____.
7. Every arthropod has a hard external skeleton called an ____.
8. An insect's body consists of head, thorax, and ____.
11. Many people are ____ to dust mites.
15. Unlike ____, which have similar segments the length of the body, most arthropods are divided into distinct groups of segments.
17. The arthropod body consists of groups of ____.
19. An insect's wings and legs are attached to its ____.
21. ____ are wormlike plant-eaters with many short legs.

Exercise 11   (Module 18.13)

Echinoderms are a unique animal phylum. Circle the statements below that relate to echinoderms, and cross out statements that are not relevant to echinoderms.

1. Adults have radial symmetry
2. Live in salt water and fresh water
3. Larvae have radial symmetry
4. Have spines embedded under the skin
5. Most closely related to cnidarians
6. Lack segmentation
7. Have an exoskeleton
8. Good at regeneration
9. Have a water vascular system

10. Move and feed with tube feet
11. Adults have bilateral symmetry
12. Examples: sea urchins and sea stars
13. Live in salt water
14. Larvae have bilateral symmetry
15. Have an endoskeleton
16. Most closely related to chordates
17. Are segmented
18. Examples: clams and snails

Exercise 12   (Module 18.14)

Humans are vertebrates, and vertebrates are chordates, classified in Phylum Chordata. Also in Phylum Chordata are some very simple marine invertebrates that do not have backbones—the lancelets and tunicates. All chordates, from tunicates to college students, share four key chordate characteristics. Identify and label these four chordate characteristics on this drawing of a lancelet.

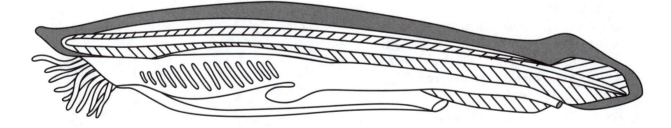

New information—especially discoveries in genetics and the new science of "evo-devo"—have shaken the animal phylogenetic tree a bit. In the revised tree, which is thought to more accurately reflect evolutionary history, characteristics such as body cavity and segmentation are not as important as other more obscure characteristics such as feeding method and details of embryological development. Study the revised tree in Module 18.15 and then see if you can identify the structural characteristics that characterize each of the following branches of the tree.

_____ 1. Sponges	A. Embryological development
_____ 2. Everything but sponges	B. Presence of true tissues
_____ 3. Cnidaria	C. Feeding structure and larva
_____ 4. All protostomes and deuterostomes	D. Radial symmetry
_____ 5. Protostomes	E. Bilateral symmetry
_____ 6. Deuterostomes	F. Molting of exoskeleton
_____ 7. Lophotrochozoa	G. Lack of true tissues
_____ 8. Ecdysozoa	

Test Your Knowledge

Multiple Choice

1. Which of the following is *not* a characteristic of all animals?
    a. They are multicellular.
    b. They have tissues, organs, and organ systems.
    c. They are eukaryotes.
    d. They ingest their food.
    e. They are heterotrophic.

2. A ____ is the simplest animal discussed in this chapter to have ____ .
    a. sponge . . . bilateral symmetry
    b. flatworm . . . a body cavity
    c. roundworm . . . a complete digestive tract
    d. jelly . . . a complete digestive tract
    e. snail . . . a body cavity

3. Which of the following animals does *not* have a body cavity?
    a. flatworm
    b. ant
    c. mouse
    d. clam
    e. earthworm

4. Which of the following phyla include numerous parasites and pests?
    a. roundworms and flatworms
    b. mollusks and roundworms
    c. annelids and flatworms
    d. annelids and roundworms
    e. mollusks and flatworms

5. Animals probably evolved from colonial protists. How do animals differ from these protist ancestors?
    a. Animals are eukaryotic.
    b. Animals have more specialized cells.
    c. Animals are heterotrophic.
    d. Animals are autotrophic.
    e. Animals are able to reproduce.

6. Which of the following animals is *not* segmented?
    a. leech
    b. human
    c. lancelet
    d. lobster
    e. snail

7. Phylum _____ includes the largest number of species.
    a. Mollusca
    b. Chordata
    c. Annelida
    d. Arthropoda
    e. Echinodermata

8. The water vascular system of a sea star functions in
    a. movement of the tube feet.
    b. circulation of nutrients around the body.
    c. pumping water for swimming movements.
    d. waste disposal.
    e. keeping all parts of the body moist at low tide.

9. A ____ is a chordate, but not a vertebrate.
   a. squid
   b. shark
   c. tunicate
   d. beetle
   e. frog

10. Zoologists have traditionally placed chordates and echinoderms on one major branch of the animal phylogenetic tree, and mollusks, annelids, and arthropods on another major branch. Which of the following is the basis for this separation into two branches?
    a. whether or not the animals have a skeleton
    b. type of symmetry
    c. whether or not the animals have a body cavity
    d. how the body cavity is formed
    e. whether or not the animals are segmented

11. Which of the following are most numerous and successful on land?
    a. mollusks and chordates
    b. annelids and arthropods
    c. arthropods and chordates
    d. annelids and chordates
    e. mollusks and arthropods

12. Which of the following has been suggested to explain the Cambrian explosion of animal diversity?
    a. increase in atmospheric oxygen levels
    b. development of more complex predator-prey relationships
    c. variation in expression of *Hox* genes
    d. evolution of hard skeletons
    e. all of the above

13. According to the information in this chapter, the largest existing invertebrate is
    a. a mollusk
    b. a whale
    c. an arthropod
    d. a cnidarian
    e. a sponge

14. Which of the following are all suspension feeders?
    a. clams, sponges, and tunicates
    b. barnacles, cnidarians, and sponges
    c. lancelets, cnidarians, and clams
    d. cnidarians, sponges, and sea stars
    e. barnacles, sea stars, and tunicates

Essay

1. Describe the characteristics that separate animals from the other groups of living things.

2. Describe some of the characteristics that biologists consider important when deciding the phylum into which an animal should be classified.

3. What kinds of animals have a body cavity? What kinds lack a body cavity? Describe some of the advantages of having a body cavity.

4. Describe how the mantle, mantle cavity, and shells of snails, clams, and squids are modified for their different ways of life.

5. In terms of numbers of individuals and numbers of species, it could be argued that insects are the most successful creatures on Earth. What are some characteristics or adaptations that have made them so successful?

Apply the Concepts

Multiple Choice

1. Compare the two phylogenetic trees in Modules 18.4 and 18.15. The tree based on molecular data most drastically revises which of the following relationships?
   a. cnidaria and all other phyla
   b. annelids and mollusks
   c. sponges and all other phyla
   d. annelids and arthropods
   e. chordates and echinoderms

2. Which of the following includes the largest number of species?
   a. animals that are segmented
   b. animals with radial symmetry
   c. animals with bilateral symmetry
   d. animals that are unsegmented
   e. animals with a notochord

3. Which of the following is radially symmetrical?
   a. a doughnut
   b. an automobile
   c. a spoon
   d. a peanut butter sandwich
   e. a wristwatch

4. A marine biologist dredged up a small animal from the bottom of the ocean. It was uniformly segmented, with short, stiff appendages and soft, flexible skin. It had a complete digestive system and a circulatory system but no skeleton. Based on this description, this animal sounds most like
   a. a lancelet.
   b. a crustacean.
   c. a mollusk.
   d. a roundworm.
   e. an annelid.

5. "Pill bugs" or "sow bugs," often found under rocks and logs in moist places, are perhaps most noticed for their ability to roll up into a ball when disturbed. Sow bugs are really crustaceans, not insects. Therefore, a sow bug does *not* have
   a. an exoskeleton.
   b. gills.
   c. three pairs of legs.
   d. antennae.
   e. jointed appendages.

6. Which of the following is thought to be most closely related to you?
   a. sea star
   b. snail
   c. earthworm
   d. jelly
   e. ant

7. Planarian worms, tapeworms, and flukes are all in the Phylum Platyhelminthes, but flukes lack sensory structures such as eyes, and tapeworms don't even have a digestive system. How can such different animals be placed in the same phylum?
   a. Phylum Platyhelminthes isn't a true clade; it is a catch-all for unrelated species.
   b. Whether or not an animal has a digestive system isn't important in classification.
   c. Roundworms are *very* different from planarian worms and are in the same phylum.
   d. Flukes and tapeworms worms may have lost some structures because they are parasites.
   e. These three worms are *not* in the same phylum.

8. A zoologist referred to a group of animals as "the insects of the sea." What group do you think she was talking about?

   a. mollusks
   b. cnidarians
   c. arachnids
   d. echinoderms
   e. crustaceans

9. Dragonflies, like most insects, have two pairs of wings. In flies, the rear pair of wings is modified into a pair of balancing structures called halteres. In beetles, the front pair of wings are more like hard protective shells. If the evolution of wings is like evolution of other parts of the insect body, what genetic differences probably account for these differences in insect wings?
   a. These three insects do not share a common ancestor, so you would not expect their wings to be similar.
   b. These insects probably have the same "wing" genes, but changes in homeotic genes alter their timing.
   c. These insects probably each have different "wing" genes, so their wings develop differently.
   d. Insect wings are actually modified legs, so mutations to "leg" genes shaped the different structures.
   e. Insect wings are shared ancestral characters; genetic changes would not affect them.

10. You would expect to find the greatest number of phyla of animals _____ and the greatest number of species of animals _____.
    a. on land . . . in the sea
    b. in fresh water . . . in the sea
    c. in the sea . . . on land
    d. in the sea . . . in fresh water
    e. on land . . . in fresh water

11. Most insects go through a process of complete metamorphosis. Which of the following "advertisements" most accurately summarizes this process?
    a. "Little and Lethal"
    b. "Feeders and Breeders."
    c. "Eaters and Skeeters"
    d. "Wingers and Stingers"
    e. "Fighters and Biters"

Essay

1. Sponges have no muscles or nerve cells and cannot move. Why are they considered animals?

2. A flattened, disk-shaped creature called *Trichoplax*, in phylum Parazoa (a small phylum not discussed in this chapter), is one of the simplest known animals. Its body consists of a single ciliated outer layer over a core of unspecialized cells. It has no digestive tract, but it crawls over food and hunches its "back" to form a temporary hollow that serves as a digestive sac. What does this animal suggest about the early evolution of animals?

3. Imagine that you are a Peace Corps volunteer assigned to a small African village where many people are infected with pork tapeworms, which are spread from pig to pig and from pigs to people by eating infected meat. Resources are scarce; the poor villagers cannot afford expensive medicines. If these worms have life cycles like other tapeworms, suggest three ways the villagers could interrupt the worm's life cycle and keep themselves from becoming infected.

4. Name what you consider to be a successful phylum of invertebrates. What are your criteria for choosing these animals? What makes them successful?

5. Zoologists have found that certain marine snails and polychaete worms have similar ciliated swimming larvae. What does this evidence suggest about the evolution of annelids and mollusks? Is this reflected in the animal phylogenetic trees given in this chapter? Explain.

6. Briefly explain how changes in homeotic genes might have led to different adaptations in insect wings, mouthparts, and legs. Would you expect a fly, a grasshopper, and a moth to have quite similar genes or quite different genes? Explain.

Put Words to Work

Correctly use as many of the following words as possible when reading, talking, and writing about biology:

annelid, anterior, arachnid, arthropod, bilateral symmetry, bivalve, body cavity, centipede, cephalopod, chelicerate, closed circulatory system, cnidarian, cnidocyte, coelom, complete digestive tract, complete metamorphosis, crustacean, cuticle, dorsal, dorsal hollow nerve cord, deuterostome, echinoderm, endoskeleton, exoskeleton, flatworm, fluke, foot, free-living flatworm, gastropod, gastrovascular cavity, horseshoe crab, hydrostatic skeleton, incomplete metamorphosis, ingestion, invertebrate, lancelet, larva (plural, larvae), leech, mantle, medusa (plural, medusae), millipede, mollusk, molting, nematode, notochord, open circulatory system, pharyngeal slit, polychaete, polyp, post-anal tail, posterior, protostome, pseudocoelom, radial symmetry, radula, segmentation, sessile, sponge, suspension feeder, tapeworm, tunicate, ventral, visceral mass, water vascular system

Use the Web

To enhance your knowledge of the invertebrates, see the activities at www.mybiology.com.

The Evolution of Vertebrate Diversity

Focus on the Concepts

This chapter continues the survey of the Animal Kingdom by exploring vertebrate phylogeny and diversity. It concludes with a discussion of human origins. In studying the vertebrates, focus on the following concepts:

- Most chordates are vertebrates. Ten major chordate clades are defined by derived characters such as the presence of a head, jaws, lungs, and legs.

- Tunicates and lancelets are chordates, but not vertebrates. The most primitive surviving vertebrates are lampreys and hagfishes, marine creatures lacking jaws or paired appendages. Both are craniates, having a defined head, but technically, only the lamprey is a vertebrate, with a backbone of vertebrae and a skull.

- Hinged jaws appear to have evolved from gill supports, and first appear in the chondrichthyans, the sharks and rays. Besides jaws, these marine fishes have flexible cartilage skeletons and paired appendages.

- Most fishes are ray-finned fishes, with a hard skeleton and paired fins supported by thin, flexible rays. A much smaller clade is the lobe-finned fishes, whose fins are supported by rod-shaped bones—precursors of the limb bones of land vertebrates. Both these fishes have the beginning of lungs—modified as a buoyant swim bladder in ray-fins, but true lungs in the lobe-fins and their tetrapod descendants.

- Adaptation to life on land was a key adaptation of vertebrates. The tetrapod ancestor was a lobe-finned fish. A series of transitional fossils shows that the first tetrapods were freshwater fishes with necks and leg-like limbs that raised their heads above the water to breathe air.

- Amphibians—frogs and salamanders—are tetrapods, vertebrates with two pairs of limbs. Early amphibians were the first vertebrates to colonize the land, but distribution of amphibians is limited by their vulnerability to dehydration and dependence on wet environments to lay their jelly-covered eggs.

- Reptiles (including birds) and mammals are amniotes; their embryos develop inside a fluid-filled sac. The reptile clade includes lizards, snakes, crocs, birds, and many extinct forms, including the dinosaurs. Besides the amniote egg, other adaptations suit reptiles for life on land—waterproof, scaly skin, better lungs, etc.

- Birds are feathered reptiles descended from dinosaurs. Besides feathers and wings, birds have many structural and functional adaptations for flight. Like mammals (and unlike other vertebrates) birds are endothermic, deriving body heat from their high rate of metabolism.

- Mammals are amniotes that have hair and produce milk. There are three main groups: Montremes lay eggs. Marsupials have a brief gestation inside the mother's uterus and continue development in a pouch. Most mammals are eutherians. They have a more complex placenta and longer gestation, and give birth to more fully developed young.

- The mammalian order Primates includes monkeys and apes. All primates have adaptations for living in the trees—flexible limbs, grasping hands, sensitive touch, and depth perception. The three main groups of primates are: (1) the lemurs, lorises and potoos, (2) the tarsiers, and (3) the anthropoids—the monkeys and apes.

- Humans are primates, anthropoids in the ape, or homonoid, group. Humans and chimps evolved from a common ancestor 5–7 million years ago. Our family tree includes at least 20 known species of hominids, or human relatives, but all are extinct except for modern humans, *Homo sapiens*.

- Fossil skulls, knees, hips, and footprints show that early small-brained hominids such as *Australopithecus* were bipedal, walking on two legs, 3.5 million years ago. *Homo habilis* had a larger brain and was making stone tools 2.4 million years ago. *Homo erectus*, with a still larger brain, spread out of Africa starting 1.8 million years ago. Neanderthals first appeared 500,000 years ago, but were a hominid offshoot, not related to modern humans.

- Fossils of modern humans, *Homo sapiens*, were found in Ethiopia and go back 160,000 to 195,000 years. Studies of mitochondrial DNA and Y chromosomes point to an African origin at that time. *Homo sapiens* left Africa and spread to all the continents. Key genetic changes, like that involving the *FOXP2* gene, may have facilitated symbolic communication. By 36,000 years ago, we were producing sophisticated cave paintings. Now we are so numerous and successful we threaten the planet.

Review the Concepts

Work through the following exercises to review the concepts in this chapter. For additional review, refer to the activities at www.mybiology.com. The website offers a pre-test that will help you plan your studies.

Exercise 1   (Module 19.1)

This module introduces the major clades of chordates and some of the derived characters that define the clades. Study the phylogenetic tree, and identify the group or groups of chordates described below.

1. The first group to branch from the chordate lineage: _____

2. Have a small brain, but no well-defined head: _____

3. Have a head, but no vertebral column: _____

4. The first group to have a skull and vertebral column: _____

5. Jawless vertebrates (careful!): _____

6. First group with "Jaws": _____

7. Five groups with lungs or derivatives: _____

8. First with muscular fins with skeletal support, making land life possible: _____

9. First tetrapods: _____

10. Two other groups of tetrapods:_____

11. Two groups with amniotic eggs: _____

12. Feed their young milk: _____

Exercise 2   (Modules 19.2–19.3)

These modules describe the major groups of fishes and fishlike vertebrates. Match each of the descriptions below with the correct group or groups.

_____ 1. Jawed fishes with a cartilage skeleton

_____ 2. Two other groups of fishes with jaws

_____ 3. Lungfishes are in this group.

_____ 4. Chondrichthyans

_____ 5. Two groups that lack hinged jaws

_____ 6. Largest and most diverse group of fishes

_____ 7. Bones in muscular fins may have enabled them to "walk."

_____ 8 Craniates but not vertebrates

_____ 9. Trout, perch, and seahorse

_____ 10. Possess a buoyant swim bladder

_____ 11. One lineage led to tetrapods.

_____ 12. Operculum helps pump water over gills.

_____ 13. Notochord is their main skeletal support.

_____ 14. Secrete slime for defense

_____ 15. Fins supported by thin, flexible skeletal rods

_____ 16. Parasites that scrape a hole in a fish to feed

_____ 17. Two groups that lack paired pectoral and pelvic fins

A. Hagfishes
B. Lampreys
C. Sharks and rays
D. Ray-finned fishes
E. Lobe-finned fishes

Exercise 3   (Module 19.4)

Adaptation to life on land—the evolution of the tetrapods—was a key event in vertebrate history. State whether each of the following is true or false (T or F) and then *correct* the false statements to make them true.

_____ 1. Tetrapods appeared in the late Devonian period, about 360–365 million years ago.

_____ 2. The first tetrapods probably lived in shallow salt water.

_____ 3. The direct ancestors of tetrapods were ray-finned fishes.

_____ 4. Gas exchange, water conservation, support, and movement are challenges of life on land.

_____ 5. The fossil record gives us little information about the evolution of land vertebrates.

_____ 6. *Eusthenopteron* was fishlike and *Ichthyostega* was adapted for life on land, but we do not know what kind of creature bridged the gap between them.

_____ 7. *Panderichthyes* was rather fishlike, but had lungs as well as gills.

_____ 8. *Tiktaalik* had a neck and limbs with fingers and toes, but fishy scales.

_____ 9. *Acanthostega* came later, and had a neck and limbs with fingers and toes.

_____ 10. *Acanthostega* and *Tiktaalik* show that the first tetrapods were fish with lungs.

_____ 11. A neck, an upturned snout, and lungs were useful adaptations in stagnant pools.

_____ 12. Modern amphibians are the only descendants of these early tetrapods.

Exercise 4   (Modules 19.5–19.8)

The tetrapods—amphibians, reptiles (including birds), and mammals—are the dominant large animals on land. Read up on these animals and then match each statement with a group or groups of tetrapods: Amphibians (A), Reptiles (R), Birds (B), or Mammals (M). Some questions have multiple answers.

_____   1. The first tetrapods

_____   2. Two groups of endothermic tetrapods

_____   3. First vertebrates with amniotic eggs

_____   4. Two other groups of amniotes

_____   5. Feathered reptiles

_____   6. Have hair

_____   7. Scales help them live on dry land.

_____   8. Mice, tigers, bats, and porpoises

_____   9. Frogs, toads, and salamanders

_____   10. Proliferated after the dinosaurs died out

_____   11. Widespread and diverse in the Carboniferous period

_____   12. Eutherians, marsupials, and monotremes

_____   13. Feed their young milk

_____   14. Two groups descended from early reptiles

_____   15. Populations are dramatically declining around the world.

_____   16. The largest animal ever is in this group.

_____   17. Two groups that are ectothermic

_____   18. Go through an aquatic tadpole larva stage

_____   19. The young of most in this group develop inside their mothers.

_____   20. *Archaeopteryx* is an extinct form.

_____   21. The largest land animals ever were in this group.

_____   22. Most live on land, but have aquatic eggs and larvae

_____   23. Have light, honeycombed bones

_____   24. Gave rise to reptiles and mammals

_____   25. Most nourish their young via a placenta.

_____   26. Adaptations include losing their teeth and their long tails.

_____   27. Lizards and snakes

_____   28. Evolved from a type of dinosaur

_____   29. Turtles, crocs, and birds

_____   30. Humans

Review the chordates by labeling this phylogenetic tree. First, identify the major steps in chordate evolution by labeling the derived characters on each clade along the left side of the tree (1–9). (Write out terms; don't just use letters to label.) Choose from: **legs, head, lobed fins, milk, lungs or derivatives, vertebral column, brain, jaws, amniotic egg**

Next identify the broad groupings of chordates by labeling the "bars" down the right side (10–15). (Again, write out the words.) Choose from: **tetrapods, chordates, amniotes, craniates, vertebrates, jawed vertebrates**

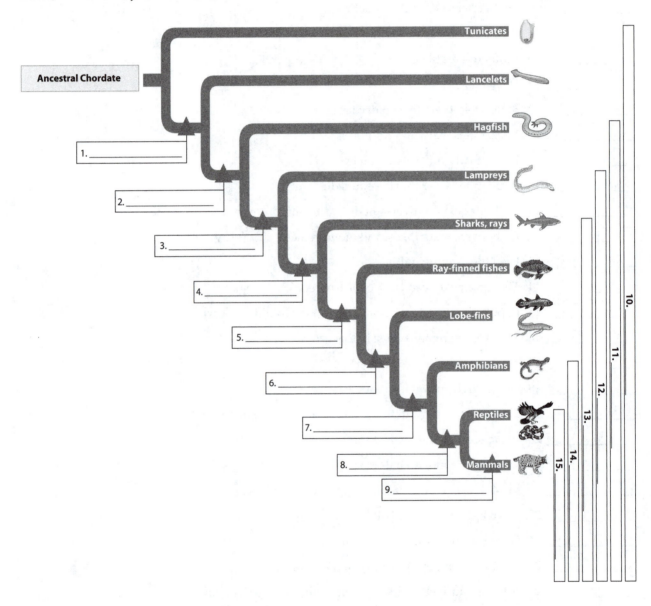

Exercise 6   (Module 19.9)

You are a primate. Review the characteristics of primates, and then explain which of these characteristics enables you to do each of the following.

1. Find a quarter among a pocketful of coins

2. Catch a ball that is thrown to you

3. Pick up a postage stamp from your desktop

4. Do a handspring

5. Thread a needle

6. Shoot a basketball

Match each of the statements below with the major modern groups of primates (A–I) shown on the phylogenetic tree. (Some questions will require more than one answer.)

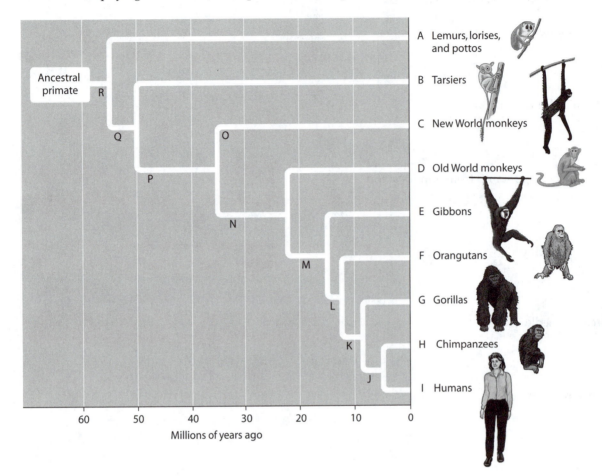

A Lemurs, lorises, and pottos

B Tarsiers

C New World monkeys

D Old World monkeys

E Gibbons

F Orangutans

G Gorillas

H Chimpanzees

I Humans

Millions of years ago

_____ 1. These groups make up the hominoids.

_____ 2. These are all anthropoids.

_____ 3. Apes

_____ 4. The largest ape

_____ 5. New World monkeys

_____ 6. Old World monkeys

_____ 7. Two groups of prosimians

_____ 8. Tarsiers

_____ 9. Lorises, potoos, and lemurs

_____ 10. Chimps

_____ 11. The sifaka, from Madagascar

_____ 12. The golden lion tamarin

_____ 13. Baboons

_____ 14. Gibbons

_____ 15. Orangutan

_____ 16. Gorillas

_____ 17. Spider monkey

_____ 18. Rhesus macaque

_____ 19. Lack tails and have relatively large brains, long arms, and short legs

_____ 20. South American monkeys

_____ 21. African and Asian monkeys

_____ 22. Ape group that is most distantly related to other apes

_____ 23. Earliest lineage of primates

_____ 24. All the groups with a fully opposable thumb

_____ 25. Monkeys with flat faces, widely spaced nostrils, and prehensile tails

_____ 26. Monkeys with downward-pointing nostrils, lacking prehensile tails

_____ 27. Closest living relatives of humans

Now match the statements below with important steps in primate evolution (J–R). (One answer each.)

_____ 28. Common ancestor of humans and chimps.

_____ 29. The earliest anthropoid

_____ 30. The ancestral primate

_____ 31. This primate seems to have rafted to South America.

_____ 32. Cro-Magnon cave-painters

_____ 33. Ancestor of all primates except lemurs, potoos, and lorises

_____ 34. In 2004, fossils of this ancestor of all apes, named *Pierolapithecus catalaunicus*, were found in Spain.

Humans and chimps are thought to have diverged from a common ancestor 5–7 million years ago. There are a dozen or more species of fossil hominids that have been identified — the human branch of the primate tree has itself branched and rebranched several times — but only our own species, *Homo sapiens*, remains. One of the most dramatic trends seen in hominid evolution is a drastic increase in the size of the brain. This graph traces change in brain size seen in various species of hominids over time. Label the graph, starting with the name of each hominid species (1–6). Then identify when other important physical and cultural changes are thought to have occurred (7–13) during the course of hominid evolution.

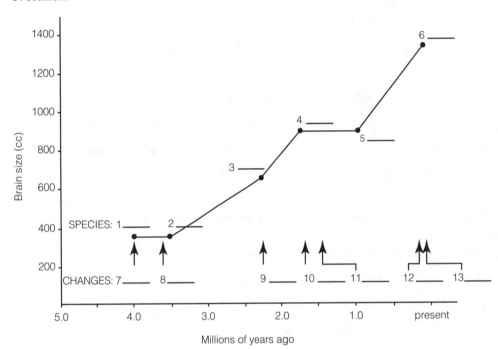

Species

A. *Homo erectus*

B. *Australopithecus anamensis*

C. *Homo habilis*

D. *Homo sapiens*

E. *Australopithecus afarensis*

F. *Homo ergaster*

Changes

G. Earliest known stone tools

H. Human *FOXP2* speech gene

I. First hominids to leave Africa

J. African footprints

K. Modern humans leave Africa

L. More sophisticated stone tools

M. Bipedalism

Exercise 9   (Modules 19.11–19.18)

Over the last few decades, paleoanthropologists have made great strides in fleshing out the story of human evolution, shattering some of our old ideas. The quotes below reflect various lingering misconceptions and fallacies about human evolution. Counter each of these false statements with a more up-to-date correction.

1. "Scientists have very little evidence of our origins. They are still looking for the 'missing link.'"

2. "Human beings are descended from chimps."

3. "The key change that set us apart from other primates was the development of a larger brain. Walking upright came later."

4. "All hominid fossils show a single line of descent, progressing from the ancestors of *Australopithecus*, through *Homo erectus*, to modern humans."

5. "Modern humans are descended from the Neanderthals."

6. "Hominids first appeared in Africa, then spread around the globe. Modern *Homo sapiens* evolved in Europe."

7. "Humans are just one species out of many on Earth. Our actions can't have much effect on the planet."

8. "Skin color is a useful characteristic for classifying humans into different races."

9. "Human speech is something we can *learn* simply because we are so intelligent. It can't be a product of our evolution because there is nothing *genetic* about it."

Test Your Knowledge

Multiple Choice

1. Jaws probably evolved from
   a. fin structures.
   b. ribs.
   c. gill structures.
   d. teeth.
   e. skull bones.

2. Which of the following is *not* thought to be in the lineage that led to human beings?
   a. an amphibian
   b. a dinosaur
   c. a jawless vertebrate
   d. a colonial protist
   e. a lobefin

3. The first vertebrates to live on land were
   a. dinosaurs.
   b. reptiles.
   c. amphibians.
   d. cartilaginous fishes.
   e. mammals.

4. There are three major groups of mammals, categorized on the basis of

   a. their diet.
   b. habitat.
   c. method of locomotion.
   d. teeth and digestive system.
   e. method of reproduction.

5. Which of the following probably appeared earliest in the story of hominid evolution?
   a. bipedal posture
   b. speech
   c. increase in brain size
   d. increase in stature
   e. symbolic thought

6. The lemurs of Madagascar are examples of
   a. Old World monkeys.
   b. anthropoids.
   c. prosimians.
   d. New World monkeys.
   e. apes.

7. Which of the following is regarded as evidence of a major change that may have led to human speech?
   a. Fossils that show that bipedal locomotion freed the hands for use of sign language
   b. Discovery of a gene that shapes the brain for use of language

c. Evidence that the tongue of *Homo sapiens* differed from that of *Homo erectus*

d. DNA studies of Neanderthals and other fossil humans

e. Changes in the nasal passages and voice box of early *Homo sapiens*

8. Biochemical evidence indicates that _____ are more closely related to humans than to other apes.
   a. the gorilla and the orangutan
   b. the chimpanzee and the baboon
   c. the orangutan and the chimpanzee
   d. the gorilla and the chimpanzee
   e. the orangutan and the gibbon

9. Fossil evidence suggests that humans evolved in
   a. Europe.
   b. South America.
   c. Asia.
   d. the Middle East.
   e. Africa.

10. Which of the following probably coexisted for a time with *Homo sapiens*?
    a. *Australopithecus robustus*
    b. *Homo erectus*
    c. *Australopithecus afarensis*
    d. *Homo habilis*
    e. *Australopithecus africanus*

11. Which of the following appear to be the best explanations so far for differences in human skin color?
    a. Darker skin shields the skin from UV; lighter skin allows synthesis of vitamin D.
    b. Darker skin cools the body; lighter skin allows synthesis of folate.
    c. Darker skin warms the body; lighter skin prevents breakdown of folate.
    d. Darker skin prevents breakdown of folate; lighter skin allows synthesis of vitamin D.
    e. Darker skin warms the body; lighter skin cools the body.

12. Which of the following correctly pairs a hominid with a "first" in hominid evolution?
    a. *Homo sapiens*—out of Africa
    b. *Homo erectus*—bipedal stance
    c. Neanderthal—cave paintings
    d. *Australopithecus*—large brain
    e. *Homo habilis*—stone tools

13. As *Homo sapiens* spread around the world, they probably reached which of the following last?

a. North America
b. South America
c. Asia
d. Europe
e. Australia

Essay

1. Describe the characteristics of lobe-finned fishes that make them the most likely aquatic ancestors of land vertebrates (tetrapods). Scientists expected the transitional forms to be fishes with lungs that gradually developed legs as they dragged themselves from pool to pool. What kinds of fossils did they find?

2. Describe adaptations of birds for flight.

3. The spadefoot toad of the southwestern United States is an unusual amphibian; it is capable of surviving in the desert. Few amphibians can tolerate dry desert conditions, but many reptiles—horned toads, rattlesnakes, and desert tortoises—thrive in hot, arid regions. In what ways are reptiles better adapted to life in the desert than amphibians?

4. List the specialized characteristics of primates.

5. Which primate is the closest living relative of humans? What is the evidence for this close relationship?

6. Describe the characteristics that separate each of the paired groups:
   a. anthropoids—prosimians
   b. Old World monkeys—New World monkeys
   c. monkeys—apes
   d. other apes—humans

7. Describe genetic evidence that points to when and where *Homo sapiens* originated. Do fossils support these data?

Apply the Concepts

Multiple Choice

1. Which of the following are all vertebrates but not amniotes?
   a. frog, trout, lancelet, lungfish
   b. lamprey, lizard, lobefin, shark
   c. shark, lamprey, salamander, trout
   d. sparrow, lizard, raccoon, turtle
   e. trout, frog, turtle, lungfish

2. There are relatively few species of cartilaginous fishes, compared with the bony fishes. Cartilaginous fishes are mostly limited to a lifestyle of swimming fast in open water. Ray-finned fishes have adapted to many different lifestyles—clinging to seaweed, hiding in crevices, even burrowing in the bottom. This could probably be attributed to the fact that ray-finned fishes
   a. have more rigid skeletons.
   b. are smaller than cartilaginous fishes.
   c. have operculums and swim bladders.
   d. have lateral line systems and paired fins.
   e. are endothermic.

3. Which of the following is *least* closely related to all the others?
   a. shark
   b. human
   c. tuna
   d. salamander
   e. garter snake

4. Imagine that you are a paleontologist (a scientist who studies fossils). In recent digs, you have unearthed bones of all of the following. Which could you have found in the oldest sediments?
   a. amphibians
   b. crocodiles
   c. dinosaurs
   d. birds
   e. marsupials

5. Which of the following is a shared derived character of reptiles (including birds) and mammals?
   a. legs
   b. amniotic egg
   c. jaws
   d. lungs
   e. all of the above

6. Which of the following is *least* closely related to the others?
   a. human
   b. baboon
   c. gorilla
   d. squirrel monkey
   e. lemur

7. At the zoo, Tom saw a species of primate he had never even heard of before. He said, "It was called a white-faced saki. It had long, dark, spiky fur, a long fluffy tail, forward-facing brown eyes, a white forehead, and yellow cheeks." Based on this information, the white-faced saki could *not* be
   a. an ape.
   b. a prosimian.

   c. a New World monkey.
   d. an Old World monkey.
   e. an anthropoid.

8. Emma saw an old Tarzan movie on television. The movie supposedly took place in Africa, but Emma easily spotted that it was not really filmed there. Which of the following could have tipped her off?
   a. Chimps like Tarzan's sidekick Cheetah do not live in Africa.
   b. The monkeys in the jungle had prehensile tails.
   c. There were no prosimians shown in the forest.
   d. Tarzan wrestled with a gorilla.
   e. Only Old World monkeys were shown.

9. Which of the following categories includes all of the others?
   a. apes
   b. Old World monkeys
   c. anthropoids
   d. hominids
   e. New World monkeys

10. Most known species of hominids
    a. were the ancestors of modern humans.
    b. evolved during the last million years or so.
    c. had large brains.
    d. lived primarily in trees.
    e. were not our ancestors.

11. Which of the following is a hominoid, but not a hominid?
    a. Neanderthal
    b. lemur
    c. *Australopithecus africanus*
    d. gorilla
    e. squirrel monkey

12. The bonobo, an ape closely related to the chimpanzee, displays behaviors very similar to those of humans, and some anthropologists have suggested that the bonobo is the living primate most closely related to humans. Which of the following would be the easiest way to try to substantiate this idea?
    a. Look for fossils of bonobos, chimps, and humans.
    b. Study the DNA of bonobos and chimps.
    c. Determine which of the species are anthropoids.
    d. Compare the DNA of bonobos and humans.
    e. Compare the DNA of bonobos, chimps, and humans.

13. Paleontologists have found fossils of several kinds of australopiths. What is their place in human evolution?
    a. They are all thought to be ancestors of modern humans.
    b. They are all extinct side branches of the human family tree.
    c. Some evolved into humans, others into apes.
    d. They are the ancestors of various modern apes.
    e. Some may have been our ancestors, others offshoots of our family tree.

Essay

1. Looking at the chordate phylogenetic tree, state the derived character that you might share with each of the following creatures: Tunicate, lancelet, hagfish, lamprey, shark, trout, lungfish, frog, rattlesnake, bear.

2. Nearly all the land vertebrates in the Arctic and Antarctic are birds and mammals—polar bears, walruses, and penguins, for example. Why do you think there are so many birds and mammals, but virtually no reptiles or amphibians, in these regions?

3. Unlike many other mammals, such as horses or deer, humans are born at a quite "undeveloped," helpless stage in development. Wouldn't it be better for humans to spend a little longer period *in utero* and enter the world a little more self-sufficient? What evolutionary compromise seems to prevent this?

4. Language, use of tools, and self-awareness are often cited as human characteristics—traits that set us apart from other members of the animal kingdom. Do these characteristics really appear to be unique to humans? Explain.

5. Finding which of the following would require anthropologists to change their current ideas about human evolution the most? Why?
    a. A large-brained, quadrepedal (walking on all fours) hominid 2.5 million years old
    b. A small-brained, bipedal hominid 5.0 million years old

Put Words to Work

Correctly use as many of the following words as possible when reading, talking, and writing about biology:

amniote, amniotic egg, amphibian, anthropoid, australopith, chondrichthyan, Chordate, craniate, ectotherm, ectothermic, endotherm, endothermic, eutherian, hominid, hominoid, lateral line system, lobefin, mammal, marsupial, operculum (plural, *opercula*), *opposable thumb, paleoanthropology, placenta, placental mammal, ray-finned fish, reptile, swim bladder, tetrapod, vertebra* (plural, *vertebrae*), *vertebral column, vertebrate*

Use the Web

Continue exploring the evolution of vertebrate diversity at www.mybiology.com.

Unifying Concepts of Animal Structure and Function

20

Focus on the Concepts

The structural and functional adaptations of an animal are related to the problems the animal must solve—feeding, movement, reproduction, etc. As you study this introduction to animal structure and function, focus on the following concepts:

- Natural selection has shaped animal "body plans" that fit the challenges of animals' environments. Structure fits function at all levels of the biological hierarchy, from the cell to the whole organism, and an animal's particular structures and abilities emerge from the arrangement of structures at each level.

- An organism's organs and organ systems are based on varied combinations of a limited set of cells and tissues. A tissue is a group of cells with common structure and function. There are four main categories of tissues. Epithelial tissue covers and lines body structures. Connective tissue forms a supporting framework. Muscle tissue functions in movement. Nervous tissue forms a communications network.

- Twelve organ systems carry out the body's functions. The systems are interdependent and work together to form a functional organism. For example, the integumentary system—the skin and its structures—encloses and protects the animal. To do its job, the skin depends on food and oxygen from the digestive and respiratory systems, transport carried out by the circulatory system, and waste disposal by the excretory system.

- Animals must exchange materials with their environments to stay alive. In most animals, the evolutionary adaptation that provides for exchange is folded surfaces that increase surface area. Food, gases, and wastes pass through these surfaces, and are exchanged with cells via the blood and interstitial fluid.

- Most animals maintain a relatively constant internal environment, a principle known as homeostasis. Homeostatic mechanisms involve negative feedback: Sensors monitor body conditions, and systems respond to changes by triggering responses that reverse the changes.

Review the Concepts

Work through the following exercises to familiarize yourself with the basics of animal structure and function. To enhance your understanding, explore the activities on the Web at www.mybiology.com. The website offers a pre-test that will help you plan your studies.

Exercise 1   (Introduction and Modules 20.1–20.2)

A gecko's feet enable it to climb walls. A bird's wings enable it to fly. The sleek shape of a dolphin allows it to zip through the water. You are probably familiar with many other examples of animal adaptations that illustrate the correlation between animal structure and function. In a sentence or two, state how each of the following illustrates this correspondence of structure and function. The first one is done for you. (A cautionary note: It is *not* a good idea to say "The gecko has specialized setae on its feet *so it can climb walls.*" The setae are the result of millions of years of natural selection, not designed on purpose.)

1. A dolphin's tail. *The tail is flattened into broad flukes, which propel the dolphin through the water.*

2. A hummingbird's beak.

3. Your hand.

4. A frog's legs.

5. A mosquito's sharp, tubular mouthparts.

6. A cow's multichambered stomach.

Exercise 2   (Modules 20.1–20.3)

Review the hierarchy of structure and function in an animal by filling in the blanks in the following paragraphs.

The body of an animal—a cat, for example—is organized on several hierarchical levels. The smallest parts of the cat that are alive are individual [1]_____, such as the muscle fibers in the wall of the stomach. Their function is to contract and move the contents of the stomach, mixing cat food with digestive juices. Many muscle cells cooperate to form a [2]_____, the second level of body structure and function. Besides muscle, there are three other kinds of tissues that make up the cat's stomach: [3]_____ tissue, [4]_____ tissue, and [5]_____ tissue. The stomach itself, formed of these four tissues, is an [6]_____, which performs the functions of storing and digesting food. The stomach, esophagus, intestines, and digestive glands make up the digestive system, which exemplifies the [7]_____ level of structure and function. It is one of a dozen or so systems that cooperate to form the cat—an [8]_____. This is the whole animal, the highest level of the hierarchy. At each level of this hierarchy, functions and abilities—such as the casts ability to catch and digest a mouse—emerge from the specific [9]_____ of specialized structures making up that level.

To review, starting from the top down: A cat is an [10]_____. It is composed of a number of [11]_____, each of which performs specific functions such as digestion or circulation. Each system is composed of [12]_____ such as the heart or stomach, which are built from four kinds of [13]_____. At the most fundamental level of the hierarchy, a tissue is composed of individual [14]_____.

Exercise 3 (Module 20.4)

Epithelial tissue covers the body and lines body organs. Complete the following chart comparing four kinds of epithelium. Refer to the illustrations in Module 20.4 to aid you with the descriptions.

Tissue Type	*Description*	*Body Locations*	*Functions*
1.	2.	3.	4.
5.	6.	7.	Absorbs and secretes fluid in kidneys
8.	Single layer of thin, flattened cells	9.	10.
11.	12.	Lines the intestine	13.
14.	15.	Lines respiratory tract	16.

Exercise 4 (Modules 20.5 and 20.8)

Complete this crossword puzzle to review the structures, functions, and locations of connective tissues and the medical uses of artificial tissues.

Across

2. Researchers recently reported the successful transplantation and functioning of lab-grown ____.
5. Artificial skin is grown from cells called human____.
7. Artificial skin may be used to treat burns and diabetic skin ____.
8. ____ tissue contains fat.
13. Bone matrix contains fibers embedded in a hard substance made of ____, magnesium, and phosphate
15. ____ is a connective tissue with a liquid matrix.
16. Connective tissue is one of ____ major categories of tissues.
17. The matrix is a web of ____ in a liquid, jelly, or solid.
18. Connective tissue cells are scattered in a non-living ____.
19. Cartilage supports the nose and ears and forms discs between the ____.
22. Adipose tissue pads and insulates the body and stores "fuel" to give the body ____.

Down

1. Organs grown from a patient's own ____ may prevent rejection.
3. ____ connective tissue holds other tissues and organs in place.
4. Artificial ____ has been successfully used to cover burns.
6. A ____ is a group of cells with a common structure and function.
9. Blood matrix is called ____.
10. Fibrous connective tissue forms ____ and ligaments.
11. ____ connective tissue has densely packed bundles of collagen fibers.
12. Some kinds of connective tissues contain rope-like ____ fibers.
14. ____ is a strong but flexible skeletal material, but it does not regenerate easily.
20. ____ is the most rigid connective tissue.
21. Inkjet technology can be used to "print" cells on a biodegradable ____.

Exercise 5   (Modules 20.6–20.7)

Review nervous tissue and the three types of muscle tissue by matching each phrase on the left with a tissue type on the right.

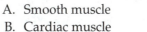

_____ 1. Figure 1 at the right	A. Smooth muscle
_____ 2. Figure 2 at the right	B. Cardiac muscle
_____ 3. Figure 3 at the right	C. Skeletal muscle
_____ 4. Figure 4 at the right	D. Nervous tissue

_____ 5. Contractile tissue of the heart
_____ 6. Forms body communications system
_____ 7. Attached to bones by tendons
_____ 8. Branching, interconnected muscle cells
_____ 9. Cells characterized by axons and dendrites
_____ 10. Carries out voluntary body movements
_____ 11. Composed of neurons and other supporting cells
_____ 12. Cells are striated, or striped
_____ 13. Muscle cells that lack striations
_____ 14. Found in walls of digestive tract, bladder, arteries, and so on
_____ 15. Involuntary muscle of internal organs other than the heart

1.

2.

3.

4.

Exercise 6   (Modules 20.8–20.10)

Review the functions of human organs and organ systems by filling in the blanks in the following paragraphs.

An animal—be it a gecko or a human being—consists of a number of cooperating organ systems, each of which performs specific life functions. Each system consists of organs, which in turn carry out particular jobs. As you read these words, your brain receives nerve impulses from your eyes. It evaluates the information received and sends out responses via the spinal cord and nerves. These parts—sense organs, brain, spinal cord, and nerves—make up the [1]_____ system, one of two systems that control and coordinate body activities.

As you read, your eyes scan the page, and your hand moves to write answers in the blanks. These responses are carried out by muscles, which make up the [2]_____ system, the system responsible for body movements. To move the body, muscles pull against bones, which make up the [3]_____ system. This system also supports the body and protects delicate internal organs such as the brain.

Muscles, like all organs, require food and oxygen to function. Food is digested and absorbed by the [4]_____ system, whose parts include the stomach, intestines, and digestive glands such as the liver and pancreas. As you inhale and exhale, oxygen enters the body via the lungs, key organs of the [5]_____ system. The [6]_____ system—the heart, blood, and blood vessels—functions to transport food and oxygen to your muscles and other organs. Some of the fluid delivered to body tissues leaves the blood and is picked up by vessels of the [7]_____ system. The fluid passes through lymph nodes, where special cells called lymphocytes attack foreign substances and microbes. The skin, or [8]_____ system, is normally able to keep most disease-causing organisms out. When they get through, they are attacked by cells of the [9]_____ system—lymphocytes, which are produced and stored in the thymus, bone marrow, and spleen, as well as lymph nodes.

As the blood delivers its cargo of food and oxygen, it also picks up waste products that must be expelled by the body. The kidneys—the key organs of the [10]_____ system—remove metabolic waste products from the blood and dispose of them via the ureters, bladder, and urethra. The kidneys also regulate the osmotic balance of blood. This activity, and many others in the body, are controlled by chemical signals called hormones. Hormones are sent out by the glands of the [11]_____ system, which acts in concert with the nervous system to coordinate the activities of all the other body systems. Endocrine glands—the pituitary, thyroid, adrenals, and others—regulate such activities as digestion, growth, metabolism, and water balance. They even help to control the process of reproduction, through their effects on the testes and ovaries, the major organs of the male and female [12]_____ systems.

Exercise 7   (Module 20.11)

Compare body imaging methods by summarizing each method on the chart.

Method	Medium Used	Used For
Conventional X-ray	1.	2.
Computerized tomography (CT)	3.	4.
Magnetic resonance imaging (MRI)	5.	6.
Positron-emission tomography (PET)	7.	8.
PET-CT scan	9.	10.

Exercise 8   (Module 20.12)

Match each of these descriptions on the left with a component of the integumentary system on the right.

_____ 1. Surface layer of skin
_____ 2. Tissue type of the outer epidermis
_____ 3. Inner skin layer
_____ 4. Main tissue type of the dermis
_____ 5. Fibrous protein of skin, hair, and nails
_____ 6. Strong fibers of the dermis
_____ 7. Sends sensory information to the brain
_____ 8. Helps to regulate temperature
_____ 9. Secretes a material that lubricates skin and hair and inhibits bacteria
_____ 10. Produces hair
_____ 11. Insulates the body; also has a sensory function
_____ 12. Made in skin; helps absorb calcium

A. Keratin
B. Epidermis
C. Hair
D. Oil gland
E. Dermis
F. Stratified squamous epithelium
G. Sweat gland
H. Touch receptor
I. Collagen
J. Hair follicle
K. Dense connective tissue
L. Vitamin D

Exercise 9   (Module 20.13)

Animals must exchange materials with their environment. The cells of saclike or flattened animals (such as *Hydra* and tapeworms) can exchange substances directly with the surrounding fluid. Larger animals, whose outer coverings are small compared with their volume, have specialized surfaces for exchanging materials with the environment. Below is a simplified diagram of an animal, similar to Figure 20.13A in the text. The circulatory system is shown. Add your own simple sketches and label the **digestive system, respiratory system, excretory system, interstitial fluid,** and **cells.** Draw and label arrows to show exchange of **food, nutrients, O₂, CO₂, metabolic wastes,** and **feces.**

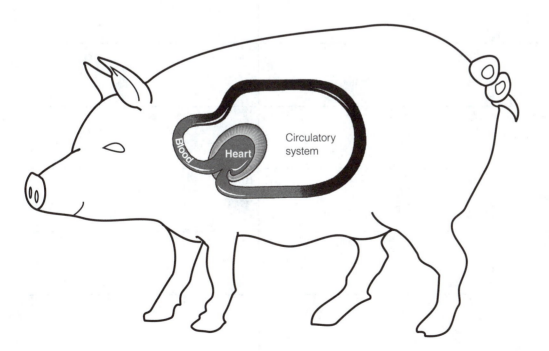

Exercise 10   (Modules 20.14–20.15)

Homeostasis is the most important principle of animal function. To review homeostasis and negative feedback, read the following paragraph and then fill in the chart that follows.

Despite changes in the external environment, an animal can keep its internal environment remarkably constant. This maintenance of a constant internal environment is called homeostasis. The text states that animals such as the ptarmigan maintain relatively constant salt and water balance and body temperature. Conditions within the human body are also more or less constant, fluctuating within narrow limits. For example, an organ called the pancreas monitors and regulates the amount of sugar in the blood. This maintains a constant supply of fuel for body cells, even though the body's intake of food varies widely during the day. After a meal, when blood sugar rises, the pancreas sends out a chemical signal, a hormone called insulin. Insulin causes body cells to take up and store sugar, which lowers blood sugar to the optimum range—70 to 110 mg of sugar per 100 mL of blood. This illustrates negative feedback: An increase in blood sugar triggers a response that counteracts the increase. Between meals, as cells consume sugar, the concentration of sugar in the blood starts to decrease. The pancreas responds by reducing its output of insulin and stepping up its secretion of a second hormone, glucagon. Glucagon signals certain cells to release sugar from storage, raising blood sugar to the optimum level. Thus, despite changes in the external environment (timing and content of meals), blood sugar usually fluctuates within the narrow range that is best for cells.

Compare the regulation of blood sugar described in the preceding paragraph with the control of body temperature outlined in Modules 20.14 and 20.15. Fill in the chart by identifying the components of each homeostatic control system.

	Body Temperature	*Blood Sugar*
Type of change in external environment	1.	2.
Control center	3.	4.
Stimulus	5.	6.
Kind of signal sent by control center to effector	7.	8.
Effector	9.	10.
Response	11.	12.
Set point	13.	14.

Test Your Knowledge

Multiple Choice

1. The four major categories of tissues are
   a. bone, muscle, blood, and adipose.
   b. nervous, epithelial, connective, and muscle.
   c. muscle, epithelial, bone, and cartilage.
   d. blood, nervous, connective, and muscle.
   e. simple squamous, simple cuboidal, simple columnar, and stratified squamous.

2. Which of the following levels of structure encompasses all the others?
   a. tissue
   b. cell
   c. organ
   d. organism
   e. system

3. How many organ systems make up your body?
   a. four
   b. hundreds
   c. twelve
   d. millions
   e. It depends on the size of the person.

4. Which of the following tissues produces voluntary body movements?
   a. smooth muscle
   b. simple cuboidal epithelium
   c. cardiac muscle
   d. skeletal muscle
   e. fibrous connective tissue

5. Neurons are specialized cells characteristic of
   a. muscle tissue.
   b. nervous tissue.
   c. connective tissue.
   d. epithelial tissue.
   e. all of the above.

6. All but one of the following systems are correctly paired with one of their parts. Which pair is *incorrect?*
   a. circulatory system—heart
   b. respiratory system—lung
   c. endocrine system—thyroid gland
   d. integumentary system—hair
   e. excretory system—intestine

7. A change in the body often triggers a response that counteracts the change. This kind of response is known as
   a. negative feedback.
   b. empowerment.
   c. cause and effect.
   d. positive feedback.
   e. adaptation.

8. Homeostasis is
   a. exchange of materials with the surrounding environment.
   b. the idea that all vertebrates are built in a similar way.
   c. the correlation of structure and function.
   d. maintaining a relatively constant internal environment.
   e. cooperation of body parts to form tissues, organs, and systems.

9. An animal's "internal environment" is
   a. the blood.
   b. the interior of compartments like the heart and stomach.
   c. anyplace beneath the skin.
   d. any fluid inside the body.
   e. the interstitial fluid that surrounds the cells.

10. Which of the following are listed in the correct hierarchical order?
    a. system-tissue-organ
    b. cell-tissue-organ
    c. organ-tissue-system
    d. tissue-cell-organ
    e. organism-organ-system

Essay

1. Name the four major types of body tissues, and briefly describe the functions of each type.

2. Name the 12 organ systems of a vertebrate. Describe the function of each system in one sentence each.

3. What are the advantages of computerized tomography (CT) over conventional X-rays?

4. Compare a thermostat controlling room temperature with your brain controlling body temperature. Describe the following for each system: stimulus sensor, set point, control center, effector, and response.

5. Some animals can "breathe" through their skins, without the aid of lungs or gills, or absorb food and expel wastes through the surfaces of their bodies, without specialized digestive tracts or kidneys. Why are all these animals rather small?

Apply the Concepts

Multiple Choice

1. Bone does *not* show which of the following correlations between structure and function?
   a. It is rigid.
   b. Its cells are packed tightly together.
   c. It contains reinforcing fibers.
   d. It can grow with the animal.
   e. It contains canals for blood vessels and nerves that keep it alive.

2. When you sprain your ankle, the "straps" of tissue that hold the bones together are stretched and torn. What kind of tissue do you think is damaged in a sprain?
   a. stratified squamous epithelium
   b. smooth muscle
   c. fibrous connective tissue
   d. adipose tissue
   e. cartilage

3. Which of the following forms a thick protective barrier that keeps bacteria out of the body?
   a. skeletal muscle
   b. fibrous connective tissue
   c. stratified squamous epithelium
   d. cartilage
   e. simple columnar epithelium

4. An organ such as the heart or intestine contains
   a. muscle tissue.
   b. nervous tissue.
   c. connective tissue.
   d. epithelial tissue.
   e. all of the above.

5. A new drug has been developed that impairs the movement of smooth muscle. It would affect the muscle
   a. that moves the arms and legs.
   b. of the heart.
   c. in the wall of the intestine.
   d. all of the above
   e. b and c only

6. Which of the following is *not* an organ?
   a. the stomach
   b. a blood vessel
   c. a neuron
   d. the heart
   e. a lung

7. A researcher wants to study the metabolic activity of various parts of exercising heart muscle. This might be accomplished by
   a. doing a CT scan.
   b. taking some X-rays.
   c. doing an ultrasound scan.
   d. using MRI.
   e. doing a PET scan.

8. Which of the following best illustrates homeostasis?
   a. All the cells in the body have much the same chemical composition.
   b. Cells of the skin are constantly worn off and replaced.
   c. When blood CO_2 increases, you breathe faster and get rid of CO_2.
   d. All organs are composed of the same four kinds of tissues.
   e. The lung has a large surface for exchange of gases.

9. Which of the following do the excretory, digestive, and respiratory systems have in common?
   a. They are present only in vertebrate animals.
   b. They contain specialized surfaces for exchange with the environment.
   c. They work independently, without any control by the nervous system.
   d. They enable the animal to absorb needed materials from its environment.
   e. They are isolated from the animal's internal environment.

10. Which of the following doctors is correctly matched with the system they specialize in treating? See if you can figure this out, based on this chapter.
    a. Gastroenterologist—digestive system
    b. Dermatologist—integumentary system
    c. Orthopedist—Skeletal and muscular systems
    d. Urologist—excretory system
    e. All of the above are correct.

Essay

1. You have probably read books or seen nature programs on television that describe adaptations of animals to their environments. Choose an animal and briefly describe how its body shows the correlation of structure and function.

2. What might interest an anatomist about each of the following: how a fish swims, how a penguin keeps warm, how an insect defends itself from its enemies? What might interest a physiologist about each of them?

3. Read the descriptions and look at the illustrations in the text of the following tissue types, and then explain how their structure correlates with their function: bone, simple squamous epithelium, and blood.

4. Briefly describe how several organ systems might cooperate in delivering food and oxygen to your brain cells.

5. The parathyroid glands regulate the amount of calcium in the blood. They send out hormone signals that control how much calcium the intestine absorbs from food and how much calcium the kidneys excrete in the urine. What do you think the parathyroids cause to happen when blood calcium gets too high? What happens when blood calcium gets too low? How does this illustrate negative feedback?

Put Words to Work

Correctly use as many of the following words as possible when reading, talking, and writing about biology:

adipose tissue, anatomy, blood, bone, cardiac muscle, cartilage, circulatory system, connective tissue, digestive system, endocrine system, epithelial tissue, epithelium (plural, *epithelia*), *excretory system, fibrous connective tissue, homeostasis, immune system, integumentary system, interstitial fluid, loose connective tissue, lymphatic system, muscular system, muscle tissue, negative feedback, nervous system, nervous tissue, neuron, organ, organism, organ system, physiology, reproductive system, respiratory system, skeletal muscle, skeletal system, smooth muscle, tissue*

Use the Web

Don't forget to check out the material on animal structure and function on the Web at www.mybiology.com.

Nutrition and Digestion 21

Focus on the Concepts

This chapter concerns nutrition and the structure and function of animal digestive systems. To get the most out of your study time, focus on the following concepts:

- Animals ingest a variety of foods in various ways. There are herbivores, carnivores, and omnivores. Most animals are bulk feeders, but there are also suspension feeders, substrate feeders, and fluid feeders. Food processing usually occurs in special compartments in four stages: ingestion, mechanical and chemical digestion, absorption of nutrients, and elimination of undigested material.

- The human digestive system consists of a tubular alimentary canal where food is processed (mouth, oral cavity, pharynx, esophagus, stomach, small intestine, large intestine, rectum, anus) and several accessory glands that produce digestive secretions (salivary glands, pancreas, liver). Adaptations accommodate different diets: Carnivores often have large, expandable stomachs that accommodate prey, and herbivores often have very long alimentary canals that allow for processing of difficult to digest plant material.

- Mechanical and chemical digestion start in the oral cavity. Swallowing delivers food from the pharynx to the esophagus, and peristalsis pushes food to the stomach, where acid and enzymes continue food breakdown. Muscular sphincters keep food material moving forward. Most chemical digestion and absorption occur in the small intestine, with help of secretions from the pancreas and liver. Villi and microvilli increase the small intestine's absorptive area. The large intestine reclaims water and compacts solid wastes, which are eliminated via the rectum and anus.

- To fulfill their nutritional needs, an animal must obtain fuel molecules (usually carbohydrates and fats) to power body activities, various organic molecules used to build the animal's own molecules, and essential nutrients the animal cannot make for itself (essential fatty acids, essential amino acids, vitamins, and minerals).

- A diet chronically deficient in calories leads to undernourishment. Malnourishment results from the long-term deficiency of one or more essential nutrients. Undernourishment and malnourishment are mainly diseases of the poor. Among the more prosperous, large portions of fattening foods, combined with sedentary lifestyle, can lead to overnutrition and obesity. Diet can influence cardiovascular disease and cancer. For good health, it is best to combine a balanced diet with exercise.

Review the Concepts

Work through the following exercises to review nutrition and digestion. For additional review, explore the activities on the Web at www.mybiology.com. The website offers a pretest that will help you plan your studies.

Exercise 1   (Module 21.1)

Test your knowledge of animal diets and feeding methods by filling in this chart. Most examples are from the module; others are animals with which you are probably familiar. For diet, choose among omnivore, herbivore, and carnivore. As for feeding methods, animals can be suspension feeders, substrate feeders, fluid feeders, or bulk feeders.

Animal	*Diet*	*Feeding Method*
1. Cow		
2. Earthworm		
3. Aphid		
4. Humpback whale		
5. Human being		
6. Fruit fly maggot		
7. Shark		
8. Female mosquito		
9. Clam		
10. Grasshopper		

Exercise 2   (Modules 21.2–21.3)

Some small, simple animals digest their food in multipurpose food vacuoles or larger gastrovascular cavities. But most animals have an alimentary canal, which processes food in four stages. Read about the four stages of food processing and kinds of digestive systems. Then connect the two by (1) labeling the drawings, and (2) coloring the drawings to show which digestive stages take place in each portion of these alimentary canals. On the drawings label **mouth, pharynx, esophagus, crop, gizzard, stomach, intestine,** and **anus.** Then color the alimentary canals to show where each stage in food processing occurs. Use yellow for **ingestion** (including swallowing and storage), red for **digestion,** green for **absorption,** and blue for **elimination.** (If two processes occur in the same area, mix the colors.)

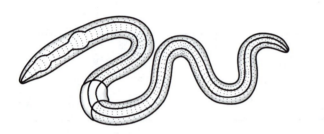

Earthworm

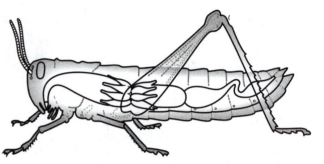

Grasshopper

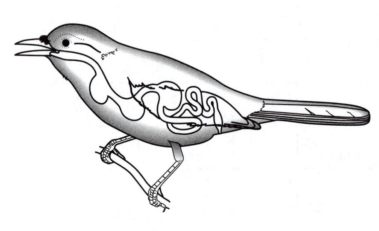

Bird

Exercise 3   (Module 21.4)

Picture the parts of the human digestive system and their relationships to one another. Label and color these parts of the digestive system on the diagram below: **stomach, oral cavity, small intestine, esophagus, rectum, pancreas, gallbladder, mouth, large intestine, tongue, salivary glands, anus, liver, pyloric sphincter,** and **pharynx.**

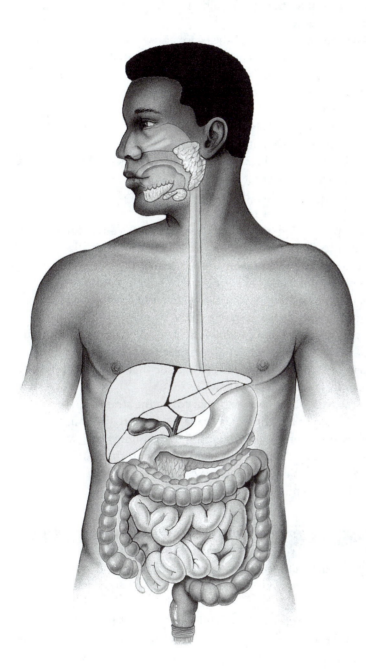

Exercise 4   (Modules 21.4–21.12)

Identify the part of the digestive tract described in each of the following statements. Start by seeing how many you know without looking up the answers. Some answers occur more than once.

_____   1. Kind of front tooth used for biting
_____   2. Largest digestive gland
_____   3. Relaxes to allow food into the esophagus
_____   4. The opening through which wastes are eliminated
_____   5. Where liver secretions are stored
_____   6. A ring of muscle that controls food leaving the stomach
_____   7. The throat
_____   8. The digestive gland above and to the right of the stomach
_____   9. Back teeth that grind and crush food
_____   10. Peristalsis in the walls of this structure pushes food to the stomach.
_____   11. Where food goes after it leaves the stomach
_____   12. Digestive gland below the stomach
_____   13. Where taste buds are located
_____   14. The portion of the tract that wraps around the small intestine
_____   15. Pointed teeth that are overdeveloped in Count Dracula
_____   16. Where the tongue and teeth are located
_____   17. The portion of the canal between the esophagus and the small intestine
_____   18. The technical name for a swallowed ball of food
_____   19. Digestive glands that secrete into the oral cavity
_____   20. Where final steps of digestion and absorption of nutrients occur
_____   21. Where water is absorbed and feces are formed
_____   22. A flap of cartilage and connective tissue that keeps food out of the trachea
_____   23. The pyloric sphincter regulates its exit.
_____   24. The muscular tube from the pharynx to the stomach
_____   25. The portion of the alimentary canal just before the anus
_____   26. The mouth opens into this portion of the alimentary canal.
_____   27. Part of the digestive tract blocked when you choke
_____   28. Secretes pepsin and hydrochloric acid
_____   29. Semiliquid, semidigested food in the stomach

Exercise 5   (Modules 21.8–21.12)

The following story will help you to visualize the movement of food through the stomach and intestine and the processes that occur there. Write the proper word in each blank.

The experimental subject has been complaining of abdominal pain, but his doctors have been unable to pinpoint its cause. Your assignment is to inspect the linings of his stomach and small intestine. You step into the Microtron and are quickly reduced to microscopic size. You enter the subject via a drink of water, and the swallowing reflex sweeps you into the [1]_____, the tube to the stomach. A wave of muscle contraction, called [2]_____, propels you forward. Ahead, the opening broadens, and you tumble into the stomach.

The subject has been on a liquid diet, so the stomach is filled with clear fluid instead of [3]_____—the normal mixture of food particles and gastric juice. Huge folds in the stomach lining look like underwater ridges and valleys. Periodic waves of peristalsis sweep across this landscape like earthquakes. As you approach the stomach wall, you note that the lining is dotted with numerous pits—the openings of tubular [4]_____, which produce [5]_____, the stomach's digestive fluid.

You enter one of the pits and swim into the gastric gland. Your instruments indicate an increase in secretion of gastric juice. The presence of food—and you—in the stomach causes cells in the stomach wall to secrete a hormone called [6]_____, which (along with nerve signals) stimulates secretion of gastric juice.

Several kinds of cells line the walls of the gastric gland. Many of the cells are undergoing [7]_____, rapidly replacing cells that are damaged in this harsh environment. Deep in the gland, large parietal cells secrete [8]_____. Nearby chief cells secrete inactive pepsinogen, which quickly changes into the enzyme [9]_____ when it comes into contact with the acid. The function of pepsin is to begin the digestion of [10]_____, preparing them for further digestion in the [11]_____. Secreting the enzyme in inactive form [12]_____ the cells where it is produced. [13]_____ cells secrete thick [14]_____, which helps protect the stomach lining.

You exit the gland and continue your inspection of the stomach lining. The opening from the esophagus, where you entered the stomach, is sealed tightly. There is little evidence of backflow of acid chyme into the esophagus, which might cause gastroesophageal reflux disease, abbreviated [15]_____, and nowhere do you see any evidence of a [16]_____—an open sore that develops in the stomach lining. It used to be thought that this happened when there was too much pepsin or acid, or not enough [17]_____. Evidence now points to [18]_____ by a bacterium—*Helicobacter pylori*—which damages the stomach lining.

The stomach's exit is guarded by a tightly closed doughnut of muscle, the [19]_____. As you approach, it opens to allow a squirt of liquid to leave the stomach. Soon you enter the next portion of the digestive tract, the [20]_____.

This portion of the alimentary tube is much narrower than the cavernous stomach. It looks like a curving, twisting tunnel. The first few inches is called the duodenum. Here you see a jagged sore, tucked into a fold in the wall of the duodenum—a duodenal [21]_____. This must be the cause of the problem. The medical team will have to deal with this.

Continuing your survey of the small intestine, you note an opening where fluid is squirting into the duodenum. This is the duct through which [22]_____ from the gallbladder and an alkaline, enzyme-rich solution from the [23]_____ enter the small intestine. The alkaline solution [24]_____ acid from the stomach. Bile, produced by the [25]_____ and stored in the gallbladder, contains bile salts that break up [26]_____ into small droplets, a process called [27]_____. This makes it easier for pancreatic enzymes to digest them. An enzyme called [28]_____ breaks fat molecules down into fatty acids and glycerol. Other pancreatic enzymes, trypsin and chymotrypsin, continue the digestion of [29]_____ that began in the stomach. They break polypeptides down to smaller polypeptides, and then peptidases break these smaller polypeptides down to [30]_____. Pancreatic amylase hydrolyzes [31]_____ to form maltose, a disaccharide. Other enzymes digest it and other disaccharides to form 32_____. Nucleases digest DNA and RNA.

Contractions in the walls of the small intestine gently propel you along. The next two portions of the small intestine are specialized for 33_____ of nutrients. The surface is highly folded, with numerous small projections called 34_____. You swim downward, and the villi surround you like the huge rubbery trunks of some inflatable forest. Their surfaces look and feel velvety, because the epithelial cells covering them bear their own tiny projections called 35_____. All these folds give the small intestine a huge surface area—over 600 square meters! You press your light against the surface of one of the villi. Inside, you can see the ghostly outlines of pinkish 36_____ and a network of yellowish, transparent 37_____ vessels. Nutrients are pumped or 38_____ through the intestinal epithelium and enter these vessels. The capillaries join to form a large blood vessel called the 39_____ that carries blood directly to the liver, where nutrients are converted to forms the body needs. For example, the liver stores excess glucose in the form of a polysaccharide called 40_____.

As you finish your inspection of the small intestine, you see the sphincter ahead that controls the movement of unabsorbed food material into the 41_____, or colon. The colon is where 42_____ is absorbed and undigested material is turned into 43_____, which are stored in the rectum and expelled through the 44_____. That part of the digestive system is not on your itinerary! You work your way through the wall and into a blood vessel, and are soon back in the lab, munching on a sandwich.

Exercise 6   (Module 21.13)

Animal diets vary—from meat to plant nectar, from plankton to bamboo leaves. Animal digestive systems display a variety of evolutionary adaptations to match. See if you can match each of the following animals with a description of its digestive system. (Use each answer once.)

_____ 1. Koala
_____ 2. Desert mouse
_____ 3. Lion
_____ 4. Cow
_____ 5. Rabbit
_____ 6. Deer

A. Re-chews "cud," which is further digested by microbes in its four-chambered stomach
B. Has a large stomach, but relatively short intestine
C. Very long intestine, and microbes in long cecum digest cellulose
D. Reingests feces to extract nutrients twice
E. Ruminant herbivore
F. Conserves water by reingesting feces

Exercise 7   (Module 21.14)

This module is short, but it is an important introduction to the subject of nutrition. See if you can summarize the module in a sentence of *exactly* 25 words.

Exercise 8   (Module 21.15)

Cellular metabolism breaks down the molecules in food and uses the energy released to build ATP, which in turn fuels body activities. The energy in food is measured in Calories (which are actually kilocalories). To get a handle on the energy contents of foods and the energy consumed by body activities, calculate how much you would need to exercise to use the calories in the foods listed below (similar to the information in Table 21.15).

First, you need to calculate your own weight in kilograms. One kilogram equals 2.2 pounds, so:

$$\text{your weight (in kg)} = \frac{\text{your weight (lbs)}}{2.2} = \underline{\qquad}$$

Next you need to calculate how many kcal/min you use performing various activities:

$$\text{kcal/min you use} = \text{kcal/kg/min (from table)} \times \text{your weight (in kg)}$$

Use the above formula to figure out how many kcal/min you would use jogging, swimming, and walking. Then write the figures in the spaces below.

Finally, use the following formula to figure out how long you would have to exercise to use up the energy in a food item, and write the figures in the table:

$$\text{exercise time} = \frac{\text{energy in food (kcal)}}{\text{kcal/min you use}}$$

Food–exercise energy equivalents:

	Jogging	*Swimming*	*Walking*
Speed	9 min/mile	30 min/mile	20 min/mile
kcal/kg/min	0.19	0.10	0.06
kcal/min for you:	_____	_____	_____

Number of minutes you would have to do the above exercises to consume the calories in:

Cheeseburger 417 kcal	_____	_____	_____
Cheese pizza—slice 280 kcal	_____	_____	_____
Soft drink—12 oz 152 kcal	_____	_____	_____
Whole-wheat bread—slice 65 kcal	_____	_____	_____

(You may want to look at the labels to get the energy contents of some of your favorite foods. You can record their exercise equivalents in the spaces below.)

_____	_____	_____	_____
_____	_____	_____	_____
_____	_____	_____	_____

Exercise 9   (Modules 21.16–21.22)

Vitamins, minerals, and certain fats and amino acids are all essential nutrients. They are substances an animal cannot make that must be obtained in food. These and other nutrients are listed on food labels. Paying attention to food labels can help us avoid obesity and other health problems. There is a lot of detail in these modules, so try to focus on overall concepts. These concepts, the names of some nutrients, principles of food labeling, problems related to obesity, and diets are all covered in this crossword puzzle.

Across

2. The hormone leptin is produced by ____ (fat) cells.

4. The minerals sodium, chlorine, and ____ are important in acid-base balance, water balance, and nerve function.

5. People in ____ countries often cannot afford animal protein

9. Excess ____-soluble vitamins are not stored, but excreted from the body.

10. A diet low in one or more essential amino acids can lead to ____ deficiency.

12. The vitamin debate is mainly about ____.

15. The hormone _____ is made by fat cells and normally suppresses appetite.

16. ____ acid (vitamin C) is important for forming connective tissue.

17. Many B vitamins function as ____ .

19. People on low-____ diets avoid sugar, fruits, bread, and potatoes.

20. Humans naturally crave foods high in _____.

21. Our modern taste for fats and ____ might be an evolutionary adaptation to an unpredictable diet.

24. We need the mineral ____ to make the hormone thyroxine.

25. There are ____ essential amino acids that we require to make proteins.

26. A ____ is an essential organic nutrient needed in very small quantities.

27. ____ bypass surgery is a treatment option for very obese individuals.

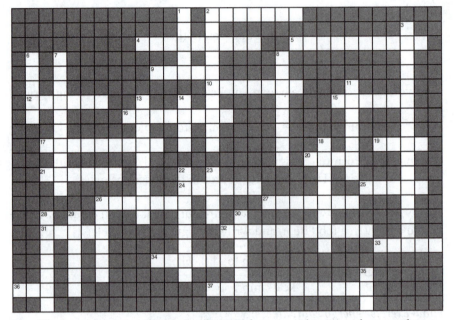

31. All essential amino acids can be obtained from a combination of corn and ____.

32. Packages list ____ and nutritional information.

33. Low-fat diets often lack adequate amounts of ____ acids and protein.

34. ____ is a mineral important in bones, teeth, muscle, nerves, and blood clotting.

36. ____-soluble vitamins (A, D, and so on) can be stored in the body.

37. An individual whose diet is lacking in one or more essential nutrients is said to be ____.

Down

1. Ingredients are listed on packages from greatest amount to ____ amount.

2. Vitamin D aids in the ____ and use of calcium and phosphorus.

3. ____ results from a diet chronically deficient in calories.

6. The body can make fats by combining fatty _____ with molecules such as glycerol.

7. Calcium and ____ are important in building the skeleton.

8. Most victims of malnutrition are ____.

11. Anorexia _____ is an eating disorder associated with compulsive starvation.

13. The amino acids that body cells cannot make are called ____ amino acids.

14. ____ is a component of hemoglobin and electron carriers of energy metabolism.

18. The energy content of foods is expressed in ____.

22. ____ are simple inorganic nutrients.

23. ____ are minimal standards of nutrients established by nutritionists for preventing deficiencies.

26. Vitamin A is important for normal ____.

28. ____ contributes to a number of human health problems, such as diabetes, cancer, and heart disease.

29. In earlier times, humans who stored more fat were able to survive _____.

30. Essential amino acids are most easily obtained from ____ sources.

35. In the United States, the ____ requires certain information on packaged-food labels.

Exercise 10   (Modules 21.22–21.23)

Maintaining a healthy diet affects not only weight and appearance, but can also reduce the risk of cancer and cardiovascular disease. If you would like to make changes that will lead to a healthier lifestyle, which of these should you try to increase (I) and which should you try to decrease or limit (D) ?

_____ 1. Fruits and vegetables
_____ 2. Vegetable oils
_____ 3. Sodium
_____ 4. Saturated fats
_____ 5. Dietary fiber
_____ 6. Exercise
_____ 7. Smoking
_____ 8. LDLs
_____ 9. Blood cholesterol
_____ 10. HDLs

_____ 11. Animal fats
_____ 12. Foods containing antioxidants
_____ 13. Unsaturated fats
_____ 14. Red meat
_____ 15. Processed and snack foods
_____ 16. Alcohol
_____ 17. Trans fats
_____ 18. Fish oils
_____ 19. Smoked, cured, and barbequed
        foods

Test Your Knowledge

Multiple Choice

1. The box elder bug is an insect that sucks plant juices. It is a
   a. suspension-feeding omnivore.
   b. substrate-feeding herbivore.
   c. suspension-feeding herbivore.
   d. fluid-feeding herbivore.
   e. fluid-feeding carnivore.

2. Which of the following lists the four stages of food processing in order?
   a. ingestion, digestion, absorption, elimination
   b. digestion, ingestion, absorption, elimination
   c. ingestion, absorption, elimination, digestion
   d. ingestion, digestion, elimination, absorption
   e. absorption, digestion, ingestion, elimination

3. Which of the answers below would *not* work in the following sentence? "In order for the body to absorb and use _____ , they must be broken down by hydrolysis into ____."
   a. polysaccharides . . . monosaccharides
   b. amino acids . . . proteins
   c. fats . . . glycerol and fatty acids
   d. disaccharides . . . monosaccharides
   e. starch . . . monosaccharides

4. How does a gastrovascular cavity differ from an alimentary canal? The gastrovascular cavity
   a. stores food but does not digest it.
   b. is usually much larger.
   c. has only one opening.

   d. functions in digestion but not absorption.
   e. can use only small food particles.

5. In humans, most nutrient molecules are absorbed by the
   a. stomach.
   b. liver.
   c. small intestine.
   d. large intestine.
   e. pancreas.

6. The largest variety of digestive enzymes function in the
   a. large intestine.
   b. oral cavity.
   c. stomach.
   d. gallbladder.
   e. small intestine.

7. After nutrients are absorbed, the blood carries them first to the
   a. brain.
   b. pancreas.
   c. kidneys.
   d. liver.
   e. large intestine.

8. Digestion of proteins begins in the _____ , and digestion of polysaccharides begins in the ___.
   a. mouth . . . stomach
   b. stomach . . . small intestine
   c. stomach . . . mouth
   d. stomach . . . stomach
   e. small intestine . . . stomach

9. In humans, a major function of the large intestine is
   a. absorption of water.
   b. digestion of food molecules.
   c. breakdown of toxic substances.
   d. absorption of nutrients.
   e. storage of food before it is digested and absorbed.

10. The energy needed to fuel essential body processes is called
    a. essential nutrient level.
    b. metabolism.
    c. recommended daily allowance.
    d. optimum energy intake.
    e. basal metabolic rate.

11. Which of the following statements about fat is *incorrect*?
    a. Craving and storing fat may have been beneficial to early humans.
    b. It is best to strive to have as little body fat as possible.
    c. Too much fat in the diet can lead to cardiovascular disease.
    d. It is best to avoid food that contain trans fats.
    e. In the U.S., the majority of people are overweight or obese.

12. Which of the following is *not* an essential nutrient?
    a. iron
    b. glucose, a monosaccharide
    c. methionine, an amino acid
    d. sodium
    e. pantothenic acid, a vitamin

13. Which of the following is an organic molecule needed by the body in small amounts?
    a. protein
    b. zinc
    c. vitamin C
    d. monosaccharide
    e. calcium

14. ____ are needed in the diet as components of teeth and bone, regulators of acid-base and water balance, and parts of certain enzymes.
    a. Amino acids
    b. Fats
    c. Nucleic acids
    d. Minerals
    e. Vitamins

15. To maintain health, it is best to try to
    a. decrease intake of unsaturated fats.
    b. decrease HDL levels in the blood.
    c. decrease intake of saturated fats.
    d. maintain a diet in which at least 50% of calories come from fat.
    e. increase LDL levels in the blood.

Essay

1. Why do many nutrient molecules contained in the food that you eat have to be digested before your body can use them?

2. In a few sentences, compare the digestive system of a grasshopper with that of a human.

3. Describe the substances in saliva and their roles in the function of the digestive tract.

4. Plants are difficult to digest. Describe how the digestive tracts and habits of three different herbivorous mammals are suited to a plant diet.

5. Briefly describe the potential problems with "fad" diets. What is the best approach to healthy, effective weight control?

6. Describe the kinds of information you can get from reading the labels on food. How might this information be useful?

Apply the Concepts

Multiple Choice

1. Which of the following do a wolf, a trout, a termite, and an elephant have in common?
   a. All are omnivores.
   b. All are substrate feeders.
   c. All are bulk feeders.
   d. All are carnivores.
   e. All are herbivores.

2. How would you expect the digestive system of a hawk, a carnivore, to compare with that of a finch, a seed-eater?
   a. The hawk would have a larger gastrovascular cavity.
   b. The finch digestive system would be longer (relative to body size).
   c. The hawk would have a gizzard, but the finch would not.
   d. The hawk digestive system would be longer.
   e. The hawk would lack both crop and gizzard.

3. Which of the following might make the most effective anti-ulcer medication? A chemical that
   a. stimulates parietal cells of the gastric glands.
   b. kills bacteria in the stomach.
   c. inhibits mucous cells of the gastric glands.
   d. stimulates secretion of bile.
   e. triggers GERD.

4. The lungs are folded into many small air sacs covered with blood vessels, which divide to form many small capillaries that increase the transfer of substances through their walls. The structures in the digestive system similar in function to these air sacs and capillaries are the
   a. villi.
   b. colon and rectum.
   c. gastric glands.
   d. high-density lipoproteins.
   e. sphincters.

5. Imagine that you have eaten a meal containing the following nutrients. Which would *not* have to be digested before being absorbed?
   a. protein
   b. polysaccharide
   c. disaccharide
   d. nucleic acid
   e. amino acid

6. Andrea had her gallbladder removed and afterward
   a. could not eat foods containing large amounts of fat.
   b. had trouble digesting proteins.
   c. could eat monosaccharides and disaccharides but not polysaccharides.
   d. had to wash her food down with large quantities of water.
   e. needed to take an amino acid supplement.

7. Laxatives work in the large intestine to relieve constipation. Which of the following would probably *not* be an effective laxative? A substance that
   a. contains lots of fiber.
   b. promotes water absorption in the large intestine.
   c. speeds up movement of material through the large intestine.
   d. decreases water absorption in the large intestine.
   e. stimulates peristalsis.

8. It is important to get some vitamin B-1 every day, but it is all right if intake of vitamin A varies a bit. Why?
   a. Vitamin B-1 is an essential nutrient, and vitamin A is not.
   b. Vitamin A can be stored by the body, but vitamin B-1 cannot.
   c. The body needs much larger amounts of vitamin B-1 than of vitamin A.
   d. The body requires vitamin B-1, but vitamin A is just an "extra."
   e. Vitamin A is water-soluble, and vitamin B-1 is fat-soluble.

9. Why does your body need grams of carbohydrates each day but only milligrams of vitamins, thousands of times less?
   a. Carbohydrates are used up, but vitamins are reusable.
   b. The body needs carbohydrates to function, but not vitamins.
   c. Vitamins contain much more energy per gram.
   d. The body makes vitamins itself but needs to get carbohydrates from food.
   e. Carbohydrates are essential nutrients, but vitamins are not.

10. Joe is trying to decide whether to buy a candy bar or some gumdrops to reward himself for losing 10 pounds. Both are 300 calories. Most of the calories in the candy bar come from fat, and most calories in the gumdrops come from carbohydrate. If Joe wants to keep his weight under control, it would be best to buy the
    a. gumdrops, because the body tends to hoard carbohydrates.
    b. gumdrops, because the body tends to hoard fat.
    c. candy bar, because the body tends to hoard carbohydrates.
    d. candy bar, because the body tends to hoard fat.
    e. It doesn't matter; calories are calories, whatever the source.

Essay

1. James and Michael are both working out to build up their muscles. Muscle is mostly protein, so they both are on high-protein diets. James is eating twice as much meat as normal. Michael is a vegetarian, so he is consuming plant foods high in protein, such as beans and peanuts. What are the pros and cons of each diet?

2. It is sometimes necessary to remove a diseased portion of the digestive system, but usually a patient can get along on a modified diet. It would be most difficult to live without which of the following, and why? Stomach, large intestine, gallbladder, small intestine, salivary gland.

3. The pancreas does not actually secrete the protein-digesting enzymes trypsin and chymotrypsin. It secretes substances called trypsinogen and chymotrypsinogen, which are converted to trypsin and chymotrypsin by a substance secreted by the walls of the small intestine. What do you think is the reason for this two-step process?

4. NASA calculates that an astronaut uses an average of 2,500 kcal of food energy per day. For every 4.83 kcal of food energy used, 1 liter of oxygen is consumed. How many liters of oxygen will the astronaut consume in a day?

5. A researcher suspects that the element scandium is an essential mineral in the human diet. How might he go about demonstrating this? Why might it be difficult?

6. Look at some food labels. How many calories are contained in one serving of one of the foods? How far would you have to jog, swim, or walk to use the calories?

Put Words to Work

Correctly use as many of the following words as possible when reading, talking, and writing about biology:

absorption, alimentary canal, anus, appendix, basal metabolic rate (BMR), bile, bolus, bulk feeder, carnivore, cecum (plural, ceca), chyme, colon, crop, digestion, duodenum, elimination, esophagus, essential amino acid, essential fatty acid, essential nutrient, feces, fluid feeder, gallbladder, gastric juice, gastric ulcer, gastrin, gastrovascular cavity, gizzard, hepatic portal vein, herbivore, high-density lipoprotein (HDL), ingestion, intestine, kilocalorie (kcal), large intestine, liver, low-density lipoprotein (LDL), malnourishment, metabolic rate, microvillus (plural, microvilli), mineral, mouth, omnivore, oral cavity, overnourishment, pancreas, peristalsis, pharynx, Recommended Dietary Allowance (RDA), rectum, ruminant, saliva, salivary glands, small intestine, sphincter, stomach, substrate feeder, suspension feeder, undernourishment, villus, villi, vitamin

Use the Web

There are a variety of activities and questions reviewing nutrition, the structure of the digestive system, and the stages of food processing on the Web at www.mybiology.com. The animations of digestion and absorption are particularly helpful.

Gas Exchange

Focus on the Concepts

Gas exchange, also called respiration, is the interchange of oxygen and carbon dioxide between an organism and its environment. This chapter describes the structure and function of animal respiratory systems. As you study this chapter, focus on these major concepts:

- Gas exchange involves ventilation (breathing), exchange of O_2 and CO_2 across moist body surfaces, transport of gases, and exchange of gases with cells. Some animals exchange gases through their skin. Most have specialized respiratory structures such as gills, tracheal tubes, or lungs.

- Several adaptations allowed early tetrapods to gulp air into primitive lungs and transition to life on land. The lungs of modern birds and mammals are more complex than those of amphibians and non-bird reptiles. In a mammal, air enters the nasal cavity, passes through the pharynx, larynx (voice box), trachea, bronchi, and branching bronchioles to tiny air sacs called alveoli, where gas exchange with blood in alveolar capillaries occurs.

- Contraction of muscles of the chest wall and the diaphragm expand the rib cage and chest cavity. This lowers the pressure in the alveoli relative to outside air, and air moves into the lungs—a mechanism called negative-pressure breathing. Breathing control centers in the brain regulate breathing rate in response to the CO_2 level of the blood.

- The blood transports respiratory gases. The heart pumps oxygen-poor blood to the lungs. Oxygen diffuses from the air in the alveoli into the blood of lung capillaries, where it is picked up and transported by hemoglobin in red blood cells. The heart then pumps this oxygen-rich blood to the capillaries in body tissues, where oxygen leaves the blood and enters cells.

Review the Concepts

Work through the following exercises to review gas exchange. For additional review, explore the activities on the Web at www.mybiology.com. The website offers a pre-test that will help you plan your studies.

Exercise 1 (Module 22.1)

In most animals, there are three phases of gas exchange: **breathing, transport** of gases by the circulatory system, and **exchange** of gases with tissues. Which of these phases do you think is blocked in each of the following scenarios?

_____ 1. In the disease cystic fibrosis, thick mucus coats the inside of the lungs, blocking passage of gases.

_____ 2. A broken neck can paralyze the muscles of the chest.

_____ 3. Babies sometimes inhale small objects that can block the windpipe.

_____ 4. Anemia is a decrease in the oxygen-carrying protein hemoglobin.

_____ 5. During a heart attack, blockage of a blood vessel causes heart muscle cells to die from lack of oxygen.

_____ 6. An asthma attack narrows air passages into the lungs.

_____ 7. Bedridden patients sometimes get bedsores when blood vessels to the skin are pinched.

_____ 8. A mountain climber is breathing rapidly and his heart is beating strongly, but in the thin air there is not enough oxygen in his blood to diffuse into brain cells.

_____ 9. In a smoker's blood, carbon monoxide prevents some oxygen from attaching to hemoglobin in red blood cells.

Exercise 2 (Module 22.2)

Match each of the following animals with a term (A–D) that describes it and a diagram (P–S) that shows its respiratory surface. Also color each respiratory surface yellow.

A. Lungs B. Gills C. Tracheae D. Body surface

Animal	*Term*	*Diagram*
1. Beetle	____	____
2. Cat	____	____
3. Earthworm	____	____
4. Trout	____	____
5. Human	____	____
6. Chicken	____	____
7. Crayfish	____	____
8. Clam	____	____
9. Cockroach	____	____
10. Flatworm	____	____

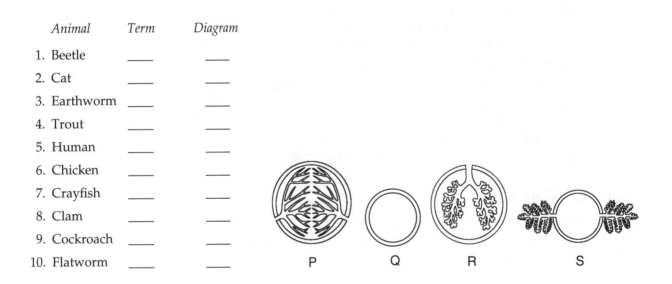

Exercise 3   (Modules 22.2 – 22.6)

Review gas exchange mechanisms of different animals by filling in the blanks below.

Although some small animals, such as earthworms, use their
[1]_____ as a gas-exchange organ, most animals have specialized organs that
enable them to obtain [2]_____ and expel waste [3]_____. The
part of an animal where gas exchange occurs is called the [4]_____ surface.
Individual molecules of O_2 and CO_2 diffuse through a membrane only if they are dissolved in [5]_____, so the respiratory surface must be [6]_____.
The surface must also be [7]_____ enough to take in sufficient oxygen for the
body's needs and [8]_____ enough for gases to diffuse through it rapidly.

Most aquatic animals obtain dissolved oxygen from the surrounding water by
means of [9]_____—outfoldings of the body surface. An advantage of exchanging gases in water is that the animal does not have to expend effort to keep the respiratory surface [10]_____. A disadvantage is that the concentration of
available oxygen is much [11]_____ in water than in the air.

The gills of a fish are efficient gas-exchange organs. Gill arches on each side of
the fish's body bear numerous elongated gill filaments, and each of these bears numerous
platelike [12]_____. The gills are red because the lamellae are filled with tiny
[13]_____ covered by a thin layer of cells. The fish opens and closes its
mouth and gill coverings to [14]_____ its gills, increasing contact between
the water and the respiratory surface. Water flows past the lamellae in a direction
[15]_____ to the flow of [16]_____ inside the lamellae. This
countercurrent increases the efficiency of the gills. [17]_____ is transfer of a
substance from a fluid moving in one direction to a fluid moving in the opposite direction. As water and blood flow past each other, [18]_____ diffuses from water
to blood. As the blood picks up more and more oxygen, it comes into contact with water
containing [19]_____ and [20]_____ available oxygen. Thus the
countercurrent flow of water and blood creates a [21]_____ gradient that
favors the diffusion of oxygen along the entire length of the lamella, greatly enhancing
the efficiency of the gill.

Land animals obtain oxygen from the [22]_____. There are advantages to breathing air: It contains a [23]_____ concentration of oxygen than
water, and it is [24]_____ to move than water. The biggest disadvantage of
breathing air is that it tends to [25]_____ the respiratory surface. The
[26]_____ system of an insect is an effective mechanism for gas exchange.
[27]_____ tubes that branch throughout the body deliver gases directly to
body cells without the help of the [28]_____ system. The tracheae branch and
rebranch and end in tiny fluid-filled tubes that touch the surface of individual body
[29]_____. [30]_____ dissolves in the fluid and diffuses into the
cells, while [31]_____ diffuses out of the cells and is expelled from the
insect's body. Because gas exchange occurs deep inside the body, the tracheal system
allows the insect to live on land without great loss of [32]_____.

The evolution of [33]_____ allowed tetrapods to move into land; this
was one of the most important steps in the history of life. Paleontologists think that the
first land vertebrates evolved in shallow fresh [34]_____. These animals

probably had both 35_____ and lungs. Adaptations of their snouts, jaws, and shoulders enabled them to gulp air in a pumping motion still used by present-day 36_____. The early tetrapods evolved into three main lineages: amphibians, reptiles (including birds), and 37_____. 38_____ have small lungs and rely partly on diffusion though their skin for gas exchange. The more complex lungs of birds and mammals reflect the fact that they have higher rates of 39_____ than amphibians and non-bird reptiles.

In humans, as in other mammals, the lungs are located in the 40_____, above a thin sheet of muscle called the 41_____ that functions in breathing. When you inhale, air enters through the 42_____ and is warmed, humidified, and 43_____ in the 44_____ cavity. The air then passes into the 45_____ (throat) and through the 46_____, or voice box. Here a pair of 47_____ make the sounds that enable us to speak. When the vocal cords are 48_____, high-pitched sounds are produced. When the cords are 49_____, they make lower-pitched sounds. From the larynx, air passes through the 50_____ into a pair of 51_____, one leading to each lung. Inside the lungs, the bronchi branch into numerous narrow tubes called 52_____. The surfaces of these respiratory passageways are covered by a film of 53_____ that traps dust and other contaminants. The beating of numerous 54_____ move the mucus and trapped particles out of the respiratory tract. The bronchioles end in clusters of tiny air sacs called 55_____. There are 56_____ of these tiny sacs in each lung, so their total surface area is enormous. The lining of each alveolus is a thin layer of epithelial cells that makes up the respiratory surface. Oxygen diffuses through a thin layer of moisture, through the epithelium, and into a network of 57_____ that covers the surface of the alveolus. 58_____ diffuses out of the blood and into the air within the alveolus. Thus the respiratory system works with the circulatory system in the process of gas exchange.

Use these diagrams to review the parts of the human respiratory system. Label and color the parts in **bold** type. On the left diagram, color the **nasal cavity** purple, the **pharynx** blue, the **larynx** green, the **trachea** yellow, the **bronchi** orange, and the **bronchioles** rust. (Coloring will help you to focus on the parts.) Also color the surrounding **lung tissue** healthy pink (this woman is definitely a nonsmoker!) and the **diaphragm** brown. On the right close-up view, color **alveoli** pink, **oxygen-poor blood** blue, **oxygen-rich blood** red, **and blood capillaries** purple.

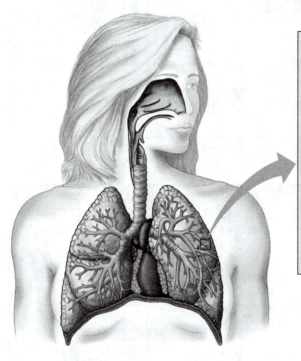
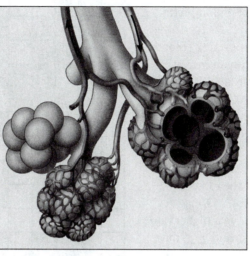

Exercise 5   (Modules 22.6 – 22.7)

What are some common respiratory diseases and problems? How do tobacco smoke and other air pollutants affect the respiratory system? Test your knowledge by completing this crossword puzzle.

Across

3. Tobacco smoke consists of small carbon particles coated with ____.

4. Quitting smoking has a significant impact on long-term ____.

7. Children exposed to second-hand smoke run an increased risk of ____, bronchitis, and pneumonia.

8. Some of the dangerous chemicals in tobacco smoke cause lung ____.

10. In ____, the alveoli become damaged and the lungs lose their elasticity.

12. Smokers have a higher rate of heart attacks and ____ than nonsmokers.

13. Smoke irritates cells lining the respiratory tract, damaging ____ that sweep out pollutants

14. Frequent ____ is a symptom of the respiratory system attempting to clear out mucus.

15. ____ kills about 440,000 people each year in the United States.

17. The ____ of a heavy smoker are black from buildup of smoke particles.

18. Secretions called ____ are lacking in babies born very prematurely.

Down

1. Chronic obstructive pulmonary disease includes emphysema and chronic ____.

2. ____ is an abbreviation for chronic obstructive pulmonary disease.

5. Defensive blood cells called ____ protect the alveoli.

6. One of the worst sources of lung-damaging pollutants is ____ smoke.

9. ____ cells line the respiratory passages.

11. Cilia lining the respiratory passages clear out pollutant-laden ____.

16. On average, adult smokers die 13 to 14 ____ earlier than nonsmokers.

Exercise 6   (Module 22.8)

State whether each of the following pertains to inhalation (I) or exhalation (E).

_____ 1. The diaphragm contracts.
_____ 2. The rib cage expands.
_____ 3. The diaphragm expands and arches upward.
_____ 4. Air enters the lungs.
_____ 5. The diaphragm moves downward.
_____ 6. Muscles between the ribs contract.
_____ 7. Muscles between the ribs relax.
_____ 8. The volume of the chest cavity decreases.
_____ 9. Air pressure in the alveoli is less than that of the atmosphere.
_____ 10. The diaphragm relaxes.
_____ 11. Air pressure in the alveoli is greater than that of the atmosphere.
_____ 12. Air is forced out of the lungs.

Exercise 7   (Module 22.9)

Rate of breathing is automatically controlled by breathing centers in the brain. Review control of breathing: State whether each of the following changes would speed up ($\uparrow$) or slow down ($\downarrow$) your rate of breathing.

_____ 1. A rise in blood CO_2 concentration
_____ 2. Purposely hyperventilating, expelling CO_2
_____ 3. A severe drop in blood oxygen concentration
_____ 4. An increase in pH of the cerebrospinal fluid
_____ 5. An increase in carbonic acid in the blood
_____ 6. A drop in blood pH
_____ 7. A decrease in blood CO_2 concentration
_____ 8. Holding your breath as long as you can, then releasing it
_____ 9. Vigorous exercise

Exercise 8   (Modules 22.10 – 22.11)

The following story summarizes the cooperation of the respiratory and circulatory systems in gas exchange and transport. Fill in the blanks to complete the story.

Your respiratory system works together with your [1]_____ system in exchange and transport of gases. Imagine that you are riding a bicycle. Your leg muscles are working hard, consuming [2]_____ and producing [3]_____ as a waste product. Blood returning from the muscles therefore has a relatively [4]_____ concentration of oxygen and a [5]_____ concentration of carbon dioxide. One side of the heart pumps this oxygen-poor blood through the capillaries covering the [6]_____ in your lungs. As your muscle cells work, the [7]_____ level of the blood increases, causing a [8]_____ in blood pH. Breathing control centers in your [9]_____ respond, speeding up the pace of nerve impulses sent to the [10]_____ and muscles between your [11]_____. You breathe more rapidly, and this helps expel the [12]_____ and meet the muscles' needs for more [13]_____.

The air in the lungs has a high partial pressure of [14]_____ and a low partial pressure of [15]_____ relative to the blood. [16]_____ diffuses out of the blood into the air inside the alveolus, moving from a region of [17]_____ partial pressure to a region of [18]_____ partial pressure. [19]_____ similarly diffuses down its pressure gradient from the air in the alveolus into the blood.

Oxygen is not very [20]_____ in water, so little oxygen is transported in dissolved form. Most oxygen is carried by a protein called [21]_____, contained within [22]_____ blood cells. A hemoglobin molecule consists of four polypeptide chains, each of which contains a heme group with an [23]_____ atom at its center. Each of these atoms can pick up one [24]_____ molecule, so the millions of hemoglobin molecules in a red blood cell can carry millions of O_2 molecules.

The oxygen-rich blood that leaves the lungs returns to the heart, which pumps it out to the exercising [25]_____ of your legs. The blood passing through the capillaries in a muscle contains a [26]_____ partial pressure of O_2 than the muscle cells where it is being used up, so O_2 leaves the blood cells and diffuses out of the blood and into the muscle cells. The exercising cells are making CO_2 at a fast pace, so the partial pressure of CO_2 is [27]_____ in the cells than in the blood. CO_2 diffuses [28]_____ the muscle cells and [29]_____ the blood.

Some CO_2 is carried dissolved in blood plasma. Most of it enters [30]_____ blood cells, but most does not combine with [31]_____. Instead, enzymes in the blood cells cause most of it to react with [32]_____ molecules, forming carbonic acid (H_2CO_3). Each carbonic acid molecule then breaks apart, forming a hydrogen ion (H^+) and a [33]_____ ion (HCO_3^-). Hemoglobin picks up most of the H^+ ions, so it does not acidify the blood much. The bicarbonate ions diffuse out into the blood [34]_____, where they are part of the blood-[35]_____ system that stabilizes the pH of the blood. If blood pH drops, the bicarbonate ions combine with H^+ ions and remove them from the plasma. If pH [36]_____, the bicarbonate releases these H^+ ions back into solution.

When the blood from the muscles returns (via the heart) to the lungs, the events that formed bicarbonate ions are reversed. Bicarbonate and H^+ form carbonic acid, which breaks up to form water and [37]_____, which in turn diffuses out of the blood into the air of the alveoli. Oxygen diffuses from the air into the blood, and the cycle is repeated. Thus the respiratory and circulatory systems continue to work in close cooperation as you continue your bicycle ride.

Exercise 9 (Modules 22.10 – 22.12)

Number the following in order from first to last to show the path an O_2 molecule must follow through the body of a mother to a cell in her fetus.

_____ A. The mother's heart pumps the oxygen-rich blood to her uterus.
_____ B. Oxygen diffuses out through the walls of capillaries in the walls of the uterus.
_____ C. The mother takes a deep breath of fresh air.
_____ D. Oxygen leaves the blood of the fetus and diffuses into a growing cell in the fetus's brain.
_____ E. Oxygen diffuses across the thin wall of an alveolus in the mother's lung and into a capillary.
_____ F. The mother's blood, now loaded with oxygen, returns from her lungs to her heart.
_____ G. Oxygen-rich fetal blood flows into the fetus through a vein in the umbilical cord.
_____ H. Oxygen diffuses through the wall of a capillary in the placenta and into the blood of the fetus.
_____ I. Oxygen attaches to hemoglobin in the mother's blood.
_____ J. Oxygen attaches to hemoglobin in fetal blood.
_____ K. The fetus's heart pumps the oxygen-rich blood out to its tissues.

Test Your Knowledge

Multiple Choice

1. Which of the following has no specialized respiratory structures?
   a. crab
   b. earthworm
   c. salmon
   d. ant
   e. snake

2. The breathing control centers are located in the
   a. heart.
   b. lungs.
   c. diaphragm and rib muscles.
   d. brain.
   e. large arteries.

3. When you exhale, the diaphragm
   a. relaxes and arches.
   b. relaxes and flattens.
   c. contracts and arches.
   d. contracts and flattens.
   e. contracts and arches, but only when you are exercising vigorously.

4. Which of the following has the simplest lungs?
   a. Grasshopper
   b. Sparrow
   c. Horse
   d. Earthworm
   e. Frog

5. Inhaled air passes through which of the following last?

   a. bronchiole
   b. larynx
   c. pharynx
   d. trachea
   e. bronchus

6. An advantage of gas exchange in water, compared with gas exchange in air, is that
   a. water usually contains a higher concentration of O_2 than air.
   b. water is easier to move over the respiratory surface.
   c. the respiratory surface does not dry out in water.
   d. ventilation requires less energy in water.
   e. the respiratory surface does not have to be as extensive in water.

7. In the blood, bicarbonate ions
   a. help transport oxygen.
   b. act as buffers to guard against pH changes.
   c. are transported by hemoglobin.
   d. attach to numerous CO_2 molecules, keeping them from solution.
   e. are poisonous and must constantly be removed.

8. Smoking destroys the cilia in the respiratory passageways. This
   a. makes it harder to move air in and out of the lungs.
   b. decreases the surface area for respiration.
   c. slows blood flow through lung blood vessels.
   d. makes it harder to keep the lungs clean.
   e. interferes with diffusion across the respiratory surface.

9. Most oxygen is carried by the blood ____.
   Most carbon dioxide is carried by the
   blood ____.
   a. attached to hemoglobin . . . in the form of
      bicarbonate ions
   b. dissolved in the plasma . . . dissolved in the
      plasma
   c. in the form of H^+ ions . . . in the form of
      bicarbonate ions
   d. attached to hemoglobin . . . attached to
      hemoglobin
   e. attached to hemoglobin . . . dissolved in the
      plasma

10. A disease called emphysema decreases
    the springiness of the lungs. This
    decreases ____ and makes it exhausting
    to breathe.
    a. the volume of air the lungs can take in
    b. respiratory rate
    c. blood flow through the lungs
    d. countercurrent exchange
    e. mucus production

11. The ____ is a structure specialized for diffusion
    of gases and nutrients between the blood of the
    mother and the fetus.
    a. uterus
    b. placenta
    c. lamella
    d. alveolus
    e. umbilicus

Essay

1. Compare the advantages and disadvantages of
   obtaining oxygen from water and obtaining it
   from the air.

2. How are gills and lungs similar? How are they
   different?

3. Describe how the structure, number, and
   arrangement of alveoli are well suited to their
   function in gas exchange.

4. Where are your vocal cords? How do they
   work? How do they produce high-pitched and
   low-pitched sounds?

5. When you inhale, does air flow into the lungs,
   causing them to expand? Or do the lungs
   expand, causing air to flow in? Explain.

6. Explain how countercurrent exchange in a fish
   gill enhances absorption of oxygen from water.

Apply the Concepts

Multiple Choice

1. Which of the following normally contains the
   highest concentration of oxygen?
   a. body cells
   b. inhaled air
   c. air in the alveoli
   d. blood entering the lungs
   e. blood leaving the lungs

2. Which of the following in a human is most
   similar in function to the gill lamellae of
   a fish?
   a. vocal cords
   b. bronchioles
   c. alveoli
   d. tracheae
   e. diaphragm

3. In which of the following does oxygen pass directly from the air, through a moist surface, to
   individual cells, without being carried by the
   blood?
   a. mouse
   b. ant
   c. shark
   d. earthworm
   e. frog

4. A fish opens and closes its mouth and gill
   covers. A dog pants. A marine worm waves
   long, filmy gills in the water. All of these
   movements
   a. are examples of ventilation.
   b. show how circulation aids respiration.
   c. are examples of breathing.
   d. slow diffusion of CO_2.
   e. enhance countercurrent exchange.

5. ____ in CO_2 in your blood, which causes
   ____ in pH, would cause your breathing
   to speed up.
   a. An increase . . . a rise
   b. An increase . . . a drop
   c. A decrease . . . a rise
   d. A decrease . . . a drop
   e. Actually, it is rise and fall of O_2, not CO_2,
      that controls breathing.

6. Which of the following would have the same O_2 content?
   a. blood entering the lungs—blood leaving the lungs
   b. blood entering the right side of the heart—blood entering the left side of the heart
   c. blood entering the tissue capillaries—blood leaving the tissue capillaries
   d. blood entering the right side of the heart—blood leaving the right side of the heart
   e. blood leaving the tissue capillaries—blood leaving the lungs

7. Patients with chronic lung disease and difficulty breathing often adapt to the high concentration of CO_2 in their blood. The breathing centers stop responding to CO_2 level. If such a patient has difficulty breathing, medical personnel are reluctant to give the patient pure oxygen. Based on what you know about control of breathing, why do you think this is the case?
   a. The patient's body would use the oxygen to make even more CO_2.
   b. The oxygen would increase concentration of bicarbonate, altering pH.
   c. Increased oxygen in the blood might slow or stop breathing.
   d. The body is not used to the oxygen, and the patient would overdose.
   e. The patient would breathe too fast and become tired out.

8. In an old science fiction movie, the hero tries to drown a giant ant by holding its head under water. Would this work? Why?
   a. Yes. Ants use lungs to breathe much as we do.
   b. Yes. The skin surface, covered with water, could not get O_2 from the air.
   c. No. Ants use gills for respiration, like crabs do.
   d. No. Ants breathe through holes in the sides of their bodies.
   e. No. The ant could get oxygen by diffusion from the water.

9. A zoologist compared the respiratory efficiency and swimming speed of different fish. He found that less efficient fish tended to have
   a. greater ventilation.
   b. a thicker respiratory surface.
   c. more hemoglobin.
   d. a faster heart rate.
   e. a more extensive respiratory surface.

10. A biochemist mixed 10 drops of acid with 100 mL of water, and the pH dropped from 7.4 to 5.0. She then mixed 10 drops of acid with 100 mL of blood. The pH dropped from 7.4 to 7.2. What is the reason for this difference?
    a. Blood is thicker than water.
    b. Blood is already very acidic, so the acid has less effect.
    c. Blood is saturated with oxygen; there is little room for acid.
    d. Blood contains buffers that reduce pH change.
    e. Water is already more acidic than blood; there is little room for more.

11. Based on what you have learned in this chapter, which of the following animals would you expect to depend on negative-pressure breathing?
    a. mouse
    b. salmon
    c. frog
    d. beetle
    e. lobster

Essay

1. Trace the path of an oxygen molecule from the air to one of your brain cells, naming all the places and structures it passes through on its way.

2. You are on the team to design a robot that will patrol the devastated terrain around the Chernobyl nuclear power plant. The robot will function like a living organism, gathering organic debris for "food" and obtaining oxygen from the surrounding air. What features would you want to include in your design of its respiratory surface?

3. Carbon monoxide molecules in cigarette smoke and automobile exhaust attach to hemoglobin molecules where oxygen normally attaches, and they hold on more strongly than oxygen. What effect would this have on the body?

4. A man smokes a pack of cigarettes (20) a day for 40 years. If each cigarette shortens his life, on average, by 11 minutes, how much "before his time" will he die? (*Note:* On average, each cigarette *does*, in fact, shorten life by *more* than 11 minutes!)

5. In a submarine, the oxygen supply was accidentally interrupted, causing the oxygen content of the air to drop. A machine that removes carbon dioxide continued to function, so there was no corresponding buildup of CO_2. None of the sailors felt short of breath or noticed anything wrong until several individuals fainted. Why do you think they did not feel short of breath?

Put Words to Work

Correctly use as many of the following words as possible when reading, talking, and writing about biology:

alveolus (plural, *alveoli*)*, breathing, breathing control center, bronchiole, bronchus* (plural, *bronchi*)*, countercurrent exchange, diaphragm, gas exchange, gill, hemoglobin, larynx, lung, negative pressure breathing, partial pressure, pharynx, surfactant, trachea* (plural, *tracheae*)*, tracheal system, ventilation, vital capacity, vocal cord*

Use the Web

To enhance your understanding of gas exchange, see the activities on the Web at www.mybiology.com.

Circulation

23

Focus on the Concepts

Most animals have a circulatory system that carries out internal transport. While studying the structure and function of circulatory systems, focus on the following concepts:

- Circulatory systems distribute nutrients and oxygen to body cells and collect carbon dioxide and wastes for disposal. Two main kinds of circulatory systems have evolved: Many invertebrates, such as insects, have open systems, where a simple heart pumps blood through open-ended vessels and it flows out among the cells. Other animals, most notably vertebrates, have closed circulatory systems, where blood is confined to vessels and the heart pumps it through a closed loop.

- The fish circulatory system consists of a single loop: A two-chambered heart pumps the blood through gill capillaries, where it is oxygenated, and then arteries carry this blood to systemic capillaries, where oxygen is delivered to tissues and back to the heart. As vertebrates moved onto land, double circulation evolved: A bird or mammal has a four-chambered heart and pumps blood twice. The right atrium and ventricle pump blood via the pulmonary circuit to the lungs and back. Then the left atrium and ventricle pump the oxygenated blood out to the tissues and back via the systemic circuit.

- The heart contracts and relaxes rhythmically, stimulated by signals from the SA node (pacemaker), relayed via the AV node. Blood returns from the systemic circuit via the vena cavae and enters the right atrium, which pumps it to the right ventricle. The right ventricle then pumps blood to the lungs via the pulmonary arteries. The pulmonary veins deliver the oxygenated blood to the left atrium and ventricle, which pumps it out to the rest of the body via the aorta. Valves at the exits of the heart chambers keep blood moving forward and create the heart sounds.

- Blood pressure drives the flow of blood through the vessels. Blood in the aorta exiting the heart is under the highest pressure, and pressure drops as the blood flows through arteries, capillaries, and veins. Blood leaving the heart flows fastest, slows almost to a stop as it spreads out in capillaries, and then speeds up again as it returns to the heart via veins.

- The structure of blood vessels is related to their functions. Arteries have thick muscular walls that contain the higher pressure of arterial blood. Smooth muscle in arteriole walls contracts and relaxes to control blood distribution. The walls of capillaries are thin, leaky epithelium, facilitating exchange via diffusion and pressure-driven fluid flow. Veins have thinner walls than arteries, reflecting the lower pressure of blood returning to the heart.

- Blood consists of several kinds of cells suspended in liquid plasma. Plasma is mostly water, but it carries various plasma proteins, salts, nutrients, hormones, gases, and waste products. Red blood cells, or erythrocytes, carry oxygen. White blood cells, or leukocytes, fight infections and cancer. Tiny platelets, along with plasma proteins such as fibrinogen, are responsible for blood clotting when vessels are damaged.

Review the Concepts

Work through the following exercises to review the concepts in this chapter. Also check out the activities on the Web at www.mybiology.com. The website offers a pre-test that will help you plan your studies.

Exercise 1 (Module 23.1)

This module gives an overview of circulatory systems—their basic functions and parts. Review the information by matching the following words and phrases. Each answer is used only once.

_____ 1. Network of small blood vessels	A. Ventricle
_____ 2. Circulatory system of vertebrates	B. Open system
_____ 3. Cannot transport molecules more than a few cell widths	C. Capillary
_____ 4. Vessel that carries blood away from the heart	D. Atrium
_____ 5. Solution in spaces between cells	E. Blood
_____ 6. Heart chamber that pumps blood out via arteries	F. Closed system
_____ 7. Carries out circulatory functions in a flatworm or cnidarian	G. Capillary bed
_____ 8. Circulatory system of insects and mollusks	H. Diffusion
_____ 9. Heart chamber that receives blood from veins	I. Interstitial fluid
_____ 10. Vessel that conveys blood from arteries to veins in tissues	J. Artery
_____ 11. Vessel that returns blood to the heart	K. Gastrovascular cavity
_____ 12. The circulatory fluid in a closed system	L. Vein

Exercise 2 (Module 23.2)

The fish heart pumps blood through two sets of capillaries. As the blood of a fish passes through the tiny gill capillaries, it loses pressure. Therefore, once it has picked up oxygen, it delivers this oxygen to the capillaries in body tissues rather half-heartedly. As vertebrates adapted to life out of the water, double circulation evolved: The mammal heart is like two fish hearts side by side. Each side pumps blood only through one set of capillaries. The right heart pumps blood only to the lungs. The left heart then raises the pressure and sends this blood on its way to body tissues. This pressure boost allows for more efficient blood flow to body tissues, an adaptation to a more vigorous terrestrial lifestyle. (Amphibians display an intermediate stage, with a three-chambered heart.) Compare fish and mammal circulatory systems. On the next page, trace the flow of blood in each animal by numbering the parts blood passes through in order.

A. Fish

____1____ a. Ventricle
_____ b. Systemic capillaries
_____ c. Atrium
_____ d. Gill capillaries

B. Mammal

____1____ a. Right ventricle
_____ b. Left atrium
_____ c. Lung capillaries
_____ d. Right atrium
_____ e. Systemic capillaries
_____ f. Left ventricle

Exercise 3   (Module 23.3)

To review the mammalian cardiovascular system, start by labeling the parts indicated in this diagram. Then color the vessels that carry oxygen-rich blood red and those that carry oxygen-poor blood blue. Finally, trace the path of blood flow by numbering the circles (1–10). (Note: Some numbers are duplicated because blood vessels split and blood goes two places at once.)

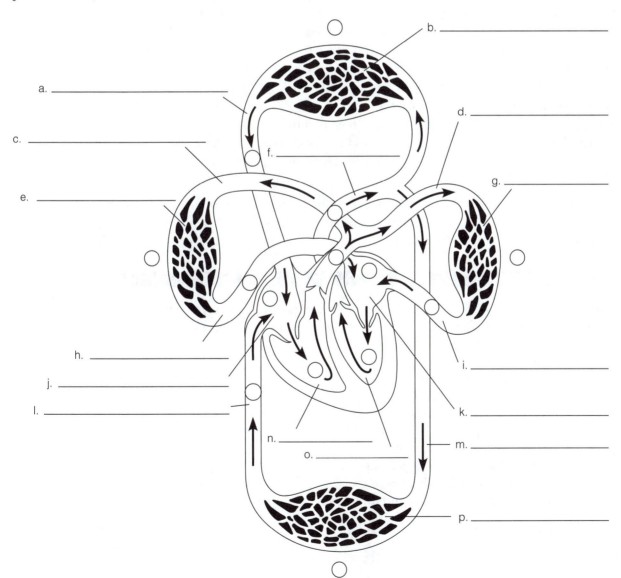

One way to learn about circulation and the parts of the heart is to trace the circulation of a drop of blood from one location in the body to another, naming all the structures that the drop of blood passes on its way. Use Figures 23.3A and 23.3B in the text. (A tip: Remember that blood always has to circulate from an artery to capillaries to a vein. There aren't any shortcuts. For example, to get from the capillaries of the big toe to capillaries of the little toe, blood must go back to the heart, to the lungs, back to the heart, and then to the little toe.)

Imagine a drop of blood starting in a brain capillary, circulating to the foot, and then circulating to the hand. The blood flows from the capillary into a vein that runs down the neck, and empties into the [1]_____, the large vein that serves the head and arms. From there, the blood enters the [2]_____ of the heart. This chamber pushes the blood into the [3]_____, which pumps it out through the [4]_____ to the lungs. In the capillaries of the lungs, the blood picks up oxygen. Then it returns to the heart via the [5]_____. It enters the [6]_____, which pumps the blood to the [7]_____. This chamber pumps blood out through the [8]_____, the largest blood vessel in the body. This vessel branches and rebranches, and finally the blood is delivered to [9]_____ in the foot, where nutrients and oxygen are dropped off at the tissues. The drop of blood travels back to the heart via leg veins, which join the [10]_____, which empties into the heart. The [11]_____ atrium and ventricle again pump blood through the [12]_____ circuit to the lungs. The [13]_____ atrium and ventricle then pump the blood out through the aorta into the [14]_____ circuit. This time the drop of blood flows down a large [15]_____ to the hand, where the blood again passes through a network of capillaries.

Try going on, tracing the flow of blood from the hand, to the intestine to pick up food molecules, or to the kidneys for filtration, and then back to the brain. Once you have done this a couple of times, try it without the diagram, or make your own sketch.

Heart valves prevent backflow of blood. Indicate which of the following statements refer to the atrioventricular (AV) valves and which refer to the semilunar (SL) valves.

_____ 1. Between atria and ventricles
_____ 2. Open during ventricular systole
_____ 3. At exits from ventricles
_____ 4. Prevent backflow from ventricles to atria
_____ 5. Prevent backflow from aorta and pulmonary artery to ventricles
_____ 6. Close during ventricular systole
_____ 7. "Lub" of "lub-dupp" heart sounds
_____ 8. "Dupp" of "lub-dupp" heart sounds
_____ 9. Possibly damaged if heart murmur sounds like "lub-ssssst"
_____ 10. Close during ventricular diastole

Exercise 6   (Module 23.4)

Cardiac output is the amount of blood that the left ventricle pumps into the aorta per minute. It is equal to the amount of blood pumped per beat multiplied by the number of beats per minute.

1. What is your cardiac output right now? An average heart pumps about 75 mL of blood per beat. To calculate cardiac output, take your pulse, then multiply 75 mL times the number of beats per minute.

   Cardiac output = 75 mL × _____ beats per minute = _____ mL of blood per minute

2. Now run in place for a minute, take your pulse, and make the same calculation.

   Cardiac output = 75 mL × _____ beats per minute = _____ mL of blood per minute

3. How much did output change with exercise?

   _____ mL of blood per minute equals _____ % change

Exercise 7   (Modules 23.5 – 23.6)

What do you know about cardiovascular health and disease? Find a word or phrase in Module 23.5 or 23.6 that goes with each of the following.

_____   1. Lowering these dietary lipids can reduce heart attack risk.

_____   2. Region that sets heart rate

_____   3. Damage or death of cardiac muscle cells due to heart vessel blockage

_____   4. Relays signal to contract from atria to ventricles

_____   5. Inserting a catheter to compress plaques in arteries

_____   6. Buildup of plaques on the inside of blood vessels

_____   7. Surgery that routes blood around blocked arteries

_____   8. Location (outside heart) of centers that control heart rate

_____   9. "Fight or flight" hormone that can speed up the heart

_____   10. A device that "shocks" the heart to restore its rhythm

_____   11. Imaging techniques used to "see" the heart

_____   12. Vessel blocked in a heart attack

_____   13. A wire mesh tube that opens an artery

_____   14. An implanted device that triggers normal heartbeat

_____   15. Anti-inflammatory drug that may prevent recurrence of heart attacks

_____   16. Recording of electrical activity of the heart

_____   17. General term for disorders of the heart and blood vessels

_____   18. Levels of this substance in blood predict cardiovascular disease.

Exercise 8   (Module 23.7)

Complete the following chart, comparing the structure and function of blood vessel types.

	Capillaries	*Arteries*	*Veins*
Carry blood from	1.	2.	3.
Carry blood to	4.	5.	6.
Thickness of walls (thick, thin, or in-between)	7.	8.	9.
Layers in walls (names)	10.	11.	12.
Valves? (yes or no)	13.	14.	15.

Exercise 9   (Modules 23.8 – 23.9)

Pay special attention to Figure 23.8A; it contains a lot of information. Included are sizes, arrangement, and names of the blood vessels and the changes that occur in pressure and velocity as blood passes through them. After reading the modules and studying the figures, match each of the following statements with one of the blood vessels listed on the right.

_____   1. Pressure is lowest here.
_____   2. Pressure and speed drop the most in these vessels.
_____   3. Pressure is usually measured in this kind of vessel.
_____   4. Blood moves fastest here.
_____   5. These vessels have the strongest pulse.
_____   6. Blood moves most slowly here.
_____   7. These vessels are the narrowest.
_____   8. Diastolic pressure here might be 70 mm Hg.
_____   9. Pressure here might be 20 mm Hg.
_____   10. Pressure is highest here.
_____   11. Systolic pressure here might be 120 mm Hg.
_____   12. Velocity of blood increases sharply as blood flows through these vessels.
_____   13. Hypertension may cause damage to the walls of these vessels, aggravating atherosclerosis.
_____   14. Blood flows fairly rapidly here, but there is no pulse.
_____   15. Muscles and breathing help propel blood through these vessels.

A. Aorta
B. Arteries
C. Arterioles
D. Capillaries
E. Venules
F. Veins
G. Vena cava

Exercise 10   (Module 23.10)

The two mechanisms that control blood distribution, discussed in Module 23.11, sound very much alike. Sometimes, smooth muscle in arterioles *leading to* a capillary bed relax or contract to allow blood into the capillaries or divert it away. The second mechanism, illustrated in Figure 23.11B, involves muscle at the beginnings of the *capillaries themselves* that control blood flow. Nerve impulses and hormones can control these mechanisms. Sometimes changes in the tissues surrounding capillaries influence increase or decrease in vessel diameter. Can you imagine two physical or chemical changes that might occur in exercising leg muscles that might trigger vessel dilation and cause an increase in blood flow through the capillaries in the muscles? Briefly describe those changes below.

Through the thin walls of capillaries, materials are exchanged between the blood and the interstitial fluid that surrounds body cells. Complete this diagram showing the movement of blood through a capillary and exchange of materials with the interstitial fluid. Draw a red arrow to show blood moving from an arteriole into the capillary. Draw a blue arrow to show blood moving out of the capillary into a venule. (Why use two different colors?) Two forces are responsible for exchange of fluid between the blood and the interstitial fluid: Blood pressure tends to push fluid out of a capillary, and osmotic pressure tends to push water in. Draw large and small arrows at the arterial end of the capillary to show the relative strengths of blood pressure and osmotic pressure. Do the same at the venous end. Now draw arrows to show the net movement of fluid at the venous and arterial ends. (Do your arrows show net movement of fluid into or out of the capillary at the arterial end? Why? Is there net movement of fluid into or out of the capillary at the venous end? Why?)

Tissue cells

Arterial end of capillary

Venous end of capillary

Interstitial fluid

Most of the information on blood is summarized in the tables in Module 23.12. Study the composition of blood, and then compare the three blood cell types by filling in the blanks in the following table.

Cell Type	Relative Size	Relative Numbers	Function
Platelets	Smallest	1.	2.
3.	4.	5.	Oxygen transport
Leukocytes	6.	Least numerous	7.

All the components necessary for blood clotting are present in blood all the time. Tissue damage activates them so that clotting occurs in a sort of "chain reaction." Try to visualize blood flow and blood clotting by filling in the missing words in the following story.

You step into the Microtron, and you are quickly reduced to a size slightly smaller than a red blood cell. The support team injects you into a small artery in the arm. Blood pressure is fairly [1]_____ here. You feel a boom of pressure on your eardrums about once per second; this is simply the [2]_____, and it will gradually disappear as the blood in the artery flows into the narrower [3]_____ that lead to the capillary beds. Bright lights on the subject's arm enable you to see what is around you: Most of the cells around you are [4]_____, flexible disks carrying [5]_____ to the body's cells. There are also a few larger, irregular [6]_____, important in body defense. They slowly crawl along the blood vessel walls. Some even move against the current. It is best to avoid them, because some are [7]_____, capable of eating bacteria and debris. All around you are tiny "blobs." These must be [8]_____, which are involved in maintaining osmotic balance, defense, and blood clotting. There are also swarms of small fragments of cell cytoplasm, called [9]_____, that assist in the clotting process.

As you enter a [10]_____, blood slows almost to a stop, and the scene brightens. The walls of these vessels are a single layer of [11]_____, only one cell thick. You can even see gaps between cells, where fluid in the capillary is exchanged with the [12]_____ fluid. There isn't much room here, though. The capillaries are so narrow that the red blood cells have to line up single file in some places.

You are just under the skin of the fingertip. The team pricks the subject's skin with a pin. You are moving directly toward the wound, so you use your gripper to hang onto the vessel wall. The clotting process is already under way. The damaged lining of the vessel exposes [13]_____ to the blood. [14]_____ stick to the exposed tissue and release a cloud of chemicals. These chemicals cause even more platelets to adhere. But in this case, the damage is too serious for a platelet plug to stop the leak. Chemicals released from the platelets and damaged cells in the vessel wall activate blood enzyme. The enzyme then causes small blobs of [15]_____ floating in the blood to change shape and form sticky strands of [16]_____. These strands stretch like a tangle of cords across the hole in the vessel, trapping red blood cells. The blood strains against the fibrin clot, but finally it holds and leakage stops.

A phagocyte has caught your leg! You break free, but in the process you loosen a big chunk of the clot. It could travel to the heart, lodge in one of the [17]_____, and cause a [18]_____! You let the flow of blood carry you along. As blood leaves the capillary bed and enters a [19]_____, blood flow speeds up. The vessel walls thicken, and it gets darker again. You enter an even larger [20]_____, and the blood slows down even more. Ahead you can dimly make out the flaps of a [21]_____, which keeps the blood moving toward the heart. Fortunately, the clot is briefly caught in an eddy downstream from the valve. You use your laser to break it into fragments small enough to pose no threat to the subject. This is a good time to make your exit, and you are soon back in the lab discussing your adventure.

Exercise 14   (Modules 23.13 – 23.15)

These modules describe several blood-related disorders and diseases, and the promise of stem cell research for curing some of them. Review these topics by matching each phrase on the left with a term from the list on the right. Each answer is used only once.

_____ 1. Cancer of white blood cells
_____ 2. Spongy bone tissue where blood cells develop
_____ 3. Unspecialized cell that can become multiple types of cells
_____ 4. A cell that divides uncontrollably
_____ 5. An inherited blood clotting disorder
_____ 6. A standard cancer treatment
_____ 7. Cell that gives rise to lymphocytes
_____ 8. A hormone that stimulates bone marrow to make red blood cells
_____ 9. A deficiency of this often causes anemia
_____ 10. Stem cells from here have potential, but little success so far
_____ 11. A general name for any white blood cell
_____ 12. Low hemoglobin or red blood cell count
_____ 13. Where stem cells first form
_____ 14. Anticoagulant drug
_____ 15. Develops from myeloid stem cell

A. Umbilical cord
B. Leukocyte
C. EPO
D. Hemophilia
E. Leukemia
F. Erythrocyte
G. Red marrow
H. Multipotent stem cell
I. Radiation
J. Iron
K. Anemia
L. Cancer cell
M. Aspirin
N. Lymphoid stem cell
O. Embryo

Exercise 15   (Summary)

This chapter introduces a lot of vocabulary connected with circulation. Circle the term that does not fit with the others in each of these groups, and briefly explain what the other terms in each group have in common.

1. venule   artery   atrium   capillary

2. fibrin   leukocyte   platelet red   blood cell

3. sphygmomanometer   systolic pressure fibrinogen   hypertension

4. pulmonary veins   right ventricle   lungs   aorta

5. fibrin   leukocyte   platelet   hemophilia

6. semilunar valve   AV node   SA node   pacemaker

7. diastole   systole   pulse   leukocyte

8. oxygen   red blood cell   hemoglobin   epithelium

9. pulmonary artery   systemic circuit   aorta   left ventricle

Test Your Knowledge

Multiple Choice

1. Rhythmic stretching of the arteries caused by heart contractions is called
   a. hypertension.
   b. heart murmur.
   c. hemophilia.
   d. pulse.
   e. diastole.

2. Which of the following animals has an open circulatory system?
   a. fish
   b. human
   c. frog
   d. jelly
   e. fly

3. Which of the following *cannot* move freely in and out of a capillary?
   a. sugar
   b. oxygen
   c. carbon dioxide
   d. water
   e. plasma protein

4. Heart valves function to
   a. keep blood moving forward through the heart.
   b. mix blood thoroughly as it passes through the heart.
   c. control the amount of blood pumped by the heart.
   d. slow blood down as it passes through the heart.
   e. propel blood as it passes through the heart.

5. Which of the following correctly traces the electrical impulses that trigger each heartbeat?
   a. ventricles, pacemaker, AV node, atria
   b. pacemaker, AV node, atria, ventricles
   c. atria, pacemaker, AV node, ventricles
   d. pacemaker, atria, AV node, ventricles
   e. pacemaker, AV node, atria, ventricles

6. A heart attack occurs when
   a. a heart valve malfunctions.
   b. a coronary artery is blocked.
   c. the heart is weakened by overwork.
   d. the aorta is blocked.
   e. a pulmonary artery is blocked.

7. The cells responsible for defense against infections are
   a. red blood cells.
   b. white blood cells.
   c. epithelial cells.
   d. platelets.
   e. pacemaker cells.

8. If cells are not receiving enough oxygen, a hormone signals the bone marrow to produce more
   a. leukocytes.
   b. fibrin.
   c. plasma.
   d. platelets.
   e. erythrocytes.

9. The primary sealants that plug leaks in blood vessels are
   a. platelets and fibrin.
   b. red blood cells and albumin.
   c. fibrin and white blood cells.
   d. white blood cells and platelets.
   e. hemoglobin and platelets.

10. Which of the following best describes an artery?
    a. carries blood away from the heart
    b. carries oxygenated blood
    c. contains valves
    d. has thin walls
    e. carries blood away from capillaries

11. Blood moves most slowly in
    a. capillaries.
    b. the aorta.
    c. veins.
    d. arterioles.
    e. venules.

12. In a fish, blood circulates through ____, while in a mammal, it circulates through ____ .
    a. two circuits . . . four circuits
    b. one circuit . . . two circuits
    c. four circuits . . . two circuits
    d. one circuit . . . four circuits
    e. two circuits . . . one circuit

13. The amount of blood flowing to skeletal muscles is greatly increased during exercise. This redirection of blood into muscles is accomplished by
    a. contraction of muscle in the walls of arteries.
    b. relaxation of muscle in walls of arterioles.
    c. opening of valves in veins.
    d. opening of valves in arteries.
    e. relaxation of muscle in the walls of veins.

Essay

1. Explain how the circulatory system changed to accommodate lung breathing and greater activity as land vertebrates evolved.

2. Cardiac output is the amount of blood the left ventricle of the heart pumps per minute. Cardiac output can be as much as four times greater when you are exercising than when you are at rest. What two things could the heart do to increase cardiac output when you are exercising?

3. How does high blood pressure contribute to cardiovascular disease?

4. Briefly describe and explain changes in blood pressure as blood flows from arteries to capillaries to veins.

5. Describe the structural characteristics of capillaries that make them well suited for exchange of materials between blood and tissues.

Apply the Concepts

Multiple Choice

1. Just after blood leaves the left ventricle of the human heart, it passes through the
   a. pulmonary artery.
   b. left atrium.
   c. aorta.
   d. superior vena cava.
   e. right ventricle.

2. In which of the following animals are blood and interstitial fluid the same?
   a. grasshopper
   b. flatworm
   c. fish
   d. dog
   e. sparrow

3. The heart specialist listened to Paul's heart through a stethoscope. Instead of the normal "lub-dupp, lub-dupp" heart sounds, he heard "siss-dupp, siss-dupp." The doctor said, "Hmm . . . I'm not sure it is anything to worry about, but I think there is something wrong with
   a. one of your coronary arteries."
   b. the pacemaker."
   c. an atrioventricular valve."
   d. your aorta."
   e. a semilunar valve."

4. Which of the following terms would be *least* useful in describing the circulatory system of a fish?
   a. capillary bed
   b. pulmonary artery
   c. ventricle
   d. atrium
   e. cardiovascular system

5. In circulating by the shortest route from the lungs to the foot, how many times would a drop of blood pass through the left ventricle?
   a. 0
   b. 1
   c. 2
   d. 3
   e. 4

6. A recording of the electrical activity of a patient's heart shows that the atria are contracting regularly and normally, but every few beats the ventricles fail to contract. Which of the following is probably not functioning properly?
   a. AV node
   b. semilunar valve
   c. coronary artery
   d. pacemaker
   e. AV valve

7. Which of the following functions most like a valve in a vein?
   a. a kitchen faucet
   b. a revolving door
   c. the volume control on a radio
   d. a subway turnstile
   e. a sliding patio door

8. In circulating from the brain to the arm, a drop of blood would *not* have to pass through which of the following?
   a. left atrium
   b. aorta
   c. superior vena cava
   d. pulmonary vein
   e. inferior vena cava

9. Emphysema damages the tissues of the lungs and slows pulmonary blood flow. This causes blood to back up, stretching and weakening the walls of the heart and blood vessels. Which of the following do you think would be most affected by this backup of blood from the lungs?
   a. aorta
   b. right atrium
   c. left atrium
   d. right ventricle
   e. left ventricle

10. The hormone erythropoetin (EPO) could be used to
    a. prepare a mountain climber for high altitude
    b. treat hemophilia
    c. increase the number of leukocytes to fight an infection
    d. prevent blood clotting in a stroke patient
    e. stimulate the growth of new blood vessels in a heart attack victim

Essay

1. A runner's heart rate is 160 beats per minute, and 90 mL of blood is pumped by the left ventricle with each beat. What is the runner's cardiac output?

2. The figures for blood pressure in an artery are usually given like this: 130/80 mm Hg. But it takes only one figure to specify blood pressure in a vein: 5 mm Hg. Why the difference?

3. Recall the forces that cause fluid to leave and reenter a capillary. How do you think high blood pressure would affect this balance of forces? How does this help explain that one of the symptoms of high blood pressure is swelling of the tissues with fluid?

4. Sometimes a baby is born with its large blood vessels reversed: The right ventricle pumps blood out through the aorta, and the left ventricle is connected to the pulmonary artery. The system is otherwise normal. How would this alter blood flow? Why would this be disastrous if not quickly corrected by surgery before or immediately after birth?

Put Words to Work

Correctly use as many of the following words as possible when reading, talking, and writing about biology:

anemia, aorta, arteriole, artery, atherosclerosis, AV (atrioventricular) node, atrium (plural, *atria*), *blood, blood pressure, capillary, capillary bed, cardiac cycle, cardiac output, cardiovascular disease, cardiovascular system, circulatory system, closed circulatory system, diastole, double circulation, erythrocyte, erythropoietin (EPO), fibrin, fibrinogen, heart, heart attack, heart murmur, heart rate, hypertension, inferior vena cava, leukemia, leukocyte, open circulatory system, pacemaker, phagocyte, plasma, platelet, pulmonary artery, pulmonary circuit, pulmonary vein, pulse, red blood cell, SA (sinoatrial) node, stem cell, stroke, superior vena cava, systemic circuit, systole, vein, ventricle, venule, white blood cell*

Use the Web

There are a variety of activities and questions reviewing circulatory system structure and function on the Web at www.mybiology.com. The animations of the heart cycle and path of blood flow are particularly helpful.

The Immune System

24

Focus on the Concepts

The immune system defends the body against invaders. While studying the immune system, focus on the following concepts:

- Both invertebrates and vertebrates have innate immunity, a first line of defense, which acts the same whether or not an invader has been previously encountered. Innate defenses include barriers such as the body covering, hairs, mucus, and cilia. If invaders get through, they can be attacked by enzymes, engulfed by phagocytes, or burst by the complement system. Interferons signal cells of virus attack. The lymphatic system is a battleground in an infection. The inflammatory response disinfects and cleans tissues.

- The acquired immune response of vertebrates is a second line of more specific defenses, activated only after exposure to pathogens. Foreign antigens trigger specific cells to attack the invader directly or secrete antibody molecules that attach to an antigen and inactivate it. Acquired immunity is specific to an invader, has a long "memory," and responds quickly and vigorously to subsequent exposures.

- Acquired immunity is usually obtained by natural exposure to—infection by—pathogens. Vaccination with a vaccine containing antigens can generate acquired immunity without disease. Either way, this is called active immunity because the individuals' own immune system actively produces antibodies. Antibodies acquired via injection or mothers' milk confer temporary passive immunity.

- B and T lymphocytes of the acquired immune response mount a dual defense: B cells deal with bacteria and viruses in body fluids by secreting antibodies into blood and lymph. Cytotoxic T cells carry out the cell-mediated immune response by attacking body cells infected with pathogens. Helper T cells mediate both responses.

- Antigens are molecules that elicit an acquired immune response—toxin molecules or surface molecules of pathogens. An antigen encounters a diverse pool of T and B lymphocytes, but a particular antigen only "selects" cells with matching receptors. These cells proliferate, forming a clone of cells specific for that antigen. This is called clonal selection.

- In the humoral immune response, an antigen stimulates specific B cells to form a clone of plasma cells, which produce antibodies that circulate in body fluids. The antigen-binding sites of Y-shaped antibody molecules attach to antigens, clumping bacteria and toxins and marking them for destruction by complement and phagocytes. The primary immune response is immediate. Memory B cells are held in reserve; they can carry out a faster, more vigorous secondary response for long-term immunity.

- In the cell-mediated immune response, a clone of cytotoxic T cells is mobilized to destroy cells infected with viruses or bacteria. Memory T cells are held in reserve.

- Both humoral and cell-mediated immunity are activated when a macrophage presents foreign antigens to helper T cells. The helper T cells then activate B and T cells. The HIV virus causes AIDS by targeting helper T cells, which cripples immune responses and leaves the body open to opportunistic infections and cancer. HIV is hard to stop because the virus evolves rapidly.

- Genes called the MHC code for key "self" proteins that identify the body's own cells. In autoimmune diseases such as MS and rheumatoid arthritis, the immune system turns against the body's own molecules. In immunodeficiency diseases such as SCID and AIDS, components of the immune system are damaged, opening the way for infection. Allergies are overreactions to antigens that are normally harmless.

Review the Concepts

Work through the following exercises to review the concepts in this chapter. For additional review, check out the activities on the Web at www.mybiology.com. The website offers a pre-test that will help you plan your studies.

Exercise 1   (Modules 24.1–24.3)

Innate immunity is a mammal's first line of defense against invaders. (It is the only line of defense in invertebrates). It is nonspecific, resisting any invader. Match each of the following components of innate immunity with a phrase from the right.

_____ 1. Natural killer cell
_____ 2. Lymph node
_____ 3. Macrophage
_____ 4. Interferons
_____ 5. Low pH (acid) in stomach
_____ 6. Complement system
_____ 7. Mucus
_____ 8. Hair
_____ 9. Neutrophil
_____ 10. Clotting proteins and platelets
_____ 11. Inflammatory response
_____ 12. Skin
_____ 13. Lymph
_____ 14. Lysozyme
_____ 15. Cilia
_____ 16. Histamine

A. Proteins that burst bacteria
B. Seal off infected region
C. Protective enzyme that digests bacteria
D. Barrier that is first line of defense
E. Large phagocytic cell of interstitial fluid
F. Chemical "alarm signal" released by damaged cells.
G. Abundant white blood cell that engulfs bacteria
H. Attacks cancer cells and cells infected by viruses
I. Proteins that help nearby cells resist viruses
J. Triggered by histamine, it disinfects and cleans injured tissues.
K. Fluid that carries microbes to lymphatic organs
L. Sweep mucus out of respiratory tract
M. May become swollen due to infection-fighting activity
N. Filters inhaled air
O. Traps microbes and dirt that get past nasal filter
P. Protects digestive system

Exercise 2   (Module 24.2)

Study this module on the inflammatory response. Then, from memory, fill in the missing words in this diagram.

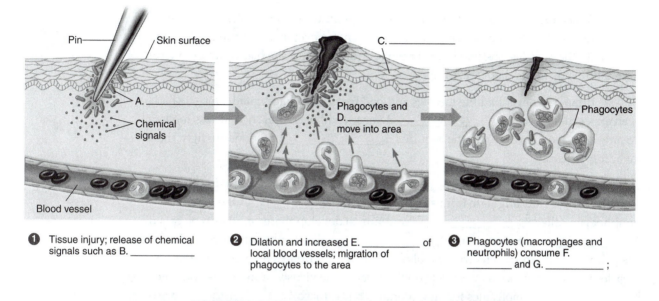

1 Tissue injury; release of chemical signals such as B. _____

2 Dilation and increased E. _____ of local blood vessels; migration of phagocytes to the area

3 Phagocytes (macrophages and neutrophils) consume F. _____ and G. _____ ;

Exercise 3   (Modules 24.3 – 24.5)

These modules introduce some basic concepts and terminology related to the immune and lymphatic systems. To test your knowledge of this material, describe the difference between the terms in each of the following pairs.

1. innate immunity—acquired immunity

2. antigen—antibody

3. T cell—B cell

4. humoral immunity—cell-mediated immunity

5. lymphatic system—circulatory system

6. active immunity—passive immunity

7. blood capillary—lymphatic capillary

8. hepatitis vaccine—plasma containing hepatitis antibodies

What is immunity? Why do you sometimes have to get sick to become immune to a disease? To answer these questions, fill in the blanks in the following story, using words and phrases from Modules 24.4 – 24.8.

Brian stopped by one afternoon to visit his friends Tom and Alicia and their 2-year-old daughter Samantha. Little did any of them suspect that Samantha was coming down with the measles. She had not yet been immunized, and she had caught the virus from a little boy in her day-care group.

Brian had never had the measles and had never been vaccinated. Once he had been exposed and his [1]_____ defenses had been breached, it was too late to do anything. Now it was up to the acquired [2]_____ response to fight the invading viruses.

White blood cells called [3]_____ carry out the acquired immune response. These cells originate from stem cells in [4]_____. One type of lymphocyte, called [5]_____, continue to develop there; another type called [6]_____ mature in the [7]_____ gland in the upper part of the chest. Both B cells and T cells were involved in defending Brian's body against the measles virus, but we will concentrate on the B cells. The viruses' outer coats contained [8]_____ molecules foreign to Brian's body; these [9]_____ are what triggered his immune response. Throughout his body were many different types of B cells, each capable of responding to a different antigen. On each of these B cells were [10]_____ that acted as receptors for various potential foreign antigens. Only a specific B cell type possessed antibodies whose [11]_____ were complementary to the shape of antigenic [12]_____ on the measles virus antigens. Eventually, the measles viruses encountered some of these B cells in a [13]_____ node. Only these cells were "chosen" to be activated to fight the invading viruses. This process is called clonal [14]_____.

The stimulated B cells began to multiply, forming a [15]_____, a population of genetically identical [16]_____ cells. When B cells do this, the activated cells are called [17]_____ cells. These cells secrete [18]_____ capable of locking on to the viral antigens and inactivating the measles viruses.

The initial phase of immunity described so far is called the [19]_____ immune response. Unfortunately, Brian had to suffer through the measles, because this response is too slow to wipe out the invaders before they cause harm. It usually takes several [20]_____ for lymphocytes to become activated and form effector cell clones, and by that time Brian was already sick.

After he recovered, however, Brian was [21]_____ to measles. When his college roommate came down with measles just before final exams, Brian was safe. A second exposure to the same antigen triggers the [22]_____ immune response, which is much quicker and stronger than the primary response and also lasts longer. The explanation for this is that the first exposure actually triggers the formation of two cell clones, the effector cells that fought the original infection and also a clone of

23 _____ cells, which are held in reserve. Whereas effector cells may live only a few days, these other cells may last for 24 _____. They are capable of mounting a quick and powerful secondary response.

Although Brian was protected from the measles, he still caught a cold the day before his math exam. But that is another story entirely.

Exercise 5   (Module 24.7)

Review the role of B cells in immunity by labeling and coloring the following diagram. First identify and label the **primary immune response** and the **secondary immune response.** Label **antigens,** and color them red. Label **B cells,** and color them yellow. Draw antigen receptors on the B cell types, and make them different colors on each type. Label **effector cells,** and color them dark blue. Label and color purple **antibodies** produced by effector cells. Label **memory cells,** and color them yellow. When you are done, explain the diagram.

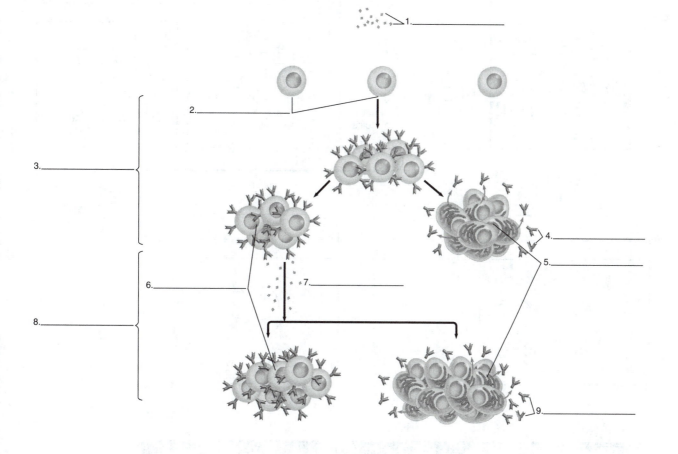

Exercise 6 (Modules 24.8–24.9)

The variable regions at the tips of an antibody's "arms" recognize antigens, and the constant region in the base of the Y helps destroy and eliminate antigens. As shown in Figure 24.9 in the text, there are a number of ways in which antibodies assist in destruction of antigens. These processes have been scrambled in the diagrams below. Identify each process, and label the diagrams accordingly. Choose from **phagocytosis, precipitation of dissolved antigens, neutralization, cell lysis, activation of complement,** and **agglutination of microbes.**

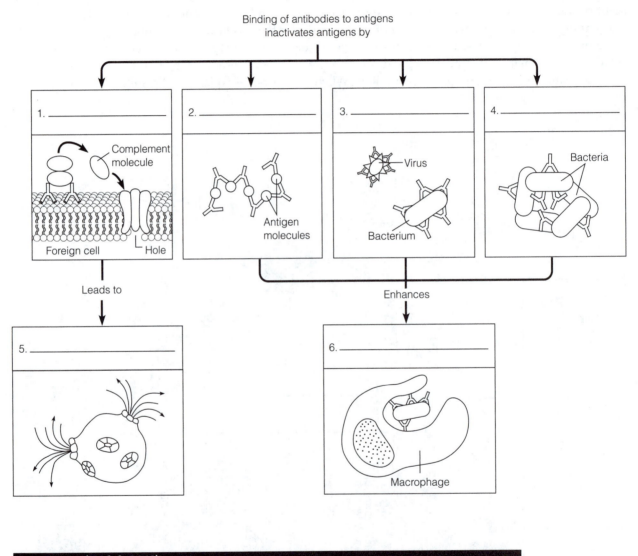

Exercise 7 (Module 24.10)

The usefulness of monoclonal antibodies stems from their ability to recognize and attach to specific molecules, which otherwise might be difficult to single out. See if you can think of several areas not mentioned in the text where this ability might be useful, and write your ideas on a separate piece of paper. Try not to limit your speculations to medicine. For example, how might monoclonal antibodies be useful in criminology, wildlife management, or archaeology?

Exercise 8   (Modules 24.11–24.14)

T cells are responsible for cell-mediated immunity. Contrast this arm of the immune system with humoral immunity, described in Module 24.9. Match each of the following components with its role in cell-mediated immunity (A–J) and then find each component on the diagram (P–Y).

Role   Diagram
(A–J)   (P–Y)

Role (A–J)	Diagram (P–Y)		
_____	_____	1. Interleukin-1	A. Protein that recognizes antigens
_____	_____	2. Antigen-presenting cell	B. Cell that secretes antibodies (humoral immunity)
_____	_____	3. Helper T cell	C. Chemical signal that activates T sand B cells
_____	_____	4. T-cell receptor	D. Macrophage that displays antigen
_____	_____	5. Self protein	E. Evokes immune response
_____	_____	6. Perforin	F. Cell that helps activate T and B cells
_____	_____	7. Interleukin-2	G. Body's own molecule that displays foreign antigens
_____	_____	8. Cytotoxic T cell	H. Chemical signal from APC to T cell
_____	_____	9. Antigen	I. Cell that attacks infected cells
_____	_____	10. B cell	J. Chemical that ruptures infected cells

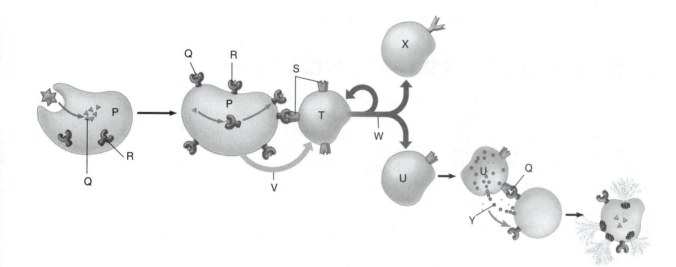

Exercise 9 (Modules 24.13–24.14)

Understanding the immune system is important and useful. Understanding HIV and AIDS could save your life. After reading about HIV and AIDS, choose the correct word or phrase to complete each sentence.

1. A virus called _____ causes AIDS.
2. HIV most often attacks _____.
3. The disease is most common in southern Asia and _____.
4. Infection hampers both the _____ and the humoral immune response.
5. The virus is transmitted to a new host via _____.
6. Inside a host cell, the _____ genome of the virus is reverse-transcribed.
7. The virus genome becomes part of the host cell's _____.
8. The damaging effects of virus _____ often kill the host cell.
9. As HIV spreads, the immune system becomes less able to fight _____.
10. A type of pneumonia and a rare _____ often occur.
11. There is no _____ for HIV or AIDS
12. Certain _____ can slow HIV reproduction, but are expensive.
13. Some HIV _____ are becoming resistant to drugs.
14. The virus has a high rate of _____ that helps it adapt.
15. The best way to stop AIDS is stop the _____ of HIV.
16. The best way to stop HIV transmission is through _____.

A. DNA
B. body fluids
C. Africa
D. transmission
E. T cells
F. reproduction
G. drugs
H. mutation
I. RNA
J. cell-mediated
K. cancer
L. infection
M. HIV
N. safe sex
O. cure
P. strains

Exercise 10 (Modules 24.13–24.17)

The body is supposed to be able to distinguish self from nonself. Genes called the major histocompatibility complex (MHC) code for self-protein labels. But sometimes the immune system ignores these labels and attacks the body's own tissues. Sometimes the system fails to respond to invaders. Sometimes it goes overboard, responding to harmless substances in the environment. Briefly describe how the normal function of the immune system goes awry in each of the following situations.

1. Severe combined immunodeficiency (SCID)

2. An allergy to cat hair

3. Radiation therapy for cancer leads to pneumonia

4. Rejection of a heart transplant

5. Rheumatoid arthritis

6. AIDS

7. Crohn's disease

Test Your Knowledge

Multiple Choice

1. An antigen is
   a. a protein molecule that helps defend the body against disease.
   b. a type of white blood cell.
   c. an invading virus or bacterium.
   d. a foreign molecule that evokes an immune response.
   e. a body cell attacked by an invading microorganism.

2. Your lymphatic system fights infection and
   a. delivers food and water to tissues.
   b. carries glandular secretions.
   c. maintains high blood pressure.
   d. allows red blood cells to approach cells more closely.
   e. drains fluid from tissues.

3. How do memory cells differ from effector cells?
   a. Memory cells are more numerous.
   b. Memory cells are responsible for the primary immune response.
   c. Memory cells attack invaders, and effector cells do not.
   d. Memory cells live longer.
   e. Memory cells are capable of producing antibodies.

4. Which of the following triggers tissue inflammation?
   a. accumulation of phagocytes in an injured area
   b. release of interferon by infected cells
   c. increased blood flow in an infected or injured area
   d. fever
   e. release of chemicals such as histamine by damaged cells

5. A cell capable of producing monoclonal antibodies is produced by fusing a lymphocyte with a
   a. tumor cell.
   b. red blood cell.
   c. bone marrow cell.
   d. T cell.
   e. macrophage.

6. A clone of lymphocytes
   a. produces a variety of antibodies.
   b. lives in a specific area of the body.
   c. consists of immature cells, incapable of carrying out an immune response.
   d. makes antibodies against the same antigens.
   e. consists of both B cells and T cells.

7. Individuals infected with HIV
   a. can live for 15 to 20 years without symptoms.
   b. have little chance of developing AIDS.
   c. often die from autoimmune reactions.
   d. suffer from increased sensitivity to foreign antigens.
   e. can die from other infections.

8. Tissues are "typed" before an organ transplant to make sure that the _____ of donor and recipient match as closely as possible.
   a. T cells
   b. antibodies
   c. MHC markers
   d. histamines
   e. B cells

9. A vaccine contains
   a. white blood cells that fight infection.
   b. antibodies that recognize invading microbes.
   c. inactivated disease-causing microbes.
   d. a hormone that boosts immunity.
   e. lymphocyte antigens.

10. When you are immune to a disease
    a. antibodies against the disease are constantly circulating in your blood.
    b. certain lymphocytes are able to make the proper antibodies quickly.
    c. your innate defenses are strengthened.
    d. B cells are stimulated to quickly engulf invaders.
    e. antigens are altered so invaders can no longer attack your tissues.

11. An antibody is a
    a. protein that attaches to an antigen.
    b. foreign substance or organism.
    c. white blood cell that attacks invading bacteria or viruses.
    d. molecule on a body cell that identifies the cell as "self."
    e. large carbohydrate molecule that helps defend the body.

12. B lymphocytes
    a. attack cells that have been infected by viruses.
    b. engulf and destroy bacteria and viruses.
    c. multiply and make antibodies that circulate in blood and lymph.
    d. are responsible for cell-mediated immunity.
    e. do all of the above.

13. HIV infects mostly
    a. cells of the nervous system.
    b. B lymphocytes.
    c. cytotoxic T cells.
    d. macrophages.
    e. helper T cells.

14. The biggest difference between cell-mediated immunity and humoral immunity is
    a. how long their protection lasts.
    b. whether a subsequent secondary immune response can occur.
    c. whether clonal selection occurs.
    d. how they respond to and dispose of invaders.
    e. how fast they can respond to an invader.

15. Viruses and bacteria in body fluids are attacked by
    a. antibodies from B cells.
    b. cytotoxic T cells.
    c. MHC markers.
    d. helper T cells.
    e. antigens.

16. What do the antibodies secreted by plasma cells (the effector cells of humoral immunity) do to attack their targets?
    a. activate complement to punch holes in them
    b. clump cells together so that phagocytes can ingest them
    c. cause antigen molecules to settle out of solution
    d. attach to antigens and detoxify them
    e. all of the above

17. Which of the following is *not* part of the body's innate defenses?
    a. interferons
    b. helper T cells
    c. complement proteins.
    d. macrophages
    e. natural killer cells

Essay

1. Explain how vaccination allows you to develop immunity to a disease without becoming ill. Include in your explanation the primary and secondary immune responses.

2. Compare humoral and cell-mediated immunity. In your comparison, discuss types of lymphocytes involved, roles of antibodies in the immune response, where invaders are attacked, and methods used to destroy invaders.

3. When biologists first started to work out the mechanisms of acquired immunity, they found that the body produced antibodies that matched the shape of invading antigens. The researchers suspected that they would find that the immune system somehow analyzed the antigens and custom-built antibodies to fit. Did they find this to be the case? Explain.

4. Andrea was telling her friend Jason why she had not been able to make it to biology class for several days: "My throat has been so sore I could hardly swallow," she croaked. "And the glands in my neck are really sore and swollen." Jason said, "We have been talking about this in class. They are not really 'glands,' you know, and the reason they are sore is . . ." Complete Jason's explanation.

5. Describe the inflammatory response and how it helps the body deal with injury or infection.

Apply the Concepts

Multiple Choice

1. Which of the following is not present until after the primary immune response occurs?
    a. memory cells
    b. macrophages
    c. helper T cells
    d. complement proteins
    e. antigens

2. The relationship between an antigen and an antibody is most like
    a. a battery and a flashlight.
    b. a hand and a glove.
    c. a hammer and a nail.
    d. a left foot and a right foot.
    e. a recipe and a cake.

3. A group of researchers have tested many chemicals and found several that have potential for use in treating malfunctions of the immune system. Which of the following would seem to have the most promise as a drug for inhibiting autoimmune diseases?
    a. Compound A13: acts like histamine
    b. Compound Q6: stimulates cytotoxic T cells
    c. Compound N98: a potent allergen
    d. Compound B55: suppresses specific cytotoxic T cells
    e. Compound M31: stimulates helper T cells

4. The body produces antibodies complementary to foreign antigens. The process by which the body comes up with the correct antibodies to a given disease is most like
   a. going to a tailor and having a suit made to fit you.
   b. ordering the lunch special at a restaurant without looking at the menu.
   c. going to a shoe store and trying on shoes until you find a pair that fits.
   d. picking out a video that you haven't seen yet.
   e. selecting a lottery prize winner by means of a random drawing.

5. Mari has been diagnosed as suffering from an immunodeficiency disease. Her doctor suspected Mari might have an immunodeficiency because
   a. Mari strongly rejected an organ transplant.
   b. Mari suffered from numerous allergies.
   c. Mari's blood showed high levels of numerous antibodies.
   d. Mari seemed to be immune to her own "self" molecules.
   e. Mari suffered from repeated, prolonged infections.

6. The idea behind vaccination is to induce _____ without the vaccinated individual having to get sick.
   a. passive immunity
   b. the primary immune response
   c. anaphylactic shock
   d. nonspecific defenses
   e. inflammation

7. Researchers found that when laboratory rats were already infected with a virus, they were better able to resist infection by a second completely different virus. The first infection apparently caused _____, which protected the rats from the second infection.
   a. increased stress
   b. secretion of interferons
   c. production of antibodies
   d. passive immunity
   e. cell agglutination

8. Which of the following would be effective in eliminating bacteria but ineffective against viruses? (Hint: Viruses are not cells.)
   a. activation of complement proteins
   b. secretion of interferon by infected cells
   c. neutralization by antibodies
   d. agglutination by antibodies
   e. perforin secretion by cytotoxic T cells

9. An allergen acts like
   a. an antigen.
   b. histamine.
   c. interferon.
   d. an antibody.
   e. complement.

10. In a series of immune system experiments, the thymus glands were removed from baby mice. Which of the following would you predict as a likely result?
    a. The mice suffered from numerous allergies.
    b. The mice never developed cancerous tumors.
    c. The mice suffered from autoimmune diseases.
    d. The mice readily accepted tissue transplants.
    e. The mice were unable to produce an inflammatory response.

Essay

1. More people die each year from bee stings than from rattlesnake bites. The individual who is stung dies of anaphylactic shock, a massive, life-threatening allergic response. Anaphylactic shock usually occurs the second time an individual is stung by a bee. Why does this occur the second time and not the first time?

2. Researchers have not yet come up with a cure for the common cold, but they have made some interesting observations, among them: (1) There are more than 50 different known kinds of rhinoviruses, the viruses that cause colds, and (2) an average 2-year-old might catch three or four colds per year, but an 80-year-old catches a cold only once every three or four years. How do these findings relate to each other and the function of the immune system?

3. Before traveling to Africa, John got a "gamma-globulin" injection containing antibodies against hepatitis. Two years later, John planned another trip to Africa. His doctor recommended another gamma-globulin injection because John was no longer immune to hepatitis. What kind of immunity did John acquire from the first injection? Why didn't the immunity last?

4. After the flu season, blood samples were obtained from flu patients, and antibodies in the blood were analyzed. Researchers found that the antibodies produced by different patients in response to the same virus were often quite different. In some cases, the antigen-binding sites of the antibodies were completely different shapes. Explain how this could occur.

5. Antibodies are Y-shaped, with antigen-binding sites at the tip of each "arm" of the Y. How does the fact that an antibody has two antigen-binding sites help the antibody inactivate invaders? (Think what would happen if they had only one antigen-binding site.)

6. Organ transplant recipients must be very careful to avoid infections, especially viral infections. Explain why.

Put Words to Work

Correctly use as many of the following words as possible when reading, talking, and writing about biology:

acquired immunity, active immunity, AIDS, allergy, allergen, anaphylactic shock, antibody, antigen, antigen receptor, antigen-binding site, antigen-presenting cell (APC), antihistamine, autoimmune disease, B cell, cell-mediated immune response, clonal selection, complement system, cytotoxic T cell, effector cell, helper T cell, histamine, HIV, humoral immune response, immune system, immunodeficiency disease, inflammatory response, innate immunity, interferon, lymph, lymphatic system, lymphocyte, macrophage, major histocompatibility complex (MHC), memory cell, monoclonal antibody, nonself molecule, opportunistic infection, passive immunity, pathogen, phagocytosis, plasma cell, primary immune response, secondary immune response, self protein, T cell, vaccination, vaccine

Use the Web

Be sure to access the activities and questions pertaining to the immune system on the Web at www.mybiology.com. The animations of immune system function are particularly helpful.

Control of Body Temperature and Water Balance

25

Focus on the Concepts

Three important aspects of homeostasis are temperature control, regulation of water and solute balance, and disposal of wastes. This chapter describes the homeostatic mechanisms of thermoregulation, osmoregulation, and excretion. Focus on these concepts:

- Internal metabolism and the outside environment provide heat for thermoregulation. Endotherms such as mammals and birds are warmed mostly by heat generated by metabolism. Most fishes, amphibians, reptiles, and invertebrates are ectotherms, who get most of their heat from external sources.

- An animal can exchange heat with the environment via conduction, convection, radiation, or evaporation. Thermoregulatory adaptations include regulation of heat production, insulation, circulatory structures and responses, evaporative cooling, and behavioral responses.

- Osmoregulation balances the uptake and loss of water and solutes. Many marine invertebrates are osmoconformers; they do not undergo net loss or gain of water, but do regulate certain ions in body fluids. Other animals are osmoregulators, maintaining water and solute concentrations different from their surroundings. A freshwater fish tends to take on water by osmosis; it excretes the excess in dilute urine and replaces salts via the gills and food. A saltwater fish tends to lose water; it gains water and salt from food and drinking seawater and excretes excess salts via the gills and scanty urine. Land animals are osmoregulators and face problems similar to saltwater fish.

- Metabolism generates toxic wastes, especially nitrogenous wastes from protein breakdown; these are dealt with via the process of excretion. Most aquatic animals excrete ammonia; it is toxic but soluble. Land animals cannot excrete ammonia; it is too toxic and must be diluted too much. Mammals and some other animals convert ammonia into urea; this takes energy, but urea is much less toxic and highly soluble, so it does not take much water to excrete. Birds, other reptiles, and insects secrete uric acid. Making uric acid requires even more energy, but it is insoluble and saves water.

- In humans, fluid homeostasis and waste excretion are the jobs of the two kidneys, the processing centers of the urinary system. Each kidney consists of a million tiny units called nephrons. Each nephron consists of a nephron tubule and associated capillaries. Blood is processed by the nephrons, and wastes, water, and other substances pass though the tubules to the ureters, the bladder, and out of the body via the urethra.

- A nephron "refines" blood in a four-step process: (1) In filtration, water and other small molecules are forced though the capillary wall, forming a fluid called filtrate. (2) In reabsorption, valuable solutes are returned to the blood. (3) Next comes secretion, where certain substances in the blood are transported into the filtrate. (3) Finally, in excretion, the products of the first three processes—now called urine—are expelled from the kidney to outside the body.

Review the Concepts

Work through the following exercises to review the concepts in this chapter. For additional review, check out the activities on the Web at www.mybiology.com. The website offers a pre-test that will help you plan your studies.

Exercise 1   (Introduction–Module 25.1)

A black bear has many adaptations that help in thermoregulation—maintenance of its internal body temperature within narrow limits. Bears are endotherms, animals that derive most of their body heat from metabolism. Many other animals are ectotherms, which absorb heat from their surroundings. Which of the following are endotherms and which are ectotherms?

_____ 1. Most fishes
_____ 2. A squirrel
_____ 3. Most invertebrates
_____ 4. A frog
_____ 5. Some insects
_____ 6. Some fishes
_____ 7. An eagle
_____ 8. A lizard
_____ 9. Some reptiles
_____ 10. A sparrow

Exercise 2   (Modules 25.2–25.3)

Does the room temperature where you are right now feel comfortable? Do you feel a bit chilly? Or are you sweating because it is hot and humid? These modules discuss thermoregulation, and they provide numerous examples of methods animals use to regulate their internal temperatures. Animals regulate temperatures two ways: (1) by changing rate of heat production and (2) by adjusting rate of heat gain or loss. In the first column, state whether each of the following is:

A. A method of warming or cooling the body by regulating heat production
B. A method of warming the body by reducing heat loss
C. A method of warming the body by increasing heat gain
D. A method of cooling the body by increasing heat loss
E. A method of cooling the body by decreasing heat gain

Then, in the second column, identify which of the following adaptations is involved in each case:

P. Metabolic heat production
Q. Insulation
R. Circulatory adaptations
S. Evaporative cooling
T. Behavioral responses

Method   Adaptation

_____ _____   1. Moisture evaporates from a lizard's nostrils.
_____ _____   2. A robin fluffs up its feathers to trap more air near the skin.
_____ _____   3. A rabbit grows a thicker coat in the winter.
_____ _____   4. A lizard comes out of its burrow and turns broadside to the sun.
_____ _____   5. An elephant sprays itself with cold water.
_____ _____   6. Bees shiver.
_____ _____   7. Blood vessels dilate in a jackrabbit's ears.
_____ _____   8. You jump up and down and swing your arms to warm up on a cold day.
_____ _____   9. A desert kangaroo rat presses itself against the cool wall of its burrow.
_____ _____   10. A countercurrent heat exchanger captures heat from the blood flowing to a duck's feet.
_____ _____   11. Hormones increase a mouse's metabolic rate.
_____ _____   12. A network of vessels in a tuna retains warm blood, so muscles function better in cold water.
_____ _____   13. A cat licks itself, and saliva evaporates from its skin.
_____ _____   14. A bee seeks flowers that focus sunlight on its body.
_____ _____   15. A snake moves out of the sun and into the shade.
_____ _____   16. A chipmunk hibernates through the winter.

To stay alive, animals must osmoregulate to maintain a correct balance of water and solutes in their body fluids. The concentration of seawater is "close enough" for many marine animals, so they are osmoconformers, simply matching the osmotic concentration of their environment. Other marine creatures, as well as freshwater and land animals, are osmoregulators. They actively move solutes and water in and out of their cells to maintain body fluid compositions different from their environments. Land animals have particular problems with loss of water and solutes. Use the information in the modules to complete this chart comparing osmoconformers and various osmoregulators.

	Marine Worm	Freshwater Fish	Saltwater Fish	Human
Osmoregulator or osmoconformer?	1.	2.	3.	4.
Tends to gain or lose water?	5.	6.	7.	8.
Tends to gain or lose solutes (ions)?	9.	10.	11.	12.
Method of compensating for gain or loss?	13.	14.	15.	16.

Breakdown of proteins and nucleic acids produces nitrogen-containing waste products. Different animals dispose of nitrogen in different ways. Summarize the kinds of animals that excrete each of the following nitrogenous wastes, and discuss the advantages and disadvantages of each.

	Ammonia	Urea	Uric Acid
Animals excreting this compound	1.	2.	3.
Advantages of excreting this compound	4.	5.	6.
Disadvantages of excreting this compound	7.	8.	9.

Exercise 5   (Module 25.6)

The human excretory system performs important functions in fluid homeostasis. This exercise and the next will help you to become familiar with the kidneys and nephrons, their functional units. First, complete this diagram of the excretory system. Label a **renal artery,** and color it red. Label a **renal vein,** and color it blue. Label the **kidneys,** and color them brown. Then label and color the **ureters** (light brown), the **bladder** (yellow), and the **urethra** (green).

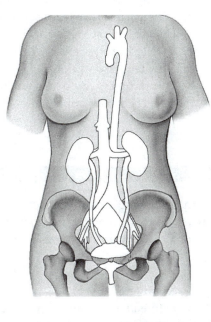

Exercise 6   (Module 25.6)

Note in Figure 25.6 in the text how many tiny nephrons are arranged in each kidney. Label and color the following diagram of a nephron. Color arterioles and arterial capillaries red; color venous capillaries and the renal vein blue; and color the entire renal tubule, from Bowman's capsule through the collecting duct, yellow. Label **arteriole from renal artery, glomerulus, arteriole from glomerulus, capillaries, Bowman's capsule, proximal tubule, loop of Henle, distal tubule,** and **collecting duct.**

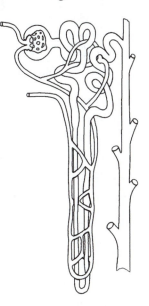

Exercise 7   (Modules 25.7–25.8)

Nephrons regulate the water and solute content of blood in a four-step process: filtration, reabsorption, secretion, and excretion. The next two exercises review nephron function, which is the heart of this chapter. After studying the modules, match each of the following parts of a nephron with its function.

_____ 1. Loop of Henle
_____ 2. Bowman's capsule
_____ 3. Proximal and distal tubules
_____ 4. Glomerulus
_____ 5. Collecting duct
_____ 6. Renal artery

A. Mainly functions in water reabsorption
B. "Refine" filtrate by reabsorption and secretion
C. Delivers blood to glomerulus
D. Carries urine to central "pelvis" of kidney; also functions in water reabsorption
E. Porous ball of capillaries where filtration occurs
F. Collects filtrate from glomerulus

Exercise 8   (Modules 25.7–25.8)

After studying kidney function, match each of the following components of blood with what happens to it in a nephron. (Hint: Ask yourself, "Is this something the body generally wants to keep, or something it wants to get rid of?") (Answers may be used more than once.)

_____ 1. Glucose
_____ 2. Water
_____ 3. Urea
_____ 4. H^+
_____ 5. Plasma protein
_____ 6. Amino acid
_____ 7. Red blood cell
_____ 8. Nicotine (drug)
_____ 9. Salt (NaCl)

A. Not filtered
B. Filtered and mostly reabsorbed
C. Filtered, then mostly *not* reabsorbed, and finally excreted in urine
D. Filtered, mostly *not* reabsorbed, *also* secreted, and finally excreted in urine

Exercise 9   (Module 25.9)

Kidney dialysis has saved many lives, often as a stopgap until a kidney transplant can be arranged.

1. What are some advantages and disadvantages of dialysis, compared with a transplanted kidney?

2. What do you think would be some advantages and disadvantages of a kidney transplant, compared with dialysis?

Exercise 10   (Modules 25.5 – 25.8)

Review by filling in the blanks in the following scenario summarizing the role of the kidneys in regulating body fluid composition.

Tom finished his breakfast, downed his fourth cup of coffee (with milk and sugar), and dashed off to work. Digestion of milk proteins began immediately. The resulting amino acids were absorbed through the intestinal lining into Tom's blood. At the same time, enzymes split the sucrose (sugar) molecules, forming glucose and fructose, which also entered the blood. Caffeine and water molecules and some trace minerals entered the blood immediately, without digestion.

Tom's heart pumped the blood from the intestine throughout his body. Some of the caffeine from the coffee had the intended "wake-up" effect on Tom's brain as he was driving to work. Like other potentially harmful chemicals, the rest of the caffeine was processed by liver cells, and the resulting less-toxic products were released back into the blood.

Some of the amino acids from the milk were carried by the blood to cells that could use them to make proteins. Other amino acids were broken down for energy or converted to other nutrients. Leftover amino groups were first changed into highly toxic ^{1}_____, but the liver cells quickly converted this into ^{2}_____, a much safer nitrogenous waste. This was also dumped into the blood exiting the liver.

The blood processed by the liver, with its cargo of glucose, amino acids, caffeine by-products, urea, and H^+ ions from the acid in the coffee, flowed via veins to the heart. The heart pumped the blood to the lungs and back and then out through the aorta, whose branches distributed the blood to body tissues. About one-fifth of the blood pumped by each heartbeat flowed through the ^{3}_____ arteries to the kidneys.

In each of Tom's kidneys, branches of the renal artery delivered blood to a million ^{4}_____, the kidney's tiny working units. As a drop of blood passed through a ^{5}_____, a porous knot of capillaries, blood pressure filtered some of the blood through the pores into a funnel-like structure called ^{6}_____. Blood cells and large protein molecules were left behind in the blood, while smaller molecules were filtered. The fluid that collected in Bowman's capsule, called ^{7}_____, contained some of the water, glucose, H^+, and amino acids from Tom's morning coffee, plus urea and caffeine by-products from the liver. But filtration was just the beginning of the work of the nephron. The processes of [8]_____ and [9]_____ had to occur before the blood could leave the kidney and waste products could be [10]_____ from the body in urine.

As Tom exited from the freeway, the walls of the proximal tubules were beginning to reabsorb the glucose and amino acids in the filtrate via the process of [11]_____. Uptake of these substances, and others such as salt, caused [12]_____ to be reabsorbed by osmosis, especially in the long [13]_____ of Henle and the collecting duct. These materials, valuable to the body, would reenter the blood and leave the kidney through the [14]_____.

Filtration is not usually enough to get rid of all the acid the body needs to excrete, so Tom's kidneys actually removed some H^+ from the blood and added it to the filtrate, a process known as [15]_____. Although some of the urea was reabsorbed, most of it, along with the caffeine by-products, remained in the refined filtrate, which is called [16]_____ when it leaves the kidney.

Because the caffeine in Tom's coffee is a stimulant that raises blood pressure, Tom's kidneys were filtering his blood at a slightly higher than normal rate as he drove into the company parking lot. Urine left the kidneys through the [17]_____ and accumulated in the [18]_____. Tom had just enough time to stop at the restroom on his way to his office.

Test Your Knowledge

Multiple Choice

1. Which of the following is an endotherm?
   a. mouse
   b. iguana
   c. frog
   d. trout
   e. all of the above except a

2. Which of the following describes the route of urine out of the body after it leaves the kidney?
   a. renal vein, bladder, urethra, ureter
   b. urethra, bladder, ureter
   c. renal vein, ureter, bladder, urethra
   d. ureter, bladder, urethra
   e. ureter, urethra, bladder

3. Some small birds can lower their body temperature for a few hours on a cold night. This is called
   a. torpor.
   b. dialysis.
   c. hibernation.
   d. ectothermy.
   e. estivation.

4. As your kidneys regulate your body fluid composition, the volume of which of the following is the largest?
   a. the volume of filtrate formed by the nephrons
   b. the volume of urine excreted
   c. the volume of blood flowing through the nephrons
   d. the volume of solutes added to the filtrate by secretion
   e. the volume of filtrate reabsorbed

5. A countercurrent heat exchanger enables an animal to
   a. produce more heat when needed.
   b. reduce loss of heat to the environment.
   c. slow metabolism when food is not available.
   d. increase heat loss by evaporation.
   e. absorb heat from the environment.

6. Uric acid is the nitrogenous waste excreted by birds, insects, and many reptiles. An advantage of excreting uric acid is that it ____, but a disadvantage is that it _____ .
   a. saves water . . . costs energy
   b. saves energy . . . is highly toxic
   c. is not very toxic . . . wastes a lot of water
   d. is much more soluble in water than other wastes . . . costs energy
   e. saves water . . . is highly toxic

7. Which of the following is the most accurate and comprehensive description of the function of the kidneys?
   a. breaking down body wastes
   b. excreting wastes
   c. regulating fluid composition
   d. filtering the blood
   e. producing urine

8. Which of the following is the first step that occurs as a nephron processes blood?
   a. excretion
   b. osmosis
   c. secretion
   d. reabsorption
   e. filtration

9. On a cold day, blood vessels in the skin
   a. dilate, allowing blood to keep the skin warm.
   b. constrict, forcing blood to flow through vessels to warm the skin.
   c. constrict, reducing heat loss from blood at the surface.
   d. dilate, causing blood to pass through the cold skin more quickly.
   e. dilate, preventing blood flow to the surface.

10. The animals in which of these pairs have similar problems regulating water balance?
    a. freshwater fish — saltwater fish
    b. land animal — freshwater fish
    c. osmoconformer — freshwater fish
    d. salmon in fresh water — salmon in salt water
    e. saltwater fish — land animal

11. The filtrate formed by the nephrons in the kidney is not the same as urine. The filtrate is first refined and concentrated by the processes of _____, forming the urine that leaves the body.
    a. filtration and secretion
    b. reabsorption and secretion
    c. reabsorption and excretion
    d. filtration and reabsorption
    e. secretion and excretion

12. Nitrogenous waste products are made from by-products of the breakdown of
    a. fats.
    b. starch.
    c. glucose.
    d. urea.
    e. proteins.

13. By definition, an ectotherm
    a. is "cold-blooded."
    b. is "warm-blooded."
    c. obtains most of its heat from its environment.
    d. derives most of its heat from its own metabolism.
    e. is none of the above.

14. Most aquatic animals excrete ammonia, while land animals excrete urea or uric acid. What is the most likely explanation for this difference?
    a. They have different diets.
    b. Land animals can get the energy needed to make urea or uric acid.
    c. Ammonia is very toxic, and it takes lots of water to dilute it.
    d. Land animals cannot afford the energy needed to make ammonia.
    e. Fish need to get rid of ammonia, but land animals need it to live.

15. A freshwater fish tends to _____ water by osmosis. As a consequence, its kidneys excrete _____ .
    a. gain . . . large amounts of dilute urine
    b. lose . . . small amounts of concentrated urine
    c. gain . . . large amounts of concentrated urine
    d. lose . . . large amounts of dilute urine
    e. gain . . . small amounts of concentrated urine

Essay

1. Describe how the skin can make adjustments that cool the body on a hot day and warm it on a cold day.

2. Explain how a goose can stand barefoot on ice without losing large amounts of body heat.

3. Compare the problems that freshwater and saltwater fish face in maintaining the water and solute balances of their body fluids. How does each kind of fish solve these problems?

4. Describe five behaviors that help different animals control their body temperatures.

5. The kidneys regulate body fluid composition by means of filtration, reabsorption, secretion, and excretion. Where does each of these processes occur? How does each contribute to the formation of filtrate and urine?

Apply the Concepts

Multiple Choice

1. Which of the following primarily involves heat transfer by convection?
   a. You roll down the car window to let the cool breeze blow through.
   b. The water in a lake is so cold that your legs become numb.
   c. You sweat profusely as you mow the lawn on a hot summer day.
   d. After sunset, you can feel heat from the warm pavement.
   e. As you lie on the sand, you can feel the sun's warm rays on your skin.

2. Which needs to drink the smallest amount of water to maintain its water balance?
   a. a sparrow
   b. a saltwater fish
   c. a freshwater fish
   d. a dog
   e. both a and b drink very small amounts

3. Humid weather makes you feel warmer because humid air
   a. interferes with heat loss by conduction.
   b. holds warm water vapor.
   c. interferes with heat loss by evaporation.
   d. prevents countercurrent heat exchange from occurring.
   e. increases metabolic heat production.

4. Pound for pound, a kidney uses as much energy as the heart. What do you think the energy is used for?
   a. to produce pressure for filtration
   b. for water reabsorption
   c. for the breakdown and detoxification of harmful substances
   d. to pump urine to the bladder
   e. for reabsorption and secretion of solutes by active transport

5. Look at the diagram of kidney dialysis, Figure 25.9, in the text. For kidney dialysis to work properly, the dialyzing solution should contain
   a. a higher solute concentration than blood.
   b. a higher concentration of urea than blood.
   c. a lower glucose concentration than blood.
   d. a lower concentration of urea than blood.
   e. a much smaller volume of fluid than the blood passing through it.

6. Which would have the toughest time surviving over the long term in the environment given?
   a. an osmoconformer in seawater
   b. an endotherm in a warm environment
   c. an ectotherm in a cold environment
   d. an osmoregulator in seawater
   e. an ectotherm in a warm environment

7. The loops of Henle in the kidneys of a desert kangaroo rat are much longer than those in a white laboratory rat because
   a. the kangaroo rat lives in an environment where water is scarce.
   b. the white rat's diet is much less varied than the kangaroo rat's.
   c. the kangaroo rat cannot always find food.
   d. the kangaroo rat produces more wastes.
   e. the kangaroo rat has less stress and lower blood pressure.

8. How does the filtrate produced by the filtration process of the glomerulus differ from urine? The filtrate
   a. contains very little water.
   b. contains a higher concentration of glucose.
   c. contains a much lower concentration of salt.
   d. contains a lower concentration of proteins.
   e. contains all of the above.

9. The kidney's filtration process is nonselective, so
   a. many valuable substances are lost in urine.
   b. the proportions of substances in urine are the same as in blood.
   c. urine is much less concentrated than blood.
   d. it really has little control over body fluid composition.
   e. useful substances must be selectively reabsorbed.

10. Which of the following would be filtered from the blood but not normally found in urine?
    a. water
    b. red blood cell
    c. H^+ ions
    d. amino acid
    e. urea

11. When you drink a lot of water, the brain signals for _____ in the amount ADH, which causes the kidney tubules to reabsorb _____ water, and the body excretes _____ urine
    a. a drop . . . more . . . dilute
    b. a drop . . . more . . . concentrated
    c. a drop . . . less . . . dilute
    d. an increase . . . less . . . concentrated
    e. an increase . . . less . . . dilute

Essay

1. In terms of heat loss, why is the wind chill factor given in a weather report always a lower temperature than the air temperature?

2. An animal behaviorist found that a large part of a horned lizard's daily routine consisted of behavior related to thermoregulation. Very little of the lizard's time was occupied by searching for food. A pocket mouse in the same environment spent most of its time seeking food but very little time thermoregulating. Explain this difference in behavior.

3. You may have noticed that you never see pictures of Antarctic crocodiles lying in ambush for unwary penguins or seals. Why are birds and mammals more successful in the polar regions than crocodiles or frogs?

4. Imagine that you have been exercising and sweating on a hot, dry day. How does your body conserve water?

Put Words to Work

Correctly use as many of the following words as possible, when reading, talking, and writing about biology:

ammonia, antidiuretic hormone (ADH), Bowman's capsule, collecting duct, countercurrent heat exchange, dialysis, distal tubule, ectotherm, endotherm, excretion, filtrate, filtration, glomerulus (plural, glomeruli), loop of Henle, nephron, osmoconformer, osmoregulation, osmoregulator, proximal tubule, reabsorption, renal cortex, renal medulla, secretion, thermoregulation, urea, ureter, urethra, uric acid, urinary bladder, urinary system, urine

Use the Web

There is more on control of body temperature and fluid homeostasis on the Web at www.mybiology.com. The illustrations and animations of kidney function are particularly helpful.

Hormones and the Endocrine System **26**

Focus on the Concepts

Chemical signals called hormones are important in coordinating body activities and maintaining homeostasis. In studying this chapter, focus on these concepts:

- There are various kinds of chemical signals. Endocrine glands secrete hormones that are carried by the blood to target cells. Local regulators are similar to hormones, but signal neighboring cells. Neurons secrete neurotransmitters to signal nearby neurons. Neurosecretory cells perform both functions by secreting their chemical signals into the blood. Pheromones are chemical signals between different individuals.

- In vertebrates, two classes of molecules function as hormones. Amino acid-derived hormones bind to receptors in the target cell membrane. This activates a set of relay molecules, which activate a protein that carries out a response. Steroid hormones enter the target cell and bind to a receptor inside. The receptor-hormone complex turns genes on or off. This has a somewhat slower, longer-lasting effect. A hormone can bind to a variety of receptors in various cells, so one hormone can have many effects.

- In vertebrates, there are more than a dozen endocrine glands. The hypothalamus connects the nervous and endocrine systems and controls many glands. Releasing and inhibiting hormones from the hypothalamus control secretion of several hormones from the anterior pituitary. Hormones of the posterior pituitary are produced in the hypothalamus, and nerve impulses from the hypothalamus trigger their secretion.

- Two hormones from the thyroid gland regulate development and metabolism. Secretion is triggered by thyroid-stimulating hormone from the anterior pituitary. Increase in thyroid hormones inhibits the hypothalamus and anterior pituitary, damping thyroid secretion. This kind of negative feedback maintains constant levels of many hormones. The thyroid and parathyroid glands regulate blood calcium.

- Antagonistic pancreatic hormones regulate blood glucose. When blood glucose rises, the pancreas increases its secretion of insulin, which causes cells to take in and store glucose. When blood glucose drops, insulin secretion drops and the pancreas puts out more glucagon. This causes the liver to release glucose from storage, maintaining blood glucose level.

- The adrenal gland mobilizes response to stress. Nerve impulses can quickly trigger the adrenal medulla to release the "fight-or-flight" hormones epinephrine and norepinephrine, which raise blood pressure, breathing rate, and metabolic rate. In response to longer-term stress, the hypothalamus signals the anterior pituitary to release ACTH, which stimulates the adrenal cortex to secrete mineralocorticoids and glucocorticoids, which act on salt and water balance and fuel mobilization.

- The gonads secrete sex hormones. The female ovaries produce estrogens and progestins, which maintain the reproductive system, stimulate the development of female characteristics, and prepare the uterus for pregnancy. The male testes secrete androgens, mainly testosterone, that shape male sexual development. Secretion of sex hormones is regulated by the hypothalamus and anterior pituitary.

Review the Concepts

Work through the following exercises to review the concepts in this chapter. For additional review, check out the activities on the Web at www.mybiology.com. The website offers a pre-test that will help you plan your studies.

Exercise 1 (Module 26.1)

This module describes the roles of chemical signals in control of body functions. Color and label the diagrams below, comparing the activities of endocrine cells, neurosecretory cells, and nerve cells. Note that the diagrams are not in the same order as those in the text. First, choose a color for each of the cells. Make hormone molecules from the endocrine cell green dots, hormone molecules from the neurosecretory cell blue dots, and neurotransmitter molecules red dots. Then label the following: **neurosecretory cell, hormone molecules, nerve signals, endocrine cell, neurotransmitter molecules, secretory vesicles, target cell, blood vessel,** and **nerve cell.**

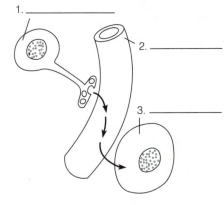
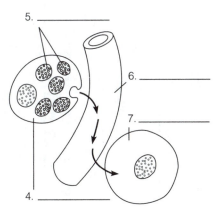
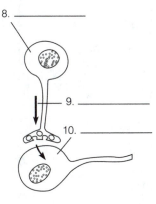

Exercise 2   (Module 26.1)

Module 26.1 contrasts the roles of the endocrine and nervous systems in controlling body activities. Check the statements that apply to each system. (Note: Some statements apply to both.)

	Endocrine	Nervous
1. A system of internal communication	_____	_____
2. Function involves electrical signals	_____	_____
3. Function involves chemical signals	_____	_____
4. Signals carried by neurons	_____	_____
5. Signals carried by body fluids	_____	_____
6. Signals carried by blood	_____	_____
7. Rapid messages	_____	_____
8. Slower messages	_____	_____
9. Split-second responses	_____	_____
10. Longer-lasting responses	_____	_____
11. Widespread effects	_____	_____
12. More pinpoint, localized effects	_____	_____

Exercise 3   (Module 26.2)

Response of target cells to hormone occurs in three steps: reception, transduction, and response. Just about everything you need to know about the mechanisms by which hormones affect their target cells is summarized in the two figures in Module 26.2. To review them, label the diagrams below. Choose from: **DNA, steroid hormone, enzyme, receptor protein, plasma membrane, target cell, new protein, signal transduction pathway, epinephrine, cellular response, nucleus, hormone-receptor complex,** and **relay molecules.**

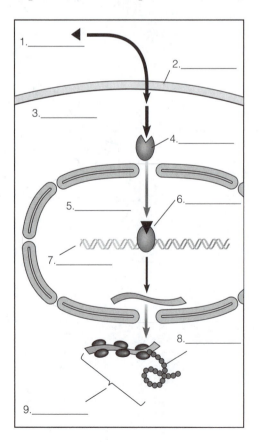

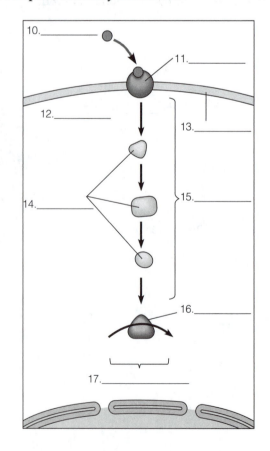

Exercise 4   (Module 26.3)

This module is an overview of the endocrine glands and hormones discussed in this chapter. You may want to refer back to the diagram and table as references as you read about glands and hormones in the modules that follow. Label the human endocrine glands on the diagram below, and color each one a different color. Choose from **thymus, testis, pineal gland, pituitary gland, adrenal glands, parathyroid glands, pancreas, thyroid gland, hypothalamus,** and **ovary.** Consulting Table 26.3 in the text, list under the name of each gland one of the hormones it produces.

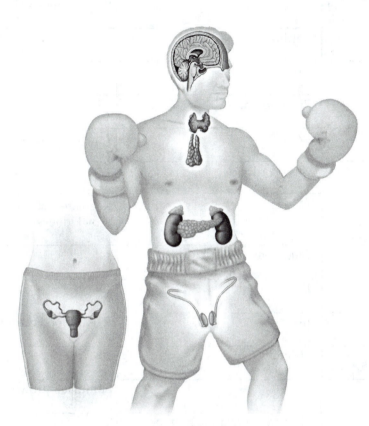

332   *Chapter 26*

Exercise 5   (Module 26.3)

The table in Module 26.3 is a useful summary of glands, hormones, and hormone actions. Use it to complete the following chart by filling in the blanks. Note that the order of the glands here is different from the table in the text. (You may want to skip this for now and come back to it as a review after you have studied the glands and hormones in the following modules in more detail.)

Gland	*Hormone(s)*	*Action*
1.	2. & 3.	Control metabolic processes
	Calcitonin	4.
Ovaries	5.	6.
	Progesterone	7.
8.	9. & 10.	Trigger "fight or flight" response
11.	Glucocorticoids	12.
	13.	Regulate reabsorption of Na and loss of K by kidneys
Pancreas	Glucagon	14.
	15.	16.
17.	18.	Raises blood calcium level
19.	Melatonin	20.
21.	22.	Stimulates growth
	Thyroid-stimulating hormone	23.
	Prolactin	24.
	25.	Stimulates adrenal cortex
	Luteinizing hormone	26.
	27.	Stimulates egg and sperm production
Posterior pituitary	Oxytocin	28.
	29.	Promotes water retention by kidneys
Thymus	30.	31.
32.	Androgens	33.
34.	Hormones that regulate anterior pituitary; hormones released by posterior pituitary	Affect pituitary

Exercise 6   (Module 26.4)

The hypothalamus and pituitary glands are closely related and tie the nervous and endocrine systems together. But the anterior and posterior lobes of the pituitary do different jobs and are controlled in different ways. State whether each of the following relates to the hypothalamus (H), anterior pituitary (A), or posterior pituitary (P).

_____ 1. Stores and secretes hormones actually made in the hypothalamus
_____ 2. Secretes releasing and inhibiting hormones that stimulate the anterior pituitary
_____ 3. Composed of non-nervous, glandular tissue
_____ 4. Part of the brain
_____ 5. Where the hormones oxytocin and ADH are released into the blood
_____ 6. Responds to releasing and inhibiting hormones from the hypothalamus
_____ 7. Secretes growth hormone, ACTH, and thyroid-stimulating hormone
_____ 8. Is the terminus of neurosecretory cells from the hypothalamus
_____ 9. Most hormones that it secretes influence other endocrine glands.
_____ 10. Exerts master control over the endocrine system
_____ 11. Consists of nervous tissue and is actually an extension of the hypothalamus
_____ 12. Hypersecretion of one of this gland's hormones can cause gigantism.
_____ 13. Negative feedback usually acts on this part of the system.
_____ 14. Blood vessels carry releasing hormones from here to the anterior pituitary.
_____ 15. Along with the brain, it secretes the body's natural painkillers.
_____ 16. Secretes a hormone that affects the kidney's reabsorption of water

Exercise 7   (Module 26.5)

Most people know at least a little about the thyroid gland. Once you understand that thyroid hormones stimulate metabolism, it is easy to remember the symptoms of over- and undersecretion and the effects of thyroid medication. Review the thyroid gland by matching each phrase on the left with a substance on the right.

_____ 1. Secreted by hypothalamus; stimulates anterior pituitary   A. Iodine
_____ 2. Secreted by anterior pituitary; stimulates thyroid   B. T_3
_____ 3. Hormone secreted by thyroid   C. TSH
_____ 4. Another hormone secreted by thyroid   D. T_4
_____ 5. Needed for the manufacture of thyroid hormones   E. TRH
_____ 6. T_3 and T_4 signal the hypothalamus to stop making this
_____ 7. Goiter occurs if there is not enough of this in the diet
_____ 8. Also called thyroxine
_____ 9. Sometimes called triiodothyronine

Two opposing, or antagonistic, hormones control the balance of calcium in the blood. To review calcium homeostasis, study Figure 26.6 in the text. Use it to help you choose the correct italicized words to complete the following paragraph.

Calcium is important for nerve impulse transmission, muscle contraction, transport of molecules through cell membranes, and (1) *digestion, protein synthesis, blood clotting.* Blood calcium concentration is held in a narrow range by the thyroid and parathyroid glands. If blood calcium drops, the (2) *thyroid gland, parathyroid glands* increase(s) secretion of (3) *parathyroid hormone, calcitonin.* This causes (4) *increased, decreased* reabsorption of calcium as the kidneys form urine, (5) *increased, decreased* absorption of calcium from food, and (6) *release of calcium from, deposition of calcium in* bone. If blood calcium concentration climbs too high, (7) *parathyroid hormone, calcitonin* is secreted by the (8) *thyroid gland, parathyroid glands.* This causes (9) *increased, decreased* reabsorption of calcium in the kidneys and (10) *release of calcium from, deposition of calcium in* bone.

To understand how the pancreas controls blood sugar, you need to remember only two things: (1) the pancreas makes two opposing hormones, insulin and glucagon and (2) diabetics have high blood sugar, so many of them have to take insulin. Given these two facts, you can figure everything else out: If diabetics have to take insulin, insulin must make blood sugar go down. It must make cells take sugar out of the blood and use it or store it. That means glucagon must make blood sugar go up by causing cells to get it out of storage and put it into the blood. Finally, an increase in blood sugar must trigger insulin secretion (to make sugar go down), so a drop in blood sugar must trigger glucagon secretion (to make blood sugar go up).

To review control of blood sugar level, choose either the word *increase(s)* or *decrease(s)* to complete each of the following statements.

1. Eating a meal rich in carbohydrates immediately causes blood glucose to _____.
2. When blood glucose _____, the pancreas secretes more insulin.
3. Insulin causes body cells to _____ their uptake and use of glucose.
4. Insulin also causes glycogen formation by the liver to _____.
5. Insulin therefore causes blood glucose to _____.
6. Between meals, blood glucose levels tend to _____.
7. When blood glucose _____, the pancreas secretes more glucagon.
8. Glucagon causes blood glucose to _____.
9. Glucagon _____ breakdown of glycogen in the liver and release of glucose to the blood.
10. Glucagon also triggers liver cells to _____ conversion of fats to glucose.
11. As blood glucose rises toward the set point, secretion of glucagon _____.
12. In type 1 diabetes, blood sugar _____ because the body is unable to produce insulin.
13. In type 2 diabetes, there is a _____ in cells' ability to respond to insulin.
14. Hypoglycemia results from an excess of insulin, which causes a sudden _____ in blood glucose.

Exercise 10   (Module 26.9)

The adrenal gland is similar to the pituitary in that it is two glands in one. The parts have different jobs and send out different hormones. One of its parts (the medulla) is stimulated by nerve impulses to secrete its hormones, while the other part (the cortex) is stimulated by hormonal signals. The hormones from the adrenal medulla and cortex help the body deal with stress. After reading the modules and studying Figure 26.9 in the text, try to match each of the phrases on the left with adrenal stress hormones, E, M, or G.

_____ 1. Increase breathing rate
_____ 2. Respond to short-term stress
_____ 3. Secreted by the adrenal cortex
_____ 4. Also secreted by the adrenal cortex
_____ 5. Increase blood volume and pressure in response to long-term stress
_____ 6. Triggered by nerve impulses from the hypothalamus
_____ 7. Secreted by the adrenal medulla
_____ 8. Cause proteins and fats to be broken down to make glucose
_____ 9. Triggered by ACTH from the pituitary
_____ 10. Also triggered by ACTH from the pituitary
_____ 11. Suppress the inflammatory response and immune system
_____ 12. Cause retention of sodium and water by kidneys
_____ 13. Dilate and/or constrict blood vessels, redirecting blood flow
_____ 14. Increase metabolic rate
_____ 15. Often prescribed to relieve pain from athletic injuries

E. Epinephrine and norepinephrine
M. Mineralocorticoids
G. Glucocorticoids

Exercise 11   (Module 26.10)

Sex hormones are discussed in more detail in Chapter 27. After reading Module 26.10, try sketching on a separate sheet of paper a concept map for sex hormones. Refer to Chapter 3, Exercise 10, for a discussion of constructing a concept map. Include the following in your concept map: **testosterone, gonads, estrogens, hypothalamus, FSH and LH, progestins, testes, releasing factor, anterior pituitary, ovaries, sex hormones,** and **steroids.**

Exercise 12   (Module 26.11)

Prolactin is an ancient hormone whose functions have diversified through evolutionary history. It is only one example of many hormones with different roles in different species. After reading this module, see if you can match each of the following functions of prolactin with a particular species or group of species.

_____ 1. Regulates salt and water balance
_____ 2. Stimulates nest building
_____ 3. Stimulates mammary glands to produce milk
_____ 4. Affects movement toward water in preparation for breeding
_____ 5. Regulates fat metabolism and reproduction
_____ 6. High levels during nursing prevents ovaries from releasing eggs
_____ 7. Affects metamorphosis from larva to adult

A. salmon
B. amphibians
C. birds
D. nonhuman mammals
E. humans

Exercise 13   (Summary)

On a separate sheet of paper, briefly describe how the items in each of the following groups are related to one another. They may affect one another, oppose one another, or have something in common.

1. luteinizing hormone   estrogens   androgens   follicle-stimulating hormone
2. oxytocin   antidiuretic hormone
3. glucagon   glucocorticoids   epinephrine
4. glucocorticoids   ACTH
5. antidiuretic hormone   mineralocorticoids   parathyroid hormone   calcitonin
6. thyroxine   calcitonin T_3
7. glucagon   insulin
8. glucocorticoids   epinephrine   mineralocorticoids   norepinephrine
9. ACTH   LH   growth hormone   FSH   TSH   prolactin
10. calcitonin   parathyroid hormone
11. oxytocin   prolactin

Test Your Knowledge

Multiple Choice

1. Another system that works closely with the endocrine system to control body processes is the
   a. circulatory system.
   b. immune system.
   c. digestive system.
   d. nervous system.
   e. muscular system.

2. Every time you eat a cookie or candy bar, your blood sugar increases. This triggers an increase in the hormone
   a. thyroxine.
   b. epinephrine.
   c. adrenocorticotropin (ACTH).
   d. glucagon.
   e. insulin.

3. Every hormone
   a. is a protein.
   b. is produced in response to stress.
   c. is under the control of the pituitary gland.
   d. enters a cell and interacts with DNA.
   e. acts as a signal between cells.

4. Researchers have found increased levels of hormones from the _____ in the blood of students preparing for final exams. These hormones are produced in response to stress.
   a. thyroid gland
   b. pineal gland
   c. posterior pituitary
   d. adrenal glands
   e. parathyroid glands

5. Which of the following hormones have antagonistic (opposing) effects?
   a. thyroxine and calcitonin
   b. insulin and glucagon
   c. growth hormone and epinephrine
   d. ACTH and glucocorticoids
   e. epinephrine and norepinephrine

6. What is the role of a receptor in hormone action?
   a. It signals a cell to secrete a hormone.
   b. It informs a gland as to whether its hormones are having an effect.
   c. It enables a target cell to respond to a hormone.
   d. It stops hormone action when it is no longer needed.
   e. It carries a hormone while it is in the blood.

7. When a boy goes through puberty, the steroid hormone testosterone "puts hair on his chest" by
   a. interacting with DNA in the nuclei of cells.
   b. causing cells to change shape.
   c. altering the permeability of plasma membranes.
   d. triggering nerve impulses in cells.
   e. turning enzymes on.

8. A hormone from the parathyroid glands works in opposition to a hormone from the _____ to regulate _____.
   a. posterior pituitary . . . metabolic rate
   b. thyroid gland . . . blood calcium
   c. pancreas . . . water reabsorption
   d. adrenal medulla . . . blood calcium
   e. thyroid gland . . . blood glucose

9. Some glands produce hormones that stimulate other endocrine glands. Which of the following hormones specifically acts to trigger secretion of hormones by another endocrine gland?
   a. thyroxine
   b. progesterone
   c. adrenocorticotropic hormone (ACTH)
   d. antidiuretic hormone (ADH)
   e. melatonin

10. How is the level of thyroxine in the blood regulated?
    a. Thyroxine stimulates the pituitary to secrete thyroid-stimulating hormone (TSH).
    b. TSH inhibits secretion of thyroxine from the thyroid gland.
    c. TSH-releasing hormone (TRH) inhibits secretion of thyroxine by the thyroid gland.
    d. Thyroxine stimulates the hypothalamus to secrete TRH.
    e. Thyroxine and TSH inhibit secretion of TRH.

11. Steroid hormones are produced only by the
    a. adrenal medulla and pancreas.
    b. thyroid gland and pancreas.
    c. anterior and posterior pituitary.
    d. thyroid gland and gonads.
    e. gonads and adrenal cortex.

12. It usually takes much longer for sex hormones and other steroids to produce their effects than it takes for most nonsteroid hormones. Why?
    a. Steroids are bigger, slower molecules.
    b. Steroids usually must be carried longer distances by the blood.
    c. Steroids cause target cells to make new proteins, which takes time.
    d. Steroids must relay their message via a receptor.
    e. It takes longer for endocrine cells to make and secrete steroids.

13. The pituitary is actually two glands. The anterior pituitary secretes its hormones when stimulated by _____, and the posterior pituitary secretes its hormones when stimulated by _____ .
    a. hormones from the adrenal cortex . . . hormones from the thyroid
    b. hormones from the hypothalamus . . . nerve impulses from the hypothalamus
    c. hormones from the hypothalamus . . . hormones from the thyroid
    d. nerve impulses from the hypothalamus . . . hormones from the hypothalamus
    e. hormones from the pineal gland . . . hormones from the pancreas

14. Injections of a hormone are sometimes given to strengthen contractions of the uterus during childbirth. What hormone might this be?
    a. adrenocorticotropin (ACTH)
    b. thyroxine
    c. oxytocin
    d. insulin
    e. follicle-stimulating hormone (FSH)

15. Which of the following hormones has the broadest range of targets in the human body?
    a. ADH
    b. prolactin
    c. TSH
    d. epinephrine
    e. calcitonin

Essay

1. The pituitary is often called the master gland. In what way is this true? In what way is it misleading?

2. Compare how the adrenal cortex and adrenal medulla deal with stress.

3. How are hormones and neurotransmitters alike? How are they different?

4. Briefly describe an example of a endocrine disease or disorder in which
   a. too much hormone is secreted.
   b. not enough hormone is secreted.
   c. hormone is secreted in a normal amount, but cells fail to respond.

5. Describe two different situations in which a pair of hormones have opposite (antagonistic) effects on the body.

Apply the Concepts

Multiple Choice

1. Jet lag occurs when a person moves rapidly from one time zone to another, causing conflict between the body's biological rhythm and the new cycle of light and dark. Some scientists suspect that jet lag may result from disruption of a daily hormone cycle. Which of the following hormones do you think is the most likely suspect?
   a. epinephrine
   b. insulin
   c. melatonin
   d. estrogen
   e. prolactin

2. Which of the following hormones triggers secretion of the other two?
   a. thyroxine
   b. thyroid-stimulating hormone (TSH)
   c. TSH-releasing hormone (TRH)
   d. Any of the above can trigger secretion of the others.
   e. None of the above can trigger secretion of the others.

3. A tumor in an endocrine gland caused Jennifer to have weakened bones and unusually high levels of blood calcium. Which of the following was affected?
   a. anterior pituitary
   b. pancreas
   c. adrenal glands
   d. parathyroid glands
   e. thymus

4. Which of the following exerts control over all the others?
   a. adrenal cortex
   b. hypothalamus
   c. thyroid gland
   d. anterior pituitary
   e. testes

5. Because only the _____ gland uses iodine to make its hormones, radioactive iodine is sometimes used as a treatment for tumors of this gland.
   a. pituitary
   b. pancreatic
   c. thyroid
   d. adrenal
   e. testicular

6. Diabetes insipidus is an inherited endocrine malfunction (unrelated to the more common diabetes mellitus) in which the kidneys fail to reabsorb normal amounts of water. Victims of this disease produce gallons of urine each day, and their kidneys soon wear out. Treatment of this disease involves replacing a missing hormone. Which of the following do you think it is?
   a. glucagon
   b. epinephrine
   c. glucocorticoids
   d. antidiuretic hormone (ADH)
   e. thyroid-stimulating hormone (TSH)

7. In an experiment, researchers removed the ____ of young mice, and as a result, these mice were able to accept organ transplants without rejection.
   a. pineal glands
   b. thymus glands
   c. thyroid glands
   d. parathyroid glands
   e. adrenal glands

8. Jake once suffered a severe allergic reaction to a bee sting. The sting caused him to suffer a near-fatal drop in blood pressure called anaphylactic shock. Now he carries a kit containing a syringe of _____, which he can inject to speed up his heart if he reacts to a bee sting.
   a. insulin
   b. thyroxine
   c. testosterone
   d. calcitonin
   e. epinephrine

9. It has been found that certain salamanders fail to go through the normal transformation from tadpole to adult if there is a shortage of iodine in the pond water in which they live. This makes it impossible for the _____ to manufacture hormones necessary for normal development.
   a. thyroid gland
   b. posterior pituitary
   c. adrenal cortex
   d. pineal gland
   e. pancreas

10. As a child, Maria suffered a head injury that damaged her pituitary. An injury to the pituitary is particularly serious because of all the functions controlled by this gland. As Maria got older, she and her doctors found that all of the following except _____ were affected.
    a. metabolic rate
    b. growth
    c. her menstrual cycle
    d. milk production
    e. blood calcium level

11. If you poke an ant nest, the ants will become stirred up and try to defend their home. Even ants underground, in the dark, get the message to come to the surface, without hearing, seeing, or touching one another. The ants must be signaling one another by means of a
    a. hormone
    b. neurotransmitter
    c. pheromone
    d. steroid
    e. local regulator

12. It has been found that male mammals have essentially the same hormones as females, but respond to some of them in slightly different ways. One hormone is secreted when a male is in contact with newborns, and it stimulates the male to protect and nurture the offspring. Which of these hormones would you expect this to be?
    a. melatonin
    b. calcitonin
    c. thyroxine
    d. prolactin
    e. glucagon

Essay

1. Hypoglycemia is a condition in which blood sugar drops to abnormally low levels. What seems to be the cause of hypoglycemia? Why would it not be a good idea to try to deal with this problem by eating more sugar?

2. One of the symptoms of severe diabetes mellitus is breath that has a sweetish, acetone smell. The smell comes from by-products of fat breakdown in the body. If another consequence of diabetes is excess sugar in the blood, why do the cells not just use the sugar instead of breaking down fat?

3. A tumor can cause enlargement of the thyroid gland. How could this result in abnormally high metabolic rate and body temperature? How could similar symptoms be produced by a tumor of the pituitary gland?

4. Some hormones, such as ADH, act on very specific targets (for ADH, the kidney). Other hormones, such as insulin, are able to affect every cell in the body. Based on what you have learned about how hormones exert their effects on target cells, speculate as to how some hormones can affect many targets, while other hormones affect only one.

5. A chemical called dioxin, or TCDD, is produced as a contaminant during some chemical manufacturing processes. Trace amounts of this substance were present in Agent Orange, a defoliant sprayed on vegetation during the Vietnam War. There has been a continuing controversy over its effects on Vietnam veterans exposed to it during the war. Animal tests have shown that dioxin can be lethal and can cause birth defects, cancer, liver and thymus damage, and immune system suppression. But its effects on humans are unclear, and even animal tests are uneven; a hamster is unaffected by a dose that can kill a guinea pig. Researchers have discovered that dioxin enters a cell and binds to a receptor protein, which in turn attaches to the cell's DNA. How might this mechanism help explain the variety of effects of dioxin on different body systems and in different animals?

Put Words to Work

Correctly use as many of the following words as possible, when reading, talking, and writing about biology:

adrenal cortex, adrenal gland, adrenal medulla, ACTH, androgen, antagonistic hormones, anterior pituitary, calcitonin, diabetes mellitus, endocrine gland, endocrine system, endorphin, epinephrine, estrogen, glucagon, glucocorticoid, goiter, gonad, growth hormone (GH), hormone, hypoglycemia, hypothalamus, inhibiting hormone, insulin, mineralocorticoid, neurosecretory cell, pancreas, parathyroid glands, parathyroid hormone (PTH), pineal gland, pituitary gland, posterior pituitary, progestin, prolactin (PRL), releasing hormone, steroid hormone, target cell, testosterone, TSH-releasing hormone (TRH), T₃, thyroid gland, thyroid-stimulating hormone (TSH), thyroxine (T₄), thymus gland

Use the Web

For further review, see the activities and questions on hormones and the endocrine system on the Web at www.mybiology.com.

Reproduction and Embryonic Development 27

Focus on the Concepts

This chapter discusses modes of reproduction, reproductive system structure and function, and the stages and mechanisms of embryonic development. Focus on these concepts:

- There are two primary modes of animal reproduction: Asexual reproduction is the creation of genetically identical offspring by a lone parent. Sexual reproduction is the creation of offspring via the fusion of haploid eggs and sperm to form a diploid zygote. Sexual reproduction may increase variation and adaptability in changing environments.

- In the human female, eggs develop in follicles in the ovaries and are released in a process called ovulation. Eggs pass down the oviducts, where fertilization and early development may occur, and an embryo implants in the endometrium, the lining of the uterus. The vagina is the entrance for sperm and serves as the birth canal.

- In the male, sperm are made in the testes, mature in the epididymis, and are ejaculated through the vas deferens and urethra. The prostate and other glands produce fluid that carries and feeds sperm. The penis has erectile tissue that can swell with blood, causing an erection. FSH from the pituitary gland regulates sperm production. LH from the pituitary controls production of androgens, mainly testosterone.

- In spermatogenesis, a diploid primary spermatocyte undergoes the two divisions of meiosis, producing four cells that mature into haploid sperm. Millions may be produced each day. In oogenesis, a diploid primary oocyte divides twice, but unevenly, forming one haploid ovum and smaller polar bodies. In humans, one follicle matures and releases one ovum each month.

- Hormones coordinate a woman's ovarian and menstrual cycles. Each month a rise in FSH and LH from the pituitary trigger follicle growth and ovulation. The growing follicle—after ovulation, the corpus luteum—makes estrogen and progesterone, which stimulate development of the uterine lining and inhibit FSH and LH secretion. A developing embryo makes HCG, which acts like LH to maintain the corpus luteum. If fertilization does not occur, the corpus luteum breaks down, causing menstruation.

- Fertilization is the union of sperm and egg to form a diploid zygote. The zygote undergoes a series of divisions—cleavage—forming a hollow ball of cells called a blastula. Gastrulation rearranges the cells, forming a gastrula with three layers of cells and the beginnings of a digestive cavity. Cell migration, differentiation, induction, and programmed cell death shape organs. Pattern formation of body limbs and regions is controlled by ancient homeotic genes.

- In humans, fertilization occurs in the oviduct. The zygote divides to form a blastocyst, which implants in the wall of the uterus. Outer trophoblast cells form extraembryonic membranes that protect and nourish the embryo. Most organs and systems take shape during the first trimester. By nine weeks, the fetus looks human. It grows dramatically during the second and third trimesters. Late hormonal changes make the uterus more sensitive to oxytocin, the main hormone that brings on labor and birth.

Review the Concepts

Work through the following exercises to review the concepts in this chapter. For additional review, utilize the activities on the Web at www.mybiology.com. The website offers a pre-test that will help you plan your studies.

Exercise 1   (Introduction and Modules 27.1–27.2)

These modules introduce some terms and concepts basic to an understanding of sexual and asexual reproduction. Write the correct term in the space next to each phrase below.

_____   1. Male and female reproductive systems in the same body
_____   2. Small gamete usually propelled by a flagellum
_____   3. When one parent produces genetically identical offspring
_____   4. Diploid cell formed by union of egg and sperm
_____   5. General name for egg or sperm
_____   6. Regrowth of lost body parts
_____   7. Discharge of gametes into water
_____   8. Large gamete that is not self-propelled
_____   9. Sexual intercourse
_____   10. Splitting of new individuals from existing ones
_____   11. When egg and sperm unite inside the female's body
_____   12. Breakup of the parent body into several pieces
_____   13. Inability to bear children
_____   14. Separation of a parent into two or more individuals of the same size

Exercise 2   (Modules 27.3 – 27.4)

These modules introduce the human reproductive systems. The diagrams below show a front view of the reproductive system of the human female and a side view of the reproductive system of the human male. See how many of the following parts you can label without looking at the modules, and then consult the modules for the remainder: **seminal vesicle, ovary, scrotum, urethra, epididymis, vagina, oviduct, vas deferens, prepuce, uterus, follicles, erectile tissue, testis, endometrium, cervix, glans, prostate gland, bulbourethral gland, corpus luteum, wall of uterus, ejaculatory duct, penis.**

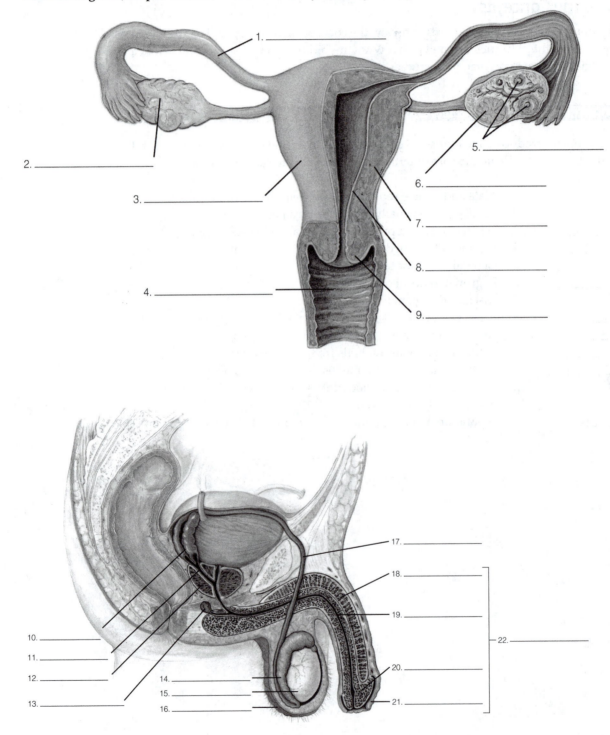

Exercise 3 (Module 27.5)

Both spermatogenesis and oogenesis produce haploid gametes, but there are several important differences in how they occur. Compare these processes in the chart below.

	Spermatogenesis	*Oogenesis*
Location of process	1.	2.
When primary cells form	3.	4.
Number of chromosomes in primary cells (humans)	5.	6.
Numbers and names of cells resulting from first meiotic division	7.	8.
Numbers and names of cells resulting from second meiotic division	9.	10.
Number of gametes produced from division of one primary cell	11.	12.
Number of chromosomes in each gamete produced (humans)	13.	14.
Total number of gametes produced per day or per month	15.	16.

Exercise 4 (Module 27.6)

This module explains the monthly ovarian and menstrual cycles that coordinate the activities of the ovaries and uterus. You can use the table and diagrams in the text to get most of the words you need to complete the following story, which integrates the ovarian and menstrual cycles. (Figure 27.6 is particularly informative.)

Jennifer wasn't consciously aware of it, but a slight change was occurring in her brain. Her menstrual period had barely started, but already her [1]_____ was starting to secrete tiny amounts of a releasing hormone, which in turn signaled the pituitary gland to boost its output of [2]_____ and [3]_____, restarting her ovarian and menstrual cycles. FSH stimulated a [4]_____, dormant in her [5]_____ since before she was born, to begin to grow and secrete its own hormones. As the follicle grew, it started to secrete small amounts of

6_____. This hormone in turn signaled the 7_____, the lining of the 8_____, to begin to grow and develop in preparation for possible implantation of a fertilized 9_____. The estrogen from the 10_____ also exerted 11_____ feedback on the hypothalamus, keeping levels of 12_____ and 13_____ low during this 14_____ phase of the ovarian cycle.

As the follicle grew in Jennifer's ovary, it secreted more and more estrogen. The higher level of estrogen caused the hypothalamus to signal the 15_____ to release bursts of 16_____ and 17_____. The surge of LH caused the mature follicle to rupture, releasing an ovum in a process called 18_____. This occurred on the 19_____ day of Jennifer's 28-day cycle. The ovum entered the 20_____, where fertilization could occur, and began to travel toward the uterus. The 21_____ phase of the ovarian cycle had begun.

One of the effects of luteinizing hormone was to transform the ruptured follicle into a yellow, glandular structure, called the 22_____, which continued to secrete estrogen, plus another hormone called 23_____. Both these hormones stimulated further growth and thickening of the uterine lining, further preparing it for the fertilized egg. These hormones also exerted strong negative feedback on the hypothalamus and pituitary, suppressing secretion of FSH and LH and so preventing additional 24_____ from growing in the ovary. Eventually, the drop in LH caused the corpus luteum to degenerate and reduce its output of estrogen and 25_____. (Because Jennifer's egg was not 26_____, there was no hormonal signal sent from a developing embryo to maintain the corpus luteum.) Estrogen and progesterone were maintaining the lining of the uterus. As levels of these hormones dropped, the 27_____ began to slough off. The monthly cycles were coming to an end, and 28_____ had begun.

Exercise 5   (Module 27.7)

Review sexually transmitted diseases by completing this crossword puzzle.

Across

4. Viral STDs are not ____, but bacterial STDs usually are.

6. Bacterial STDs are treated with ____.

8. ____ is short for "sexually transmitted disease."

9. A drug called valacyclovir is used to treat symptoms of genital ____.

11. Chlamydial infections, syphilis, and ____ are caused by bacteria.

12. Chlamydial infections and gonorrhea may cause no symptoms in ____.

14. A fungus called *Candida* causes ____ infections.

15. Pelvic inflammatory disease, or ____, may damage the oviducts and uterus.

16. The use of ____ can usually prevent the spread of STDs

Down

1. STDs are spread by sexual ____.

2. _____ warts can be removed by freezing.

3. Trichomoniasis is caused by a ____.

5. Genital warts have been linked to cervical ____.

7. AIDS and genital herpes are both caused by ____.

10. A genital discharge, sores, or painful ____ can be symptoms of STDs.

13. A virus called HIV is responsible for ____.

17. ____ usually show no symptoms of trichomoniasis.

Exercise 6   (Module 27.8)

Various methods of contraception can prevent pregnancy. Review them by matching each of the statements on the left with a contraceptive method listed on the right. Some answers are used more than once.

_____ 1. Refraining from intercourse around the time of ovulation
_____ 2. A barrier method used by women
_____ 3. Another barrier method used by women
_____ 4. High-dose birth control pills used after intercourse
_____ 5. Sterilization of the male
_____ 6. Removing the penis from the vagina before ejaculation
_____ 7. Hormones administered via injection
_____ 8. A foam or jelly that kills sperm
_____ 9. Cutting the vas deferens to interrupt sperm path
_____ 10. Also called oral contraceptives
_____ 11. The only method that also reduces STDs
_____ 12. Least effective method as typically used
_____ 13. A barrier method used by men
_____ 14. Sterilization of the female
_____ 15. Blocks progesterone and terminates pregnancy
_____ 16. Overall, the most effective method of contraception
_____ 17. A daily tablet containing synthetic estrogen and progesterone

A. Birth control pill
B. Vasectomy
C. Tubal ligation
D. Depo-provera
E. RU-486 (mifepristone)
F. Rhythm method
G. Withdrawal
H. Male condom
J. Cervical cap
K. Spermicide
L. Morning after pill (MAP)
I. Diaphragm

Exercise 7   (Module 27.9)

Embryonic development begins with fertilization. After reading the module and studying the illustrations, answer each of the following questions with one sentence.

1. How is the sperm able to get to the egg?

2. How does the sperm penetrate the egg?

3. In species whose eggs are fertilized externally, what keeps sperm of other species from fertilizing the egg?

4. Why is it important that only one sperm penetrate the egg?

5. What prevents more than one sperm from penetrating the egg?

6. In what ways does the egg change after fertilization?

Exercise 8   (Modules 27.10 – 27.12)

Modules 27.10–27.12 trace the development of an embryo from the zygote to development of rudimentary organs. The three stages in this process are cleavage, gastrulation, and organ formation. The illustrations below (combining development of a sea urchin and a frog) are in random order. (Two intermediate stages of frog gastrulation not shown in the text are illustrated here, to show how this occurs step-by-step). Number the stages in the correct sequence (blanks A–K), and then label the following (blanks L–Z): **ectoderm, blastula, zygote, blastopore, digestive cavity, blastocoel, endoderm, notochord, mesoderm, gastrula, neural tube, body cavity (coelom),** and **somite.** Some terms are used more than once.

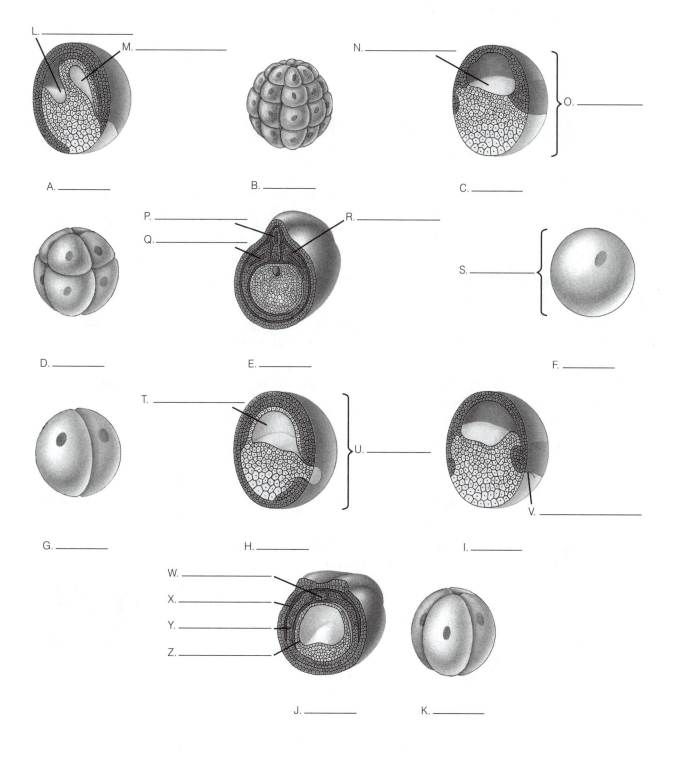

L. _____
M. _____
N. _____
O. _____

A. _____   B. _____   C. _____

P. _____
Q. _____
R. _____
S. _____

D. _____   E. _____   F. _____

T. _____
U. _____
V. _____

G. _____   H. _____   I. _____

W. _____
X. _____
Y. _____
Z. _____

J. _____   K. _____

Gastrulation produces a three-layered embryo. Each of the three tissue layers gives rise to particular structures. After reading the module and studying the table, name the embryonic tissue layer from which each of the following structures develops—**ectoderm, mesoderm,** or **endoderm.** (You can figure out most of them just by remembering that ectoderm is on the outside of the embryo, endoderm is on the inside, and mesoderm is in between.)

_____   1. Lining of digestive tract
_____   2. Circulatory system
_____   3. Pancreas and liver
_____   4. Nervous system
_____   5. Skeleton
_____   6. Muscles
_____   7. Lining of respiratory system
_____   8. Skin epidermis

These three modules describe some of the processes that shape an embryo. Each of the following statements describes an experiment or observation that demonstrates one of the processes. Match the process with the experiment. Choose from **pattern formation, cell migration, programmed cell death, induction, effects of homeotic genes, differentiation,** and **changes in cell shape.**

1. Ectoderm from the side of a frog gastrula normally develops into skin. If a piece of this ectoderm is removed from the side and placed on the back of the embryo in contact with the notochord, it turns into neural tissue. This experiment illustrates _____.

2. A certain chemical is painted on a limb bud of a mouse. An additional set of toes develops at the site, with the little toe where the chemical is most concentrated. This experiment illustrates

    _____.

3. The reproductive system develops from mesoderm, except for the germ cells that will actually divide to form eggs and sperm. Embryologists have dyed these cells and followed them as they "crawl" from a distant site to the developing gonads, an example of _____.

4. When a tadpole changes into a frog, cells in its tail break down and the tail shrinks and eventually disappears, an example of _____.

5. If cells of a blastula are treated with chemicals that interfere with changes in the cells' internal cytoskeletons, the cells fail to fold inward properly to start the process of gastrulation. This experiment shows the importance of _____.

6. Researchers have discovered mutant fruit flies that "don't know which end is up." These "bicoid" flies have no heads and tails on both ends! This suggests a change in _____.

7. As a muscle cell matures and becomes specialized, it is no longer able to divide. Researchers are attempting to determine how genes are activated and deactivated in developing muscle cells. If turned-off genes could be turned back on, perhaps a cell could be made to return to a dividing, embryonic state. This might be useful in repairing, for example, damage from a heart attack. These experimenters are attempting to reverse the process of _____.

Exercise 11   (Modules 27.15 – 27.17)

Many of the stages and processes seen in the development of other animals are also seen in human development and birth. Match the words and phrases below to review human development.

_____ 1. Outer layer that allows embryo to implant
_____ 2. Carrying developing young in reproductive tract
_____ 3. Name for embryo after several weeks' gestation
_____ 4. Stage of labor in which baby is delivered
_____ 5. Hormone secreted by chorion; prevents spontaneous abortion
_____ 6. Structure that taps the mother's blood supply
_____ 7. Period when embryo grows rapidly and prepares for birth
_____ 8. Membrane that becomes the embryo's part of the placenta
_____ 9. Embedding of the blastocyst in the wall of the uterus
_____ 10. Forms fluid-filled sac around embryo
_____ 11. Trimester in which the biggest changes occur
_____ 12. When placenta starts making progesterone and corpus luteum degenerates
_____ 13. The first stage of labor
_____ 14. In birds and reptiles it is important in waste disposal.
_____ 15. Portion of the blastocyst that actually forms the baby
_____ 16. A pituitary hormone that initiates labor
_____ 17. Produces the embryo's first blood cells and germ cells
_____ 18. This process begins development.
_____ 19. Outgrowths of the chorion that absorb nutrients and oxygen
_____ 20. Contractions of the uterus that bring about birth
_____ 21. Equivalent of frog blastula
_____ 22. Help oxytocin induce labor

A. HCG
B. First trimester
C. Chorionic villi
D. Inner cell mass
E. Yolk sac
F. Fertilization
G. Prostaglandins
H. Blastocyst
I. Trophoblast
J. Fetus
K. Placenta
L. Labor
M. Gestation
N. Dilation
O. Oxytocin
P. Third trimester
Q. Allantois
R. Implantation
S. Amnion
T. Chorion
U. Second trimester
V. Expulsion

Exercise 12   (Module 27.18)

In vitro fertilization, surrogate motherhood, and the possibility of sperm and embryo donation open a multitude of reproductive possibilities to infertile couples. A woman's eggs could be fertilized with her husband's sperm and implanted in her uterus. Alternatively, donated sperm or eggs could be used. Or the eggs could be implanted in another woman's uterus. How many combinations are possible? List as many as you can.

Test Your Knowledge

Multiple Choice

1. Which of the following is a form of sexual reproduction?
   a. budding
   b. fission
   c. fragmentation
   d. regeneration
   e. hermaphroditism

2. Which of the following correctly traces the path
   of sperm from their site of production out of a
   man's body?
   a. seminiferous tubule, vas deferens, epi-
      didymis, urethra
   b. epididymis, urethra, seminiferous tubule, vas
      deferens
   c. seminiferous tubule, epididymis, vas defer-
      ens, urethra
   d. epididymis, seminiferous tubule, vas defer-
      ens, urethra
   e. vas deferens, epididymis, urethra, seminifer-
      ous tubule

3. A peak in _____ triggers ovulation on about the
   _____ day of the monthly cycle.
   a. progesterone . . . fourteenth
   b. LH . . . seventh
   c. FSH . . . second
   d. LH . . . fourteenth
   e. estrogen . . . twentieth

4. External fertilization occurs mostly in
   a. land animals.
   b. insects.
   c. aquatic animals.
   d. animals that reproduce asexually.
   e. mammals.

5. Human eggs are usually fertilized in the
   a. vagina.
   b. oviduct.
   c. ovary.
   d. cervix.
   e. uterus.

6. On its way to fertilize a human egg, a sperm cell
   does *not* have to pass through which of the following?
   a. oviduct
   b. vagina
   c. ovary
   d. vas deferens
   e. cervix

7. Which of the following hormones is the first to
   increase significantly every 28 days or so and
   initiates the ovarian cycle?
   a. progesterone
   b. follicle-stimulating hormone
   c. estrogen
   d. luteinizing hormone
   e. human chorionic gonadotropin

8. The fertilization envelope
   a. prevents an egg from being fertilized by more
      than one sperm.
   b. attracts sperm to an egg from some distance
      away.
   c. enables an egg to obtain nourishment from its
      environment.
   d. ensures that an egg is fertilized by sperm of
      the proper species.
   e. enables a fertilized egg to implant in the uter-
      ine wall.

9. After ovulation occurs, the empty follicle
   a. can be recycled to produce more eggs.
   b. changes into the corpus luteum and makes
      hormones.
   c. quickly degenerates.
   d. immediately initiates menstruation.
   e. becomes the site of implantation of a fertil-
      ized egg.

10. The first stage of embryonic development is
    _____ . This process produces _____ .
    a. gastrulation . . . a three-layered embryo
    b. gestation . . . a gastrula
    c. ovulation . . . a zygote
    d. cleavage . . . a hollow ball of cells
    e. blastulation . . . a fetus

11. The recent increase in multiple births—twins
    and "super twins"—is attributed to
    a. couples having children earlier.
    b. better nutrition.
    c. use of contraceptives.
    d. fertility drugs.
    e. environmental chemicals.

12. There is no cure for which of the following
    STDs?
    a. candidiasis
    b. chlamydia
    c. genital herpes
    d. gonorrhea
    e. syphilis

13. Which of the following correctly matches a body structure with the embryonic layer from which it develops?
    a. stomach lining—endoderm
    b. spinal cord—mesoderm
    c. epidermis of skin—mesoderm
    d. rib—endoderm
    e. liver—ectoderm

Essay

1. Some animals, such as aphids, alternate between sexual and asexual reproduction. Describe environmental conditions under which each form of reproduction might be most advantageous to the animal.

2. Name and describe the functions of the extraembryonic membranes that surround a human embryo.

3. Compare sperm formation in the testes with egg formation in the ovaries. When in life does each process begin? What triggers each to occur or be completed? How many cells result from division of a primary oocyte? A primary spermatocyte?

4. Draw a series of simple sketches that show how a fertilized egg develops into a three-layered gastrula.

Apply the Concepts

Multiple Choice

1. In an early stage of development of the frog nervous system, the bits of tissue that will form the backs of the eyes bulge from the growing brain. If one of these bits of tissue is removed and transplanted under the ectoderm in the area that will become the frog's belly, a lens develops in front of it. This experiment illustrates the importance of _____ in shaping an embryo.
   a. pattern formation
   b. homeotic genes
   c. induction
   d. cell migration
   e. cleavage

2. Birth control pills contain synthetic estrogen and progesterone. How might these hormones prevent pregnancy?
   a. They trigger premature ovulation, before an egg is mature.
   b. They cause the lining of the uterus to be sloughed off.
   c. They cause the corpus luteum to degenerate.
   d. They keep the pituitary from secreting FSH and LH, so ovulation does not occur.
   e. They prevent monthly development of the uterine lining.

3. Which of the following are hermaphrodites?
   a. sea anemones
   b. frogs
   c. earthworms
   d. mammals
   e. insects

4. Pregnancy tests detect a hormone in a woman's urine that is present only when an embryo is developing in her uterus. This hormone is secreted by
   a. the ovary.
   b. the chorion.
   c. a follicle.
   d. the fetus.
   e. the endometrium.

5. One difference between the blastula and gastrula stages of development is that
   a. blastula cells are more differentiated than gastrula cells.
   b. there are many more cells in a blastula.
   c. the blastula consists of more cell layers.
   d. the blastula is a solid ball of cells, but the gastrula is hollow.
   e. there is an opening from the cavity inside the gastrula to the outside.

6. Which of the following systems begin to take shape first in a frog?
   a. digestive and nervous systems
   b. nervous and muscular systems
   c. circulatory and digestive systems
   d. nervous and skeletal systems
   e. skeletal and circulatory systems

7. Doctors hoping to increase the chances of implantation of embryos fertilized in vitro might treat the recipient with
   a. progesterone.
   b. follicle-stimulating hormone.
   c. oxytocin.
   d. prolactin.
   e. all of the above.

8. Diseases, drugs, alcohol, and radiation have their more drastic effects on a human embryo if they are present during the first two months of gestation, because at this time
   a. substances can most easily enter the embryo through the placenta.
   b. the embryo has not yet implanted in the wall of the uterus.
   c. the mother's hormone levels are highest.
   d. the most drastic and rapid changes are occurring in the embryo.
   e. the cells of the embryo are not yet fully activated.

9. If you wanted to examine the extraembryonic membranes surrounding a lizard or mouse embryo, you would have to cut through which of the following to see all the others?
   a. yolk sac
   b. amnion
   c. endoderm
   d. allantois
   e. chorion

10. The trick to developing a successful male contraceptive seems to be in figuring out a way to block sperm production without affecting androgen (testosterone) levels. (The latter tends to inhibit sexual desire.) Looking at Figure 27.4D, which of the following would seem to be the best approach for developing a male "pill"?
    a. block the hypothalamus
    b. block the anterior pituitary
    c. block releasing hormone
    d. block LH
    e. block FSH

11. By treating a fruit fly embryo with certain chemicals, it can be made to develop into a fly with missing or extra eyes, wings, or body segments. These chemicals apparently interfere with
    a. homeotic genes
    b. apoptosis
    c. cleavage
    d. induction
    e. fertilization

Essay

1. Some populations of tropical fish called mollies consist of males and females that reproduce sexually, the way most fish do. Other populations consist only of females, whose eggs develop without fertilization. Which of these populations do you think would adapt most successfully to a change in its environment? Why?

2. A man who soaked in a hot tub for an extended period every day was found to be suffering from temporary sterility. It seems that his sperm production had been slowed down considerably. Explain why this might have occurred. (Note: This phenomenon cannot be relied on as a method of birth control!)

3. In general, land animals engage in more elaborate courtship behaviors before mating than do aquatic animals. Why might this be more important for land animals?

4. An embryologist carried out an experiment where tissue from the belly of a "red fish" embryo was transplanted onto the back of a "blue fish" embryo. The tissue from the red fish would not normally grow into a dorsal fin, but it did in its new location, on the blue fish. Interestingly, it grew into the kind of dorsal fin normally found on a red fish! Briefly explain (a) what caused the dorsal fin to develop in the right spot, and (b) why it was a red fish fin, and not the fin of a blue fish.

5. It has recently been discovered that genetic changes resulting from prolonged starvation can be passed on and affect the health of children and grandchildren. For example, if a man suffers starvation during adolescence, when sperm start to form, his children are more likely to suffer from diabetes. If a woman is subjected to starvation, her *grandchildren* are more likely to suffer from diabetes. Why her grandchildren?

Put Words to Work

Correctly use as many of the following words as possible when reading, talking, and writing about biology:

acrosome, amnion, asexual reproduction, birth control pill, blastocyst, blastula, budding, cervix, chlamydia, chorion, cleavage, clitoris, conception, contraception, copulation, corpus luteum, ectoderm, egg, ejaculation, embryo, endoderm, endometrium, epididymis, external fertilization, extraembryonic membranes, fetus, fertilization, fission, follicle, fragmentation, gamete, gametogenesis, gastrula, gastrulation, genital herpes, gestation, hermaphroditism, homeotic gene, human chorionic gonadotropin (HCG), impotence, in vitro fertilization (IVF), induction, infertility, internal fertilization, labia, labor, menstrual cycle, menstruation, mesoderm, morning after pill (MAP), neural tube, notochord, oogenesis, oral contraceptive, orgasm, ovarian cycle, ovary, oviduct, ovulation, ovum (plural, ova), pattern formation, penis, placenta, primary oocyte, primary spermatocyte, programmed cell death, prostate gland, regeneration, reproduction, scrotum, secondary oocyte, secondary spermatocyte, semen, seminal vesicle, spermatogenesis, seminiferous tubule, sexual reproduction, sexually transmitted disease (STD), sperm, spermicide, testis (plural, testes), trimester, trophoblast, tubal ligation, uterus, vagina, vas deferens, vasectomy, zygote

Use the Web

Learn more about reproduction and development by accessing the activities and questions on the Web at www.mybiology.com. The videos and animations are particularly helpful and informative.

Nervous Systems **28**

Focus on the Concepts

This chapter explores the structure, function, and evolution of nervous systems, emphasizing the vertebrate nervous system and the human brain. As you study the chapter, focus on the following concepts:

- Nervous systems receive sensory input, integrate it, and formulate and direct responses. Sensory neurons within nerves of the peripheral nervous system (PNS) conduct impulses to the central nervous system (CNS)—the brain and spinal cord. Interneurons within the CNS integrate data and signal motor neurons, which carry impulses via the PNS to effectors—muscles and glands.

- The functional units of the nervous system are neurons and supporting glial cells. Dendrites carry impulses to the neuron cell body, and axons carry them away from the cell body and to receiving cells. Certain glial cells produce an insulating myelin sheath that insulates neuron fibers and speeds nerve impulses.

- The sodium-potassium pump maintains a resting membrane potential across the neuron membrane. Na^+ is pumped out of the cell. K^+ is pumped in, but easily diffuses back out, leaving negatively charged ions behind. This results in net positive charge just outside the neuron and negative charge inside, storing energy like a battery.

- A stimulus causes Na^+ channels in the membrane to open, letting Na^+ enter. If a certain threshold is reached, more Na^+ channels open, letting more sodium in, and causing the cell to depolarize—to become $+$ inside and $-$ outside. Meanwhile, K^+ channels open, letting K^+ out and repolarizing the cell. This quick depolarization causes the next part of the membrane to depolarize, and the next, propagating a nerve impulse, or action potential, along the length of the neuron.

- Neurons communicate at synapses. When an action potential reaches the end of an axon, the cell dumps a chemical neurotransmitter into the synaptic cleft between the sending cell and the receiving cell. The neurotransmitter diffuses across the synapse and binds to receptors on the receiving cell. A variety of chemicals act as neurotransmitters; some trigger action potentials and some inhibit them. The receiving cell responds to the sum of its inputs. Many drugs and poisons act at synapses.

- Simple radial animals have weblike nervous systems; bilateral animals move actively, and their nervous systems are more centralized and cephalized, with a linear nerve cord or cords. The vertebrate nervous system consists of a CNS, composed of the brain and spinal cord, acting through cranial and spinal nerves of the PNS. The somatic portion of the PNS carries signals to skeletal muscles in response to external stimuli. The autonomic nervous system regulates the internal environment.

- In the course of vertebrate evolution, the brain grew larger and became subdivided into regions with specialized functions—hindbrain, midbrain, and forebrain. In birds and mammals, the portion of the forebrain called the cerebrum became much larger and more complex, making these animals capable of more complex behaviors.

- In humans various areas of the brainstem relay data to and from higher centers, control sleep and arousal, and regulate breathing and circulation. The cerebellum coordinates movement. The thalamus is a sensory relay center. The hypothalamus regulates homeostasis. The cerebrum, especially the folded cerebral cortex, is responsible for higher activity—memory, learning, language, emotions, creativity, etc. The limbic system, connecting several regions, links emotions, memory, and learning.

Review the Concepts

Work through the following exercises to familiarize yourself with the basics of nervous system structure and function. To enhance your understanding, explore the activities on the Web at www.mybiology.com. The website offers a pre-test that will help you plan your studies.

Exercise 1   (Module 28.1)

This module introduces the overall structure and function of the nervous system, using the example of a knee-jerk reflex, one of the simplest automatic behaviors. Review the parts and processes involved in a reflex by matching each in the diagram at the top of the next page with its name (A–P) and description (a–p).

A. motor neuron
B. ganglion
C. peripheral nervous system
D. spinal cord
E. quadriceps muscles
F. flexor muscles
G. reflex
H. interneuron
I. sensory input
J. sensory neuron
K. integration
L. nerve
M. brain
N. motor output
O. sensory receptor
P. central nervous system

a. a bundle of neurons wrapped in connective tissue
b. conduction of signals from sensory receptors in integration centers
c. nerve cell that conducts signals from sensory receptors to the CNS
d. nerves to and from the CNS
e. main control center of the CNS
f. conduction of signals from integration centers to effector cells
g. brain and spinal cord
h. a cluster of neuron cell bodies
i. nerve cell that carries signals from CNS to effector cells
j. contract to kick leg
k. picks up stimuli
l. relay neuron within the CNS
m. interpretation of sensory signal and formulation of response
n. works with brain as part of CNS
o. inhibited when knee jerks
p. an automatic response to stimuli

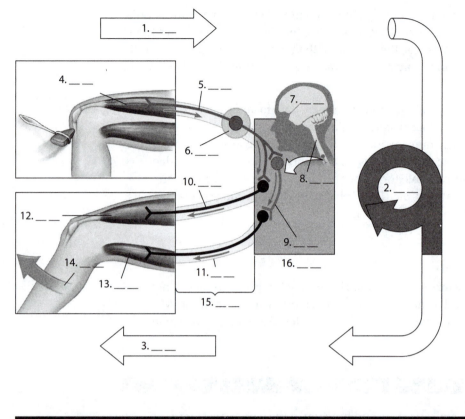

Review the structure of a neuron by labeling and coloring this diagram. Label the **cell body, axon, myelin sheath, dendrites, synaptic terminals,** and a **node of Ranvier.** Color the dendrites green, the cell body blue, the axon red, and the myelin sheath yellow.

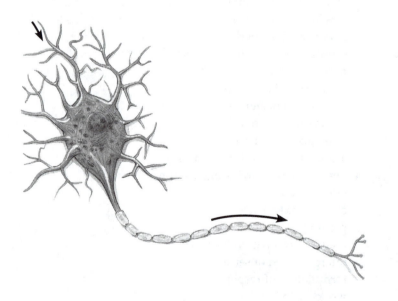

Exercise 3 (Modules 28.1–28.2)

Before moving on, it wouldn't hurt to review nervous system structure and function one more time. See how well you know your nervous system vocabulary by completing this crossword puzzle.

Across

2. Supporting cells, or _____, help neurons do their jobs.

4. A _____ is a cell specialized for carrying signals.

7. Motor _____ is conduction of signals from integration centers to effectors.

8. _____ is the interpretation of sensory signals and formulation of responses.

11. The _____ nervous system (PNS) consists of nerves outside the CNS.

13. _____ neurons carry information from sensory receptors in the CNS.

17. _____ receive messages and carry them to the neuron cell body.

18. The CNS consists of the _____ and spinal cord.

19. A Schwann cell is one type of _____ cell.

20. An axon has many branches, each with a synaptic _____ at its end.

22. A _____ sheath insulates a neuron.

Down

1. _____ in the central nervous system integrate data.

3. Sensory _____ is conduction of signals from sensory receptors to the brain.

5. The _____ nervous system is abbreviated "CNS."

6. _____ neurons carry signals from the CNS to effectors.

9. A _____ is a bundle of neuron extensions wrapped in connective tissue.

10. The nucleus of a neuron is in the cell _____.

12. Interneurons are entirely within the _____.

13. The myelin sheath is destroyed in multiple _____.

14. Signals go faster when they jump along a neuron, between _____ of Ranvier.

15. The site of contact between two nerve cells is called a _____.

16. _____ are clusters of cell bodies belonging to neurons making up a nerve.

21. An _____ carries signals toward another neuron or an effector.

Exercise 4   (Modules 28.3–28.5)

How do neurons carry signals? What exactly is a nerve signal? These three modules explain how nerve cells transmit nerve signals, which are called action potentials. Number in sequence each of the following steps (A–E) in the generation and transmission of a nerve signal. Then match each of the steps with an explanation (P–T).

Sequence   *Explanation*

_____	_____	A. Action potential occurs
_____	_____	B. Resting potential
_____	_____	C. Stimulus affects neuron
_____	_____	D. Action potential propagates along axon
_____	_____	E. Cell repolarizes

P. Na^+-K^+ pump moves Na^+ out of cell, K^+ into cell; K^+ leaks out but Na^+ can't get in. Membrane is $+$ outside and $-$ inside.

Q. Na^+ channels open, Na^+ leaks into cell, and cell depolarizes a bit; membrane becomes somewhat less $+$ outside and less $-$ inside.

R. Na^+ spreads out and causes adjacent area of membrane to depolarize.

S. Na^+ channels close, K^+ channels open, and K^+ leaves cell. Membrane becomes $+$ outside and $-$ inside.

T. Threshold reached and more Na^+ channels open, allowing Na^+ to rush into cell. Membrane becomes $+$ inside and $-$ outside.

Exercise 5   (Module 28.6)

A synapse is a relay point where nerve impulses are transmitted from the synaptic terminal of a neuron to another cell. The following paragraph describes how this occurs. Complete the description by filling in the blanks.

Synapses are junctions between [1]_____ and receiving cells. Some are [2]_____ synapses, where action potentials pass from one neuron to the next. Most are chemical synapses, where chemical signals called [3]_____ pass between cells. A transmitting neuron and a receiving neuron are separated by a narrow gap, called the synaptic [4]_____. Neurotransmitter is contained in small vesicles in the synaptic terminals at the end of the [5]_____ of the transmitting cell. When an action potential arrives at the end of the transmitting cell's axon, chemical changes occur that cause the vesicles to fuse with the [6]_____. The neurotransmitter molecules are released into the synaptic cleft, rapidly diffuse across the cleft, and bind to [7]_____ molecules in the plasma membrane of the [8]_____. Commonly, this opens ion [9]_____ in the receiving cell's membrane, allowing ions to diffuse through the membrane and initiating new [10]_____ in the receiving cell. The neurotransmitter is quickly broken down by an [11]_____ or transported back into the signaling cell, and the ion channels close. This ensures that the transmission of a signal from cell to cell will be brief and precise.

Exercise 6   (Module 28.7)

Some neurotransmitters cause receiving cells to transmit signals, and other neurotransmitters inhibit the receiving cell's transmission of signals. The receiving cell "adds up" incoming signals and responds, or not, allowing for complex information processing. Neuron X, diagrammed below, adds up all the excitatory and inhibitory impulses it receives. This particular neuron transmits its own impulses only when at least *two* neurons send it excitatory impulses at the same time. Each neuron sending it inhibitory impulses can cancel out the effect of one neuron sending excitatory signals. Given the inputs outlined below, state whether neuron X would or would not transmit nerve impulses in each case.

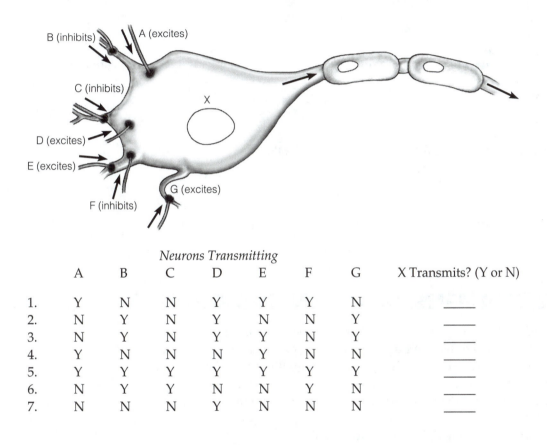

	Neurons Transmitting							
	A	B	C	D	E	F	G	X Transmits? (Y or N)
1.	Y	N	N	Y	Y	Y	N	_____
2.	N	Y	N	Y	N	N	Y	_____
3.	N	Y	N	Y	Y	N	Y	_____
4.	Y	N	N	N	Y	N	N	_____
5.	Y	Y	Y	Y	Y	Y	Y	_____
6.	N	Y	Y	N	N	Y	N	_____
7.	N	N	N	Y	N	N	N	_____

Exercise 7   (Modules 28.8–28.9)

These modules describe the functions of various neurotransmitters and the effects of stimulants, tranquilizers, depressants, and antidepressants on the nervous system. Imagine a neuron in your brain that transmits impulses to other neurons that elevate mood. This "feel good" neuron transmits when it is stimulated by serotonin from nearby neurons. It is inhibited by GABA from its neighbors. With this in mind, explain how each of the following might affect your mood.

1. Hereditary shortage of serotonin

2. A drug, such as alcohol, that enhances the effects of GABA

3. A drug that stimulates serotonin receptors

4. A chemical that blocks the enzymes that normally break down GABA after it has docked with receptors in the receiving neuron

5. A drug that blocks receptors so that they cannot respond to GABA

6. A drug that blocks the enzymes that normally break down serotonin after it has docked with receptors in the receiving neuron

7. A drug that enhances the release of serotonin by sending cells.

8. A drug that blocks the reuptake of serotonin from the synapse by sending cells.

Exercise 8   (Module 28.10)

Evolution of nervous systems reflects changes in lifestyle. The nervous systems of radially symmetrical animals are quite different from the nervous systems of animals with bilateral symmetry. Radial animals are fixed in place or slow-moving. Bilateral animals move actively through their environment. Start by defining some terms used to describe nervous systems (questions 1–6), then compare the nervous systems of three animals (questions 7–9).

Match the descriptive terms on the right with their definitions.

_____ 1. Concentration of the nervous system at the head end   A. Bilateral symmetry
_____ 2. A headless, circular body   B. Nerve cords
_____ 3. Presence of a central nervous system   C. Centralization
_____ 4. A body with a head and tail and left and right sides   D. Radial symmetry
_____ 5. A weblike network of nerve cells   E. Cephalization
_____ 6. Lengthwise groupings of nerve cells   F. Nerve net

Now match each of the animals below (7–9) with the correct terms from the list above (A–F). Multiple terms apply to each animal.
_____ 7. Which terms describe a leech?
_____ 8. Which terms describe a cnidarian (hydra)?
_____ 9. Which terms describe a human?

Exercise 9 (Modules 28.11–28.12)

Vertebrate nervous systems vary in complexity, but they all have certain features in common. Write the name of the part or division of the vertebrate nervous system that goes with each of the phrases below.

_____ 1. Master control center

_____ 2. Component of the peripheral nervous system (PNS) that transmits information to and from skeletal muscles

_____ 3. Component of the PNS that regulates internal activities

_____ 4. Fluid-filled spaces in the brain

_____ 5. Layers of connective tissue around the brain and spinal cord

_____ 6. Nerves that originate in the brain and terminate mostly in the head and upper body

_____ 7. Myelinated axons and dendrites in the CNS

_____ 8. Fluid that circulates around and through the CNS

_____ 9. All nerves and ganglia outside the brain and spinal cord

_____ 10. Nerve tissue of the brain and spinal cord consisting mainly of cell bodies and dendrites

_____ 11. Brain and spinal cord

_____ 12. Part of the CNS inside the vertebral column

_____ 13. Nerves that originate in the spinal cord and extend to parts of the body other than the head

_____ 14. Maintains a stable chemical environment for the brain

Exercise 10 (Module 28.13)

What is the overall effect of activity of the parasympathetic nervous system on the body? The sympathetic nervous system? One way to keep their actions straight is to remember that the sympathetic system triggers the "fight or flight" response to stress, gearing up the body for action. Effects of the parasympathetic system are best summarized in the phrase "rest and digest." It slows the body down, for the most part, but stimulates the digestive organs. With these effects in mind, see if you can state the effect of each system on the structures in the chart below.

Body Structure	Parasympathetic Effect	Sympathetic Effect
Stomach (via enteric nervous system)	1.	2.
Bronchi (lungs)	3.	4.
Genitals (male)	5.	6.
Heart	7.	8.
Salivary gland	9.	10.
Pupil	11.	12.

Exercise 11   (Modules 28.14–28.15)

Name each of the structures identified by numbers on this diagram of the human brain. Then match each structure or area with its function (A–G).

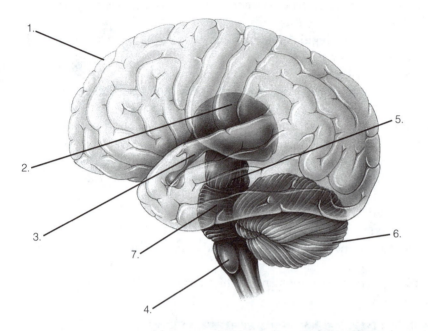

Name   Function

1. _____   _____
2. _____   _____
3. _____   _____
4. _____   _____
5. _____   _____
6. _____   _____
7. _____   _____

A. Coordination, balance, motor memory
B. Controls breathing
C. Visual reflexes, integrates auditory data
D. Controls breathing, circulation, swallowing, and digestion
E. Memory, learning, speech, emotions
F. Sorts and relays information to higher brain centers
G. Body temperature, hunger, thirst, sex, biological clock

Exercise 12   (Modules 28.15–28.19)

These modules outline activities of various parts of the cerebrum. Review these activities and other aspects of the nervous system by filling in the blanks in the following story.

Liz appeared to be sleeping quietly, but she was dreaming, and her mind was active. Beneath her closed eyelids, her eyes darted back and forth rapidly, showing that she was in a state called [1]_____ sleep. Sleep is an active state, important in the consolidation of [2]_____ and [3]_____. Two areas in the brainstem, the [4]_____ and [5]_____, contain centers that cause sleep when stimulated.

Outside her window, a car squealed around a corner. The [6]_____, a diffuse network of neurons in the brainstem, reacted, and an area in Liz's [7]_____ triggered arousal. Liz woke up. She opened her eyes and looked at the clock—3:00 A.M. Then she remembered her dream. She had been riding her bike between two rows of trees as fluffy clouds passed overhead. "That's it! The idea for my art project," she thought. She fumbled for the light switch, grabbed a pencil and sheet of paper from the bookcase, and scribbled *clouds*. As she wrote, the movement of her hand was controlled by the [8]_____ cortex in the [9]_____ lobe of her cerebrum, which sent nerve impulses to the muscles of her arm via [10]_____ neurons. The [11]_____ helped the cerebrum by making the movements smooth and coordinated. Liz clicked off the light and soon went back to sleep.

The next morning, when she saw her scrawled *clouds* note, Liz felt happier and more energized than she had in days. She hadn't realized how the looming deadline for her term art project had been getting her down. (She also didn't realize that the [12]_____ system, a group of centers in the forebrain, was actively involved in shaping these emotions, as well as in learning and memory.) Now she knew what she was going to paint for the project—a design based on clouds! She could picture the painting perfectly in her mind (actually, mostly in the [13]_____ cerebral hemisphere, which in most people specializes in spatial relations and pattern recognition). Liz started to make a mental list of the supplies she would need to do the painting.

In the art department of the campus bookstore, Liz gathered the materials she needed. She tested the texture of several watercolor papers with her fingertips. The nerve impulses from her fingers traveled along [14]_____ neurons to her [15]_____, which transmitted them up to her brain. In the brain, the [16]_____ sorted the sensory impulses and relayed them to Liz's cerebral cortex. The [17]_____ cortex in the [18]_____ lobe of Liz's cerebrum interpreted the nerve impulses from her fingers, inferring from them the textures of the papers. The sensory information from her right hand was sent to the [19]_____ side of the brain, but Liz could check it against her mental image because the nerve impulses were able to cross from one side of the brain to the other via the [20]_____, a thick band of nerve fibers connecting the two cerebral hemispheres.

One of the shelves was empty, and Liz thought the missing paper might be just the one she needed. She read the stock number on the shelf, repeated it to herself a couple of times, and kept the number in her [21]_____-term memory just long enough to ask a clerk whether the store had more in stock. No luck. Liz recalled the paper she had used for a couple of assignments the previous term. This information was stored in ^{22}_____-term memory. The ^{23}_____, part of the limbic system, had "labeled" the name of the paper for storage in memory and linked it with the emotion of happiness that Liz was feeling. Another limbic system center, the ^{24}_____, assisted in this memory-formation process, as well as later recall.

Liz found the paper, paints, and a brush that she needed and headed for the checkout line, adding up the prices in her head (actually on the ^{25}_____ side of her head, because ^{26}_____ areas in the ^{27}_____ cerebral hemisphere seem to be primarily responsible for mathematical calculations). The total was a bit more than she had anticipated, and a moment of anxiety about her budget caused Liz's heart and breathing to speed up a bit, under orders from the ^{28}_____ and ^{29}_____, two areas of the hindbrain.

With her purchases tucked under her arm, Liz set off for home on her bicycle. She crossed the park and turned onto a bike path flanked by tall maple trees. Fluffy clouds sailed overhead, and Liz thought the scene seemed very familiar.

Exercise 13 (Module 28.20)

Review your knowledge of central nervous system disorders by matching each of the phrases on the left with one or more disorders from the list on the right. Answers may be used more than once, and three questions require more than one answer.

_____ 1. Sadness, loss of interest, sleeplessness, low energy, etc.	A. Parkinson's disease
_____ 2. Two conditions occurring in about 1% of the population	B. Schizophrenia
_____ 3. Hearing voices, delusions, blunted emotions and speech	C. Bipolar disorder
_____ 4. Two conditions whose risk increases with advancing age	D. Alzheimer's disease
_____ 5. Also called manic-depressive disorder	E. Major depression
_____ 6. Often caused by a low level of the neurotransmitter serotonin	
_____ 7. Low levels of the neurotransmitter dopamine	
_____ 8. Characterized by neurofibrillary tangles and senile plaques	
_____ 9. Symptoms result from death of neurons in the midbrain.	
_____ 10. Extreme mood swings	
_____ 11. Occurs in about 1% or 65-year-olds and 5% of 85-year-olds.	
_____ 12. A motor disorder causing slowness in movement, rigidity, muscle tremors	
_____ 13. Treated with SSRIs such as Prozac, Paxil, and Zoloft	
_____ 14. May eventually be treated by transplant of embryonic dopamine-secreting neurons	
_____ 15. In a mild form, it is linked with the creativity of several famous artists, writers, and musicians.	
_____ 16. Progressive loss of memory, confusion, hostility, inability to care for self	
_____ 17. If one identical twin has it, there is a 50% chance the other will too.	

Test Your Knowledge

Multiple Choice

1. Which of the following best describes an action potential?
   a. flow of electricity along a neuron
   b. passage of ions through the membrane of a neuron
   c. flow of neurotransmitter chemical along a neuron
   d. movement of tiny filaments of protein inside a neuron
   e. change in a neuron so that the inside becomes more negatively charged

2. A part of a neuron that carries signals toward the cell body is called
   a. a nerve.
   b. white matter.
   c. a neurotransmitter.
   d. a dendrite.
   e. an axon.

3. Which of the following maintains resting potential, the difference in electrical charge inside and outside a neuron membrane that enables the cell to transmit a signal?
   a. charges that pull sodium and potassium through the membrane
   b. opening of sodium and potassium channels in the membrane
   c. the myelin sheath, which prevents ions from entering or leaving
   d. transport and leakage of sodium and potassium into and out of the cell
   e. the mutual repulsion of sodium and potassium ions

4. The _____ contains association areas for speech, language, and calculation.
   a. right cerebral hemisphere
   b. medulla oblongata
   c. midbrain
   d. left cerebral hemisphere
   e. cerebellum

5. A stimulus triggers an action potential by
   a. causing sodium ions to leak into the neuron.
   b. triggering the release of neurotransmitter.
   c. causing potassium ions to leak out of the neuron.
   d. activating the sodium-potassium pump.
   e. causing sodium ions to leak out of the neuron.

6. The ____ of a primate, dolphin, or whale is much larger than this brain region in other mammals.
   a. brainstem
   b. hypothalamus
   c. cerebral cortex
   d. limbic system
   e. medulla

7. The autonomic nervous system
   a. is the part of the nervous system outside the brain and spinal cord.
   b. controls and coordinates voluntary movements.
   c. regulates the internal environment.
   d. integrates all sensory information from the environment.
   e. consists of the brain and spinal cord.

8. Which of the following correctly matches a part of the brain with its function?
   a. thalamus—responsible for learning and memory
   b. hypothalamus—relays sensory information to cerebrum
   c. cerebrum—controls breathing and circulation
   d. cerebellum—coordinates movements
   e. medulla oblongata—interprets visual information

9. The limbic system is involved in
   a. emotions, memory, and learning.
   b. speech and hearing.
   c. vision.
   d. sleep and wakefulness.
   e. control of heartbeat and respiration.

10. Which of the following is part of the central nervous system?
    a. cranial nerve
    b. spinal nerve
    c. spinal cord
    d. sympathetic nerve
    e. ganglion

11. Which of the following disorders is characterized by low levels of the neurotransmitter serotonin?
    a. Alzheimer's disease
    b. multiple sclerosis (MS)
    c. schizophrenia
    d. Parkinson's disease
    e. major depression

12. Which of the following correctly pairs a brain structure with the embryonic structure from which it develops?
    a. cerebrum—midbrain
    b. cerebellum—forebrain
    c. hypothalamus—midbrain
    d. pons—forebrain
    e. thalamus—forebrain

Essay

1. Sketch a chemical synapse and use your sketch to explain how a nerve impulse crosses a chemical synapse from the transmitting neuron to the receiving neuron.

2. Compare the functions of the parasympathetic and sympathetic divisions of the autonomic nervous system.

3. Briefly describe three major trends in the evolution of the vertebrate brain.

4. What are the three major functions of nervous systems? How are they illustrated as you answer this question?

5. A neuron can receive inputs from thousands of transmitting neurons, but it can respond in only one of two ways—it can either transmit action potentials or not transmit action potentials. How do transmitting cells signal the receiving cell, and what determines whether the receiving cell will or will not transmit action potentials itself?

Apply the Concepts

Multiple Choice

1. The axons of which of the following end in the central nervous system?
   a. motor neurons
   b. sensory neurons
   c. interneurons
   d. a and c
   e. b and c

2. A man was admitted to the hospital suffering from abnormally low body temperature, loss of appetite, and extreme thirst. A brain scan showed a tumor located in the
   a. hypothalamus.
   b. cerebellum.
   c. pons.
   d. right cerebral hemisphere.
   e. corpus callosum.

3. A drug that causes potassium to leak out of a neuron, increasing the positive charge on the outside, would
   a. make it easier to trigger action potentials in the neuron.
   b. cause the cell to release its neurotransmitter.
   c. speed up action potentials traveling the length of the cell.
   d. act as a stimulant.
   e. inhibit transmission of action potentials by the neuron.

4. What is the difference between a neuron and a nerve?
   a. One is sensory in function, the other motor.
   b. Nerves are found only in the central nervous system.
   c. They consist of different numbers of cells.
   d. Neurons are made of white matter, nerves of gray matter.
   e. Neurons are found only in vertebrates.

5. Which of the following animals is *least* cephalized?
   a. clam
   b. ant
   c. flatworm
   d. human being
   e. fish

6. After an accident, a young man began exhibiting bizarre psychological symptoms: He was convinced his parents had been replaced by lifelike robots because he experienced no feelings when he saw or remembered them. Apparently, the accident had damaged the man's
   a. medulla oblongata.
   b. cerebellum.
   c. basal ganglia.
   d. limbic system.
   e. reticular formation.

7. Alex became so dehydrated while playing tennis that his blood pressure started to drop. His _____ detected the pressure drop and sent signals via _____ to speed up the heart, compensating for the drop in pressure.
   a. hypothalamus . . . parasympathetic neurons
   b. medulla oblongata . . . sympathetic neurons
   c. cerebellum . . . sympathetic neurons
   d. medulla oblongata . . . parasympathetic neurons
   e. cerebellum . . . parasympathetic neurons

8. The gray matter of the cerebral cortex, where most higher-level "thinking" occurs, is composed mostly of
   a. interneuron cell bodies.
   b. myelinated axons of neurons.
   c. sensory neuron cell bodies.
   d. Schwann cells.
   e. motor neuron cell bodies.

9. Which of the following neurons would both receive nerve signals from other neurons and transmit signals to other neurons?
   a. a sensory neuron from the fingertip to the spinal cord
   b. a motor neuron from the spinal cord to the leg
   c. an interneuron in the brain
   d. all of the above
   e. both the sensory and motor neurons

10. Sensory receptors in the ear can detect a whisper or a shout and can transmit nerve signals about these sounds to the brain. How would sensory neurons relay information to the brain about the *loudness* of loud and soft sounds?
    a. The action potentials would travel at different speeds.
    b. The size of action potentials would vary.
    c. The number of action potentials would vary.
    d. The frequency of action potentials would vary.
    e. Action potentials would be routed to different parts of the brain.

11. A patient experiencing numbness of the right thumb might have damage to
    a. motor neurons or the left frontal lobe
    b. sensory neurons or the left parietal lobe
    c. sensory neurons or the right parietal lobe
    d. motor neurons or the left parietal lobe
    e. sensory neurons or the right frontal lobe

12. An artist suffered a brain injury, but his artistic ability was unimpaired. An injury to the same area of the brain ended the career of a mathematician. Injury to which area might result in such different outcomes?
    a. the cerebellum
    b. the left cerebral hemisphere
    c. the hypothalamus
    d. the right cerebral hemisphere
    e. the medulla

Essay

1. Many nervous system drugs and poisons act at synapses. Explain how each of the following would alter the transmission of nerve impulses.
   a. Curare, a substance used on poison arrows by native South Americans, competes with acetylcholine for receptor sites on receiving neurons.
   b. Botulism toxin inhibits the release of acetylcholine.
   c. Diisopropyl fluorophosphate (DF) is used in warfare as a nerve gas. It blocks the enzyme that breaks down acetylcholine after it has crossed the synapse to the receiving neuron.

2. Why is transmission of a nerve signal along a neuron sometimes compared with the toppling of a row of dominoes? How does it differ from the toppling of dominoes?

3. Andrew's back was broken in an auto accident, and he was paralyzed below the shoulders. Explain why he is able to move his arms but not his legs. Like many people with such injuries, Andrew cannot feel his legs, but he does respond when the doctor tests his knee-jerk reflex. Explain how.

4. A victim of a severe head injury may live for years in a "persistent vegetative state"—unconscious, but still alive and breathing. Explain how a person can continue to live even though the cerebrum, the largest part of the brain, ceases to function.

Put Words to Work

Correctly use as many of the following words as possible when reading, talking, and writing about biology:

acetylcholine, action potential, Alzheimer's disease, autonomic nervous system, axon, biological clock, bipolar disorder, blood-brain barrier, brain, brainstem, cell body, cephalization, central nervous system (CNS), centralization, cerebellum, cerebral cortex, cerebral hemisphere, cerebrospinal fluid, cerebrum, circadian rhythm, cranial nerve, dendrite, effector cell, forebrain, ganglion (plural, *ganglia*), *glia, gray matter, hindbrain, integration, interneuron, lateralization, limbic system, long-term memory, major depression, medulla oblongata, membrane potential, memory, meninges, midbrain, motor neuron, motor output, myelin sheath, nerve, nerve cord, nerve net, nervous system, neuron, neurotransmitter, parasympathetic division, Parkinson's disease, peripheral nervous system (PNS), pons, reflex, resting potential, schizophrenia, sensory input, sensory neuron, short-term memory, sodium-potassium (Na-K) pump, somatic nervous system, spinal cord, spinal nerve, stimulus* (plural, *stimuli*), *sympathetic division, synapse, synaptic cleft, synaptic vesicle, thalamus, threshold, ventricle, white matter*

Use the Web

There is a lot of material on the nervous system on the Web at www.mybiology.com. The animations of action potentials and synaptic transmission are particularly informative.

The Senses

29

Focus on the Concepts

An animal's senses enable it to detect and respond to changes in its external and internal environment. Many animals have senses that humans do not—the echolocation of bats and electroreception of sharks, for example. In studying this chapter on the senses, focus on the following concepts:

- Sensory receptors detect stimuli, transduce them into receptor potentials, and transmit action potentials to the CNS. Stimulus type is determined by the pattern of neurons stimulated. Stimulus intensity is determined by the frequency of action potentials. Once the brain is aware of sensations, it interprets them to produce perceptions.

- Specialized receptors detect five kinds of stimuli: Thermoreceptors detect heat and cold. Mechanoreceptors, in joints, skin, or the ear, respond to mechanical energy. Chemoreceptors in the nose and taste buds pick up chemicals. Electromagnetic receptors, such as the photoreceptors of the eye, respond to magnetism, electricity, or light. Pain receptors respond to excess stimuli or tissue damage.

- The ear converts pressure waves into action potentials perceived as sound. The outer ear channels sound to the eardrum. Bones of the middle ear conduct vibrations to the coiled cochlea, where fluid is set in motion. This moves a membrane, which bends hair cells, which transmit action potentials to the brain. Different hair cells pick up different pitches; frequency of action potentials signals volume. Similarly, bending of hair cells in the semicircular canals, utricle, and saccule sense body movement and position.

- Several types of eyes have evolved among animals. Flatworms have simple eye cups that sense direction and intensity of light. Insects have compound eyes, which form images and are good at sensing movement. Vertebrates have single-lens eyes. Light enters though the pupil. The iris opens and closes the pupil to let in more or less light. The lens focuses an image on the retina, which contains many photoreceptor cells.

- In squids and many fishes, the lens is rigid, and focusing is accomplished by moving the lens back and forth. In humans, the lens changes shape to accommodate to near or distant objects. For distant vision, a muscle relaxes, allowing a ligament to tighten and flatten the lens. For close-up vision, the muscle contracts, allowing the lens to become more rounded. Artificial lenses compensate for focusing problems.

- The human retina has two kinds of photoreceptors: rods and cones. Both contain visual pigments that absorb light. Rods are more numerous and more sensitive. They can detect dim light, but only sense shades of gray. Cones are concentrated near the center of focus and can see color and detail. Different cones detect red, green, and blue light, and the colors we perceive depend on the mix of cones stimulated.

- The senses of taste and smell depend on chemoreceptors that detect molecules in the environment. A variety of olfactory receptors in the nasal epithelium can detect thousands of odors in air. Molecules in solution elicit five taste perceptions from taste receptors in the tongue's taste buds—salty, sweet, sour, bitter, and umami.

Review the Concepts

Work through the following exercises to review the concepts in this chapter. For additional review, check out the activities on the Web at www.mybiology.com. The website offers a pre-test that will help you plan your studies.

Exercise 1   (Introduction)

Some animals are able to sense stimuli not detectable by humans. Match each of the following descriptions of non-human senses with its name (A–C) and the animals or animals (L–N) that can do it.

Sense   *Animal(s)*

_____   _____   1. Ability to detect objects by listening to reflected sounds

_____   _____   2. Used as a "compass" for navigation

_____   _____   3. Can detect hidden prey or objects in low-visibility environments

A. magnetoreception
B. electroreception
C. echolocation

L. birds, fish, turtles, amphibians, bees
M. porpoises, bats
N. fish, platypus, sharks

Exercise 2   (Modules 29.1–29.2)

As you read these words, several different processes occur in your eyes and brain before you understand what the words mean. Match each of these processes (A–F) with its description (L–Q), and number the processes in the order in which they might occur.

Order   *Description*

_____   _____   A. Transduction
_____   _____   B. Sensation
_____   _____   C. Transmission
_____   _____   D. Perception
_____   _____   E. Reception
_____   _____   F. Adaptation

L. Conscious understanding of sensory data
M. Conversion of a stimulus into electrical signals
N. Sending action potentials to the brain
O. Drop in sensitivity of receptors when stimulated repeatedly
P. Awareness of sensory stimuli by the brain
Q. Detection of stimuli by sensory cells

There are five general kinds of sensory receptors—pain receptors, thermoreceptors, mechanoreceptors, chemoreceptors, and electromagnetic receptors. To review them, complete this chart. Give the general category of receptors to which each example belongs, the kind of stimulus it responds to, and where you might find one of the receptors.

Example	General Receptor Type	Stimulus It Responds To	Locations
Infrared receptor	1.	Infrared radiation	2.
Touch receptor	3.	Contact and movement	4.
5.	6.	Nearby obstacles and animals in murky water	7.
8.	9.	10.	Eyes
Pain receptor	11.	Danger, injury, disease	Many body locations
12.	Mechanoreceptor	13.	Human ear
Heat receptor	14.	15.	16.
17.	18.	Body position	19.
Smell receptor	20.	21.	Nose
Magnetic orienting mechanism	22.	23.	Heads of certain birds, fishes, mammals

The ear is a mechanical marvel, catching and amplifying sound waves and transforming them into action potentials. The inner ear also contains structures that sense body position and movement As you review the structure and function of the ear, label the following parts on the diagram: **auditory nerve, eardrum, cochlea, semicircular canals, oval window, utricle, anvil, inner ear, auditory canal, utricle (and saccule), stirrup, outer ear, pinna, hammer, Eustachian tube,** and **middle ear.** Color all structures of the outer ear red, the middle ear yellow, and the inner ear blue.

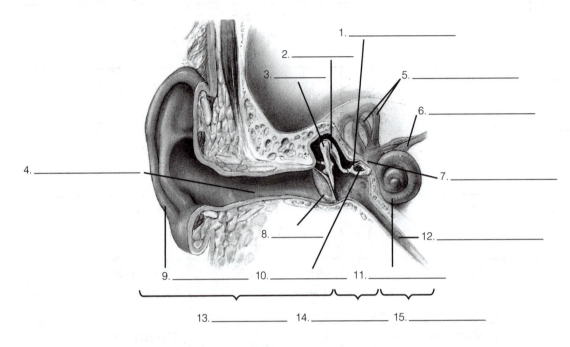

Now match each of the following functions with a part of the ear by placing its letter next to the correct label on the diagram. (Not all parts get letters.)

A. Its hair cells transduce motion of the basilar membrane into action potentials.

B. Equalizes pressure outside and inside eardrum

C. Other animals, such as dogs and horses, can turn it toward sounds.

D. Transmits action potentials from the ear to the brain

E. Transmits vibrations to middle-ear bones

F. Sense changes in the head's rotation or angular movement

G. Collects and channels sound waves to the eardrum

H. Transmits vibrations from the eardrum to the cochlea

I. Detect the position of the head relative to gravity

J. Transmits pressure waves from the stirrup to fluid within the cochlea

Exercise 5   (Modules 29.7–29.8)

Eyes can be simple or complex. The eye cups of flatworms enable them to sense light intensity and direction. The compound eyes of insects can form images, but are best at detecting movement. The single-lens vertebrate eye is a complicated and sensitive sensory organ that operates like a camera. To familiarize yourself with the major parts of the eye, label and color the diagram below. Include the **sclera, choroid, retina, pupil, iris, lens, cornea, blind spot, vitreous humor, aqueous humor, artery and vein, fovea, optic nerve,** the **ciliary body** and the **ligament** attached to the lens.

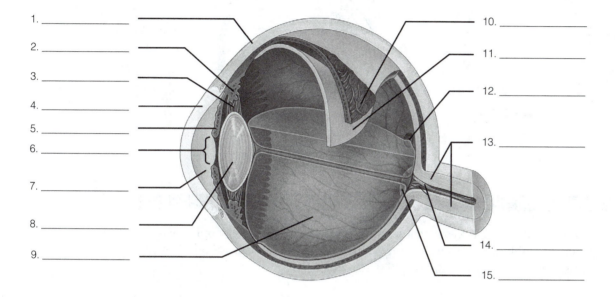

1. _____
2. _____
3. _____
4. _____
5. _____
6. _____
7. _____
8. _____
9. _____

10. _____
11. _____
12. _____
13. _____
14. _____
15. _____

Now match each of the following functions with a part of the eye by placing its letter next to the correct label on the diagram.

A. Carries nerve signals to the brain

B. Covers the front of the eye but lets light in

C. Supplies nutrients and oxygen to the lens, iris, and cornea

D. Photoreceptor cells here transmit action potentials to the brain.

E. Regulates the size of the pupil to let more or less light into the eye

F. Bends light rays and focuses them on the retina

G. Muscles in this structure change the shape of the lens.

H. Photoreceptors are highly concentrated at this center of focus.

I. Works with the ciliary body in accommodation

J. Jellylike "filling" that maintains the shape of the eyeball

K. The fibrous "white" of the eye

L. Hole in the iris that admits light

M. Supply the eye with blood

N. Pigmented layer inside the sclera

O. Area of retina with no photoreceptors

The lenses in the eyes of a fish or squid move back and forth to bring images into focus, but the lenses in human eyes focus by changing shape, a process known as accommodation. After studying Module 29.8, sketch the path of light rays and the shape of the lens when the eye focuses on a nearby object and a distant object.

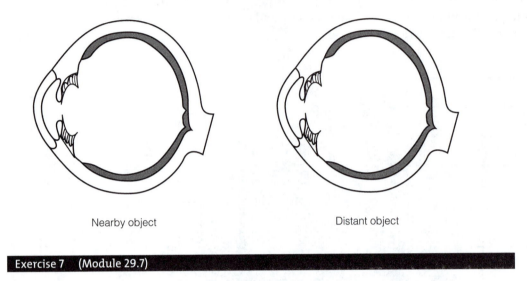

Nearby object   Distant object

There are a number of possible reasons for less-than-perfect vision. Some people are near-sighted, others farsighted. Some suffer from astigmatism. After reading about these vision problems, sort them out by completing the following chart.

Diagram			
Problem name	1.	2.	3.
Nearby or distant objects clear	4.	5.	6.
Nearby or distant objects unclear	7.	8.	9.
Cause(s)	10.	11.	12.
Lens used for correction	13.	14.	15.

Exercise 8   (Module 29.10)

This module discusses rods and cones, the sensory receptors of the retina. Indicate below the statements that relate to rods and those that relate to cones by placing checkmarks in the appropriate column. (Some statements relate to both.)

	Rods	Cones
1. Three different types		
2. Stimulated by bright light		
3. Function in night vision		
4. Most numerous in the fovea		
5. Located in the back layer of the retina		
6. Contain a visual pigment called rhodopsin		
7. Can see different colors		
8. Stimulus transducers		
9. Absent from the fovea		
10. Most numerous at the outer edges of the retina		
11. Contain pigments called photopsins		
12. See only shades of gray		
13. Transmit signals to the visual cortex		
14. Deficient in color blindness		
15. Synapse with retinal neurons		
16. Most numerous receptors		
17. Most sensitive receptors		
18. Concentrated in fovea of a hawk		
19. More numerous in animals active at night		

Exercise 9   (Modules 29.11–29.13 and Summary)

Modules 29.11 and 29.12 discuss the senses of smell and taste, and Module 29.13 reviews the role of the senses in the overall function of the nervous system. Review all the senses (including vision, hearing, and equilibrium) by matching each of the phrases on the left with a sensory structure from the list on the right. Some answers are used more than once.

_____ 1. Receptors here detect changes in head movement.
_____ 2. Receptors here sense sour, salty, sweet, bitter, umami.
_____ 3. Odor molecules bind to proteins on its cilia.
_____ 4. Site of receptors for hearing
_____ 5. Light receptor of the retina
_____ 6. Receptors here sense position of head relative to gravity.
_____ 7. Detects molecules in solution
_____ 8. Site of photoreceptors
_____ 9. Balance receptor
_____ 10. Chemoreceptor of the nasal cavity
_____ 11. This kind of receptor may lose sensitivity as we age.
_____ 12. Receptor of hearing
_____ 13. If you have more of the structures with these, you could be a "supertaster."

A. Retina
B. Semicircular canal
C. Utricle and saccule
D. Hair cell
E. Olfactory receptor
F. Taste bud
G. Organ of Corti
H. Cone
I. Taste receptor

Test Your Knowledge

Multiple Choice

1. Insect eyes are much better at seeing _____ than are our eyes.
   a. colors
   b. in dim light
   c. movement
   d. fine detail
   e. distant objects

2. A thermoreceptor in the skin converts heat energy into action potentials. This conversion process is called
   a. sensation.
   b. transduction.
   c. reception.
   d. integration.
   e. perception.

3. Which of the following correctly traces the energy of sound waves into the ear?
   a. auditory canal–eardrum–ear bones–cochlea
   b. eardrum–auditory canal–cochlea–ear bones
   c. auditory canal–ear bones–eardrum–cochlea
   d. eardrum–auditory canal–ear bones–cochlea
   e. eardrum–ear bones–auditory canal–cochlea

4. Which of the following are *not* correctly paired?
   a. mechanoreceptor—stretch receptor
   b. electromagnetic receptor—photoreceptor
   c. chemoreceptor—taste bud
   d. mechanoreceptor—touch receptor
   e. electromagnetic receptor—hair cell

5. When you focus your eyes on a nearby object, the lenses
   a. decrease in diameter.
   b. move forward.
   c. become more rounded.
   d. become more flattened.
   e. move backward.

6. As light passes into the eye, it goes through which of the following first?
   a. lens
   b. pupil
   c. aqueous humor
   d. cornea
   e. vitreous humor

7. The brain determines the loudness of a sound from
   a. the part of the organ of Corti stimulated by the sound.
   b. the frequency of action potentials received.
   c. the size of air pressure changes in the middle ear.
   d. the part of the brain receiving nerve impulses from the ear.
   e. the size of the action potentials received.

8. Josh is color blind, so he has a lot of trouble picking out clothes. Which sensory structures below are affected in a color-blind person?
   a. organ of Corti
   b. cones
   c. hair cells
   d. rods
   e. utricle and saccule

9. Which of the following is *not* a mechanoreceptor?
   a. taste bud
   b. hair cell
   c. touch receptor
   d. pressure receptor
   e. stretch receptor

10. Eating carrots really is good for your eyes. Carrots contain vitamin A, which is used to make a substance called rhodopsin, which
    a. is a visual pigment that absorbs light.
    b. provides energy for the function of rods and cones.
    c. colors the iris of the eye.
    d. stimulates the neurons in the retina to form branches and connections.
    e. keeps the lens clear and transparent.

Essay

1. Bats can navigate in total darkness by listening to the echoes of their high-pitched clicks. In a few sentences, describe three other examples of animal sensory capabilities beyond the range of human senses.

2. Explain the difference between sensation and perception.

3. What is farsightedness? What eye defect causes it? How is farsightedness corrected?

4. Explain how a stimulus (light, chemical, temperature change, movement) triggers a nerve impulse in a sensory receptor.

5. How are the sense of smell and sense of taste similar? What are the differences between the senses of smell and taste?

Apply the Concepts

Multiple Choice

1. You look all over for your glasses and then find them on your forehead. You couldn't feel them because of
   a. perception.
   b. transduction.
   c. sensation.
   d. adaptation.
   e. accommodation.

2. Michael and Laura are both looking at an illustration in her psychology textbook. Michael says, "It is an old woman looking down toward the left." Laura says, "I see a young woman looking away from us." How can they look at the same picture and see different things?
   a. Perhaps one of them has astigmatism.
   b. Their sensations differ.
   c. They sense the same thing, but their perceptions differ.
   d. It may take some time before their eyes accommodate to the picture.
   e. They perceive the same thing, but their sensations differ.

3. Whales and dolphins are known to send out clicking sounds and listen to the echoes. This suggests that they might find their prey in the same way
   a. a rattlesnake finds a mouse in total darkness.
   b. a salmon locates its home stream.
   c. an owl locates a mouse in a dark barn.
   d. an electric eel finds its prey in a muddy river.
   e. bats navigate in dark caves.

4. As you read this sentence, how do the photoreceptors of your retina tell the brain whether an area is light (the paper) or dark (the print)?
   a. Either rods or cones send signals to the brain.
   b. Nerve signals are sent to different areas of the brain.
   c. Signals are sent to the brain at different frequencies.
   d. Accommodation adjusts for light and dark areas.
   e. Larger (white) or smaller (dark) action potentials are sent to the brain.

5. A fish detects vibrations in the water around it by means of its "lateral lines," rows of sensory receptors along each side of the body. Based on what you know about sensory receptors, the lateral line receptors are probably most similar to
   a. receptors in the organ of Corti.
   b. rods and cones.
   c. receptors in taste buds.
   d. receptors in the retina.
   e. olfactory receptors.

6. Damage to the nerve from the saccule and utricle to the brain could result in
   a. loss of sense of taste.
   b. blindness.
   c. dizziness.
   d. loss of sense of smell.
   e. deafness.

7. There may be as many as 50 different kinds of smell receptors, each activated by different molecules. We can differentiate among thousands of different smells because each substance stimulates a different combination of these receptors. In much the same way,
   a. our eyes can distinguish many different colors.
   b. our ears can differentiate between sounds of many different pitches.
   c. pain receptors can distinguish between mild and severe pain.
   d. our eyes can adjust to light of differing intensities.
   e. our ears can determine how loud a sound is.

8. It is said that if you are seasick, it is better to look out at the water than at the boat. Why?
   a. It fools your brain into thinking that you are not really moving.
   b. This stimulates the saccule and utricle, organs of equilibrium.
   c. Seeing that you are moving reduces conflict between vision and equilibrium.
   d. Keeping your head level reduces activity of the semicircular canals.
   e. Actually, looking at the water just makes seasickness worse.

9. The eyes of a nocturnal animal, such as an owl, could be expected to have a larger proportion of _____ than do our eyes.
   a. chemoreceptors
   b. cones
   c. rods
   d. hair cells
   e. mechanoreceptors

10. The carotid body is a structure in the wall of the main artery carrying blood to the head. In it are special receptors that monitor blood pressure. These receptors would belong to which of the following groups?
    a. chemoreceptors
    b. thermoreceptors
    c. photoreceptors
    d. mechanoreceptors
    e. electromagnetic receptors

11. There is a small muscle attached to the inner ear bones that tightens them up in response to loud sounds, so the inner ear is not damaged. This muscle adjusts to sound in much the same way as the _____ in the eye adjusts to light.
    a. retina
    b. iris
    c. lens
    d. ciliary body
    e. fovea

Essay

1. Theresa is suffering from a mysterious disorder called Ménière's disease. Her symptoms are quite debilitating: loud roaring sounds and dizziness. Why do these symptoms occur together? What part of the body must be affected by Ménière's disease?

2. The eyes contain three kinds of cones—green, blue, and red. Which are stimulated when we see blue? Red? White? Black?

3. Some researchers believe that human beings might have an ability to sense direction that is distinct from the traditional "five senses." The human brain contains magnetite, a magnetic mineral, and it has been suggested that perhaps we have some ability to find our way by sensing the Earth's magnetic field. Describe an experiment that could test this suggestion.

4. Most sensory receptors are sensitive to a sudden change in a stimulus, but soon adapt—become less sensitive—to a continuing stimulus. For example, thermoreceptors quickly adapt, and a very hot bath soon feels comfortable. How might this be useful? Pain receptors are an exception; they do not readily adapt. Why do you think this is the case?

5. There are many different kinds and causes of deafness. Explain how each of the following would result in deafness: damage to hair cells from repeated loud sounds, brain injury, arthritis of the middle-ear bones, a torn eardrum, buildup of earwax in the auditory canal.

6. Most sensory organs come in pairs—two ears, two eyes, even two infrared detecting pits on the head of a rattlesnake and two eye cups on a flatworm. Why do you think it is adaptive for animals to have paired sensory structures?

Put Words to Work

Correctly use as many of the following words as possible when reading, talking, and writing about biology:

accommodation, astigmatism, auditory canal, chemoreceptor, cochlea, compound eye, cone, cornea, eardrum, electromagnetic receptor, Eustachian tube, farsightedness, fovea, hair cell, inner ear, iris, lens, mechanoreceptor, nearsightedness, middle ear, organ of Corti, outer ear, pain receptor, perception, photoreceptor, pupil, receptor potential, retina, rod, sclera, semicircular canals, sensation, sensory adaptation, sensory receptor, sensory transduction, stretch receptor, thermoreceptor, visual acuity

Use the Web

Continue your exploration of the senses on the Web at www.mybiology.com.

How Animals Move

30

Focus on the Concepts

This chapter explores animal locomotion, skeletons, and muscles. Focus on the following concepts:

- Animal movement involves muscles pulling against a firm skeleton. Water supports the body of an aquatic animal, but friction slows movement. Many aquatic species are streamlined. Air offers little resistance, but a land animal must support itself against gravity. Various adaptations facilitate hopping and running. Animals that crawl must fight friction. Snakes undulate, and worms move by wavelike peristalsis. Wings are curved airfoils that generate lift, enabling flyers to overcome gravity.

- Skeletons function in support, movement and protection. An earthworm has a hydrostatic skeleton, consisting of fluid held under pressure in body compartments. An exoskeleton, like that of an insect, covers the body and offers good protection, but must be molted as the animal grows. An endoskeleton, consisting of hard elements situated among soft tissues, grows with the animal.

- Vertebrates have endoskeletons made of cartilage and/or bone. Bones are complex living organs. The axial skeleton consists of the skull, vertebrae, and ribs. The appendicular skeleton consists of the appendages and supporting girdles. Genetic changes—especially changes in *Hox* genes—have modified appendages and regions of the vertebral column. For example, the giraffe has elongated cervical vertebrae, a snake has no legs and mostly thoracic vertebrae, and the human sacrum consists of five fused bones.

- Muscles pull against bones, which act as levers, to produce movement. Muscles work in antagonistic pairs to produce opposing movements. A muscle consists of numerous muscle cells, called muscle fibers. Each fiber contains thousands of myofibrils, which are bundles of thick and thin filaments, composed of the proteins myosin and actin. The thick and thin filaments overlap and are arrayed in numerous repeating longitudinal segments called sarcomeres. Sliding of the filaments along each other shortens the muscle.

- A nerve impulse triggers depolarization of the muscle fiber membrane. This causes the E.R. to release calcium ions, which bind to the thin filaments. This opens up binding sites for myosin cross-bridges, projecting from the thick filaments. Breakdown of ATP causes the myosin heads to move. The heads attach to the actin of the thin filaments and bend. This slides the thick and thin filaments along each other, so they overlap more. Sliding shortens the sarcomeres, the muscle, fiber, and the whole muscle.

- During exercise, muscle cells first use stored ATP, then convert CrP into ATP. Most ATP is made via cellular respiration. Glucose is the main fuel. Glycogen in muscle and the liver may be tapped for more glucose, and fats can also be used for fuel. Myoglobin stores some oxygen, if needed. If ATP demand outstrips oxygen supply, cells may switch to anaerobic glycolysis, which makes a smaller amount of ATP, and lactate as a waste product.

- Muscles consist of a mix of slow, intermediate, and fast fibers, specializing in different jobs. Slow fibers depend on aerobic respiration and contract slowly, but because they contain much myoglobin, more mitochondria, and have a good blood supply, they are slow to fatigue. Fast fibers depend on anaerobic metabolism to contract quickly. They have less myoglobin and fewer capillaries and mitochondria, so they have less endurance. Sprinters have more fast fibers, and distance runners have more slow fibers.

Review the Concepts

Work through the following exercises to review animal movement. For additional review, explore the activities on the Web at www.mybiology.com. The website offers a pre-test that will help you plan your studies.

Exercise 1   (Introduction and Module 30.1)

Animals have evolved many means of locomotion. Test your knowledge of locomotion by filling in the blanks.

Mammals range in size from tiny shrews to great whales. They have become adapted to a variety of lifestyles on land, in the water, and in the air.

Most mammals, from mice to musk oxen, walk or run on land. The surrounding air offers little resistance to movement, so [1]_____ is not much of a problem, even for the fastest runners. But a mammal that walks or runs must expend considerable effort supporting itself against the force of [2]_____. To maintain balance and stability when it walks, an animal employs the principle of the tripod, keeping [3]_____ feet on the ground most of the time. Fast runners, such as the horse, have bodies built for speed. The horse's spleen stores extra [4]_____, and its legs are lengthened by modified foot bones called [5]_____ bones. The fastest land animal, the cheetah, is of course a mammal. When it runs, all of its feet may leave the ground at once, but its [6]_____ stabilizes its position.

Some mammals crawl or burrow—moles and gophers, for example. Unlike the earthworm, which moves by waves of muscular contractions called [7]_____, burrowing mammals use their legs for digging. The mole's coat is velvety smooth, reducing [8]_____ as it tunnels underground.

Millions of years ago, a group of shore-dwelling mammals took up life in the sea. Their descendants are the whales and dolphins. Because of their buoyancy, [9]_____ is no problem for these aquatic animals, and some have grown to enormous size. Water is much more dense than air, but a whale's tapered, [10]_____ shape enables it to slip through the water with little effort. The

whale's tail forms a pair of broad, flat flukes, and the whale bends its body
[11]_____ to push it through the water.

     Bats are the only flying mammals. A bat's wings are modified arms, with skin
stretched between elongated fingers. The bat's wings, like those of airplanes and birds,
are [12]_____. As a bat beats its wings, air travels farther over the top surface
than the bottom surface, making the pressure underneath the wings
[13]_____ than the pressure on top. This generates [14]_____
and allows the bat to overcome the pull of [15]_____.

     No matter how an animal moves, whatever its environment or adaptations, all
movement boils down to one basic mechanism—the movement of microfilaments in
[16]_____ pulling against a firm [17]_____. Mammals illustrate
this principle in many different ways.

Exercise 2   (Module 30.2)

There are three main kinds of skeletons. Summarize your knowledge of skeletons by completing this chart.

	Hydrostatic Skeleton	*Exoskeleton*	*Endoskeleton*
Example animals	1.	2.	3.
Description of skeleton	4.	5.	6.
Materials skeleton is made of	7.	8.	9.
Functions of skeleton	10.	11.	12.
Drawbacks of skeleton	13.	14.	15.

The human skeleton is unique in several ways, but it has the same overall structure as the skeleton of a mouse or a chicken. It is useful to know the names of some of the major bones of the skeleton. Label these bones on the diagram below: **humerus, sternum, femur, tarsals, tibia, carpals, radius, phalanges, ulna, vertebrae, skull, clavicle, metacarpals, ribs, scapula, fibula, pelvic girdle, metatarsals, shoulder girdle,** and **patella.** (Note: The letters relate to Exercise 6, below.) Before you leave this exercise, color the bones of the axial skeleton green and the appendicular skeleton yellow.

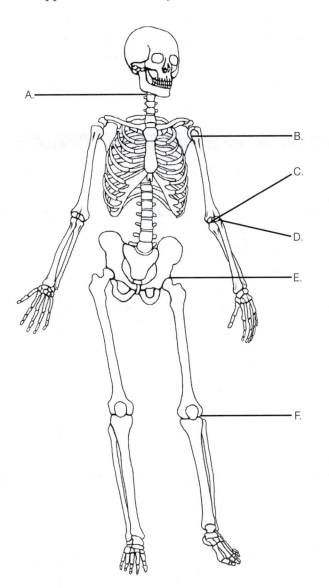

Exercise 4   (Module 30.4)

Vertebrates have a vertebral column, which is composed of different regions. Modifications of the vertebral column and the limbs, as seen in humans, chickens, and snakes, result from differences in gene expression, particularly differences in *Hox* genes. Match each of the following statements with the correct region of the vertebral column.

_____ 1. Vertebrae of the chicken neck
_____ 2. Fused together in humans
_____ 3. In all vertebrates, forelimbs start between these and cervical vertebrae.
_____ 4. Vertebrae of the lower back
_____ 5. Development of these vertebrae is controlled by *Hoxc6* and *Hoxc8*.
_____ 6. In humans, these vertebrae are immediately behind the head.
_____ 7. In a python, these vertebrae are immediately behind the head.
_____ 8. Make up the human tailbone
_____ 9. Humans have 12 of these.
_____ 10. Vertebrae usually bearing ribs
_____ 11. A cow has 7 of these.
_____ 12. Vertebrae of a monkey's chest
_____ 13. Vertebrae behind the thoracic vertebrae in a rabbit
_____ 14. Human vertebrae homologous with vertebrae in a cat's tail
_____ 15. Nearly all of the vertebrae of a snake

A. cervical
B. thoracic
C. lumbar
D. sacral
E. coccygeal

Exercise 5   (Modules 30.4–30.5)

Review bone structure, growth, disease, and injury by matching each phrase on the left with a word or phrase from the right.

_____ 1. Fat stored in central cavity
_____ 2. Made of protein fibers, calcium, and phosphorus
_____ 3. Outer hard layer of bone, forms shafts of long bones
_____ 4. Cushions ends of bone at joints
_____ 5. Decrease after menopause may cause osteoporosis
_____ 6. A break in a bone
_____ 7. Site of red marrow
_____ 8. Delivers food and hormones to bone
_____ 9. Produces blood cells
_____ 10. Bone break that occurs from long-term repeated stresses
_____ 11. Deficiency of this in the diet may contribute to osteoporosis.
_____ 12. Covers outside of bone, helps repair fractures
_____ 13. Crystals of this and calcium harden the matrix.
_____ 14. Characterized by low bone mass and deterioration of bone
_____ 15. Protein fibers that make matrix flexible

A. Spongy bone
B. Cartilage
C. Calcium
D. Red bone marrow
E. Blood vessel
F. Fibrous connective tissue
G. Yellow bone marrow
H. Compact bone
I. Osteoporosis
J. Collagen
K. Stress fracture
L. Matrix
M. Estrogen
N. Phosphorus
O. Fracture

Exercise 6   (Module 30.6)

Which of the letters on the diagram of the skeleton in Exercise 3 identifies each of the following types of joints? (Three are from the module; you will have to figure out the other three yourself.)

1. Ball-and-socket joints: _____
2. Hinge joints: _____
3. Pivot joints: _____

Exercise 7   (Modules 30.7–30.8)

Many muscles work in opposing pairs to move the skeleton. Each muscle is composed of numerous muscle cells, called muscle fibers. Each fiber contains smaller parts, which are responsible for muscle contraction. Review the contractile machinery of muscle by labeling these parts in the diagram below: **biceps muscle, triceps muscle, thin filament, muscle fiber, myofibril, tendon, thick filament, sarcomere,** and **Z line.**

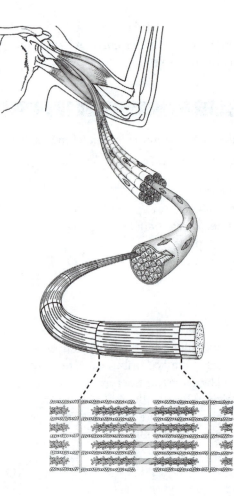

Try to visualize the inner workings of a muscle by filling in the blanks below.

Your new mission is to view contracting muscle cells close up. You step into the Microtron, and a moment later, you are working your way against the current in a small arm vein. Ahead are smaller, thin-walled [1]_____ that exit from the muscle and join to form the vein. You stay close to the wall and enter the smallest vessel.

Through the translucent wall of the capillary, you can make out a nearby surface that seems to be made of shiny white ropes. This is the [2]_____ that connects the end of the biceps to the lower arm bones. Also visible is a smoother, shinier "wire"—one of the [3]_____ that controls contraction of the muscle. Ahead, you see the wrinkled blob of a white blood cell slipping through a gap in the vessel wall, and you follow it into the intercellular fluid. Close up, the nerve you saw a moment ago appears frayed—its neurons branch and rebranch to control every muscle cell. You follow one of the neurons; it branches to about 100 cells. The neuron and the cells it controls make up a [4]_____—a group of muscle fibers that work together.

Following one of the axons to its end, you finally arrive at a [5]_____, where the axon synapses with a muscle cell. Each cell is called a muscle [6]_____. You are deep inside the muscle now, so you radio for more light and ask for the subject to tense her arm.

Suddenly a cloud of particles sprays from the end of the axon, the membrane of the muscle fiber seems to shimmer, and you are tossed by a surge of pressure into a nearby tangle of connective tissue fibers. She's moving the arm just a little bit, but to you it feels like an earthquake! You contact your colleagues: "Hey, take it easy! Try it again, but gently!" This time you hang on tightly. Again there is a spray of particles—the release of [7]_____ from the end of the axon, signaling the muscle fiber to contract. When the neurotransmitter molecules contact the muscle fiber membrane, the shimmer you saw before is repeated—an [8]_____ spreading across the cell. Another earthquake, but this time you are ready.

You grip the membrane of the muscle fiber. It is pocked with numerous openings, where infoldings of the membrane form [9]_____ that carry action potentials deep into the cell. You swim inward, following the action potentials, as the cell continues to receive impulses and contract. You can press your powerful spotlight against the tubule wall and illuminate the inside of the fiber. Inside are numerous [10]_____, large bundles of parallel protein filaments. Transverse [11]_____ lines, made of protein, separate repeating [12]_____, the basic contractile units of the muscle fiber. The filaments themselves are of two types: The [13]_____ filaments look like twisted strings of beads, the beads themselves being globular [14]_____ molecules. The [15]_____ filaments are made of elongated [16]_____ molecules, each with a head that can reach out and pull against the thin filaments.

You watch the thick and thin filaments [17]_____ along each other with each volley of nerve impulses. As each action potential travels into the cell via the tubules, you see a cloud of [18]_____ ions released from storage in the endoplasmic reticulum. The calcium ions attach to troponin molecules and shift strands of tropomyosin that wrap around the thin filaments, freeing up binding sites for the myosin [19]_____ of the adjacent thick filaments. Meanwhile, the myosin molecules of the thick filaments are writhing like a bundle of worms. Their heads use energy from [20]_____ molecules to detach from the thin filaments, straighten, attach, and bend, several times per second. The hundreds of myosin heads in one thick filament pull on all the surrounding thin filaments. This creates the pull that causes the thick and thin filaments to slide together, [21]_____ the sarcomere, and on a larger scale, the muscle fiber.

The nerve impulses cease. You can see [22]_____ ions being pumped out of the cytoplasm and back into the [23]_____. As the binding sites on the thin filaments close up, the myosin heads let go, and the muscle stops contracting. Time to make your exit. Your muscles propel you out of the muscle fiber, and you are soon back in the lab.

Exercise 9   (Module 30.11)

Muscles rely on various chemicals and chemical processes to supply energy for contraction. Match each of the following statements with the correct chemical. Some answers are used more than once.

_____   1. Used to break cross-bridges between actin and myosin molecules

_____   2. Ten to twenty-five seconds of ATP can be made by transferring P from this molecule.

_____   3. Blood vessels supply this and glucose to exercising muscles.

_____   4. The main energy-rich sugar consumed in aerobic respiration

_____   5. The main by-product of aerobic respiration

_____   6. Carries oxygen in the blood

_____   7. A muscle protein that stores oxygen

_____   8. Lactic acid fermentation can use this as an energy source.

_____   9. Broken down in muscle and liver to provide glucose for respiration

_____   10. Can be used as fuel, but a slower process than getting ATP from glucose

_____   11. Anaerobic respiration can make only a fraction as much as aerobic respiration.

_____   12. Lactic acid fermentation occurs when this is no longer available.

_____   13. Oxygen debt helps metabolize this after anaerobic exercise.

_____   14. Muscles typically have about 4–6 seconds' supply of this on hand.

A. hemoglobin
B. lactate
C. ATP
D. fats
E. oxygen
F. myoglobin
G. glucose
H. PCr
I. glycogen
J. carbon dioxide

Muscles contain a mix of fast and slow fibers, and the mix of fibers present in muscle affects athletic performance. Study the two tables below, and then answer the questions.

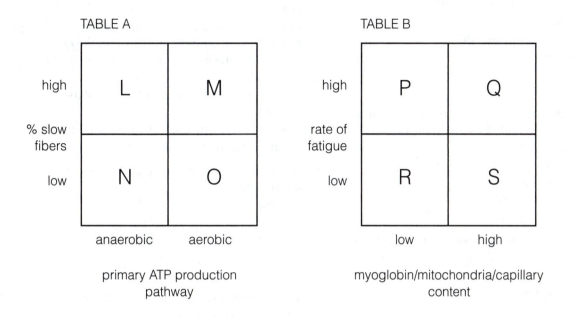

TABLE A	TABLE B

_____ 1. Which quadrant of table A (L, M, N, or O) describes the muscles of a sprinter?
_____ 2. Which quadrant of table A (L, M, N, or O) describes the muscles of a marathoner?
_____ 3. Which quadrant of table B (P, Q, R, or S) describes the muscles of a sprinter?
_____ 4. Which quadrant of table B (P, Q, R, or S) describes the muscles of a marathoner?
_____ 5. Which quadrant of table B (P, Q, R, or S) describes fast fibers?
_____ 6. Which quadrant of table B (P, Q, R, or S) describes slow and intermediate fibers?

Test Your Knowledge

Multiple Choice

1. _____ connects a muscle to a bone.
   a. Cartilage
   b. A neuromuscular junction
   c. A tendon
   d. A myofibril
   e. A motor unit

2. Inside a muscle fiber, _____ trigger(s) contraction and _____ provide(s) the energy.
   a. myosin . . . actin
   b. calcium ions . . . ATP
   c. actin . . . myosin
   d. calcium ions . . . myosin
   e. ATP . . . calcium ions

3. Which of the following is associated with osteoporosis?
   a. age
   b. calcium in the diet
   c. exercise
   d. estrogen
   e. all of the above

4. Muscles are arranged in pairs,
   a. so if one is injured, the other can take over.
   b. doubling their strength.
   c. because one pulls while the other pushes.
   d. enabling them to perform opposing movements.
   e. so they can take turns contracting and resting.

5. A marathoner's muscles would be expected to have more _____ than the muscles of a sprinter.
   a. mitochondria
   b. capillaries
   c. myoglobin
   d. all of the above
   e. none of the above

6. Which of the following correctly pairs an example of a joint with its general type?
   a. hinge—elbow
   b. pivot—shoulder
   c. hinge—hip
   d. ball and socket—elbow
   e. ball and socket—knee

7. Which of the following skeletons works poorly on land?
   a. endoskeleton
   b. hydrostatic skeleton
   c. exoskeleton
   d. a and b
   e. b and c

8. A stronger muscle contraction occurs when the brain
   a. activates muscle cells more quickly.
   b. sends stronger nerve impulses to the muscle.
   c. activates the motor units of the muscle one at a time.
   d. signals a larger number of motor units to contract.
   e. sends nerve impulses to the muscle one at a time instead of in bursts.

9. Which of the following animals has an endoskeleton?
   a. clam
   b. sea star
   c. sponge
   d. a and b
   e. b and c

10. Which of the following is part of the human appendicular skeleton?
    a. vertebral column
    b. scapula
    c. rib
    d. sternum
    e. skull

Essay

1. How do the shape of its wings and their movement through the air lift a bird off the ground?

2. What are the three major functions of a skeleton? Describe each, using parts of your own skeleton as illustrations.

3. What kinds of animals have jointed exoskeletons? What are the major advantages and disadvantages of exoskeletons, compared with endoskeletons?

4. What two forces must be overcome by a moving animal? Which is more of a problem on land? In the water?

5. Without looking it up, make a sketch showing the arrangement of thick and thin filaments in a muscle cell when the cell is relaxed and when the cell is contracting.

6. How does a nerve impulse trigger the movement of a muscle fiber? Briefly describe the sequence of events, step by step.

Apply the Concepts

Multiple Choice

1. Which of the following shortens when a muscle fiber contracts?
   a. thin filament
   b. myosin molecule
   c. sarcomere
   d. actin molecule
   e. thick filament

2. Which of the following ranks the parts in order, from largest to smallest?
   a. muscle, myofibril, filament, fiber
   b. fiber, muscle, myofibril, filament
   c. muscle, fiber, filament, myofibril
   d. myofibril, muscle, fiber, filament
   e. muscle, fiber, myofibril, filament

3. His doctor suspects that Jon might be suffering from leukemia. The doctor orders a sternal puncture, a test that samples red marrow. This test is performed to determine
   a. whether the marrow holds sufficient fat reserves.
   b. how fast Jon is growing.
   c. the rate of blood flow through the bone tissues.
   d. whether blood cell production is occurring normally.
   e. how fast cartilage is forming in the bones.

4. For every bone in the arm and hand, there is a corresponding bone in the leg and foot. Which of the following matches corresponding bones?
   a. humerus—femur
   b. metacarpals—tarsals
   c. radius—femur
   d. carpal—patella
   e. humerus—tibia

5. An animal that crawls or burrows faces problems similar to those faced by an animal that
   a. swims.
   b. runs.
   c. flies.
   d. walks.
   e. hops.

6. A scallop escapes from danger by clapping its shells together, which shoots out a stream of water and causes the scallop to hop backward. This is most like the movement of
   a. an earthworm.
   b. a water beetle.
   c. a jelly.
   d. a bird (but under water instead of in the air).
   e. a whale.

7. Smooth muscle is a type of muscle found in the internal organs—the walls of blood vessels and the intestine, for example. It is called "smooth" because the cells of smooth muscle lack the alternating light and dark bands seen in the muscle fibers that move the skeleton. Which of the following do you think might best explain this difference? Smooth muscle
   a. does not require nervous stimulation to contract.
   b. does not contain the regular, repeating filaments of skeletal muscle.
   c. cells are much larger than the cells of skeletal muscle.
   d. motor units are much smaller than those of skeletal muscle.
   e. requires much less blood flow than skeletal muscle.

8. Which of the following drugs would cause muscle spasms or cramps (uncontrolled contractions)? A drug that
   a. blocks the release of calcium ions from endoplasmic reticulum.
   b. prevents the release of acetylcholine from motor neurons.
   c. blocks neurotransmitter receptors on muscle fiber membranes.
   d. blocks the enzyme that breaks down acetylcholine after contraction.
   e. prevents attachment of myosin heads to thin filaments.

9. If you were to cut through a muscle fiber and look at the cut end under a powerful microscope, what would the contractile parts of the cell look like?
   a. many crisscrossing lines
   b. thousands of tiny dots
   c. thousands of overlapping circles
   d. numerous irregular splotches
   e. many overlapping, parallel lines

10. When a muscle is exercising hard, it quickly uses up its store of ATP and soon after runs out of oxygen. It can make more ATP aerobically by getting stored glucose from _____ and stored oxygen from _____. Finally it has to break down glucose anaerobically, making _____.
    a. lactate . . . hemoglobin . . . CrP
    b. glycogen . . . hemoglobin . . . lactate
    c. glycogen . . . myoglobin . . . lactate
    d. glycogen . . . lactate . . . CrP
    e. myoglobin . . . hemoglobin . . . lactate

11. Which of the following shortens when a muscle fiber contracts?
    a. sarcomeres and myofibrils
    b. thick and thin filaments
    c. thin filaments and sarcomeres
    d. thick filaments and myofibrils
    e. actin and myosin

Essay

1. Look at the skeleton of the frog in Module 30.2. In what ways is the frog's skeleton adapted for its way of life?

2. Both roundworms and earthworms possess hydrostatic skeletons. The body of a roundworm is a single elongated sac, filled with fluid. The roundworm is able to move only by thrashing from side to side. Compare this with the structure and movement of an earthworm, and explain the difference.

3. The wings of birds come in several different shapes. Many small songbirds have short, stout wings. The wings of an eagle are long and broad, while those of a gull are long and narrow. How are these different wing shapes (structures) well suited to the activities and environments (functions) of these different birds?

4. A map of the human cerebral cortex shows that the area of the brain that controls muscle activity of the hands is equal in size to the area that controls all the muscles below the neck. The total size of the muscles moving the hands is much smaller than all the muscles of the trunk, legs, and so on. Why do you think so much nerve tissue is required to control the hands? Explain in terms of the motor units that make up the muscles.

5. Early reconstructions of dinosaur skeletons showed them with their legs splayed out to the sides, like present-day alligators and lizards. Most dinosaur experts now think that the legs of dinosaurs were directly under the torso, like those of a dog or elephant. Why do you think they consider this important?

Put Words to Work

Correctly use as many of the following words as possible when reading, talking, and writing about biology:

actin, airfoil, appendicular skeleton, axial skeleton, ball-and-socket joint, cross-bridge, endoskeleton, exoskeleton, fast fiber, hinge joint, hydrostatic skeleton, ligament, locomotion, motor unit, muscle fiber, myofibril, myoglobin, myosin, osteoporosis, pivot joint, red bone marrow, sarcomere, sliding-filament model, slow fiber, tendon, thick filament, thin filament, yellow bone marrow

Use the Web

There are a variety of activities and questions reviewing locomotion, skeletons, and muscle contraction on the Web at www.mybiology.com. The activities and animations on the structure of muscle fibers and the sliding filament model of muscle contraction are particularly helpful.

Plant Structure, Reproduction, and Development

31

Focus on the Concepts

This chapter introduces the tissues, organs, and systems of flowering plants, how plants grow, and how flowering plants reproduce. As you study the chapter, focus on the following concepts:

- Plants are essential to human life, and our use of plants parallels the growth of civilization. Most plants are flowering plants, or angiosperms. There are two main groups of angiosperms—monocots, such as grasses and lilies, and eudicots such as roses and maple trees. Dicots and eudicots differ in seed structure, leaf venation, stem layout, number of flower parts, and root structure.

- The body of an angiosperm consists of three basic organs—root, stems, and leaves—which are modified in many ways. The root system anchors the plant, absorbs and transports nutrients, and stores food. The shoot system consists of stems, leaves, and flowers. Leaves are photosynthetic organs. They attach to stems at nodes. Stems support leaves and flowers. Terminal and axillary buds develop into stems or flowers.

- A plant has three tissue systems: The dermal tissue system consists of an outer covering of epidermis, which can be covered by a waxy cuticle. The ground tissue system fills the space between the epidermis and the vascular tissue system. In a root vascular tissue forms a central vascular cylinder, with phloem between rays of xylem. In a stem the xylem and phloem form numerous vascular bundles. The bulk of a leaf is ground tissue called mesophyll, with vascular bundles called veins.

- There are several kinds of plant cells and tissues. Parenchyma cells are the most abundant cells, performing most metabolic and storage functions. Collenchyma and sclerenchyma cells provide support. Xylem is a water-conducting tissue composed of hollow tracheids or vessel elements. Phloem is a food-conducting tissue composed of living sieve-tube members and companion cells.

- Plants display indeterminate growth; they grow as long as they live. Annuals complete their life cycles in a single year, perennials in two years, and perennials live for many years. Growth is made possible by meristems, layers of cells that continue to divide. Apical meristems at the tips of roots and shoots elongate the plant; this is primary growth Lateral meristems, such as the vascular cambium that produces new xylem and phloem cells, are responsible for grow in girth; this is secondary growth.

- Flowers are the reproductive shoots of angiosperms. They contain four kinds of floral organs: Sepals protect the flower. Petals are often colorful and attract pollinators. Stamens are the male parts; anthers at their tips produce pollen. A carpel is the female part; it has a sticky stigma, a tubular style, and encloses ovules in the ovary at its base.

- The diploid plant is a sporophyte. Anthers produce haploid spores, which develop into male gametophytes—pollen grains. Ovules produce haploid spores that develop into female gametophytes—embryo sacs, containing eggs. After pollination, the pollen grain grows a pollen tube. A pollen cell divides to form two sperm nuclei that travel down the tube. One sperm fertilizes the egg, and the zygote becomes an embryo sporophyte. The other sperm fertilizes a diploid cell that develops into endosperm, food for the embryo. The ovule becomes a seed, and the ripened ovary becomes a fruit.

- Seed germination begins when a seed swells with water, rupturing the seed coat. The inflow of water causes metabolic changes: the embryo starts growing, enzymes break down endosperm, and food is transported to growing regions of the embryo. The embryonic root pushes downward and the shoot upward.

- Most plants are capable of asexual, or vegetative, reproduction, which produces clones that are genetically identical to the parent plant. Bulbs can fragment to form several individuals. Some plants spread by means of root sprouts or horizontal runners. Many plants can be propagated from cuttings or grown from single cells in cell culture.

Review the Concepts

Work through the following exercises to review the concepts in this chapter. For additional material, refer to the activities on the Web at www.mybiology.com. The website offers a pre-test that will help you plan your studies.

Exercise 1 (Modules 31.1 and 31.2)

Look around you and reflect a moment on all the products and "services" supplied by angiosperms—flowering plants. For example, as I write this I'm referring to some pages printed on paper (perhaps made from fiber from a flowering tree, plus some cotton), sitting at a maple desk, drinking coffee (made from beans of the coffee plant). In the space below, list four specific examples from your surroundings. Name the plants if you can.

1.

2.

3.

4.

The figures below show differences between the two major groups of flowering plants— monocots and eudicots—but the figures are scrambled. List the figures characteristic of monocots and eudicots, and briefly describe the characteristic shown in each.

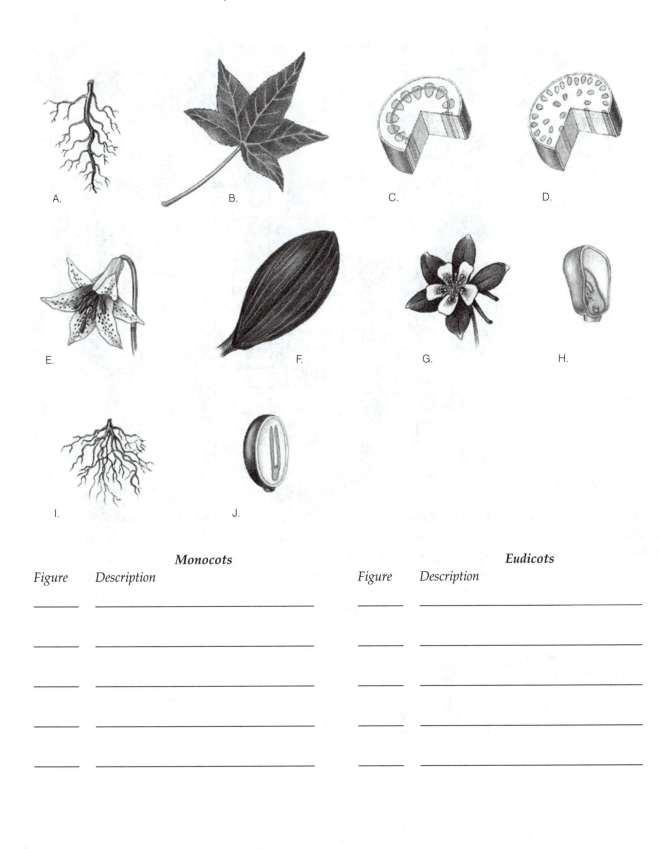

A.   B.   C.   D.

E.   F.   G.   H.

I.   J.

	Monocots		**Eudicots**
Figure	*Description*	*Figure*	*Description*
_____	_____	_____	_____
_____	_____	_____	_____
_____	_____	_____	_____
_____	_____	_____	_____
_____	_____	_____	_____

These modules introduce plant structure—cells, tissues, organs, and systems. Review the vocabulary of plant structure by completing this crossword puzzle. All answers are found in Modules 31.3–31.6.

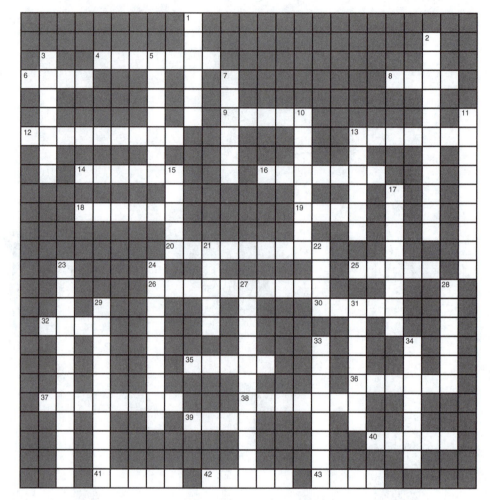

Across

4. Eudicots have one main _____, with many smaller secondary roots.

6. A root ____ is an extension of an epidermal cell that enhances absorption.

8. A ____ is an underground shoot with swollen leaves that store food.

9. The ____ system consists of stems, leaves, and reproductive structures.

12. Sieve-tube members are served by ____ cells.

13. A leaf "stalk" is called a ____.

14. The ____ is ground tissue that makes up most of a root.

16. The middle ____ is a sticky layer between walls of adjacent cells.

18. ____ dominance is inhibition of axillary buds by the terminal bud.

19. A ____ is the point where a leaf is attached to a stem.

20. ____ is the parenchyma cells in a leaf specialized for photosynthesis.

25. Plant cells are surrounded by rigid ____.

26. ____ cells are the most abundant cells in most plants.

30. A ____ is a long, slender sclerenchyma cell.

32. A ____ is a vascular bundle in a leaf.

35. A ____ is an enlarged rhizome that stores food.

36. Hard ____ is the main chemical component of wood.

37. A ____ is an irregular cell that makes nutshells or seeds hard.

38. A waxy coating called the ____ helps the plant retain water.

39. ____ fills the central portion of a stem and may function in food storage.

40. A sharp cactus ____ is a modified leaf.

41. A ____ is a tiny pore in a leaf.

42. ____ cells regulate air flow through stomata.

43. The ____ system anchors the plant and absorbs water and minerals.

Down

1. The ____ tissue system makes up much of a young plant and has many functions.
2. The cell structure that carries out photosynthesis is called a ____.
3. A plant cell contains a fluid-filled central ____.
5. A ____ is a horizontal stem.
7. A root, stem, or leaf is composed of three ____ systems.
10. The ____ bud is the growth point at the tip of a stem.
11. Plant cell walls are made of ____.
13. ____ carries sugar from the leaves to other parts of the plant.
15. ____ carries water up from the roots.
17. ____ buds form where leaves join a stem and are usually dormant.
21. A ____ plate is the perforated end of a food-conducting cell.
22. The ____ is the main site of photosynthesis in most plants.
23. ____ cells are the dead, hardened supporting cells that form the plant "skeleton."
24. The ____ covers and protects a plant.
27. ____ cells support the growing parts of a plant.
28. A ____ , or "runner," is a stem that grows along the surface of the ground.
29. The ____ forms a barrier that regulates flow into root vascular tissue.
31. Vascular tissues of stems and leaves are arranged in vascular ____.
33. The ____ tissue system is composed of xylem and phloem.
34. A ____ is a modified leaf that helps a plant cling and climb.

The cross-sectional diagrams below will help you to visualize the internal structures of roots, stems, and leaves. Label the following: **eudicot stem, monocot stem, root, leaf, cortex, cuticle, endodermis, stoma, pith, epidermis, guard cell, mesophyll, vascular bundle, phloem, xylem,** and **vein.** Color the dermal tissue system red, the vascular tissue system blue, and the ground tissue system yellow.

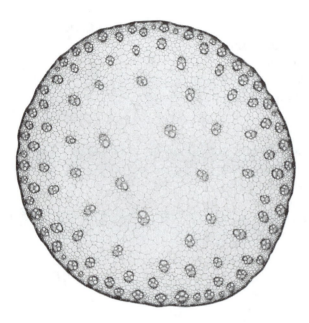

a. _____

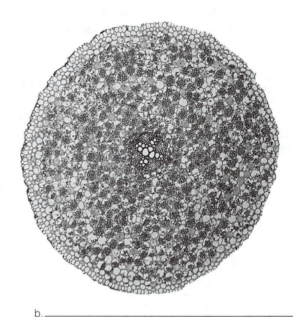

b. _____

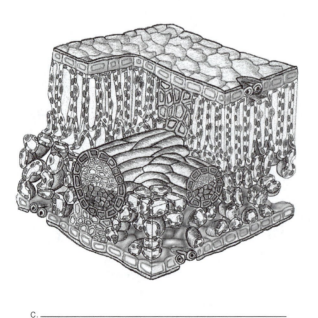

c. _____

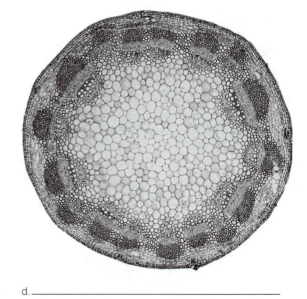

d. _____

Exercise 5   (Modules 31.7–31.8)

These modules discuss plant growth. There are different kinds of growth and different kinds of plant life cycles. After reading the modules, match each term on the left with a phrase on the right.

_____ 1. Primary xylem	A.	Lengthwise growth
_____ 2. Biennial	B.	Makes up the wood of a tree or shrub
_____ 3. Secondary growth	C.	Growth point at the tip of a shoot or root
_____ 4. Indeterminate growth	D.	Thick, waxy cells that protect woody stem
_____ 5. Perennial	E.	A plant that completes its life cycle in a year or less
_____ 6. Primary phloem	F.	Tissue just outside the vascular cambium of a tree trunk
_____ 7. Sapwood	G.	Ceasing to grow after reaching a certain size
_____ 8. Primary growth	H.	Meristematic tissue that forms cork
_____ 9. Cork	I.	A plant that takes two years to complete its life cycle
_____ 10. Secondary phloem	J.	A plant that can grow in girth
_____ 11. Vascular cambium	K.	Older layers of secondary xylem in a tree trunk
_____ 12. Annual	L.	Pushed outward by development of secondary phloem
_____ 13. Determinate growth	M.	A plant that continues to live for many years
_____ 14. Heartwood	N.	Young secondary xylem that still conducts water
_____ 15. Secondary xylem	O.	Rapidly dividing cells between xylem and phloem
_____ 16. Cork cambium	P.	Dividing cells along the length of a root or stem
_____ 17. Apical meristem	Q.	First water-conducting vascular tissue of a plant
_____ 18. Woody plant	R.	Growth in diameter
_____ 19. Lateral meristem	S.	Growing as long as a plant lives

Primary growth is growth in length. Label the following on this diagram of a growing root tip: **root cap, root hair, vascular cylinder, cortex, epidermis,** and **apical meristem.** Color the root cap red, the zone of cell division orange, the zone of cell elongation yellow, and the zone of cell maturation blue.

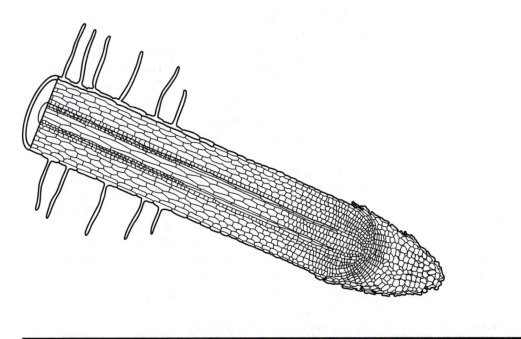

Try to picture the growth and structure of a maturing twig by identifying each of the layers in the stem. Choose from the following: A. cortex, B. secondary xylem, C. vascular cambium, D. primary phloem, E. epidermis, F. primary xylem, G. cork cambium, H. secondary phloem, I. pith, J. cork.

Near the tip of the twig:

1. Name the layers, from outside to inside:_____
2. Which layers are lateral meristems?_____

After a few years' growth:

3. Name the layers from outside to inside:_____
4. Which layers are lateral meristems?_____
5. Which layers make up the bark?_____
6. Which layer makes up wood?_____
7. Which layers present near the tip have now disappeared?_____

Exercise 8   (Module 31.8)

The diagram below will also help you visualize the secondary growth of a tree trunk. Label each of the following on the diagram: **vascular cambium, secondary phloem, sapwood, secondary xylem, heartwood, cork, cork cambium,** and **bark.** Color the vascular cambium green. Color the oldest xylem red, the newest xylem pink, the oldest phloem dark blue, and the newest phloem lighter blue.

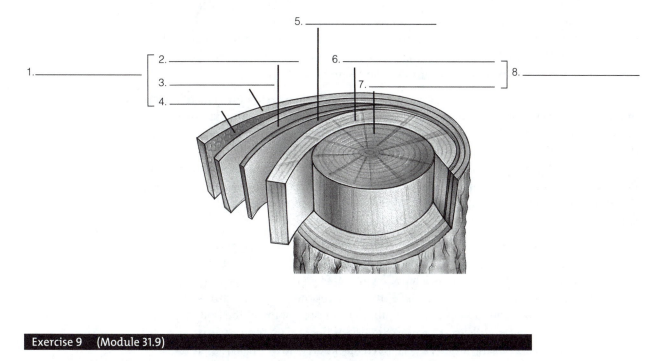

Exercise 9   (Module 31.9)

In the space below, sketch a flower, and label the following parts: **sepal, stigma, stamen, style, filament, carpel, anther, petal, ovary, ovule,** and **pollen grain.** Color the female parts of the flower red and the male parts blue.

The following story will help you review plant reproduction and development. Complete the story by choosing the correct italicized word or phrase in each set.

The spring rains have begun. A blackberry seed has lain undisturbed on the ground near a rail fence since September. Now (1) *sunlight, warmth, water* causes the seed to swell. The process of (2) *pollination, germination, fertilization* is beginning. The protective (3) *endosperm, carpel, seed coat* splits open, a tiny root emerges, and the leaves of the tiny (4) *ovule, pollen grain, embryo* unfurl toward the sun. At first the plant's leaves are pale; its seed leaves, or (5) *cotyledons, anthers, sepals,* have absorbed the (6) *pollen, endosperm, sperm* stored within the seed to obtain food for the plant's first few days of growth.

As the days grow longer and warmer, apical meristems add new cells to the tips of the blackberry's roots and stems. By midsummer, the plant has produced several branches and dozens of leaves and has begun to climb up on the fence. Clusters of white flowers, resembling small wild roses, sprout at lateral buds. The blackberry plant represents the diploid (7) *gametophyte, sporophyte* phase of the blackberry's life cycle. The flowers are the reproductive structures where (8) *mitosis, germination, meiosis* produces haploid (9) *seeds, spores, fruits,* which undergo mitosis to become haploid (10) *gametophytes, sporophytes.*

Each flower begins as a bud, enclosed by green, protective (11) *petals, carpels, sepals.* As a flower opens, white (12) *petals, carpels, sepals* attract bees. The bees swarm over the flowers, collecting nutrient-rich pollen grains from the (13) *carpels, sepals, anthers* at the tips of the (14) *stamens, carpels, stigmas,* the male reproductive structures of the flowers. As they buzz from flower to flower, the bees deposit some of the pollen grains on the sticky (15) *stigmas, anthers, stamens* at the tips of the female reproductive structures, called (16) *anthers, carpels, sepals.* Each blackberry flower has numerous stamens and carpels, so the bees are busy.

Each pollen grain is actually a tiny (17) *male, female* gametophyte, containing a tube nucleus and two (18) *egg, seed, sperm* nuclei. (19) *Fertilization, Pollination, Meiosis* occurs as a bee rubs a pollen grain onto a stigma. The pollen grain germinates, and a pollen tube grows down through the carpel into the swollen (20) *anther, ovary, sepal* at its base. The two (21) *egg, sperm, tube* nuclei now travel down the pollen tube to (22) *a sepal, a carpel, an ovule,* which contains the (23) *stigma, stamen, embryo sac,* the tiny haploid female (24) *gametophyte, sporophyte.* The sperm nuclei enter the ovule, and one of them joins with the (25) *egg, anther, pollen grain* within, forming a diploid (26) *ovule, embryo sac, zygote.* This is fertilization. The other sperm nucleus combines with a large diploid cell, forming a triploid cell. This cell will divide to form the (27) *pollen, endosperm, ovule,* which will become enclosed in the seed and nourish the developing embryo. The joining of two sperm nuclei with an egg cell and another diploid cell is called double (28) *fertilization, germination, pollination* and is unique to flowering plants.

In the ovary at the base of each carpel, each ovule with its zygote now begins to develop into a (29) *seed, stamen, fruit.* The zygote divides, and one of the cells formed divides to become the (30) *flower, endosperm, embryo.* Because the blackberry is a eudicot, the embryo sprouts two (31) *sepals, cotyledons, seed coats,* which will later help the embryo absorb nutrients stored in the seed. The outer coating of the ovule dries and thickens and becomes the seed coat.

Each blackberry flower contains multiple carpels. The flower petals shrivel and fall away, and the (32) *sepal, anther, ovary* at the base of each carpel grows and ripens. Eventually, the seeds end up in a cluster of ripened ovaries, which form a (33) *fleshy fruit, dry fruit* called a blackberry.

The blackberries are hard, green, and sour at first, but they become soft, dark purple, and sweet as (34) *starch, organic acid, sugar* accumulates in them. The fruits protect the seeds and aid in their dispersal. A blackbird perches on a swaying cane, twists off a fat berry, and gulps it down. The soft flesh of the blackberry is easily digested, but the seeds pass through the bird's digestive system unharmed, protected by their tough (35) *endosperm, seed coats, sepals.* A few hours after eating the berry, the bird perches on a fence and deposits the seeds in its droppings. Over the winter, the seeds will undergo a period of (36) *dormancy, differentiations, fertilization.* In the spring, the seeds will germinate, and the fence will make an ideal trellis for the next generation of blackberries.

Exercise 11   (Modules 31.14–31.15)

Plants are a lot easier to clone than animals. Meristems are essentially composed of "stem cells," (no pun intended) and they persist through the life of a plant, even a 4,600 year-old tree! Review plant cloning by matching each phrase with a type or method of vegetative reproduction or propagation.

_____ 1. Grass slowly spreads into a garden bed.

_____ 2. Pieces of a houseplant grow roots in water.

_____ 3. A clump of tulips expands each year.

_____ 4. Many small trees grow from a stump.

_____ 5. Tree seedlings grow from cells in a test tube.

A. Cuttings
B. Root sprouts
C. Cell culture
D. Fragmentation
E. Runners

Test Your Knowledge

Multiple Choice

1. Which of the following is *not* a characteristic of eudicots?
   a. two seed leaves
   b. parts of flowers in fours or fives
   c. taproot
   d. vascular bundles arranged in a ring
   e. veins in leaves usually parallel

2. Most of the photosynthesis in a plant is carried out by _____ in the leaves.
   a. collenchyma cells
   b. water-conducting cells
   c. parenchyma cells
   d. sclerenchyma cells
   e. food-conducting cells

3. How do cells in a meristem differ from other cells in a plant?
   a. They continue to divide.
   b. They photosynthesize at a faster rate.
   c. They are growing.
   d. They are differentiating.
   e. They store food.

4. "Angiosperm" is another name for
   a. a plant with seeds
   b. pollen.
   c. stored food in a seed.
   d. a flowering plant.
   e. fruit.

5. In the process of pollination, pollen grains are transferred from the _____ to the _____ .
   a. ovary . . . anther
   b. stigma . . . ovary
   c. anther . . . sepal
   d. carpel . . . stigma
   e. anther . . . stigma

6. Which of the following is *not* a method of asexual reproduction?
   a. runners
   b. germination
   c. stump sprouts
   d. cuttings
   e. fragmentation

7. What is endosperm?
   a. male reproductive cells in plants
   b. stored food in a seed
   c. cells that make up the bulk of a pollen grain
   d. the fleshy part of a fruit such as an apple or strawberry
   e. plant chromosomes

8. After fertilization, the _____ develops into a seed, and the _____ develops into a fruit.
   a. ovule . . . ovary
   b. pollen grain . . . ovule
   c. ovary . . . ovule
   d. egg . . . ovule
   e. egg . . . ovary

9. Lengthwise growth of a root into the soil results mainly from
   a. cell division in the apical meristem.
   b. elongation of cells.
   c. cell division in the vascular cambium.
   d. differentiation (specialization) of root cells.
   e. pulling by root hairs.

10. Which of the following is closest to the center of a daisy stem?
    a. cortex
    b. phloem
    c. pith
    d. epidermis
    e. xylem

11. Wheat was first domesticated in
    a. Mexico
    b. China
    c. the Andes of South America
    d. southern Africa
    e. the Middle East

12. Cell division would be slowest in which of these tissues?
    a. apical meristem of root
    b. cork cambium
    c. epidermis
    d. vascular cambium
    e. apical meristem of terminal bud

Essay

1. Briefly describe four functions of a plant's roots. Which of these is primarily a job of the epidermis? The vascular cylinder? The cortex?

2. Sketch a cross section of a young eudicot stem, showing pith, xylem, phloem, cortex, and epidermis. What do these layers have to do with the three tissue systems in the stem? How does the arrangement of tissues in a eudicot root differ from their arrangement in a stem?

3. What is the sporophyte stage of a flowering plant? What reproductive structures does it produce? What is the gametophyte stage? What reproductive cells do gametophytes produce? Which of the preceding are haploid and which are diploid?

4. What are meristems, and where would you find them in an apple tree? Which are responsible for primary growth? Secondary growth?

5. Briefly explain how the shape and structure of the following cells are well suited to their functions: sclerenchyma cells in a stem, water-conducting cells in a root, epidermal cells in a leaf.

6. Explain how the vascular cambium enables a tree to grow in diameter.

7. Starting with pollination, briefly explain how fertilization occurs in a flowering plant. What is the major difference between this process and fertilization in animals?

8. Briefly describe six different ways in which roots, stems, or leaves are modified for storage, reproduction, or other specialized functions.

Apply the Concepts

Multiple Choice

1. The vascular cambium in the trunk of a large, woody rhododendron shrub lies between
   a. secondary phloem and secondary xylem.
   b. secondary xylem and pith.
   c. primary xylem and primary phloem.
   d. secondary phloem and cortex.
   e. primary phloem and pith.

2. Which of the following is correctly matched with its tissue system?
   a. xylem—ground tissue system
   b. phloem—dermal tissue system
   c. cortex—ground tissue system
   d. pith—vascular tissue system
   e. All of the above are correctly matched.

3. The shoot system of a beavertail cactus consists of broad paddlelike structures covered with spines. The spines are modified _____ , so the flat green paddles must be modified _____ .
   a. buds . . . leaves
   b. buds . . . stems
   c. leaves . . . stems
   d. stems . . . roots
   e. stems . . . leaves

4. Cell division in the vascular cambium adds to the girth of a tree by adding new _____ on the inside of the cambium layer and _____ on the outside.
   a. phloem . . . xylem
   b. xylem and phloem . . . bark
   c. pith . . . xylem and phloem
   d. xylem . . . phloem
   e. xylem . . . cortex

5. A vandal killed a historic oak tree on the village green by "girdling" it with a chain saw. He cut through the bark and into the sapwood all the way around the tree. Why did the tree die?
   a. The leaves could not get water.
   b. Oxygen could not get to the roots.
   c. The roots could not get food.
   d. The leaves could not get food.
   e. The roots could not absorb water.

6. Artichoke hearts are tender and tasty. The leaves are tasty too, but most of an artichoke leaf is fibrous and impossible to chew. The leaves must contain lots of
   a. parenchyma cells.
   b. phloem.
   c. meristematic tissue.
   d. sclerenchyma cells.
   e. epidermal cells.

7. Based on your own experience with these plants, which do you think is correctly paired with the word describing its life cycle?
   a. apple tree—perennial
   b. rose bush—annual
   c. marigold—perennial
   d. oak tree—biennial
   e. tulip—annual

8. Plants growing in harsh environments such as deserts, sand dunes, and arctic tundra often reproduce asexually. This is because
   a. there are few animals available to pollinate them.
   b. they are members of plant families that only reproduce asexually.
   c. fruits would freeze or dry out in these environments.
   d. vegetative reproduction is not as risky or costly as making seeds.
   e. seeds would be eaten by hungry animals in these environments.

9. Which of the following would be *least* useful in figuring out whether a plant is a monocot or a dicot?
   a. life cycle—perennial, biennial, annual
   b. numbers of flower parts
   c. pattern of veins in leaves
   d. number of seed leaves
   e. arrangement of vascular tissue in stem

10. A cross section of part of a plant exposes epidermis, a thick cortex, and a central cylinder of xylem and phloem. This part is a
    a. fruit.
    b. seed.
    c. stem.
    d. root.
    e. bud.

11. When you cut through a carrot, the cross section reveals a lighter inner circle surrounded by a darker outer ring. The darker outer ring is called
    a. cortex
    b. xylem
    c. pith
    d. phloem
    e. sclerenchyma

Essay

1. Amy told Aaron that a giant sequoia is the largest living thing, bigger than ten blue whales. Aaron replied, "Yes, the sequoia is bigger, but it is an unfair comparison because *most of a tree is not alive*." Do the facts back up Aaron's assertion? Explain.

2. Imagine pounding a large spike into a tree. Name (in order) all the tissue layers it would pass through as it penetrates into the heartwood.

3. What part of a plant are you eating when you consume each of the following: garlic, walnut, cabbage, carrot, cauliflower, artichoke, asparagus, cucumber, broccoli, rice, and potato?

4. When he was a boy, Grandpa nailed a horseshoe to a maple tree, about 5 feet from the ground. The tree has since grown much taller, and the diameter of its trunk has quadrupled, but the horseshoe is still about 5 feet up. Explain this in terms of how trees grow.

5. Palms are monocots. Do you think that you could kill a palm tree by girdling it (as in multiple choice question 5 in the "Applying Your Knowledge" section)? Why or why not?

6. Examine the diagram of a growing root, Figure 31.7B in the text. Why does it make sense that lateral roots must start growing from the inside of the main root (the area of the endodermis, actually) rather than simply projecting from the outside?

7. Most food for humans is derived from the seeds of plants. Why does it make sense that seeds are a good source of nutrition?

Put Words to Work

Correctly use as many of the following words as possible when reading, talking, and writing about biology:

annual, anther, apical dominance, apical meristem, axillary bud, bark, biennial, carpel, clone, collenchyma cell, cork, cork cambium, cortex, cotyledon, cuticle, dermal tissue system, double fertilization, dicot, embryo sac, endodermis, endosperm, epidermis, eudicot, fiber, fragmentation, fruit, food-conducting cell, gametophyte, germinate, ground tissue system, guard cell, heartwood, indeterminate growth, internode, lateral meristem, leaf, meristem, mesophyll, monocot, monoculture, node, organ, ovary, parenchyma cell, perennial, petal, phloem, pistil, pith, pollination, primary growth, primary phloem, primary xylem, rhizome, root system, root hair, sapwood, sclerenchyma cell, secondary growth, secondary phloem, secondary xylem, seed coat, seed dormancy, sepal, shoot system, sporophyte, stamen, stem, stigma, stoma (plural, stomata), tendril, terminal bud, tissue, tissue system, tuber, vascular bundle, vascular cambium, vascular cylinder, vascular tissue system, vein, water-conducting cell, wood, xylem

Use the Web

For more review of plant structure, reproduction, and development, see the activities and questions on the Web at www.mybiology.com.

Plant Nutrition and Transport

32

Focus on the Concepts

What substances do plants need to grow? How do plants obtain and transport these substances? These questions and others are answered in this chapter. In studying the material, focus on these concepts:

- Plants acquire nutrients from soil and air. Root hairs increase surface area. Mineral nutrients enter the root dissolved in water. The solution moves along cell walls (the extracellular route) and from cell to cell (the intracellular route) to the vascular cylinder. The waxy Casparian strip forces water and solutes to pass through endodermal cells, and the selectivity of these cells controls the mix of solutes and water absorbed.

- Xylem carries water and minerals to the leaves via a transpiration-cohesion-tension mechanism: Guard cells open stomata in the leaves, allowing water to evaporate. Hydrogen bonds cause water molecules to stick together, or cohere. Evaporation of water molecules from the leaves—transpiration—pulls a string of water molecules up from the roots, while adhesion to xylem cell walls keeps gravity from pulling the water back down.

- Phloem transports sugars via a pressure-flow mechanism: Sieve tubes carry sugar both up and down, but always from a sugar source, such as photosynthesizing leaves, to a sugar sink, such as metabolizing root cells. Sugar is transported into phloem cells at the source, causing water to follow by osmosis, creating a buildup of pressure. At the sink, sugar is unloaded and water follows, lowering pressure. Phloem sap moves from cell to cell, from the higher-pressure source to the lower-pressure sink.

- Plants require 17 essential elements. Six macronutrients—C, O, H, N, S, and P—are used in large quantities to build organic molecules. Three other macronutrients—K, Ca, and Mg—have various functions. Eight micronutrients, such as Cl, Fe, and Zn, function mainly as cofactors and are used in small amounts.

- Topsoil is a mix of rock particles, humus, and organisms. Negative ions are not held tightly in the soil, and may be lost. Positive ions adhere to negatively charged clay particles, so plants take them up by cation exchange, releasing H^+ to the soil. Soils are commonly deficient in N, P, or K, but fertilizers can supply missing nutrients. Plant scientists improve crops, while many farmers try to practice sustainable agriculture.

- Most plants depend on bacteria for nitrogen. Ammonifying bacteria break down organic matter, and nitrogen fixers (often in root nodules of legumes) use N_2 from the air to make ammonia. Nitrifying bacteria convert ammonia to nitrate. Mycorrhizae—mutualistic fungi—assist with absorption. Parasitic plants may tap other plants for nutrition, while carnivorous plants trap insects.

Review the Concepts

Work through the following exercises to review the concepts in this chapter. For additional review, refer to the activities on the Web at www.mybiology.com. The website offers a pre-test that will help you plan your studies

Review nutrients required by plants by filling in the blanks in the paragraph below.

Plants need four main kinds of nutrients to survive and grow:
[1]_____, [2]_____, [3]_____, and
[4]_____. Most of a plant's mass is made of organic molecules built from
[5]_____, which is obtained from the [6]_____. Hydrogen for
organic molecules comes from [7]_____, which is obtained from the
[8]_____ and split during the process of photosynthesis. Many substances
manufactured by the plant require elements other than carbon, hydrogen, and oxygen.
Proteins, for instance, contain [9]_____. Chlorophyll contains
[10]_____ and [11]_____, and ATP and nucleic acids contain
[12]_____ and [13]_____. These [14]_____ are ab-
sorbed from the soil by the plant's roots. Finally, the plant takes in
[15]_____, which is necessary for aerobic respiration from the air and soil.

On the diagram below, trace the uptake of water and minerals by a root. Show water moving from cell to cell (the intracellular route) with a blue arrow. Trace water moving along cell walls, between cells (the extracellular route), with a red arrow. The Casparian strip forces all water and solutes to pass through endodermal cells, allowing the endodermis to control movement of water and solutes into the vascular tissues. Color the endodermis and Casparian strip yellow. Label the following: **endodermis, root hair, epidermis, cortex, xylem vessels, Casparian strip, intracellular route, extracellular route,** and **plasmodesma.**

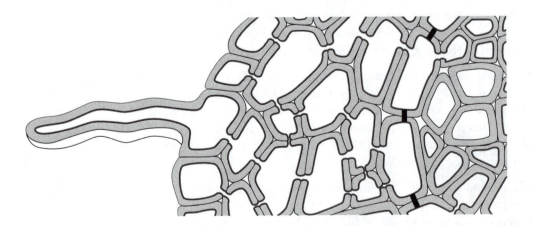

Exercise 3 (Modules 32.3–32.4)

Review the movement of water and ions up a plant stem by filling in the missing words in the phrases below, then numbering each of the phrases in order. Start with what happens in the roots and continue with what pulls water up from the top of the plant.

_____ A. Water molecules _____ from cells bordering the air spaces within the leaf.

_____ B. Inorganic ions are pumped into the _____ of the roots; this causes water to enter these cells by _____.

_____ C. Water molecules leaving leaf cells are stuck to nearby water molecules, a phenomenon known as _____. They also stick to the cellulose molecules in the walls of xylem cells, a phenomenon known as _____. As water molecules diffuse from leaf cells, they tug on adjacent water molecules.

_____ D. Water accumulating in the xylem of roots pushes xylem sap upward a few meters, a phenomenon known as _____.

_____ E. As water molecules in the leaf pull upward on adjacent water molecules, they in turn tug on water molecules in the xylem, all the way down to the roots. This occurs because water molecules are attracted to each other by _____ bonds.

_____ F. _____ cells in leaf epidermis take up K^+ ions. Water enters by _____, and the cells swell. This causes the _____ to open, and water evaporates from the leaf, a process called _____.

_____ G. Water molecules are pulled up from the roots, bit by bit, as water molecules evaporate from the leaves. This explanation for the ascent of xylem sap is called the _____ mechanism.

Exercise 4 (Module 32.5)

Phloem transports sugar from where it is produced to where it is used via a pressure flow mechanism. Review this mechanism, paying particular attention to Figure 32.5B in the text. Then test your understanding by stating whether each of the following refers to a sugar **source** or a sugar **sink**.

_____ 1. Where phloem sap has its highest sugar concentration

_____ 2. A beet root or potato tuber during the summer

_____ 3. Any photosynthesizing part of a plant

_____ 4. Phloem sap flows away from this area.

_____ 5. Where sugar is pumped into the phloem by active transport

_____ 6. Where water enters phloem sap by osmosis

_____ 7. A bulb in early spring

_____ 8. Where water (hydrostatic) pressure is highest

_____ 9. Where sugar leaves the phloem by active transport

_____ 10. A leaf on a summer day

_____ 11. Where water (hydrostatic) pressure is lowest

_____ 12. Where phloem sap has its lowest sugar concentration

_____ 13. Phloem sap flows toward this area

_____ 14. A growing bud in early spring

Seventeen essential elements are required by plants: iron (Fe), manganese (Mn), nitrogen (N), hydrogen (H), calcium (Ca), phosphorus (P), zinc (Zn), potassium (K), molybdenum (Mo), copper (Cu), carbon (C), oxygen (O), magnesium (Mg), boron (B), chlorine (Cl), sulfur (S), and nickel (Ni). Which of these nutrients are the answers to each question below?

_____ 1. Which nine nutrients are macronutrients, needed in large quantities?

_____ 2. Which six macronutrients are the major ingredients of organic compounds?

_____ 3. Which eight nutrients are micronutrients, needed in smaller amounts as cofactors that help enzymes?

_____ 4. Shortage of which nutrient is the most common deficiency of plants?

_____ 5. Which macronutrient combines with proteins to "glue" cells together?

_____ 6. Which macronutrient is the main solute for osmotic regulation, causing opening and closing of stomata by guard cells, for example?

_____ 7. Which micronutrient is a metallic component of the cytochromes involved in electron transport?

_____ 8. Deficiency of which nutrient is the second most common plant deficiency?

_____ 9. Which macronutrient is part of chlorophyll and a cofactor of several enzymes?

_____ 10. "5-6-7" on a bag of fertilizer is the ratio of which three nutrients?

Match each of the phrases on the left with a component of soil from the list on the right.

_____ 1. Fine clays and nutrients dissolved in water accumulate in this soil layer.

_____ 2. Roots exchange these for nutrient ions bound to clay particles.

_____ 3. Not bound tightly by soil particles

_____ 4. Partially decomposed organic matter

_____ 5. Their negative charges keep positive ions from draining away.

_____ 6. Where most decomposing organic material and organisms are found

_____ 7. Mainly composed of partly broken-down rock.

_____ 8. Adhere to clay particles

_____ 9. Diffuses into soil water from air spaces in the soil

_____ 10. Held in tiny spaces between soil particles

_____ 11. Break down organic matter

A. positive ions, such as Ca^{2+}

B. B horizon

C. C horizon

D. Clay particles

E. Air

F. Water

G. Bacteria and fungi

H. H^+ ions

I. Topsoil

J. Humus

K. Negative ions, like NO_3^-

Exercise 7   (Modules 32.9–32.11)

State how each of the following would be harmful or helpful to the soil or the environment.

1. Applying an excess of commercially produced inorganic fertilizer to a field

2. Crop rotation; alternating different crops

3. Plowing and planting in rows up and down hills

4. Flooding fields with irrigation water and blocking flow to prevent runoff

5. Using fertilizers that contain organic material

6. Using perforated pipes to drip water onto soil

7. Planting "super" varieties of wheat, maize, and rice

Exercise 8   (Modules 32.12–32.14)

Review the adaptations of plants for obtaining nitrogen by filling in the blanks in the following story.

[1]_____ is often the hardest to obtain of all plant nutrients but is essential to plants for making [2]_____, [3]_____, and other organic molecules. Ironically, the [4]_____ is 80% N_2 gas, but plants cannot use nitrogen in that form. Most plants must obtain nitrogen from the soil in the form of [5]_____ (NH_4^+) or [6]_____ (NO_3^-) ions. Certain soil bacteria, called [7]_____ bacteria, are able to convert N_2 to ammonium. Other bacteria, called [8]_____ bacteria, decompose organic matter in the soil and produce additional ammonium. Soil bacteria called [9]_____ bacteria then convert ammonium to nitrate, which is absorbed. Plant enzymes then convert nitrate back to ammonium, which is used to build [10]_____.

In many plants, roots form associations with fungi, called [11]_____, which assist in absorption of nutrients. This is a [12]_____beneficial arrangement—both plant and fungus gain. The fungus is good at absorbing water and inorganic ions, some of which are transferred to the plant. The fungus may secrete 13_____, which stimulate roots to grow, and 14_____, which help protect the plant from certain diseases. In return, some of the plant's 15_____ nourishes the fungus.

Other specializations for obtaining nutrients have evolved in various groups of plants. Some plants, such as 16_____ and 17_____, are parasitic. They tap into the vascular systems of other plants and steal food and nutrients. Some plants, such as orchids, grow on other plants but are not parasites. These plants get their water and minerals from rain and are called 18_____.

19_____ plants, such as the sundew and 20_____, often grow in bogs where the soil is poor in 21_____ and other minerals. These plants obtain the minerals they need by trapping and digesting 22_____.

The roots of plants in the 23_____ family, such as peas, beans, and alfalfa, have swellings called 24_____ that house 25_____ bacteria. Again, the arrangement is mutually beneficial. The legume plant provides the bacteria with 26_____, and the bacteria convert atmospheric 27_____ into ammonium. Excess ammonium may actually enter the soil, and for this reason, farmers often 28_____ crops, alternating between legumes and nonlegumes such as 29_____.

Test Your Knowledge

Multiple Choice

1. Nitrogen fixation is
   a. using nitrogen to build molecules such as proteins and nucleic acids.
   b. converting nitrogen in the air into a form usable by plants.
   c. recycling nitrogen from organic matter in the soil.
   d. absorbing N_2 from the soil.
   e. an unhealthy interest in nitrogen.

2. Guard cells
   a. control the rate of transpiration.
   b. push water upward in a plant stem.
   c. protect the plant's roots from infection.
   d. control water and solute intake by roots.
   e. protect nitrogen-fixing bacteria in root nodules.

3. Which of the following is a macronutrient?
   a. hydrogen
   b. nitrogen
   c. calcium
   d. phosophorus
   e. all of the above

4. Soil can easily become deficient in _____, because these ions are negatively charged and do not stick to negatively charged clay particles.
   a. potassium
   b. calcium
   c. magnesium
   d. nitrate
   e. ammonium

5. Which of the following would trigger opening of stomata?
   a. extreme heat
   b. loss of potassium by guard cells
   c. nightfall
   d. swelling of guard cells due to osmosis
   e. all of the above

6. The sundew plant has to digest insects because
   a. it obtains nitrogen from their flesh that it cannot get from the soil.
   b. it has lost the ability to perform photosynthesis.
   c. it lives in a dry environment and needs the moisture in their bodies.
   d. it needs to get rid of insects that accidentally get stuck in its hairs.
   e. its flowers are fertilized by pollen in their digestive tracts.

7. Ammonifying bacteria in the soil
   a. convert ammonium to nitrate.
   b. fix nitrogen.
   c. convert nitrogen in organic molecules into ammonium.
   d. change nitrate into ammonium.
   e. use nitrate to make amino acids that plants can use.

8. Mycorrhizae are
   a. nutrients required by plants in relatively small amounts.
   b. plants such as mistletoe that parasitize other plants.
   c. medium-sized soil particles.
   d. cells that control evaporation of water from leaves.
   e. associations of roots with beneficial fungi.

9. Which of the following is a sugar source?
   a. a photosynthesizing leaf
   b. a developing fruit
   c. a growing root
   d. a growing shoot
   e. a tree trunk

10. Phytoremediation is
    a. supplying plants with the proper nutrients
    b. evaporation of water from plant leaves
    c. sharing of nutrients with mycorrhizae
    d. adding organic or natural fertilizer to the soil
    e. using plants to clean the environment

11. What is the main source of energy that moves water upward in the trunk of a tree?
    a. muscle-like contraction of xylem cells
    b. evaporation of water by the sun
    c. pressure exerted by root cells
    d. breakdown and release of energy of sugar molecules
    e. osmotic changes caused by alterations in salt content

12. All water and solute molecules must pass through _____ before they can enter the vascular system and move upward to the leaves.
    a. a stoma
    b. a root hair cell
    c. an endodermal cell
    d. an epidermal cell
    e. a cortex cell

13. A plant does *not* obtain which of the following substances from soil?
    a. magnesium
    b. nitrogen
    c. carbon
    d. oxygen
    e. phosphorus

Essay

1. Explain what causes water to enter the xylem of a tree root and move upward in the xylem of the trunk to the leaves.

2. Describe an experiment to find out whether the element molybdenum is an essential plant nutrient.

3. In a growing potato plant, what causes sugar to flow from a leaf to the potato? Early in the spring, what causes sugar to flow from the potato to the growing shoot?

4. Explain how root hairs obtain ions that cling to the surfaces of clay particles in the soil.

5. Organic fertilizers such as manure and compost and inorganic commercial fertilizers contain the same nutrient ions needed by plants. In addition to this, what does the humus in organic fertilizer contribute to the soil?

Apply the Concepts

Multiple Choice

1. The roots of many aquatic plants have special structures that project above the surface of the water. For example, cypress trees (which grow in swamps) have "knees" that extend upward above water level. Which of the following is the most logical function of these structures?
   a. obtaining carbon dioxide for photosynthesis
   b. nitrogen fixation
   c. obtaining oxygen for the roots
   d. transpiration
   e. absorbing trace minerals

2. Helen had a terrarium on her windowsill containing various houseplants. She wondered why the glass was often fogged with water droplets. Her friend Sara, who has had a biology class, tried to impress Helen by explaining, "The water evaporates from the leaves—it's a process called ____."
   a. root pressure
   b. adhesion
   c. photosynthesis
   d. pressure flow
   e. transpiration

3. In an apple tree, sugar might flow from ____ to ____ .
   a. a developing apple . . . a leaf
   b. the trunk . . . a leaf
   c. a growing root . . . a growing shoot tip
   d. a leaf . . . a developing apple
   e. a growing shoot tip . . . the trunk

4. Soil could be deficient in any of the following nutrients. If you had to supply one of them, which would be needed in the smallest amount?
   a. sulfur
   b. phosphorus
   c. nitrogen
   d. potassium
   e. iron

5. Which of the following is a difference between transport by xylem and transport by phloem?
   a. Active transport moves xylem sap but not phloem sap.
   b. Transpiration moves phloem sap but not xylem sap.
   c. Phloem carries water and minerals; xylem carries organic molecules.
   d. Xylem sap moves up; phloem sap moves up or down.
   e. Xylem sap moves from sugar source to sink, but phloem sap does not.

6. What keeps the force of gravity from pulling water molecules down from the leaves?
   a. upward pressure from the roots
   b. high water pressure in the leaves
   c. the Casparian strip blocks them from moving out
   d. movement of water toward a sugar sink
   e. cohesion and adhesion of water molecules

7. Jon is performing a chemical analysis of xylem sap. He should not expect to find much of which of the following?
   a. nitrogen
   b. sugar
   c. phosphorus
   d. water
   e. potassium

8. A botanist discovered a mutant plant that was unable to produce the material that forms the Casparian strip. This plant would be
   a. unable to fix nitrogen.
   b. unable to transport water or solutes to the leaves.
   c. able to exert greater root pressure than normal plants.
   d. unable to control the amounts of water and solutes it absorbs.
   e. unable to use its roots as a sugar sink.

9. Normally when an aphid feeds by puncturing plant tissues, it does not have to suck the sap out. An inexperienced aphid, however, accidentally inserted its feeding tube in the wrong place and found the fluid in its gut being sucked out through the feeding tube. It had punctured
   a. the Casparian strip.
   b. a root nodule.
   c. a xylem cell.
   d. a phloem tube.
   e. a stoma.

10. Professor Timothy Schmidlap claims to have discovered a new macronutrient required for plant growth. Most of his colleagues are skeptical of his claim. Why might they consider it unlikely?
    a. All the nutrients required for plant growth have already been found.
    b. It is very difficult to prove that a plant needs a certain nutrient.
    c. Plants need thousands of nutrients; a new one is not significant.
    d. Any nutrient needed in large amounts has probably already been found.
    e. His colleagues are jealous and want to claim credit for the discovery.

11. You would be unlikely to find an organic farmer
    a. practicing crop rotation
    b. planting a GM crop
    c. adding humus to the soil
    d. encouraging the growth of mycorrhizae
    e. using drip irrigation

Essay

1. A farmer found swellings full of bacteria on the roots of his bean plants, so he sprayed the plants with antibiotics. The swellings disappeared, but the plants no longer grew well. Examining the plants more closely, the farmer concluded that the cause of the difficulty was fungi associated with the roots. He treated the plants with a fungicide, but then the plants did even worse. Explain what is happening.

2. Sometimes on very hot days, water evaporates so quickly from the leaves of a plant that empty "bubbles" occur in xylem tubes. This can be harmful to the plant. Explain why.

3. The jade plant, a familiar houseplant, has a peculiar adaptation to the hot, dry environment where it grows naturally. Its stomata open at night, carbon dioxide enters the leaves, and the carbon dioxide is stored in a chemical form. The stomata remain closed during the day, when the stored carbon dioxide is used in photosynthesis. How does this differ from most other plants? How might it be advantageous in the jade plant's environment?

4. A plant does not use potassium for a building material or energy source, but it stops doing photosynthesis and dies if it runs out of potassium. Why?

5. Denise and Roger dreamed of starting an organic homestead in southeast Alaska. They bought a plot of forested land, cut down the fir and spruce trees, built a cabin, and then had the soil tested. An accumulation of acid from decaying conifer needles, in combination with the heavy rainfall, had caused the soil to become quite deficient in a particular macronutrient. What nutrient was most likely to be in short supply? Why was it and not other nutrients washed from the soil? How might they enrich the soil with this nutrient without using commercial fertilizers?

Put Words to Work

Correctly use as many of the following words as possible when reading, talking, and writing about biology:

adhesion, Casparian strip, cation exchange, cohesion, compost, essential element, fertilizer, humus, macronutrient, micronutrient, mycorrhiza (plural, *mycorrhizae*), *nitrogen fixation, nodule, phloem sap, pressure flow mechanism, root, pressure, sugar sink, sugar source, sustainable agriculture, transpiration, transpiration-cohesion-tension mechanism, topsoil, xylem sap*

Use the Web

For further review of plant nutrition and transport, see the activities on the Web at www.mybiology.com. The animations of xylem and phloem transport are particularly useful.

Control Systems in Plants

33

Focus on the Concepts

A variety of mechanisms, most involving hormones, control how a plant responds to its environment. The chapter explores these plant control systems. In studying the chapter, focus on the following concepts:

- Experiments on phototropism, the bending of plants toward light, led to the discovery of plant hormones—substances produced in one part of the body and transported to another area, where they evoke responses in target cells. There are five major types of plant hormones. Hormones work together, and each may have multiple effects, depending on the site of action, concentration, and developmental stage of the plant.

- Auxins are produced by apical meristems, and they stimulate growth of stems, roots, and fruits. They may make stems elongate by weakening cell walls so cells can stretch. Cytokinins are a class of hormones produced by roots, embryos, and fruits. They promote cell division and retard aging. Auxins from the terminal bud and cytokinins from the roots interact to suppress axillary buds—a phenomenon called apical dominance.

- Gibberellins from roots and young leaves cause stems to elongate by stimulating cell division. Gibberellins in seeds promote germination. ABA promotes seed dormancy. The balance of gibberellins and ABA determines whether a seed will germinate. ABA inhibits other plant processes as well; it causes stomata to close during drought.

- Ethylene triggers a variety of aging responses. A burst of ethylene in a fruit causes ripening. In the fall, shorter days and cooler temperatures slow growth and alter the balance between decreasing auxins and increasing ethylene. This triggers the formation of an abscission layer, where a leaf separates from the stem.

- Plant hormones have many uses in agriculture—controlling fruit ripening, thinning of fruits, fruit drop, seed production, branching, etc. The synthetic auxin 2,4-D is used as a weed killer. Modern agriculture depends on such synthetic chemicals, but there are environmental costs.

- Tropisms orient plant growth toward or away from stimuli. Phototropism is response to light. Light cause auxin to migrate to the opposite side of the stem, which grows faster. Similarly, gravitropism, response to gravity, causes roots to grow downward. Thigmotropism, response to touch, causes tendrils to wrap around support.

- Plants display daily cycles of activity called circadian rhythms, which are independent of environmental cues such as light. Plants seem to possess a cellular biological clock. The clock may be a genetic mechanism that manufactures a protein on a daily cycle.

- Seasonal events are important in the plant life cycle, and plants detect the time of year by measuring photoperiod, the length of day and night. Long-day plants flower in spring, when nights are shorter than a specific length, and short day plants flower in fall, when nights exceed a certain length. Light-absorbing proteins called phytochromes change form in the dark, allowing the biological clock to measure the night.

- A variety of defenses against herbivores have evolved in plants: There are physical defenses, such as thorns, and chemical defenses, such as distasteful or toxic compounds. Some plants attract predatory insects to attack grazers. Defenses against pathogens include the epidermis, microbe-killing chemicals, and toughening of cell walls. R proteins at the site of infection recognize Avr molecules of pathogens, triggering a defense response, and also alarm hormones that warn the rest of the plant.

Review the Concepts

Work through the following exercises to review the concepts in this chapter. For additional review, refer to the activities on the Web at www.mybiology.com. The website offers a pretest that will help you plan your studies.

Exercise 1 (Module 33.1)

A series of experiments demonstrated that a hormone (named auxin) formed in the tip of a plant shoot causes the shoot to bend toward light. The illustrations on the next page illustrate some of those experiments and others. Predict whether each of the numbered grass shoots will grow to the left, right, straight up, or show no growth in response to the experimental conditions shown. Assume that each seedling has germinated in the dark and has just been placed under the experimental conditions.

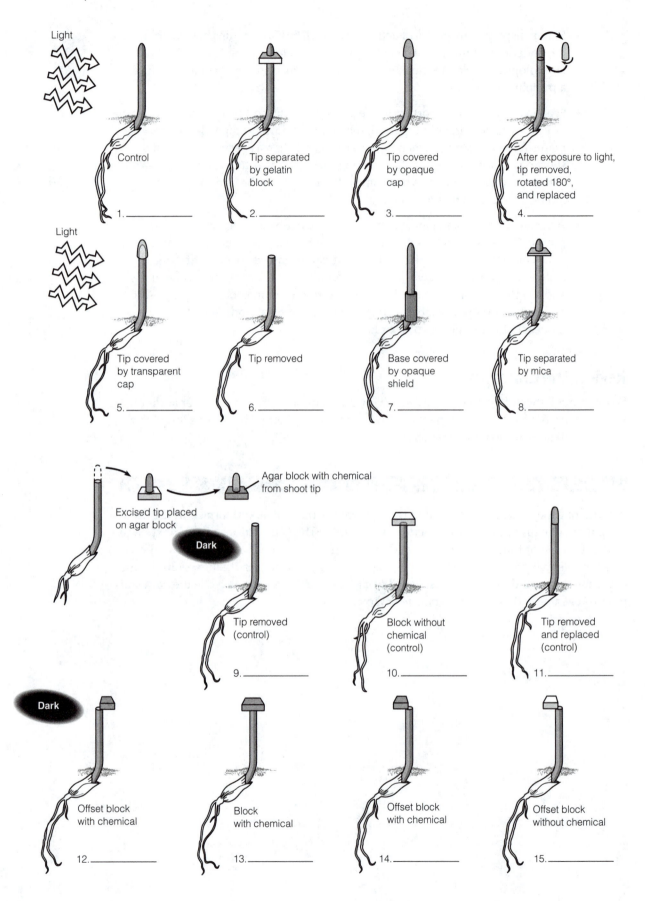

Light

Control

1. _____

Tip separated by gelatin block

2. _____

Tip covered by opaque cap

3. _____

After exposure to light, tip removed, rotated 180°, and replaced

4. _____

Light

Tip covered by transparent cap

5. _____

Tip removed

6. _____

Base covered by opaque shield

7. _____

Tip separated by mica

8. _____

Excised tip placed on agar block

Agar block with chemical from shoot tip

Dark

Tip removed (control)

9. _____

Block without chemical (control)

10. _____

Tip removed and replaced (control)

11. _____

Dark

Offset block with chemical

12. _____

Block with chemical

13. _____

Offset block with chemical

14. _____

Offset block without chemical

15. _____

Exercise 2   (Modules 33.2–33.4)

Module 33.2 introduces the major types of plant hormones, while the next two modules focus on auxins and cytokinins, two important hormone types. Review the effects of auxins and cytokinins by filling in the blanks in the following statements.

A. A number of different [1]_____, which promote elongation of stems (among other functions), have been identified in plants. Others have been synthesized artificially.

B. [2]_____ promote cell division. Several are natural; some are produced in the lab.

C. The best-known natural auxin is [3]_____.

D. Most auxin synthesis occurs in the [4]_____ at the tip of a shoot. It moves down the stem and stimulates cells to elongate.

E. Cytokinins are produced in growing tissues, mainly [5]_____, [6]_____, and [7]_____. Those that originate in roots affect stems by traveling upward in [8]_____.

F. [9]_____ promotes stem elongation, perhaps by weakening [10]_____ and allowing cells to stretch.

G. The same concentration of auxin that causes shoots to elongate [11]_____ elongation of roots. A lower concentration [12]_____ root elongation.

H. Auxin transported down the stem from the terminal bud [13]_____ growth of axillary buds, causing the plant to grow straight up without branching.

I. If the terminal bud is pinched off, [14]_____ buds become active, forming lateral branches, and the plant becomes [15]_____. [16]_____ moving upward from the roots are thought to activate the axillary buds.

J. The ratio of [17]_____ and [18]_____ probably shapes the complex growth pattern of most plants. Auxin from the terminal buds [19]_____ branching, while cytokinins from the roots [20]_____ it. The [21]_____ axillary buds usually begin to grow before those closer to terminal buds, reflecting the higher ratio of [22]_____ to auxins in the lower parts of the plant.

K. Essentially, the ratio of auxin and cytokinins helps to coordinate root and shoot growth. A growing [23]_____ system produces cytokinins, which cause the [24]_____ system to branch.

Exercise 3   (Modules 33.5–33.6)

Gibberellins and abscisic acid each have a number of functions, sometimes opposing one another. State whether each of the following might refer to a gibberellin (G) or abscisic acid (ABA).

_____ 1. Slows growth
_____ 2. Stimulates cell division and elongation in stems
_____ 3. Breakdown by winter cold may trigger spring germination.
_____ 4. Responsible for maintaining seed dormancy
_____ 5. Produced by fungi, it causes "foolish seedling disease."
_____ 6. Causes dwarf plants to grow tall
_____ 7. Opposes the effects of gibberellins in seeds
_____ 8. Primary signal that enables plants to withstand drought
_____ 9. When released from an embryo, it signals a seed to germinate.
_____10. If it is washed from seeds, they will germinate.
_____11. Acts antagonistically to ABA in seeds
_____12. With auxin, it causes fruit to develop, even without fertilization.
_____13. Causes leaf stomata to close
_____14. Causes grapes to grow larger and farther apart

Exercise 4   (Module 33.7)

Review the roles of hormones in maturation and ripening of fruit and leaf abscission by filling in the blanks in the following paragraphs.

Although auxin and gibberellins cause fruits to develop and mature, a hormone called [1]_____ triggers ripening. It is also involved in other processes related to [2]_____. Ethylene diffuses through the fruit in the air spaces between cells and brings on changes in the fruit such as [3]_____ and changes in color. As cells age, they produce more ethylene, speeding up the [4]_____ process. Ethylene can also diffuse through the air to a fruit. [5]_____ are often picked green and ripened by exposure to ethylene. Ripening of apples is slowed by flooding storage areas with [6]_____, which prevents buildup of [7]_____.

Ethylene probably triggers autumn color changes, drying, and fall of leaves from [8]_____ trees such as maples and oaks. Dropping its leaves allows a tree to save [9]_____. Fall colors are [10]_____ made in the fall and other substances revealed by the breakdown of green [11]_____.

The signals for leaf drop are the [12]_____, [13]_____ days of autumn. As a leaf ages, it produces less [14]_____. Cells in the [15]_____ layer, at the base of the leaf stalk, begin to produce ethylene. The ethylene causes enzymes to digest cell walls in the abscission layer. The wind and the weight of the leaf cause the stalk to break away, and the leaf falls.

Exercise 5 (Modules 33.3–33.8)

Review use of plant hormones in agriculture by writing the name of the hormone next to the summary of its uses.

_____ 1. Can make fruits develop without fertilization of seeds; with auxin, used for producing seedless fruits. Induce biennials to seed during the first year of growth.

_____ 2. Triggers ripening of stored fruit. May be used to thin fruit or loosen it for picking.

_____ 3. In low doses, prevent trees from dropping fruit. In larger amounts, used to cause fruit drop for thinning. A synthetic variant, 2,4-D, used as herbicide.

_____ 4. Hormone spray used to keep cut flowers fresh and make Christmas trees branch.

Exercise 6 (Modules 33.2–33.8)

Review the actions of the five classes of plant hormones by matching the hormones with their overall functions.

_____ 1. Abscisic acid

_____ 2. Cytokinins

_____ 3. Gibberellins

_____ 4. Ethylene

_____ 5. Auxins

A. Stimulate cell division and growth, affect root growth; oppose auxins by stimulating branching

B. Stimulate stem elongation; affect root growth, fruit development, apical dominance, tropisms

C. Inhibits growth, maintains dormancy, closes stomata during stress

D. Promote seed germination and bud development, stem elongation, leaf growth, flower and fruit development

E. Promotes fruit ripening, leaf drop, aging; opposes some auxin effects

Exercise 7 (Module 33.9)

Tropisms are directed growth responses that enable plants to adapt to environmental circumstances. Complete this chart comparing tropisms.

Name	Description	Mechanism
Phototropism	1.	2.
3.	Growth movement in response to gravity	4.
5.	6.	Contact triggers greater growth on opposite side, causing tendril to bend toward support

Exercise 8   (Module 33.10)

What is a circadian rhythm? A biological clock? See if you can explain the role of biological clocks in plant circadian rhythms in *exactly* 30 words. Choose your words wisely!

Exercise 9   (Modules 33.11–33.13)

Research with *Arabidopsis* is helping plant scientists understand phenomena such as flowering. Short-day plants flower when the night exceeds a certain minimum length. They are inhibited from flowering when red light (or white light, which includes red) shortens a long night. (Far-red light following red or white light reverses this effect.) Long-day plants flower when night length is shorter than a certain maximum. They flower when red or white light shortens a long night. The graphs below are similar to those in the text. Test your understanding of flowering and day (night) length: Under each graph, write "yes" if the plant will flower under the conditions described. Write "no" if the plant will not flower.

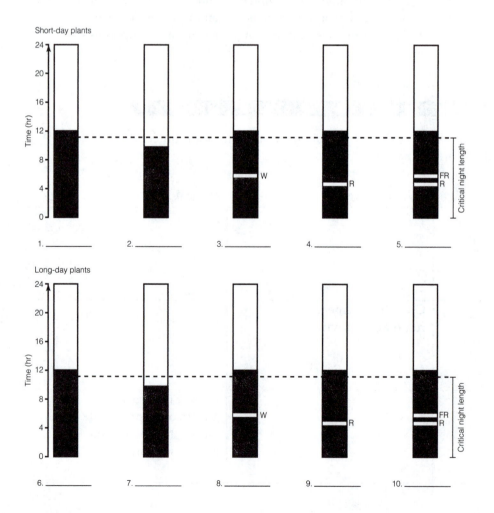

Plants protect themselves with a variety of defenses. Match each plant part or process on the left with a description of its role in defense.

_____ 1. Epidermis
_____ 2. Chemical that attracts wasps
_____ 3. Canavanine (an amino acid)
_____ 4. Cell wall
_____ 5. Alarm hormone
_____ 6. *R* gene
_____ 7. Thorn
_____ 8. *Avr* gene

A. Triggers defensive reaction distant from infection
B. Physical barrier that keeps pathogens out of plant
C. Physical defense that injures or repels herbivores
D. Lures predators that kill harmful caterpillars
E. Toughens to slow spread of pathogen in plant
F. Codes for a "signal" molecule that warns a plant of infection
G. Codes for receptor that matches and inactivates pathogen
H. Alters insect proteins so insect dies

Test Your Knowledge

Multiple Choice

1. Most plants flower when
   a. the soil reaches a certain temperature.
   b. they deplete soil nutrients.
   c. the days are the right length.
   d. a certain number of days have passed since they last flowered.
   e. the nights are the right length.

2. Plant tropisms appear to be controlled by
   a. auxins.
   b. tropogens.
   c. gibberellins.
   d. cytokinins.
   e. phytochromes.

3. The abscission layer
   a. causes a shoot to bend toward light.
   b. secretes cytokinin.
   c. is the location of the biological clock in a plant.
   d. detects light and measures photoperiod.
   e. is where a leaf separates from a stem.

4. _____ stimulate fruit development, so they are sprayed on crops to produce seedless fruits or vegetables.
   a. Ethylene and auxin
   b. Auxin and gibberellins
   c. Gibberellins and abscisic acid
   d. Ethylene and cytokinins
   e. Abscisic acid and auxin

5. Branching is inhibited by _____ from the tip of a growing shoot, but this effect is countered by _____ from the roots.
   a. cytokinins . . . auxin
   b. gibberellins . . . ethylene
   c. auxin . . . cytokinins
   d. gibberellins . . . abscisic acid
   e. auxin . . . abscisic acid

6. A synthetic auxin called 2,4-D is used
   a. to make fruit ripen.
   b. as a weed killer.
   c. to produce seedless fruits.
   d. to make plants produce larger numbers of seeds.
   e. to keep fruit from ripening.

7. A chemical change in a substance called phytochrome
   a. causes a plant to bend toward light.
   b. triggers fruit drop.
   c. enables a plant to measure day length.
   d. is responsible for gravitropism.
   e. allows a plant to deal with stresses like shortage of water.

8. A biological cycle with a period of about 24 hours is called
   a. thigmotropism.
   b. a circadian rhythm.
   c. a photoperiod.
   d. abscission.
   e. a biological clock.

9. Seeds of many desert plants will not germinate until a heavy rain washes away their
   a. abscisic acid.
   b. phytochrome
   c. gibberellins.
   d. auxin.
   e. ethylene.

10. Plant chemicals called brassinosteroids and isoflavones have effects on humans similar to the effects of
    a. aspirin.
    b. vitamin C.
    c. cholesterol.
    d. growth hormone.
    e. estrogen.

11. Researchers have discovered that plant chemicals such as the salicylic acid of aspirin
    a. attract pollinators.
    b. stimulate plant growth.
    c. help prevent or fight plant infections.
    d. can be used as stomach remedies.
    e. protect plants from grazers.

Essay

1. When a plant is tilted, it bends back toward a vertical position, even in the dark. How does a plant "know" which way is up, and what do hormones have to do with this?

2. How do we know that circadian rhythms are internal and not responses to environmental cycles? In what way are environmental cues important in circadian rhythms?

3. A farmer's almanac says, "To keep your stored apples crisp all winter, don't put bruised or wormy apples in the same barrels with the good ones." State the scientific reason why this is good advice.

4. Describe the changing balance of hormones that causes the leaves of deciduous trees to fall in the autumn. What does this change do to the structure of the leaf?

5. Explain the difference between long-day and short-day plants. How do plants "know" how long the day is?

Apply the Concepts

Multiple Choice

1. As leaf lettuce matures, the basal edible leaves suddenly send up a tall flowering shoot. After the plant "bolts" like this, it no longer produces broad, tasty leaves. Suppose you wanted to prevent bolting so that you could harvest lettuce longer. You might look for some way to interfere with the effects of
   a. abscisic acid.
   b. gibberellins.
   c. cytokinins.
   d. ethylene.
   e. gravitropism.

2. If the seeds in a cantaloupe are poorly fertilized and fail to develop, the cantaloupe itself will develop poorly and will probably drop off the vine. How does lack of seeds produce this effect?
   a. There is not enough ethylene in the cantaloupe for ripening to occur.
   b. The auxin concentration of the cantaloupe is too high for fruit growth.
   c. Abscisic acid causes the cantaloupe to shrivel and drop off.
   d. There is not enough auxin to stimulate development and stop fruit drop.
   e. Increased cytokinins cause the vine to break at the abscission layer.

3. According to the graph in Figure 33.3B in the text, an auxin concentration of 10^{-4} g/L would
   a. inhibit stem elongation and stimulate root elongation.
   b. stimulate stem elongation and inhibit root elongation.
   c. inhibit both stem and root elongation.
   d. stimulate both stem and root elongation.
   e. It is impossible to predict this from the graph.

4. If you wanted your plants to branch more, you might try spraying them with
   a. cytokinins.
   b. auxin.
   c. gibberellins.
   d. ethylene.
   e. abscisic acid.

5. When a plant structure such as a leaf is injured, it produces _____, which may cause the part to age and drop off.
   a. cytokinins
   b. phytochrome
   c. auxin
   d. abscisic acid
   e. ethylene

6. An Alaskan trapper was worried about being attacked by grizzly bears, so he left the lights in his cabin on all the time. Plants near the cabin flowered a month early. Which of the following best explains this?
   a. It was due to phototropism.
   b. They must have been long-night plants.
   c. The lights must have emitted far-red light.
   d. They must have been long-day plants.
   e. They must have been short-day plants.

7. Once a flower is pollinated, changes occur that make it less attractive to insects. Its petals, for example, shrivel and fall off. Pollination must
   a. increase the output of cytokinins in the flower.
   b. block the flow of auxin from the roots.
   c. trigger the release of ethylene in the flower.
   d. increase the formation of phytochrome, which sets the biological clock.
   e. trigger the formation of gibberellins in the flower.

8. Which of the following grass seedlings will probably bend toward the light?
   a. tip covered with a cap made of black plastic
   b. tip separated from base by a gelatin block
   c. tip cut off
   d. tip separated from base by aluminum foil
   e. tip cut off; agar with auxin placed on cut surface on side toward light

9. A biologist growing plant cells in a laboratory dish wanted to cause them to _____, so he treated them with cytokinins.
   a. enlarge
   b. become dormant
   c. grow roots
   d. produce auxin
   e. divide

10. A plant flowers only if days are shorter than 10 hours. Which of the following would cause it to flower?
    a. 8 hours light, 8 hours dark, flash of white light, 8 hours dark
    b. 12 hours light, 6 hours dark, flash of white light, 6 hours dark
    c. 8 hours light, brief dark period, 8 hours light, 8 hours dark
    d. 8 hours light, 8 hours dark, flash of white light, flash of far-red light, 8 hours dark
    e. 6 hours light, 6 hours dark, 6 hours light, 6 hours dark

Essay

1. During the 1800s, when natural gas was used for lighting, it was found that leaking gas from gas mains sometimes caused shade trees to drop their leaves. Based on what we now know about plant hormones, explain why this probably occurred.

2. A plant scientist hoped to make asparagus shoots grow faster by spraying them with auxin. She found in the lab that a 0.0001 M concentration of auxin increased elongation slightly. To increase its effect, she increased the auxin concentration to 0.01 M (100 times stronger) for spraying in the field. The results were not at all what she expected. The growth rate was less than normal! Explain why.

3. Predict the effects on a shoot of grafting it onto a much larger root system. What would be the role of hormones in these effects?

4. If you pinch off a leaf but leave the leaf stalk, the stalk soon falls from the stem. Pinching off the leaf must upset the balance between which two hormones? Which hormone, now in higher concentration in the stalk, causes the stalk to drop? Which hormone could you apply to the stalk to test whether this hypothesis is correct? What do you predict would happen?

5. In an experiment, soybean plants were planted at weekly intervals from the beginning of May through June. All the plants flowered at the same time—early September. Why?

6. Sometimes when grains are exposed to moisture they begin to germinate and are no longer usable. If you wanted to develop a hormone treatment to prevent this, what would you start with? Why?

7. Shoots exhibit positive phototropism and negative gravitropism. How could you find out which effect is stronger?

Put Words to Work

Correctly use as many of the following words as possible when reading, talking, and writing about biology:

abscisic acid (ABA), auxin, biological clock, circadian rhythm, cytokinin, ethylene, gibberellin, gravitropism, herbivore, hormone, long-day plant, photoperiod, phototropism, phytochrome, short-day plant, systemic acquired resistance, thigmotropism, tropism

Use the Web

For more questions and activities relating to plant control mechanisms, see the activities on the Web at www.mybiology.com.

The Biosphere: An Introduction to Earth's Diverse Environments

34

Focus on the Concepts

Living things interact with their environments in numerous ways. This chapter is an introduction to ecology, the study of these interactions. Focus on these concepts:

- Ecologists study the distribution and abundance of living things at several levels: Individual organisms are influenced by and adapted to the environment around them. Populations are groups of organisms of the same species. A community is an assemblage of interacting populations. An ecosystem includes the biotic factors that make up the community, plus their abiotic environment. Landscapes are large arrays of ecosystems. All of Earth inhabited by life comprises the biosphere.

- In recent decades, it has become apparent that human activities have large and long-term effects on the biosphere. The science of ecology offers insights into environmental problems.

- Abiotic—physical and chemical—factors influence living things. Some examples are energy source, temperature, water, nutrients, oxygen concentration, water salinity, wind, and fire. Organisms are also shaped by the biotic environment—factors such as food supply and predation. Each species adapts to particular abiotic and biotic factors via natural selection.

- Climate influences distribution of terrestrial communities. The tilt of Earth's hemispheres toward or away from the sun determines the seasons. The sun strikes Earth more directly near the equator, causing greater heating, rising air, and evaporation of moisture. Rising (cooling) and sinking (warming) air and rotation of Earth drive the winds (and ocean currents) and determine rainfall patterns.

- Sunlight and depth shape marine biomes. Near shore, the benthic realm (bottom) is in the photic (lighted) zone. Here intertidal communities, estuaries, wetlands, and coral reefs are rich with life. In most of the open sea (pelagic realm), the bottom is in the aphotic (lightless) zone, and life is sparse or dependent on debris from above or chemicals from undersea vents.

- Freshwater biomes include lakes and ponds, rivers and streams, and various wetlands. Light, temperature, and nutrient availability are key factors affecting lakes and ponds. A river is cold, clear, and swift at its source, and warmer, slower, and richer in nutrients near its mouth, affecting species living there. Wetlands—marshes, swamps, and bogs—have light, nutrients, and water, and are among the richest biomes.

- Distribution of terrestrial biomes reflects differences in climate—mainly temperature and rainfall. The frigid tundra circles the poles. Coniferous forest lies south of the tundra. Broadleaf forests occur in temperate areas with sufficient rainfall. Drier interior areas are characterized by temperate grasslands and savanna. Chaparral occurs in some dry coastal areas. Tropical forests cluster near the equator. The driest areas, hot or cold, are deserts.

- The global water cycle links water and land and shapes terrestrial biomes. The sun evaporates water from the sea, producing water vapor. Most of this water is returned to the sea via precipitation, but the wind carries some over land, where the water falls as rain and snow. Some of this water evaporates or is transpired by plants. Some pools as surface water or groundwater. Eventually it all returns to the sea.

Review the Concepts

Work through the following exercises to review the concepts in this chapter. For additional review, refer to the activities on the Web at www.mybiology.com. The website offers a pre-test that will help you plan your studies.

Exercise 1   (Introduction and Modules 34.1–34.2)

These modules introduce the subject of ecology. Review the relationships between the various levels of study in ecology by completing the concept map on the next page. Place the following terms on the map: **abiotic factors, biosphere, ecology, groups belonging to same species, populations, interactions, habitats, ecosystems, organisms, biotic factors, landscapes,** and **nutrients.**

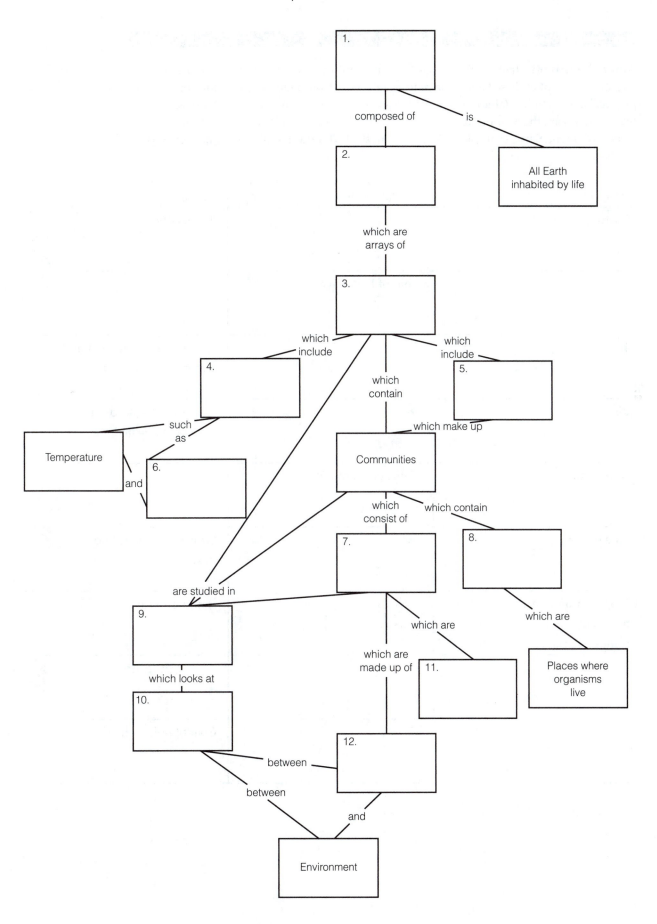

Exercise 2 (Modules 34.3–34.4)

Abiotic and biotic factors influence the distribution and abundance of organisms. Organisms adapt to these factors via natural selection. Complete the following chart listing physical and chemical (abiotic) factors that shape organisms and ecosystems. In each case, list an organism, the factor that affects it, and how the organism is adapted to that aspect of its environment. Some examples are taken from the text modules; others are new examples.

Organism	*Abiotic Factor*	*Adaptation*
Cactus	1.	Thick cuticle
2.	Cold of Himalayas	3.
4.	5.	Enzymes that function in very hot water
Pine tree	6.	Cones release seeds only after being heated
Noble rhubarb	Ultraviolet radiation	7.
Blue poppy	8.	Grows low to ground, next to rocks
Tropical rain forest tree shades out competing plants	9.	Rapid growth
Carp	10.	Able to gulp air
11.	12.	Coat of hollow hairs

Exercise 3   (Module 34.5)

Practice using concepts and terminology of climate by filling in the blanks in the following paragraphs.

The continent of North America stretches almost from the North Pole to the equator. Its life forms—from polar bears to iguanas—reflect this diversity in climate. The climate is warmest nearest the equator, where the sun's rays strike the Earth most 1_____. Near the 2_____, the sun's rays strike at more of an angle, so their energy is spread out over a larger area. The 3_____ of the Earth in relation to the sun causes the seasons. In 4_____, the Northern Hemisphere is tilted toward the sun. In 5_____, North America tilts away from the sun.

In addition to temperature differences, wind and rain are important aspects of climate. The sun's heat 6_____ air and evaporates 7_____ near the equator. The warm air is less dense than cold air, so it 8_____. The rising air near the equator creates an area of calm, light winds called the 9_____. The moist air cools as it ascends, and its moisture condenses and falls as 10_____, watering the tropical forests of Guatemala, Costa Rica, and Panama. After losing their moisture, the high-altitude air masses spread away from the equator. Cool air is denser than warm air, so the air 11_____ around 12_____ degrees north and south of the equator. This dry air is responsible for the 13_____ landscape of Mexico and the southwest United States. Some of the dry air spreads back toward the equator, and some picks up 14_____ and spreads toward temperate latitudes. This air will cool and drop its moisture as it spreads farther from the equator, into areas such as the eastern United States.

The rise and fall of air masses and their resulting spread from place to place is deflected by the 15_____ of the Earth to create the 16_____ winds. Wind rushes back to replace rising air at the equator, creating the 17_____, which dominate the tropics. The Earth's rotation causes the trade winds to blow from the 18_____ in the Northern Hemisphere.

Ocean 19_____ are created by the combined effects of unequal heating of surface waters, 20_____, the Earth's 21_____, and the locations of the 22_____. Currents profoundly affect local climate. The cold Japanese current creates the foggy conditions that water the redwood forests of northern California. The 23_____ warms the southeastern United States, as well as western Europe. Thus, 24_____ moderate the climate of nearby land.

Landforms shape local climate. Altitude influences temperature and 25_____. Moisture may be captured as air masses move over mountain ranges, creating dry 26_____ downwind. The sagebrush landscape of Nevada lies behind the Sierra Nevadas, for example. The forests of the California mountains and the Nevada desert are just two examples of 27_____, major types of ecological associations that cover large regions. As this example illustrates, these large ecosystems are shaped mostly by 28_____ and 29_____.

All aquatic communities are shaped by similar environmental factors. Review words and concepts related to marine and freshwater environments by completing this crossword puzzle.

Across

2. A ____ is an ecosystem intermediate between aquatic and terrestrial ones.
4. The bottom of the sea, a lake, or a river is called the ____ realm.
7. The ____ zone is flooded by tides every 12 hours.
8. Light does not reach the ____ zone.
12. Motile animals like fishes and whales live in the ____ realm.
13. ____ is only possible in the photic zone.
14. Nitrogen and ____ may be limiting nutrients in lakes and ponds.
16. ____ reefs are diverse communities in warm shallow tropical waters.
17. The oceans are threatened by their use as a ____ ground for waste.
18. ____ of an estuary may vary, depending on the mix of salt and fresh water.
19. ____ are drifting animals of the pelagic realm.
20. Wetlands include swamps, bogs, and ____.

Down

1. Drifting algae and cyanobacteria make up the ____.
3. The availability of ____ has a major effect on organisms in the sea.
5. An ____ is a coastal area where fresh and salt water meet.
6. Seasonal ____ in a lake brings nutrients to the surface and oxygen to the bottom.
9. The ____ shelves are the submerged edges of continents.
10. ____ vents are densely populated communities of the deep sea.
11. Near a river's mouth, the water may be murky with ____.
12. The ____ zone is the area near the surface where light can penetrate.
15. Decomposition on a lake bottom may deplete the water of ____.
16. Near a river's source, the water is usually cool and ____.

Exercise 5   (Modules 34.8–34.17)

As you read about the biomes, a good way to learn their characteristics is to make a chart that compares them. Using a large sheet of paper, list the *name* of each biome in the left column, and lay out columns for each of their characteristics across the top: *location, description, temperature, rainfall, special environmental conditions* (fire, etc.), *organisms,* and *human impact.* Once you have covered all the biomes, look at the website and do the next few exercises, which compare biomes in various ways.

Exercise 6   (Modules 34.8–34.17)

Temperature and precipitation are the most important factors shaping biomes. The figure below is a "climograph" outlining the temperature and rainfall limits within which each of the biomes occurs. Label the biomes on the graph. You do not need to know absolute temperatures and rainfall amounts; just ask yourself which biomes are hotter or colder, wetter or drier. Choose from **desert, tundra, temperate grassland, coniferous forest, temperate forest,** and **tropical forest.** You can reinforce your learning by coloring the areas on the graph to match the colors on the map in Module 34.8.

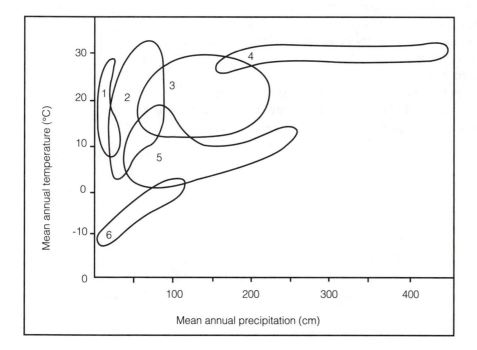

Match each of the phrases on the left with one of the biomes from the list
on the right.

_____ 1. Grows around the Mediterranean

_____ 2. Among the most complex and diverse of all biomes

_____ 3. North American prairies and Asian steppes

_____ 4. The taiga of Canada, Alaska, Siberia

_____ 5. May be hot or cold but is always dry

_____ 6. Permafrost occurs here.

_____ 7. May occur in rain shadows

_____ 8. Characterized by deciduous trees such as maples and oaks

_____ 9. Richest farmland in the United States

_____ 10. Antelope, zebras, lions, and cheetahs live here.

_____ 11. Where kangaroos live

_____ 12. Unlike other biomes, this one is growing in size.

_____ 13. Characterized by mild, rainy winters and hot, dry summers with fires.

_____ 14. Some characteristics of savannas but in colder areas than savannas

_____ 15. Straddles the equator

_____ 16. Grassland with scattered trees

_____ 17. The biome of most of the northeastern United States

_____ 18. Scrubland of dense, spiny shrubs

_____ 19. Moose, elk, snowshoe hares, beavers, bears, and wolves live here.

_____ 20. The warmest, rainiest biome

_____ 21. Closest to the North Pole

_____ 22. Rainfall can vary, but not much change in temperature or day length

_____ 23. In North America, nearly all of this biome was cleared, but it's coming back.

_____ 24. Includes temperate rain forests of California, Pacific NW, Alaska

A. Tundra

B. Desert

C. Tropical forest

D. Temperate
broadleaf forest

E. Chaparral

F. Coniferous
forest

G. Temperate
grassland

H. Savanna

Exercise 8   (Modules 34.8–34.17)

Color and label this map to show the distribution of these major biomes: **tropical forest, savanna, desert, chaparral, temperate grassland, temperate broadleaf forest, coniferous forest,** and **tundra.** (If you want to go into more detail, you can include polar and high-mountain ice areas.)

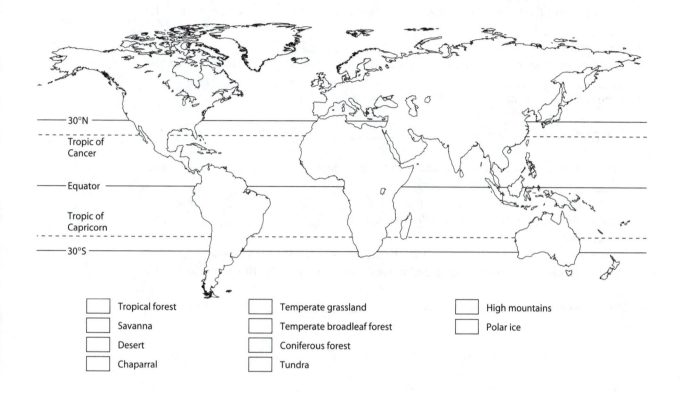

☐ Tropical forest	☐ Temperate grassland
☐ Savanna	☐ Temperate broadleaf forest
☐ Desert	☐ Coniferous forest
☐ Chaparral	☐ Tundra

☐ High mountains
☐ Polar ice

Exercise 9   (Module 34.17)

The global water cycle connects water and land, and is one of the major factors shaping the distribution of terrestrial biomes. Review the water cycle by completing the following statements.

1. Energy from the _____ drives the water cycle.

2. The biggest part of the water cycle is evaporation of water from the _____.

3. Water in the air is in the form of _____.

4. Most of the water that evaporates from the sea falls as precipitation over the _____.

5. Some of the water vapor evaporated from the sea is carried by _____ over land.

6. Precipitation over land is much _____ than precipitation over the sea.

7. Most of the rainfall over land returns to the air via _____ and _____.

8. Cutting tropical forests changes the amount of _____ in the air.

9. Some of the precipitation over land "pools" a while as _____ water and groundwater.

10. Pumping _____ for irrigation may alter the water cycle.

11. Much of the precipitation that falls over land eventually flows back to the _____.

12. Water that flows down rivers may carry _____ and chemicals.

Test Your Knowledge

Multiple Choice

1. Which of the following is the most complex?
   a. community
   b. individual
   c. species
   d. ecosystem
   e. population

2. Change of seasons is caused by
   a. change in the distance of the Earth from the sun.
   b. the tilt of the Earth.
   c. variations in output of solar energy.
   d. changes in wind patterns.
   e. gravitational pull of the sun and moon.

3. The kind of tropical forest in a given area—thorn forest, deciduous forest, or rain forest—depends mostly on
   a. rainfall.
   b. average temperature.
   c. temperature difference between day and night.
   d. distance from the equator.
   e. prevalence of fire.

4. The tundra biome gets little precipitation, but during the short summer season it is very wet, with many marshy areas, ponds, and bogs. Why?
   a. Water cannot soak into the frozen ground.
   b. It does not rain much, but the tundra gets a lot of snow.
   c. The ground is so flat that the water cannot run downhill.
   d. It is flooded by melting snow from the surrounding mountains.
   e. The tundra does not look wet; it is a cold desert.

5. If you were to compare the source (beginning) of a river to its mouth (end), you would probably find that
   a. the water is colder near the mouth.
   b. there are more nutrients in the water near the source.
   c. there are more phytoplankton in the water near the mouth.
   d. the current is swifter near the mouth.
   e. silt accumulates near the source.

6. Which of the following occurs in areas too warm for coniferous forest and too dry for temperate forest?
   a. desert
   b. tropical thorn forest
   c. temperate grassland
   d. chaparral
   e. tundra

7. Which of the following is dominated by plants that drop their leaves in the winter to conserve water?
   a. desert
   b. coniferous forest
   c. tropical rain forest
   d. temperate broadleaf forest
   e. savanna

8. Surface winds tend to blow toward the equator because
   a. this is a very rainy area.
   b. equatorial air is heated by the sun and rises.
   c. air flows "downhill" from the North Pole.
   d. the Earth's rotational speed is fastest at the equator.
   e. the oceans are broadest at the equator.

9. Which of the following describes the climatic conditions characteristic of the coniferous forest?
   a. mild with occasional drought and fires
   b. cold and dry
   c. hot or cold but always rainy
   d. cold and snowy
   e. mild, rainy winters and hot, dry summers

10. The primary ecological factor determining the distribution of deserts is
    a. windiness.
    b. elevation.
    c. moisture.
    d. temperature.
    e. fertility of soil.

11. Which of the following is *not* an abiotic factor that shapes ecosystems?
    a. soil minerals
    b. predators
    c. fire
    d. rainfall
    e. storms

12. Which of the following would you expect to find in a rain shadow?
    a. a desert
    b. an ocean
    c. a rain forest
    d. a river system
    e. tundra

Essay

1. In the post–World War II period (the 1940s through the 1960s) there was a widespread belief in the United States that technology could provide a simple solution to many of our human problems. How and why has that attitude changed?

2. Explain why rain forests occur along the equator, while deserts are found 30 degrees north and south of the equator.

3. Which biomes in North America have been most affected by human activities? How have they been changed, and what kinds of activities have changed them?

4. Using two specific biomes as examples, explain the conditions that determine why a particular biome occurs in a specific area.

5. Briefly describe the global water cycle, and explain how it affects the distribution of biomes on land.

Apply the Concepts

Multiple Choice

1. Which of the following shows how biotic environmental factors can affect an organism?
   a. Mice have the highest reproductive rate of any common mammal.
   b. Maple trees will not grow in waterlogged soil.
   c. Some shrubs grow only where forest fires scorch their seeds.
   d. Trout cannot live in shallow, warm water.
   e. Monarch butterflies live only where there are milkweed plants for food.

2. An ecologist studying how cattle grazing affects population dynamics of native animals like pronghorns and prairie dogs is focusing on
   a. populations.
   b. the biome.
   c. the ecosystem.
   d. the community.
   e. individuals.

3. Which of these biomes has been most altered by human activities in North America?
   a. temperate broadleaf forest
   b. coniferous forest
   c. desert
   d. tundra
   e. None of the above has been altered much in North America.

4. Which of the following might eat phytoplankton?
   a. great white shark
   b. beaver
   c. seal
   d. shrimp
   e. seagull

5. Which of the following lists only abiotic environmental factors?
   a. food, temperature, fire, wind
   b. soil minerals, oxygen level, light, predators
   c. food, parasites, predators, competitors
   d. wind, temperature, soil minerals, light
   e. light, rainfall, food, temperature

6. In the Sierra foothills of California, between the flat valley floors and mountain forests, there is a zone of grassland with scattered oak trees. Although it is too narrow to show on our biome map, this "oak woodland" would probably be classed as an example of
   a. coniferous forest.
   b. savanna.
   c. temperate deciduous forest.
   d. chaparral.
   e. tropical deciduous forest.

7. Which of the following countries is *not* paired with a biome that occurs there?
   a. Ecuador—tropical forest
   b. Australia—desert
   c. Argentina—temperate grassland
   d. Panama—coniferous forest
   e. France—temperate broadleaf forest

8. Which of these biomes generally lies between tundra and temperate broadleaf forest?
   a. savanna
   b. coniferous forest
   c. desert
   d. tropical forest
   e. chaparral

9. Most _____ must live in both the benthic realm and photic zone.
   a. sharks
   b. seaweeds
   c. clams
   d. whales
   e. phytoplankton

10. Bob was reading the newspaper. "More people burned out of their homes in Southern California," he remarked. His wife, Barb, looked up from her reading. "That's a shame," she said. "But I guess you have to expect a fire once in a while if you live in the _____ ."
    a. temperate grassland
    b. tundra
    c. chaparral
    d. coniferous forest
    e. savanna

Essay

1. Name five abiotic factors that might be important to the life of a tree in a mountain forest. Name one biotic factor that might affect the tree. How does the tree adapt to changes in these factors on a day-to-day basis? How has the tree species become adapted to these factors over many generations?

2. In general, temperature is an important abiotic factor shaping biomes on land, but it is less important in the ocean. In the ocean, light is an important abiotic factor, but it is less important on land. Explain why these factors differ in importance on land and in the sea.

3. Sailing ships going from Europe to North America generally sailed south, then west, and on the way back to Europe first sailed north, then east. From the information given in Module 34.5, can you explain why?

4. What effects might global warming have on the distribution of biomes? Which might shrink? Which might expand? In which directions will they probably expand or shrink?

5. A marine biologist was asked to speak on the topic of "Feeding the Growing Human Population by Harvesting the Sea." She said, "We can't do it. Marine sources are already overused. You see, the ocean is about as productive as a desert. It's because in most places the photic zone and the nutrients in the benthic realm are too far apart." Explain what she meant.

6. Ecology is the study of the interaction between organisms and their environment. We tend to think in terms of the environment affecting organisms. But interaction works both ways. Briefly describe two situations in which organisms (other than humans) might change their environment.

Put Words to Work

Correctly use as many of the following words as possible when reading, talking, and writing about biology:

abiotic factor, aphotic zone, benthic realm, biome, biosphere, biotic factor, chaparral, community, coniferous forest, continental shelf, desert, desertification, doldrums, ecology, ecosystem, estuary, habitat, intertidal zone, landscape, ocean current, organism, pelagic realm, permafrost, photic zone, phytoplankton, population, prevailing winds, savanna, temperate broadleaf forest, temperate grassland, temperate zones, trade winds, tropical forest, tropics, tundra, westerlies, wetland, zooplankton

Use the Web

For more practice using the concepts discussed in this chapter, see the activities on the Web at www.mybiology.com.

Behavioral Adaptations to the Environment   35

Focus on the Concepts

Behavior is the sum of an animal's actions, carried out in response to internal and external environmental cues. In studying animal behavior, focus on these concepts:

- Behavioral ecologists seek the proximate causes and ultimate causes of behavior. Proximate causes are *how* behaviors occur—the immediate mechanisms that underlie responses to stimuli. Ultimate causes are *why* behaviors occur—evolutionary explanations concerning the adaptive value of behaviors.

- Innate behaviors are under genetic control. A fixed action pattern (FAP) is an innate behavior sequence triggered by a specific stimulus. Once triggered, the response is carried out in full. FAPs are often involved in feeding, mating, and care of young. Researchers have found the genes responsible for certain behaviors, and interaction with the environment—an animal's experience—can affect gene expression.

- Learning is modification of behavior as a result of experiences. Habituation is learning to ignore a repeated stimulus. Imprinting is critical learning about home, parents, or signals, limited to a specific sensitive period. Spatial learning is establishing memories of landmarks. Some animals develop symbolic cognitive maps, and some can migrate long distances. Associative learning links stimulus with outcome. Social learning is acquired by observing others. Some animals have complex cognitive abilities and engage in problem-solving.

- Foraging is food-obtaining behavior. According to optimal foraging theory, an animal's feeding behavior should provide maximum energy gain with minimum energy expense and risk. Using a search image is one way to find food efficiently.

- Communication is sending and responding to signals—stimuli transmitted between animals. Animals may communicate by means of scents, sounds, or visual signals. Communication is more important and complex in social species.

- Mating behaviors and parental care enhance reproductive success. Many species engage in courtship rituals that reduce aggression and confirm species, sex, and readiness to mate. Mating systems may be promiscuous, monogamous, or polygamous. Natural selection favors mating systems and parental care that enhance survival of young and assure paternity.

- Social behavior is interaction among animals, usually of the same species. Sociobiology applies evolutionary theory to behavior. Territories and dominance hierarchies parcel out resources and may be maintained by agonistic behavior—threats, rituals, and combat.

- Altruism is behavior that reduces an individual's fitness while increasing the fitness of others. Natural selection that favors altruistic behavior toward relatives is called kin selection; genes may be passed on via relatives. Altruism toward nonrelatives is termed reciprocal altruism; a beneficiary may someday return the favor.

- Human behavior is shaped by both genetic and environmental factors. Studies of identical twins raised apart and nontwins raised together suggest that about half of complex human behavior is genetic and half the product of environment. Natural selection has equipped humans with flexibility and intelligence.

Review the Concepts

Work through the following exercises to review the concepts in this chapter. For additional review, refer to the activities on the Web at www.mybiology.com. The website offers a pretest that will help you plan your studies.

Exercise 1   (Introduction and Modules 35.1–35.3)

We can approach any animal behavior in terms of proximate (How?) questions—those related to environmental stimuli and the mechanisms of an animal's response. ("How does the bee know where to go?") We can also approach behavior in terms of ultimate (Why?) questions—those related to natural selection, the behavior's contribution to reproductive success. ("Why does the bee do that? How does it help the bee survive and reproduce?") In addition, we can also ask to what extent a behavior is genetically programmed (innate) and to what extent the behavior is modified by environment or experience. For each behavior described below, briefly describe what appear to you to be its proximate and ultimate explanations, and to what extent you think the behavior seems to be genetic (innate) or modified by experience.

1. A frog will automatically attempt to catch any small, moving, insect-sized object. It will starve if given live insects that cannot move.

2. Male and female prairie voles sniff each other, mate repeatedly, and strongly remember and bond to their mates.

3. On its first flight each morning, a bee circles the hive. After foraging, it can find its way back. If the hive is moved even a short distance after the "orientation flight," the bee will not be able to locate the hive when it returns.

4. Male red-winged blackbirds defend territories (with food and nest sites) and respond aggressively—with threats and attacks—to the red wing patches of other males. If a male's wing patches are painted over, it can enter territories of other males and is ignored by them.

5. A monkey can soon figure out how to unbolt a door and escape from a cage.

6. After it has been stung by a bee, a toad will not try to catch other bees.

7. Monarch butterflies breed in the United States, but they are basically a tropical species, unaccustomed to much cold weather. Monarchs that hatch from eggs in the eastern United States unerringly fly to a small wintering ground in the mountains of Mexico.

8. Newly hatched chicks will run for cover if a cardboard cutout of a hawk "flies" over the barnyard.

9. Gulls that nest close together in large colonies are able to remember the intricate patterns of spots on their own eggs and return only their own eggs to the nest if they roll out or are removed.

Exercise 2 (Module 35.3)

An innate (genetically programmed) behavior often takes the form of an automatic, stereotyped response to a specific stimulus. The response is called a fixed action pattern, or FAP. The following chart lists some examples of innate behaviors from Modules 35.2 and 35.3 and from the previous exercise. Fill in the blanks for species, stimulus, and fixed action pattern.

Species	*Stimulus*	*Fixed Action Pattern*
Kittiwake chick	1.	2.
3.	Flying shape	4.
5.	Open mouth of chick	6.
7.	Red wing patches	8.
9.	10.	Catches insect with tongue
11.	12.	Smile
13.	Egg near nest	14.

Exercise 3   (Modules 35.4–35.11)

There are several kinds of learning, from simple to complex. State whether each of the following describes an instance of **habituation, imprinting, spatial learning, using a cognitive map, associative learning, social learning,** or **problem solving**. Some examples are from the modules; others are new, but you can figure them out.

_____   1. A bird pecks at a caterpillar and gets a beakful of stinging hairs. After a couple of such experiences, the bird leaves this kind of caterpillar alone.

_____   2. A salmon is found stuck in a pipe, trying to return to a hatchery pond.

_____   3. During a sensitive period following hatching, gulls learn to identify their own young and keep them separate from the chicks in nearby nests.

_____   4. A hermit crab will at first withdraw into its shell if you tap on the glass of its aquarium. After a few taps it no longer responds.

_____   5. A bird called a gray jay buries food in hundreds of locations, then months later returns unerringly to dig it up and eat it.

_____   6. A chimp learns to reach bananas by either hitting them with a stick or stacking up boxes. When the bananas are placed too high for either the stick or the stacked boxes, the chimp uses both the boxes and the stick.

_____   7. If a bird does not hear its species' song shortly after hatching, it will not be able to sing it later.

_____   8. A dolphin learns the commands "Throw ball" and "Retrieve ring." The first time the trainer says, "Retrieve ball," the dolphin does it correctly.

_____   9. A monarch butterfly hatches and grows up in Ohio, then finds its way alone to wintering grounds in Mexico.

_____   10. A cat learns to come running into the kitchen when it hears the electric can opener.

_____   11. A monkey washes dirt from his food. Other monkeys observe this and start washing their food, too.

_____   12. At first, young ground squirrels dash into their burrows at any unusual sight or sound. Later they learn to run only from real danger.

_____   13. When their mother was killed, ducklings on a farm learned to follow a tractor.

_____   14. Bees circle before going out foraging and remember landmarks to find their way back to the hive.

_____   15. Young vervet monkeys see and hear their parents give a warning call when they see a snake. Next time the young monkeys give the warning call, too.

Exercise 4   (Modules 35.2–35.11)

To help you better understand innate and learned behaviors, sketch a concept map on a separate sheet of paper. Constructing a concept map is discussed in Chapter 3, Exercise 10.

Exercise 5   (Modules 35.2–35.11)

How do biologists figure out whether a worm displays negative phototaxis or a mouse has a bioclock? These modules describe some of the ways in which behaviors enable animals to survive in their environments—avoiding predators, locating food, finding their way, and attracting mates. The modules also describe some experiments and observations that have given us evidence about the causes, mechanisms, and reproductive value of various animals' behaviors. The following scenarios describe some observations and hypotheses about animal behavior. Briefly describe the kind of observations or experiments that could be done to test each hypothesis. Some examples are from the chapter, and others are new.

1. Turtles seen off the coast of Brazil are thought to lay their eggs on Ascension Island, in the middle of the Atlantic Ocean.

2. A kind of flatworm lives under rocks in streams and ponds and comes out at night to feed. This is probably an example of negative phototaxis.

3. A crab is most active at high tide and becomes inactive at low tide, roughly 6 hours later. It is thought that this is not due to an internal "bioclock" but rather that the crab can somehow sense the tidal pull of the moon.

4. Homing pigeons travel even on cloudy nights. It is thought that they do not need to use vision to find their way to the vicinity of their home.

5. Birds become restless in the fall. It is thought that an environmental change other than a drop in temperature makes them ready to migrate.

6. The kangaroo rat is a nocturnal desert rodent. It may possess an internal bioclock that signals it when to emerge from its burrow.

7. Fruit flies don't live long enough to learn much. Their courtship "dance" seems to be innate.

8. Hummingbirds may find it "worth their while" to fight more fiercely over a patch of flowers when the flowers contain more nectar (food).

Review foraging, communication, mating, and social behaviors by filling in the blanks in the following story.

The gray wolf is the top predator of forests, grasslands, and tundra of North America. Wolves are social animals, cooperating in packs to hunt large mammals such as deer, moose, and caribou. Thus they are neither food [1]_____ nor "generalists" but somewhere in between. A typical wolf pack consists of 5 to 20 animals, led by a [2]_____ "alpha" male and his mate. Unlike lone hunters such as the cougar, wolves do not ambush their prey. The pack tracks an animal over long distances, finally exhausting it, attacking from all directions, and eventually dragging their victim down for the kill. The wolf pack is successful in about one chase in ten. Deer often outrun the pack, and moose often fight wolves off.

In the summer, when larger prey are harder to corner, wolves often change their [3]_____ and seek out smaller animals, such as beaver and hares. These smaller prey are easier to catch, and when wolves are alone, they must pursue smaller prey. Why then do wolves work in a pack to go after (usually unsuccessfully) prey sometimes much larger than they are, only to share their kill with other members of the pack? Apparently it is more [4]_____ to cooperate in a pack and tackle large prey than to catch smaller prey working alone. This is an example of [5]_____, feeding behavior that provides maximum energy gain for minimal energy expense.

The wolf pack constantly roams and defends a [6]_____ of from 100 to 1,000 square kilometers. Like cheetahs and many other large territorial mammals, wolves use [7]_____ markers to stake their territorial claim. When on the move, they stop to urinate every few minutes. The territory assures a [8]_____ supply for the wolf pack and a safe place to raise their [9]_____.

The dominant feature of social behavior within the pack is a [10]_____ hierarchy. The alpha male leads the pack in chasing prey and in territorial defense. A dominance hierarchy may help to [11]_____ the wolf population. By allocating resources unequally, the hierarchy ensures that at least some of the wolves will receive an adequate food supply. It also minimizes the amount of energy spent in [12]_____ within the pack, energy that could more efficiently be used in hunting, defense, and reproduction.

An essential feature of social behavior is [13]_____, the exchange of information between individuals. Wolves use a varied repertoire of barks, yips, whines, and howls, similar to those of dogs. Body postures signal status in the pack hierarchy. The ears and tail of the alpha male are held erect. An ears-back, tail-down posture signifies submissiveness and lower rank. Thus wolves use three major channels of communication: [14]_____ signals to mark territory and [15]_____ displays and [16]_____ within the pack.

The mating season, from late fall to late winter, is a time of turmoil for the wolf pack. In most cases, the dominant male and female are the only individuals in the pack who will mate, and only with each other; they are therefore [17]_____, maintaining a [18]_____ that may last for years. Keeping other individuals from mating increases certainty of the dominant male's [19]_____. During mating season, there is an increase in [20]_____ behavior, both threats and actual fighting, as wolves jockey for position in the hierarchy. Much of the behavior is in the form of [21]_____—mostly growls and posturing. Even if the dominant male and female are not challenged, lower-ranking individuals may test each other, seeking an opportunity to move up in rank.

Unlike many species that engage in a lengthy [22]_____ ritual, mating between the dominant male and female in the pack is performed with the minimum of fuss required to neutralize [23]_____ behavior. The pair may mate several times a day over a 2-week period. In the spring, the female gives birth to four to seven pups. They are given much food and attention by all members of the pack, not just their parents. This helping or self-sacrificing behavior is an example of [24]_____. Many of the individuals in the pack are aunts, uncles, even siblings of the pups. Their caring behavior could be explained in terms of [25]_____ selection. Even though most of the adults in the pack do not get to mate, their sharing and helping improve the chances that copies of their [26]_____ will be passed on. The pups nuzzle and lick the muzzles of adults, a signal that causes the adults to regurgitate food. (This behavior is also seen between adults, even those who are not related. This might be [27]_____ altruism, where the favor of food sharing may be repaid later by the beneficiary.) In 3 to 5 months, the cubs are able to travel with the pack. Later some will leave the pack, but most stay. Thus, social ties in the wolf pack are also family ties.

Exercise 7   (Module 35.13)

There are three main channels of animal communication: chemicals (scents), movements (visual displays or touch), and sounds (vocalizations, etc.). Communication in a number of species is described in several of the modules in this chapter. A list of species is given below. Most are from the chapter; you are probably familiar with most of the others. Place the name of each species on the chart to show how you think that species communicates. For example, grasshoppers communicate almost totally by means of sound; "grasshopper" goes in the "sounds" corner. House cats use sounds, movements, and scents; they are placed near the center of the triangle, roughly equally distant from the names of all three communication channels. Add the following to the chart: **honeybee, ring-tailed lemur, wolf, cricket, human, domestic dog, common loon, frog,** and **white-crowned sparrow.** It is not important in this exercise for you to be sure of the "right" answer. The important thing is to think about the variety of forms of animal communication.

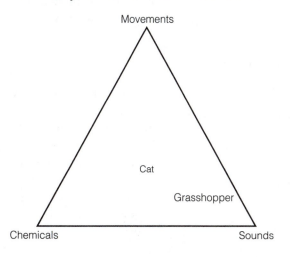

Review some of the vocabulary of animal behavior by matching each of the phrases on the left with a term from the list on the right. Each answer is used once.

_____ 1. Study of the evolutionary basis of social behavior
_____ 2. Sending and response to signals
_____ 3. A stimulus transmitted from one animal to another
_____ 4. Helps an animal look for food
_____ 5. Can result in altruistic acts that help relatives
_____ 6. A type of dominance hierarchy
_____ 7. Most efficient feeding behavior
_____ 8. Nature and nurture
_____ 9. One incentive for males to care for young
_____ 10. Interaction between individuals, usually of same species
_____ 11. Area defended against members of same species
_____ 12. Self-sacrifice that enhances fitness of others
_____ 13. May be promiscuous, monogamous, or polygamous
_____ 14. Research that helps sort out effects of genes and environment
_____ 15. Ranking by social interactions
_____ 16. Communicates that male and female are ready to mate
_____ 17. Threats, rituals, or combat over resources
_____ 18. Signals surrender
_____ 19. Self-sacrifice that helps unrelated individuals

A. Reciprocal altruism
B. Optimal foraging
C. Sociobiology
D. Kin selection
E. Courtship ritual
F. Social behavior
G. Twin study
H. Territory
I. Agonistic behavior
J. Dominance hierarchy
K. Appeasement display
L. Search image
M. Communication
N. Pecking order
O. Genes and environment
P. Altruism
Q. Signal
R. Mating system
S. Certainty of paternity

Test Your Knowledge

Multiple Choice

1. Which of the following *incorrectly* pairs an animal with a signal it might use to communicate?
   a. Ring-tailed lemur—scent
   b. Chimp—vocalization
   c. Common loon—scent
   d. Bee—movement
   e. Sparrow—vocalization

2. Scientists think that some birds may find their way during migration by
   a. imprinting on faint odors.
   b. listening to high-pitched sounds.
   c. observing the stars.
   d. seeing wavelengths of light that we cannot see.
   e. sensing the gravitational pull of the sun and moon.

3. Kineses and taxes are simple, automatic movements in response to stimuli. Movement of a moth toward a light is positive phototaxis. What is the difference between kineses and taxes?
   a. Taxes are directed toward a stimulus, kineses away.
   b. Kineses are innate, and taxes are learned.
   c. Taxes are innate, and kineses are learned.
   d. Kineses are performed much faster than taxes.
   e. Kineses are more random, and taxes are more directed.

4. Which of the following is a fixed action pattern?
   a. A goose rolls an egg back into its nest.
   b. A sparrow defends its territory.
   c. A wolf tracks its prey.
   d. A robin eats a distasteful bug, spits it out, and never eats one again.
   e. A gull returns to the same island breeding grounds each year.

5. Ducklings learn to follow their mothers. This is an example of
   a. habituation.
   b. problem solving.
   c. altruism.
   d. imprinting.
   e. kin selection.

6. Which of the following appears to be the primary method by which worker bees communicate the location of nectar to other members of the colony?
   a. They secrete a particular combination of chemical signals.
   b. They produce complex, high-pitched sounds.
   c. They fly to the nectar, and other bees follow.
   d. They perform a sequence of movements in the hive.
   e. They lay down a chemical trail to the source of food.

7. Birds sing, squirrels chatter, and wolves urinate on bushes to
   a. warn other members of the group of danger.
   b. mark their territories.
   c. attract mates.
   d. assert dominance.
   e. defend themselves from predators.

8. In _____ , an animal learns to associate one of its behaviors with reward or punishment.
   a. social learning
   b. trial-and-error learning
   c. fixed-action-pattern learning
   d. habituation
   e. imprinting

9. The ultimate question about any animal behavior is
   a. "How much does it cost?"
   b. "How does it increase reproductive success?"
   c. "How does an animal do it?"
   d. "What is the stimulus that triggers it?"
   e. "What's in it for me?"

10. When animals engage in _____ , they often perform displays that make them look as large and dangerous as possible.
    a. courtship rituals
    b. altruism
    c. kin selection
    d. kineses
    e. agonistic behavior

11. Chemicals such as DDT, PCBs, and dioxin can
    a. disrupt reproductive behavior
    b. interfere with imprinting
    c. exacerbate agonistic behavior, making animals overly aggressive
    d. disrupt optimal foraging
    e. interfere with migratory behavior

12. The key idea of sociobiology is that
    a. human social behavior is the same as animal social behavior.
    b. social behavior has an evolutionary basis.
    c. little behavior is inherited; most is learned in social situations.
    d. almost all human and animal behavior is preprogrammed.
    e. animal behavior and human behavior are so different they cannot be compared.

13. Some animals carry out their mating rituals one-on-one, while others congregate to choose mates. What advantage might there be in animals carrying out such a *group* mating ritual?
    a. It reduces aggression.
    b. It prevents individuals from accidentally mating with relatives.
    c. It saves energy.
    d. It enables individuals to compare potential partners.
    e. A large group attracts more attention.

Essay

1. Describe the functions of imprinting in the lives of birds. How does it increase fitness?

2. What are the advantages of a dominance hierarchy to a high-ranking individual? What are the advantages of the hierarchy to the group?

3. Describe four functions of courtship rituals.

4. Explain why communication is essential to social behavior, and give five examples of animal communication.

5. Describe two methods birds use to find their way during long-distance migrations.

6. What are the advantages of fixed action patterns over learned behavior? In what situations do fixed action patterns seem to be especially important?

7. Name three species of animals that cooperate in social groups, and give three general benefits or advantages of social cooperation in these animals.

Apply the Concepts

Multiple Choice

1. A frog may at first be startled by tree branches swaying in the wind, but it soon gets used to these kinds of unimportant changes in its environment. This is an example of
   a. a fixed action pattern.
   b. imprinting.
   c. altruism.
   d. habituation.
   e. trial-and-error learning.

2. Bees can see colors we cannot see, and they can detect minute amounts of chemicals we cannot sense. But unlike many insects, bees cannot hear very well. A behavioral ecologist would probably give which of the following as the ultimate explanation of their poor hearing?
   a. Bees are too small to have functional ears.
   b. Good hearing must not contribute much to a bee's reproductive success.
   c. If a bee could hear, its tiny brain would be swamped with information.
   d. This is an example of optimal foraging.
   e. If bees could hear, the noise of the hive would drive them crazy.

3. Chimpanzees usually maintain their rank in the dominance hierarchy with charging displays. Sometimes actual combat occurs briefly, but usually aggression is ritualized. Why don't chimps fight more?
   a. Chimps are not strong enough to inflict much injury.
   b. Dominant females keep order and prevent fights from escalating.
   c. Natural selection favors avoiding injury and saving energy.
   d. Chimps spend most of their time mating; they don't have time to fight.
   e. Chimps are too hard-pressed finding food to spend much time fighting.

4. Western fence lizards are sometimes called bluebellies because the undersides of the males are bright blue. If one male invades another's territory, the territory owner protests vigorously, doing "push-ups" to display his belly and, if necessary, pushing the invader out. If an intruder's blue belly is first painted brown, he is completely ignored by the territory owner. The blue belly of the intruder apparently acts as a _____ , triggering _____ .
   a. search image . . . optimal foraging
   b. stimulus . . . altruism
   c. stimulus . . . agonistic behavior
   d. search image . . . altruism
   e. fixed action pattern . . . imprinting

5. An aquaculture facility hatched salmon eggs and released young fish into a river leading to the ocean. The fish fed and grew in the ocean, and in a few years they returned to the facility. Because the number of returning fish was low, a scientist suggested that the facility add a harmless chemical to the water draining from the fishponds into the river. She hoped the chemical would increase the number of returning fish in a few years by
   a. decreasing their agonistic behavior.
   b. enabling them to imprint on the facility.
   c. stimulating habituation in the fish.
   d. improving the salmons' search image.
   e. promoting optimal foraging.

6. Every morning Brian turns on the light in the room and then feeds the fish in his aquarium. After a couple of weeks of this routine, he noticed that the fish came to the surface to feed as soon as he turned the light on, whether or not any food was present. This illustrates
   a. habituation.
   b. positive phototaxis.
   c. imprinting.
   d. associative learning.
   e. spatial learning.

7. Wildlife biologists raising condors in captivity have found that it is important not to let condor chicks see their human "parents," so they feed the chicks with a puppet that looks like the head of an adult condor. This is to allow normal _____ in the chicks.
   a. imprinting
   b. habituation
   c. agonistic behavior
   d. social learning
   e. kin selection

8. Which of the following best illustrates optimal foraging?
   a. A robin will repeatedly attack any red object near its territory.
   b. Musk oxen will form a circle to fend off a wolf attack.
   c. Bats emerge to feed at about the same time each night.
   d. A blackbird will warn others in the flock if it senses danger.
   e. A sunbird will more fiercely defend flowers that produce more food.

9. Remembering facts that you read in a textbook would probably be considered
   a. optimal foraging.
   b. imprinting.
   c. spatial learning.
   d. habituation.
   e. social learning.

10. Which of the following sayings best summarizes the idea of reciprocal altruism?
   a. "A rolling stone gathers no moss."
   b. "You scratch my back, and I'll scratch yours."
   c. "A penny saved is a penny earned."
   d. "A bird in the hand is worth two in the bush."
   e. "Beauty is in the eye of the beholder."

11. Which of the following best summarizes the relative roles of genes and environment in human behavior?
   a. Genes have little effect; virtually everything people do is a result of experience.
   b. Genes have some effects on specific behaviors, but everything else is learned.
   c. Genes and experience both have important effects on behavior.
   d. Much what we learn is based on genetic foundation, so genes are more important.
   e. Virtually all human behavior is genetically programmed; learning has a small effect.

12. A female bear risks her own life to save the lives of her cubs. This is best explained in terms of
   a. optimal foraging.
   b. inclusive fitness.
   c. agonistic behavior.
   d. imprinting.
   e. dominance.

Essay

1. Some animals, such as wolves and lions, hunt in groups, while others, like foxes and house cats, are loners. Every method of getting food involves tradeoffs; there is no one best way to do it. What do you think are the advantages of hunting in a group? The disadvantages?

2. Bats are the major predators of a species of moth. When one of these moths hears the high-pitched sound of a bat's sonar, it folds its wings and dives to the ground. How would a behavioral biologist describe the proximate cause of the moth's behavior? The ultimate cause?

3. European birds called wagtails eat insects. During the winter, when food is scarce, a wagtail may defend a territory, on which it captures an average of 20 insects per hour. Or the wagtail may join a flock that ranges widely over the countryside. A bird in a flock averages 25 insects per hour. If this is the case, why don't the birds always travel in flocks? Discuss in terms of energy gain and expenditure.

4. Cancer patients often suffer nausea and loss of appetite as a side effect of the powerful drugs used in chemotherapy. They have a particularly tough time eating foods that they have in the past eaten just before drug treatment, even if the foods are subsequently given to them days or weeks after the effects of the drugs have worn off. Explain this "food aversion" in terms of your knowledge of learning in animals. How might this kind of learning be helpful to an animal?

5. Chickens are polygamous and sparrows are monogamous. Explain how these different mating systems might have been shaped by the needs of their young.

6. Many bird species use special alarm calls to signal the approach of danger. Biologists ask: What possible advantage to a bird is there in warning other members of the flock? Isn't the bird that gives the warning call putting itself in danger? If so, how could this behavior possibly continue to exist? How could it improve a bird's fitness? Wouldn't the genes for warning other members of the flock be eliminated by natural selection? The fact is, alarm calls and other forms of altruism—self-sacrifice—continue to exist. How might altruism improve the alarm-giver's fitness? List the two general kinds of explanations that biologists give for altruistic behavior.

7. Most people think babies are "cute." They like to hold them, play with them, talk to them, and protect them. How might a sociobiologist explain this?

Put Words to Work

Correctly use as many of the following words as possible when reading, talking, and writing about biology:

agonistic behavior, altruism, associative learning, behavior, behavioral ecology, cognition, cognitive map, communication, dominance hierarchy, fixed action pattern (FAP), foraging, habituation, imprinting, inclusive fitness, innate behavior, kin selection, kinesis, learning, migration, monogamous, optimal foraging theory, polygamous, problem solving, promiscuous, proximate cause, reciprocal altruism, search image, sensitive period, signal, social behavior, social learning, sociobiology, spatial learning, stimulus, taxis (plural, *taxes*)*, territory, trial-and-error learning, ultimate cause*

Use the Web

Continue exploring animal behavior on the Web at www.mybiology.com.

Population Ecology

Focus on the Concepts

Population ecology is concerned with such characteristics of populations as age structure, birth and death rates, changes in population size, and factors that regulate populations over time. As you study this chapter, focus on these concepts:

- A population is a group of individuals of the same species, living in the same general area. Population density is the number of individuals per unit of area or volume. Dispersion pattern is the way individuals are spaced; it may be clumped, uniform, or random, depending on their degree of interaction and distribution of resources.

- A life table tracks the chances of an individual in a given population surviving to various ages. Data from a life table can be used to plot a survivorship curve, which tracks the percentage of individuals remaining alive at each age. In populations with Type I survivorship, most young survive a long time and die of old age. Species with Type II survivorship stand an equal chance of living and dying throughout life. In Type II survivorship, most individuals die young, and only a few make it to old age.

- Using idealized models, ecologists can predict population growth. With unlimited space and resources, exponential growth can occur. $G = rN$, where G is the number of new individuals added, N is the number of individuals in the population, and r is the per capita rate of increase. When resources are limited, the logistic model describes a population that levels off at carrying capacity: $G = rN(K-N)/K$, where K is the carrying capacity, the number of individuals the environment can support.

- Most populations are limited by a combination of density-dependent factors, such as predation or food supply, and abiotic factors, such as weather. Density-dependent factors regulate population density by exerting greater effects as the population increases. Abiotic factors are not related to density and may cause drastic growth and decline. Some predators and prey undergo boom and bust cycles.

- Evolution shapes life histories. *R*-selection shapes species that are small, short-lived, and produce numerous offspring that receive little care. They tend to be regulated by abiotic factors and thrive in disturbed areas where resources can suddenly become abundant. *K*-selection shapes larger organisms that live longer and produce smaller numbers of offspring that receive more care. *K*-selected species tend to be regulated by density-dependent factors in environments where resources are more limited.

- For centuries, a high birth rate and high death rate kept the human population small. Better sanitation and health care brought a high birth rate and lower death rate, and the population exploded to its current 6.5 billion. Now we are going though a demographic transition to a low birth rate and low death rate. Developing countries with youthful age structures continue to grow. Developed countries are closer to zero growth, but consume more resources per capita and have bigger ecological footprints.

Review the Concepts

Work through the following exercises to review the concepts of population ecology. For additional review, refer to the activities on the Web at www.mybiology.com. The website offers a pre-test that will help you plan your studies.

Exercise 1 (Modules 36.1–36.2)

This exercise will allow you to work with the concepts of population density, dispersion pattern, and sampling. The map on the next page represents a meadow on the edge of the city of Mapleton. It is surrounded by developed and farmed land but has remained relatively undisturbed. Developers plan to build a subdivision that would cover the meadow. The Mapleton Open Space Alliance would like the meadow to remain as public open land. They note that the dwarf hawthorn, an uncommon shrub, is found in the meadow. It is considered a "sensitive species" by the state conservation department. The city council has asked for a construction delay until the status of the shrub is determined. You have been sent to determine the density of the hawthorn population in the meadow, as well as that of a deer mouse that may also be present. Use the map of hawthorn distribution on the next page for your survey, and answer the following questions.

The area of the meadow is 16.8 hectares. (A hectare is a metric unit of area equal to about 2.2 acres, so the meadow totals about 37 acres.) This is too big an area to count every shrub, so you will have to look at sample plots. On the ground, this would be done with GPS and measuring tapes. You can choose random samples by merely dropping a penny on the map, drawing a circle around it, and counting the "shrubs" inside. On the scale of the map, the area covered by a U.S. penny equals 0.2 hectare.

1. Take ten samples. How many hectares does this total? _____
2. What is the total number of shrubs in the ten samples? _____
3. What is the density of hawthorns in shrubs per hectare? _____
4. What is the total number of hawthorns in the meadow? _____
5. How could you make your count more accurate? Why not do this?
6. Look at the map again. What is the pattern of dispersion of the shrubs? What might cause this pattern of dispersion?

You would also like to know the number of deer mice in the meadow. For this, it will probably work best to use the mark-recapture method. To learn about this method, you will have to look at the Web Activity entitled "Techniques for Estimating Population Density and Size."

7. Why does the mark-recapture method work better for mice than the method used to count the plants?

8. One night, you trap 40 mice, mark them, and let them go. Two nights later, you again trap 40 mice, and 10 of them are marked. What is the total number of mice in the meadow? _____

9. What is the population density of mice in the meadow, in animals per hectare? _____

10. What do you have to assume about the mice and your method for your results to be valid? Could you be wrong? Why or why not?

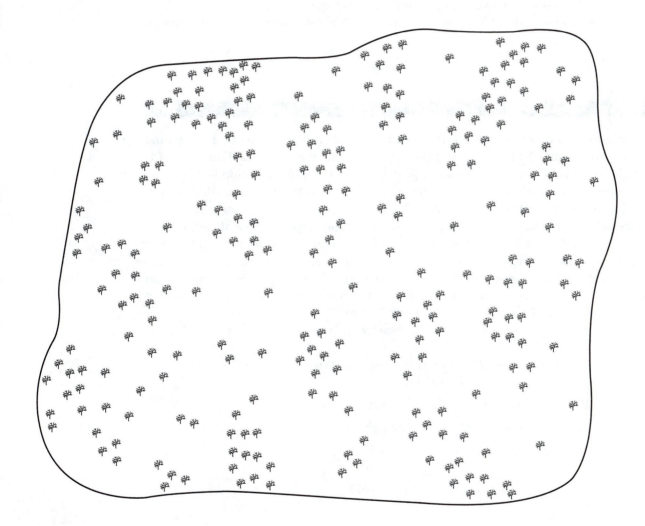

Exercise 2   (Module 36.3)

Life tables and survivorship curves enable population ecologists to describe and understand life cycles. Check your understanding of life tables and survivorship curves by matching each phrase on the left with a term on the right. Answers may be used more than once.

_____ 1. Graph of percent alive at the end of each age interval	A. Life table
_____ 2. Tabulation of deaths and chance of surviving	B. Survivorship
_____ 3. Most young die, but a few live to old age.	curve
_____ 4. Originally used to set life insurance rates	C. Type I
_____ 5. Characteristic of oysters	survivorship
_____ 6. Death rate constant over life span	D. Type II
_____ 7. Characteristic of lizards and squirrels	survivorship
_____ 8. Most offspring live a long life and die of old age.	E. Type III
_____ 9. Characteristic of humans and many other large mammals	survivorship

Exercise 3   (Module 36.4)

Models devised by ecologists describe two kinds of population growth. Exponential growth is described by this equation: $G = rN$. The number of individuals added to the populaton per unit of time, G, depends on N, the size of the population, multiplied by r, the population's intrinsic rate of increase. Intrinsic rate of increase, r, is calculated by subtracting the death rate from the birth rate. Exponential growth is unregulated. The bigger the population, the faster it grows. This cannot be sustained for long in real populations, but it is interesting as a theoretical possibility. Populations of fast reproducers such as bacteria and insects can grow at near-exponential rates for short periods.

   Let's calculate and graph the exponential growth of a population of aphids for which $r = 40\%$ (or 0.4) per week. Remember that $G = rN$. If there are 10 aphids to start with, the number of aphids added by the end of the first week (G) is equal to rN, or 0.4×10, which equals 4. So the total population (N) after one week is $10 + 4 = 14$.

1. Starting with the new total (N) of 14, how many aphids will be added (G) in the second week? _____ (Round off fractions.)

2. What will the total population (N) be at the end of the second week? _____

3. Aphids added in the third week? _____ Total after third week? _____

4. Aphids added in the fourth week? _____ Total? _____

5. Total at end of the fifth week? _____ sixth week? _____ seventh week? _____ eighth week? _____ ninth week? _____ tenth week? _____

6. Graph the size of the aphid population *(N)* versus time (in weeks) below. Population size was 10 at time = 0. Label the axes of the graph.

7. How would you describe the shape of this graph?

8. Could this kind of growth continue indefinitely? Why or why not?

Real environments will not support exponential growth. Populations are limited by space, food supply, or other factors that slow growth. The population may level off at a density the environment can maintain—carrying capacity. This so-called logistic growth is described by the equation $G = rN(K - N)/K$, where K represents carrying capacity.

The following data chart the growth of a population of deer on a small protected island off the coast of British Columbia, recorded over a 50-year period:

1945	92	1975	814
1950	151	1980	765
1955	295	1985	688
1960	603	1990	740
1965	861	1995	729
1970	920	2000	738

9. Graph the growth of the deer population below. Label the axes. How would you describe the overall shape of the graph?

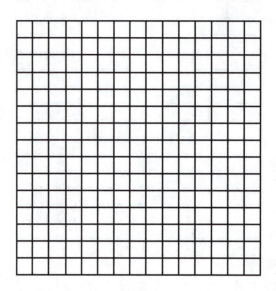

10. What happened to the population during the 1970s?

11. What may have caused the population density to level off?

12. What is your estimate of the carrying capacity of the island for deer?

13. What is the biological term for this kind of population growth?

Exercise 4　(Modules 36.5–36.6)

Population growth is limited by both density-dependent (biotic) and abiotic environmental factors. Density-dependent and abiotic factors affect birth rates and death rates in different ways. State whether each of the following words or phrases relates more to density-dependent factors (DD) or to abiotic (A) factors.

_____ 1. Have more effect when the population is larger
_____ 2. Have less effect when the population is smaller
_____ 3. Effect not dependent upon density of population
_____ 4. Competition for food
_____ 5. Fire
_____ 6. Predation
_____ 7. Stress produced by crowding
_____ 8. Competition for nest sites
_____ 9. Storms
_____ 10. Drought
_____ 11. Disease
_____ 12. Heat and cold
_____ 13. Habitat disruption by humans
_____ 14. Cause populations to stabilize in size, presumably near carrying capacity
_____ 15. Cause rapid population growth followed by unpredictable crashes
_____ 16. Seem to cause boom and bust cycles among predators and prey
_____ 17. Limit the snowshoe hare population
_____ 18. Limit the lynx population
_____ 19. Effects of the nonliving environment
_____ 20. Effects of organisms
_____ 21. Reduce clutch size as song sparrow population grows
_____ 22. Responsible for lemming boom and bust population cycles
_____ 23. Cause periodic drastic declines in song sparrow population

Natural selection shapes different life history traits under different environmental conditions. Some populations exhibit *r*-selection and others *K*-selection. Compare these contrasting life histories by completing this chart.

Characteristic	*r-Selection*	*K-Selection*
Life history emphasis	1.	Stability near carrying capacity
Relative body size	2.	3.
Number of offspring per reproduction	4.	5.
Age at first reproduction	Younger	6.
Emphasis on ____ of offspring	7.	Quality and care
Population near *K*?	No; resources abundant	8.
Limited mostly by	Abiotic factors	9.
Predators or pests	10.	11.
Examples of organisms	12.	13.

Review the dynamics of human population growth by filling in the blanks in the following essay.

The world's human population now totals more than [1]_____ billion people. We are the most numerous large animal on Earth, and our huge numbers are a result of a recent burst of population growth unprecedented in the history of life.

The human population grew relatively slowly and steadily until about 1500, when there were about [2]_____ million people on earth. After that, population growth gradually accelerated. The population doubled to 1 billion by 1850, again to 2 billion by 1927, and again to 4 billion by 1975. Current projections suggest that the human population could reach more than [3]_____ billion by the year 2025.

Population growth rate depends on two factors—per capita rate of increase, symbolized as [4]_____ , and population size, or [5]_____ . Throughout most of human history, *r* was close to [6]_____ . Times were tough; although [7]_____ rate was relatively high, it was balanced by an equally high [8]_____ rate.

The Industrial Revolution changed all this. Humans did something that no other species has ever done—they took *r* into their own hands. Economic development in the United States and [9]_____ led to improved nutrition, health care, and sanitation. Death rate [10]_____, while [11]_____ remained high. Starting in the developed world, the human population began to grow faster and faster. By the mid-1900s, improvements in sanitation and [12]_____ had spread to the developing world, and there birth rates began to outpace death rates as well.

Although the world population continues to grow, growth *rate* peaked in 1962 as health care improved [13]_____ in the developing countries and widespread used of [14]_____ started to hold down birth rate. The developing world is catching up, and the world human population finds itself undergoing a demographic [15]_____, a shift from high birth rates balanced high death rates to [16]_____ birth rates balanced by [17]_____ death rates. Currently, in the developed countries, birth rate and death rate are approximately [18]_____. In the developing world, the ratio of births to deaths is roughly 2:1. Because most people live in the developing countries, this is the biggest factor in world population growth. In fact, out of the more than 77 million people added to the global human population each year, more than [19]_____ million are born in the developing countries!

Reduced [20]_____ size is the most important change leading to demographic transition. When women's lives are improved, they [21]_____ reproduction and have [22]_____ children. Access to [23]_____ enables them to practice family planning.

The [24]_____ structure of a population—the proportion of individuals in different age groups—can help us predict growth trends and social changes. In a [25]_____ nation such as Mexico, the age structure is a broad-based pyramid, with a large percentage of prereproductive individuals. Couples tend to have large families, and each generation of children outnumber their parents. Even if birth rate is curtailed, such a population will continue to grow for many years because so many young people are coming into their reproductive years. This phenomenon is known as population [26]_____, and it makes it difficult to put the brakes on population growth. In the developing countries, about [27]_____ % of the population is in this prereproductive group (under the age of 15). In the developed countries, this figure is only [28]_____%.

The age structure of a [29]_____ nation such as the United States is shaped more like a barrel than a pyramid. Couples average about two children, so the population tends to stabilize. In the U.S., an anomaly is the post-World War II [30]_____, a population "bulge" that echoes through the years. The boomers swelled school enrollments, competed for jobs, and will soon reach [31]_____, placing pressure on younger workers and taxpayers as they apply for Medicare and [32]_____.

The global population is expected to grow to about [33]_____ billion by the middle of the 21st century. Just to accommodate all the people expected to live on the Earth by 2025, we will have to [34]_____food production. But we have already strained the Earth's resources and severely impacted our environment.

Technology allows us to grow more food, but many people are starving and undernourished. Agricultural lands are depleted. Fresh ³⁵_____ for drinking and irrigation is becoming scarce. Overgrazing turns grassland into ³⁶_____. As we turn more and more land to our own use, many species face ³⁷_____.

Although populations in the developing world are growing ³⁸_____ than those of the developed countries, the per capita environmental impact of developed countries is vastly greater. The ³⁹_____—the amount of land needed to support an individual's demands on the environment—is one way to measure and compare our ecological impact. An American has an ecological footprint of about 8.4 hectares, while a resident of India gets by on about 0.8 hectares. If the area of productive land on the Earth is divided by population, we each have a share of about ⁴⁰_____ hectares. In 2003, the average ecological footprint of the world's population was 2.23 hectares. This means that our ecological footprint already ⁴¹_____ available resources. The problem is not just overpopulation, but overconsumption. The world's richest countries, which account for only ⁴²_____ % of global population, consume 86% of the world's resources. A child born in the United States has an impact on the environment greater than 40 African children.

Will the Earth's human population eventually reach 9.5 billion, or even more? Whatever our numbers, there is no question that the overall human birth rate will eventually come into line with the death rate, and globally, r will equal 0. But the questions that remain are as follows: Will the balance come about by informed choice, or will it be harshly imposed upon us? Will it occur through a decrease in the ⁴³_____ rate or an increase in the ⁴⁴_____ rate? Finally, what kind of life will there be in the future for all the people, and other species, who share an increasingly crowded planet?

Test Your Knowledge

Multiple Choice

1. In wild populations, individuals most often show a _____ pattern of dispersion.
   a. random
   b. density-dependent
   c. type II
   d. uniform
   e. clumped

2. A population would grow exponentially
   a. if it were limited only by density-dependent factors.
   b. until it reaches carrying capacity.
   c. if there were no limiting factors.
   d. if it were a *K*-selected population.
   e. if it showed logistic growth.

3. Which of the following would be most likely to have density-dependent effects on population growth?
   a. fire
   b. storms
   c. drought
   d. food supply
   e. cold

4. The effects of which of the following environmental factors would probably *not* change as a population grows?
   a. disease
   b. weather
   c. competition for nesting sites
   d. limited food supply
   e. predation

5. Human population growth was slow and gradual for a long period, but it turned sharply upward
   a. when death rates increased.
   b. when birth rates increased.
   c. when birth rates decreased.
   d. when death rates decreased.
   e. when the population reached carrying capacity.

6. A broad-based, pyramid-shaped age structure is characteristic of a population that is
   a. growing rapidly.
   b. at carrying capacity.
   c. stable.
   d. limited by biotic factors.
   e. shrinking.

7. A population grows rapidly at first and then levels off at carrying capacity if it is
   a. limited by density-dependent factors.
   b. limited by abiotic factors.
   c. an *r*-selected species.
   d. growing exponentially.
   e. characterized by uniform dispersion.

8. Which of the following is the most accurate comment on the Earth's carrying capacity (*K*) for people?
   a. *K* is smaller now than it was a thousand years ago.
   b. The human population is still a long way from *K*.
   c. Our technology allows us to increase *K*, but not indefinitely.
   d. When it comes to humans, the concept of *K* is irrelevant.
   e. The human population has already vastly exceeded *K*.

9. Which of the following would be true of a species whose life history is shaped by *r*-selection?
   a. Members of the species take a relatively long time to reach reproductive age.
   b. They are regulated mostly by density-dependent factors.
   c. They produce large numbers of offspring.
   d. The population usually stabilizes near carrying capacity.
   e. They give their young lots of care.

10. In the models that describe population growth, *r* stands for
    a. population density.
    b. a time interval.
    c. total number of individuals in the population.
    d. growth rate.
    e. carrying capacity.

11. According to Figure 36.9B, when did the population of Mexico show the highest rate of growth?
    a. 1895
    b. 1912
    c. 1950
    d. 1970
    e. 2000

12. When needed resources are unevenly distributed, organisms often show a(n)_____ dispersion pattern.
    a. density-dependent
    b. clumped
    c. exponential
    d. random
    e. uniform

Essay

1. Describe the exponential and logistic population growth curves, explain their shapes, and explain why they are different.

2. Describe some factors that might affect carrying capacity, *K*.

3. Describe three density-dependent factors and three abiotic factors that might restrict population growth. Explain how the effects of these two kinds of factors on populations differ.

4. What has enabled Earth's human population to continue to grow for a long period at near-exponential rates? How does this differ from other species?

5. What are population cycles? What kinds of hypotheses have been suggested to explain them?

6. Compare life histories shaped by *r*-selection and *K*-selection, and give examples of each.

Apply the Concepts

Multiple Choice

1. A particular species of tropical fish has only a few offspring and takes care of them for an extended period. We might also expect the fish population
   a. to be controlled mostly by abiotic factors.
   b. to show exponential growth.
   c. to live in a harsh environment.
   d. to start reproducing very young.
   e. to be relatively stable, near carrying capacity.

2. Gorillas have a relatively low birth rate. They take good care of their young, and most gorillas live a long life (if unmolested by humans!). The gorilla survivorship curve would look like
   a. a line that slopes gradually upward.
   b. a relatively flat line that drops steeply at the end.
   c. a line that drops steeply at first, then flattens out.
   d. a line that slopes gradually downward.
   e. a horizontal line.

3. Locust populations go through periods of sudden explosive growth, followed by a sudden decline in numbers. Their numbers are probably regulated by
   a. predation.
   b. biotic factors.
   c. logistic growth.
   d. random dispersion.
   e. abiotic factors.

4. Seagulls fiercely defend the areas around their nests in their cliff-top breeding colonies. Within the colony they would show a _____ dispersion pattern.
   a. uniform
   b. random
   c. dense
   d. density-independent
   e. clumped

5. A wildlife biologist is trying to predict what will happen to a bear population if bear hunting is banned. He had the equations all worked out but then realized that he had grossly underestimated the amount of food available to the bears. To make his predictions more accurate, he will have to go back to his equations and
   a. decrease N.
   b. increase N.
   c. decrease K.
   d. increase K.
   e. increase r.

6. An ecologist would suspect a population is growing rapidly if it
   a. has a broad-based age structure.
   b. is near its carrying capacity.
   c. is limited only by density-dependent factors.
   d. shows a clumped pattern of dispersion.
   e. is far below its carrying capacity.

7. An oak tree produces thousands of acorns, but very few grow into mature oak trees. The oak tree exhibits a _____ survivorship curve.
   a. Type I
   b. Type II
   c. Type III
   d. Type I or II
   e. Type I or III

8. When birth rate equals death rate
   a. a population grows rapidly.
   b. population density levels off.
   c. density-dependent limiting factors do not affect the population.
   d. a population is in danger of extinction.
   e. a population goes through up and down cycles.

9. If you wanted to see what percentage of the population of Thailand is under 10 years old, you could look at
   a. a logistic curve for the population.
   b. a plot of population density
   c. a life table for the population.
   d. the population's age structure.
   e. the population's survivorship curve.

10. To determine the density of a rabbit population, you would need to know the number of rabbits and
    a. the factors that limit population growth.
    b. their birth rate.
    c. the area in which they live.
    d. their population growth rate.
    e. their pattern of dispersion.

11. Sometimes, applying a pesticide can actually make a pest problem worse, because
    a. pests are usually clumped, while their natural predators are uniformly distributed.
    b. pests often have a greater rate of increase, *r*, than their predators.
    c. the carrying capacity, *K*, of pests is generally greater than that of their predators.
    d. both predators and pests are usually limited by abiotic factors.
    e. the population density of predators is always greater than that of pests.

Essay

1. A fisheries biologist is interested in determining the population density of smallmouth bass in a lake. Using a net, she captures and tags 100 bass. A week later, she again catches 100 bass, and out of these fish, 5 are tagged. How many smallmouth bass are there in the lake?

2. Tim said, "It says here that Japan has a much greater population density than the United States." Carl replied, "That can't be right. There are a lot of people in Japan, but surely the population of the United States is much larger." What would you say to clear this up for them?

3. Biologists figure that the carrying capacity of a particular river system is 100,000 salmon. They want to manage the catch of salmon so that the salmon population is at a size where replacement of the fish that are caught will happen at the fastest rate. Applying the concept of maximum sustained yield, would you recommend that they catch fish until the salmon population stands at 80,000, or 50,000, or 10,000? Explain why. (Hint: They are assuming logistic growth of the salmon population.)

4. The temperature seldom dips below freezing along the Pacific coast, so robins can stay there all winter long. But during rare spells of sub-freezing weather, the birds have a hard time finding food and shelter. Birds that find a warm place sheltered by evergreen branches have the best chance of making it through a cold snap, but there are only a limited number of such places. After a week of cold, only a fraction of the robin population may remain. Explain how both density-dependent and abiotic factors affect the robin population in this case.

5. Population explosions are often seen when animals are introduced into a new area. Examples include the spread of starlings across North America, a plague of rabbits in Australia, and the spread of cane toads in Florida. Why are these kinds of population explosions more likely in areas where the animals are not native?

6. The most intensive and rigorous studies of populations have been undertaken on islands and in lakes. Why are populations easier to study on islands or in lakes?

7. Figure 36.11B shows that China has an ecological footprint comparable in size to that of the United States. The population of China is about 1.3 billion, and the population of the U.S. about 300 million. How can their footprints be similar if China has more than four times as many people as the U.S.?

Put Words to Work

Correctly use as many of the following words as possible when reading, talking, and writing about biology:

age structure, carrying capacity, clumped dispersion, demographic transition, density-dependent, dispersion pattern, ecological footprint, exponential growth model, K-selection, life history, life table, limiting factors, logistic growth model, maximum sustained yield, per capita rate of increase, population, population density, population ecology, population momentum, random dispersion, r-selection, survivorship curve, sustainable resource management, uniform dispersion, G, K, N, r

Use the Web

For further review, see the activities and questions on Chapter 36 on the Web at www.mybiology.com. Some of the simulations of population growth and sampling are particularly informative.

Communities and Ecosystems 37

Focus on the Concepts

The chapter explores the structure and function of communities and ecosystems. While studying the chapter, focus on these concepts:

- A biological community is made up of all the interacting populations living in an area. A community might be small or global, defined by ecologists according to their research interests. Understanding community ecology is important in conservation, disease control, and agriculture.

- Several kinds of interspecific interactions are fundamental to community structure: interspecific competition (negative for both populations, − −), mutualism (+ +), predation (+ −), herbivory (+ −), and parasitism and disease (+ −). Reciprocal evolutionary adaptation, called coevolution, shapes these interactions and tunes each species' ecological niche.

- Every community has a trophic structure, a pattern of feeding relationships consisting of different levels. The sequence of food transfer is called a food chain. Food chains interconnect to form food webs. Producers—usually photosynthetic organisms—form the base of the food web. Primary consumers eat producers, and they in turn feed secondary and tertiary consumers. Wastes—called detritus—are consumed/recycled by detritivores (scavengers) and decomposers (mainly bacteria and fungi).

- The species diversity of a community is defined in terms of species richness, the number of species, and relative abundance, the proportional representation of species. Keystone species, even though not abundant themselves, may impact community diversity. Diversity is an important consideration in conservation and pest management.

- Disturbances, such as fires, floods, and human activity, may alter biological communities. A disturbed area is colonized by a mix of species, which are gradually replaced by other species in a process called ecological succession. Primary succession occurs in a barren area with no soil, secondary succession where soil is intact.

- An ecosystem consists of all the organisms in a community plus the abiotic environment. Energy flow and chemical cycling are important features of ecosystems. Energy flows through an ecosystem, entering as sunlight and exiting as heat. Chemical nutrients are reused, cycling within the ecosystem or biosphere.

- The amount of solar energy converted to chemical energy is called primary production and depends on the ecosystem. Forests and wetlands have greater primary production than deserts or open ocean. Energy supply limits length of food chains. At each level, much energy is lost to detritus and much is used in cellular respiration. Only 5–20% is passed on to the next level, supporting few tertiary or quaternary consumers.

- Chemical nutrients, such as carbon, phosphorus, and nitrogen, are recycled. Details vary but each biogeochemical cycle has (an) abiotic reservoir(s)—phosphate in rocks and soil, CO_2 and N_2 in the air. Producers assimilate nutrients and pass them on to consumers. Organisms excrete wastes and die, and detritivores and decomposers return nutrients to soil, water, or air. Deforestation, fossil fuel use, and sewage runoff can disrupt these cycles.

Review the Concepts

Work through the following exercises to review the concepts in this chapter. For additional review, refer to the activities on the Web at www.mybiology.com. The website offers a pretest that will help you plan your studies.

Exercise 1   (Modules 37.1–37.7)

The structure of a community is shaped by interactions among the populations making up the community. The most important kinds of interactions are **competition**, **predation**, **herbivory**, **parasitism**, and **mutualism**. State which of these five interactions is described in each of the examples below. Then put a + or – in each of the parentheses ( ) to indicate whether the relationship is positive or negative for each organism.

_____ 1. Sheep liver flukes ( ) feed on bile and can weaken or kill their hosts ( ). They are passed on to other sheep in the animals' droppings.

_____ 2. Grazing by introduced mountain goats ( ) has reduced the numbers of alpine wildflowers ( ) in Olympic National Park.

_____ 3. Pest control specialists have brought in a moth ( ) to eat tansy ragwort, ( ) a poisonous weed.

_____ 4. Mistletoe ( ) obtains nutrients from an oak tree ( ).

_____ 5. Mycorrhizal fungi ( ) associated with roots obtain carbohydrates from a tree, while enabling the tree ( ) to absorb water and minerals more efficiently.

_____ 6. In many parts of North America, the starling ( ) has displaced the bluebird ( ) from its nest sites.

_____ 7. A bee ( ) pollinates a tropical orchid ( ) by being tricked into "mating" with the flower; the bee uses a perfume from the flower to attract a mate.

_____ 8. The influenza virus ( ) attacks the lining of the respiratory tract and is passed from person ( ) to person by contact or airborne droplets.

_____ 9. Red-winged blackbirds ( ) arrive earlier on the breeding grounds but are forced to the edges of a marsh by larger, later-arriving yellow-headed blackbirds ( ).

_____ 10. Lions ( ) hunt large herbivorous mammals such as zebras and wildebeest ( ).

_____ 11. A tropical acacia tree ( ) spouts under the taller canopy trees ( ) and grows up toward the light.

_____ 12. The acacia ( ) is infested by ants ( ) that live in enlarged tree thorns and feed on special leaves. As the tree grows, the ants prune away leaves and branches of surrounding trees.

Exercise 2   (Modules 37.2–37.7)

Each species in a community has an ecological niche, its "role" or "job" in the community. The niche includes the sum of the organism's needs, functions, abilities, and tolerances. It is possible to describe the niche as a sort of "job description" for a species, as you might see in a classified ad. An ad that reads "Applicant will be required to travel in herd, drink through nose, and knock down trees for food" could only describe the job of an elephant!

Identify the organism whose niche is outlined in each of the following job descriptions. Some are from the text; others are not but will probably (like the elephant) be familiar to you.

_____ 1. "Must be skilled at building traps to catch flying insects. May or may not need to devour mate."

_____ 2. "Ability to build and maintain reef. Must be able to work closely with dinoflagellates."

_____ 3. "Will be traveling and working outdoors in cold weather. Must have the patience to wait long periods to catch and eat seals through hole in ice. Some swimming ability and camouflage helpful."

_____ 4. "Passionflower specialist wanted. Must be able to break down toxins."

_____ 5. "Must be able to withstand coastal storms and forest fires while maintaining species' reputation as world's tallest tree."

_____ 6. "Must live in South American rain forest, eat insects, and have poison glands and bright coloration."

_____ 7. "Will work closely with legume. Will be required to fix nitrogen in exchange for daily carbohydrate allowance."

_____ 8. "Ability to eat insects important. May be called upon to compete with orange-crowned warblers."

_____ 9. "Must be able to climb trees and eat acorns."

Exercise 3   (Modules 37.8–37.9)

The trophic structure of an ecosystem is the pattern of feeding relationships by which energy and chemicals flow through the system from trophic level to trophic level. Name the trophic level of each of the organisms in the following description of a freshwater marsh food web: producer (P), primary consumer (1C), secondary consumer (2C), tertiary consumer (3C), quaternary consumer (4C), detritivore or scavenger (S), or decomposer (D). (Note that a consumer can function on more than one level, depending on what it eats.)

Marshes and other wetlands are among the most endangered of habitats. They are productive "nurseries" for many wildlife species, but many of our wetlands have been drained for agriculture or filled for development.

The freshwater marsh food web starts with plants like cattails, arrowleaf, and various floating or submerged "water weeds" (^{1}_____). They provide food for muskrats (^{2}_____) and mallard ducks (^{3}_____), both of which may in turn be eaten by hawks (^{4}_____) or mink (^{5}_____). Microscopic algae (^{6}_____) make much of the food in the marsh. Small shrimplike crustaceans (^{7}_____) and insect larvae (^{8}_____) graze on the algae. The crustaceans and insects are eaten by ducks (^{9}_____), frogs (^{10}_____), and sunfish (^{11}_____). A frog or sunfish might be eaten by a larger yellow perch (^{12}_____), a great blue heron (^{13}_____), a water snake (^{14}_____), or a mink (^{15}_____). The heron (^{16}_____) also eats perch and snakes, and the hawk (^{17}_____) will also occasionally devour a snake.

This is a highly simplified description of a marsh food web. There might be hundreds of species of large plants and animals making up the community of a marshy roadside pond, not to mention a swarm of microscopic creatures. In addition to this grazing food web, many inconspicuous worms, insect larvae, and snails ([18]_____) get their food from the dead material produced by the other plants and animals of the marsh, and in the mud a myriad of bacteria ([19]_____) recycle nutrients.

On a separate sheet of paper, sketch the marsh food web. Write the names of the organisms at their appropriate trophic levels, and connect the names with arrows. It is not necessary to include detritivores or decomposers. (Where in your diagram should the producers be? The top consumers? Which way do the arrows point? Why?)

Exercise 4   (Modules 37.10–37.11)

What is species diversity, and what effect can a single species have on the diversity of a community? A desert ecologist sets up the following experiment to find out.

The ecologist wanted to understand the effects of grazing on desert wildflowers by small rodents called pocket mice and kangaroo rats. She fenced off three small plots of desert. All the plots had essentially the same mix of species at the beginning of the experiment. Plot A had gaps in the fence so all species could come and go. Plots B and C were securely fenced. She relocated all the kangaroo rats out of plot B and all the pocket mice out of plot C. At the end, the mix of species in plot A was virtually unchanged, but there were changes in plots B and C. Study the data below and answer the following questions.

Flower Species	Plot A Control	Plot B w/o Kangaroo rats	Plot C w/o Pocket mice
P	23%	8%	16%
Q	16%	0	9%
R	20%	0	13%
S	19%	18%	47%
T	22%	74%	15%

1. At the end of the experiment, which of the plots had the greatest species richness? _____
2. At the end of the experiment, which of the plots had the lowest species richness? _____
3. At the end of the experiment, which of the plots had the greatest relative abundance? _____
4. At the end of the experiment, which of the plots had the lowest relative abundance? _____
5. If you took a look around, which of the plots would appear to be most diverse? _____
6. Which of the plots comes closest to being a monoculture? _____
7. Which of the plots might be most vulnerable to a disease that attacks one species of plant? _____
8. Which of the plots would offer the most food and habitats for species *other than* kangaroo rats and pocket mice? _____
9. Which of the plots would offer the least food and habitats for species *other than* kangaroo rats and pocket mice? _____
10. How would you describe the role of kangaroo rats in this community?
11. How would you describe the role of pocket mice in this community?
12. How would you describe the role of flower species T in this community?

Exercise 5   (Module 37.12)

Gradual transition in the species composition of a community that occurs after a disturbance is called ecological succession, described in Module 37.12. In this exercise, first state whether each of the following represents a relatively early (E) stage in succession or a relatively late (L) stage. (Hint: Ask yourself if the community were left untouched, whether it would look the same or different in a hundred years. If it would look different, it is in an early stage—there are later stages to come.) Next, for those communities in an early stage, state whether those examples represent primary (P) or secondary (S) succession.

Early or Late?	*Primary or Secondary?*	
_____	_____	1. Lichen-covered rocks near a melting glacier in Alaska
_____	_____	2. "Old growth" conifer forest in the Pacific Northwest
_____	_____	3. A weedy vacant lot near your college
_____	_____	4. An oak-maple-beech forest in Ohio
_____	_____	5. A lava flow on the island of Hawaii
_____	_____	6. A lawn in a suburb in New Jersey
_____	_____	7. A cornfield in Virginia

Exercise 6   (Module 37.13)

Invasive species can drastically alter communities. The introduction of rabbits to Australia is a dramatic example. Dutch elm disease (Module 37.7) is another. Several other examples are described in Module 19.18.

1. What are three kinds of circumstances that allow an introduced species to spread unchecked? In other words, what kinds of controls might be relaxed in a new environment?

2. What is biological control? Why is biological control a good way to deal with introduced species?

3. Why does biological control not always work? Give three ways it can go wrong.

Sunlight, moisture, and nutrient availability determine how much food the producers of an ecosystem can make—primary production. After studying the bar graph, rank the following ecosystems in terms of primary production per square meter per year (highest pp = #1, lowest pp = #8).

_____ A. Desert
_____ B. Tropical rain forest
_____ C. Cornfield
_____ D. Beech/maple forest
_____ E. Open ocean
_____ F. Estuary
_____ G. Tundra
_____ H. Kelp bed

The flowchart on the following page illustrates the movement of energy through an ecosystem. The boxes represent the biomass of organisms at each trophic level. The arrows show the amount of energy passing through each trophic level. Energy enters the producer level as sunlight. Some of this energy is stored in molecules produced in photosynthesis. Energy enters each of the consumer trophic levels when the consumers feed on the level below. Much of the energy in the food entering any level is used to power life processes; the food is used as fuel in cellular respiration, and its energy ends up as heat. Some energy is wasted; it is lost to the detritus food web in the form of dead leaves or droppings. A small portion of the energy is stored in tissue when organisms grow or reproduce; this production—roughly 5–20% of energy intake at any trophic level—is the only energy available to the next level. Label and color the trophic levels on the diagram: **producers** (green), **primary consumers** (orange), **secondary consumers** (blue), and **tertiary consumers** (purple). Label and color the pattern of energy flow: **sunlight** (yellow), **production energy** (red), **energy used in cellular respiration** (pink), and **energy in wastes** (brown). (Note: Can you see from this diagram why quaternary consumers are quite rare?)

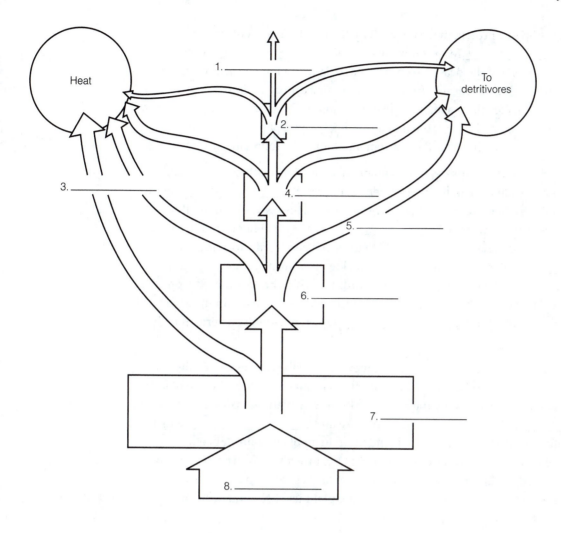

Heat

To detritivores

1. _____

2. _____

3. _____

4. _____

5. _____

6. _____

7. _____

8. _____

The biosphere receives a constant supply of energy from the sun, uses this energy for a while, then loses it to space as heat. Unlike energy, the chemicals necessary for life are present on Earth in fixed amounts, and these chemicals are used over and over. These chemicals, such as carbon, nitrogen, and phosphorus, occur in various forms and are changed from one form to another by various physical and chemical processes. The story below traces a nitrogen atom as it moves through the various reservoirs and processes of the nitrogen cycle. Fill in the blanks as you follow its journey.

The nitrogen atom, N, had been in the atmosphere for more than two years. It was paired with another identical nitrogen atom, forming a molecule of
[1]_____ gas, which makes up about 80% of the air. This is by far the largest
[2]_____ reservoir of nitrogen. During its time in the atmosphere, N had circled Earth several times, from the skies over the Philippines, to Africa, to the Antarctic, to South America, and now over a sand dune in North Carolina. There it was captured by [3]_____ in a nodule on the root of a legume called a beach pea. There N was split away from its partner and combined with hydrogen atoms, eventually ending

up in an amino group in a protein molecule built by the plant. This N entered (or reentered!) the [4]_____ reservoir of this global [5]_____ cycle. The protein was stored in one of the peas, which ripened, dried, and fell on the ground, and it was eaten by a mouse. The amino acid from the pea was incorporated into a [6]_____ molecule in a leg muscle of the mouse. On a moonlit night a month or so later, a great horned owl caught the mouse, and N became part of a protein molecule in one of the owl's feathers. And there it remained until the following spring.

In the spring, as every spring, the owl molted some of its feathers. The small feather containing N ended up in the litter under a pine tree. [7]_____ in the soil broke down the feather over a period of several months, and N was eventually released into the soil, in the form of [8]_____, NH_4^+. Other bacteria, called [9]_____ bacteria, then attached N to three oxygen atoms, forming [10]_____, which was taken up by a huckleberry bush and used to build another protein. The huckleberry was eaten by a cardinal, which broke down the protein and eventually excreted N in the uric acid in its droppings. Decomposers in the forest soil changed this waste product to [11]_____, which nitrifying bacteria quickly converted into [12]_____.

When the patch of forest was cut for lumber, most of the nitrate ions in the soil were washed downhill. A flood deposited the nitrate ion containing N in the soil of a flat river valley. N was soon absorbed by the roots of a buttercup plant—a process called [13]_____. The buttercup used N to make an [14]_____, which was used to build a protein molecule, which ended up in a pollen grain. A beetle collected some of this pollen, and soon N was part of the insect's body. The beetle flew into a bog and landed on the leaf of a Venus flytrap, an insect-eating plant. The leaf snapped shut, and the beetle was slowly digested; the plant used N to make its own proteins.

About a year later, the rotting trunk of a dead tree toppled into the swamp and buried the Venus flytrap. As the plant decomposed, bacteria again incorporated N into ammonium and then nitrate. But this time, in the low-oxygen conditions of the mud, [15]_____ bacteria were able to break down the nitrate ions, releasing N to the air as part of an N_2 molecule. The wind swept the molecule out over the ocean, where . . .

You might find it informative and fun to continue the story from here. You might also find it helpful to make up similar stories for a carbon atom or a phosphorus atom. This is a good review to do in a study group.

Exercise 10   (Modules 37.19–37.23)

These modules contrast some processes that occur in natural, "healthy" ecosystems with processes that take place in "unhealthy" systems that have been damaged by human activities. Each of the statements below describes the situation in a healthy ecosystem. Briefly describe the corresponding situation in an unhealthy system affected by human disturbance.

1. The amount of CO_2 released into the atmosphere by respiration equals the CO_2 used in photosynthesis.

2. Forest organisms and soil contain a large store of mineral nutrients such as calcium and phosphorus, which are continuously recycled. Small amounts of nutrients are lost each year, but this is balanced by the flow of nutrients into the ecosystem.

3. Over time, nutrients from the surrounding land gradually accumulate in a lake, and the lake becomes more productive, a process called eutrophication.

4. Organic fertilizers release nitrogen and phosphorus gradually, so they are absorbed by crop plants and do not run off to pollute rivers and lakes.

5. Runoff from a cattle feed lot passes through treatment lagoons where microorganisms and plants recycle nutrients.

Test Your Knowledge

Multiple Choice

1. All of the populations of organisms in a particular area make up
   a. a food chain.
   b. a biogeochemical cycle.
   c. an ecosystem.
   d. a niche.
   e. a community.

2. When you eat an apple, you are a
   a. primary consumer.
   b. secondary producer.
   c. producer.
   d. secondary consumer.
   e. tertiary consumer.

3. An organism's "trophic level" refers to
   a. the rate at which it uses energy.
   b. where it lives.
   c. what it eats.
   d. whether it is early or late in ecological succession.
   e. the intensity of its competition with other species.

4. The relationship between species A and species B is described as mutualism. This means that
   a. both species suffer.
   b. one species benefits and the other species suffers.
   c. both species benefit.
   d. one species benefits and the other species is unaffected.
   e. any of the above is possible in mutualism.

5. Most plants get nitrogen from
   a. nitrates in the soil.
   b. N_2 gas in the air.
   c. proteins.
   d. ammonium in the soil.
   e. rainfall.

6. The energy for nearly every organism in nearly every ecosystem ultimately comes from
   a. minerals in the soil.
   b. the sun.
   c. food.
   d. respiration.
   e. decomposition.

7. Why is a diagram of energy flow from trophic level to trophic level shaped like a pyramid?
   a. Organisms at each level store most of the energy and pass little on.
   b. There are more producers than primary consumers, and so on.
   c. Organisms eventually die as they get older.
   d. Most energy at each level is lost, leaving little for the next.
   e. There are always fewer secondary consumers than primary consumers, and so on.

8. The main decomposers in an ecosystem are
   a. plants and animals.
   b. bacteria and viruses.
   c. fungi and bacteria.
   d. bacteria and plants.
   e. plants and fungi.

9. In a forest, bacteria are especially important in
   a. photosynthesis.
   b. the nitrogen cycle.
   c. ecological succession.
   d. the phosphorus cycle.
   e. recycling of energy.

10. The biggest difference between the flow of energy and the flow of chemical nutrients in an ecosystem is that
    a. the amount of energy is much greater than the amount of nutrients.
    b. energy is recycled, but nutrients are not.
    c. organisms always need nutrients, but they don't always need energy.
    d. nutrients are recycled, but energy is not.
    e. organisms always need energy, but they don't always need nutrients.

11. An animal's niche is
    a. the number of individuals of the species the environment will support.
    b. the same as its habitat.
    c. the way the animal fits into its community and environment.
    d. the specific place in the habitat where the animal lives.
    e. its position in the food chain.

12. Which of the following ranks three ecosystems, from highest to lowest, in terms of primary production per square meter?
    a. open ocean—temperate grassland—desert
    b. cultivated land—estuary—deciduous forest
    c. tropical rain forest—cultivated land—coral reef
    d. cultivated land—tropical rain forest—temperate grassland
    e. coral reef—deciduous forest—open ocean

13. Gobis (fish) that eat hippo dung in Mzima Springs are best described as
a. primary consumers.
b. decomposers.
c. secondary consumers.
d. herbivores.
e. detritivores.

14. Over time, a barren sand dune is covered by shrubs and finally a forest. This process is best described as
a. primary succession.
b. coevolution.
c. secondary succession.
d. herbivory.
e. a biogeochemical cycle.

Essay

1. Explain in ecological terms why a given area of farmland can support more people if they eat plants rather than meat.

2. What is decomposition? What are the two major kinds of decomposers in most ecosystems, and how is their role important to the ecosystem?

3. Describe an example of coevolution between predator and prey. Are there ecological relationships other than predator/prey interactions that are shaped by coevolution between two species? Give an example.

4. State whether each of the following is a producer, primary consumer, secondary (or higher) consumer, or detritivore: squirrel, oak tree, mosquito, great white shark, moose, cheetah, mushroom, spider, phytoplankton, grass, and vulture.

5. Trace a carbon atom through the carbon cycle. In what chemical form is carbon in the air? How does a carbon atom enter the food chain? In what chemical form might the carbon atom be obtained by a consumer? What chemical process would put the carbon atom back into the atmosphere?

6. Agricultural ecosystems—cornfields or pear orchards, for example—have low species diversity. What are some potential negative consequences? How might these problems be reduced without harm to the environment?

7. Why can the productivity of an ecosystem be limited by a shortage of phosphorus or nitrogen, but seldom by a shortage of carbon? Briefly explain in terms of cycles.

Apply the Concepts

Multiple Choice

1. A lichen is actually composed of two organisms—a fungus and an alga. They depend on each other for survival. The most specific term that describes their relationship is
a. parasitism.
b. competition.
c. herbivory.
d. symbiosis.
e. mutualism.

2. Which of the following describes mimicry?
a. An insect's bright colors warn a predator that it tastes bad.
b. The mottled pattern on a fish looks like dead leaves on the bottom of a pond.
c. Two species of mice live in the same area and eat the same kinds of seeds.
d. A harmless frog resembles a poisonous frog.
e. Both kangaroo rats and jackrabbits hop erratically when escaping from predators.

3. When goats were introduced to an island off the California coast, the goats lived in the same areas and ate the same plants as the native deer. The deer population dwindled, and the deer finally disappeared. This is an example of
a. herbivory.
b. succession.
c. a food chain.
d. coevolution.
e. competition.

4. Suppose you wanted to establish a self-sustaining ecosystem by sealing some sterilized soil, water, and air in a glass container with a few organisms. You would be most likely to succeed with which of the following?
a. aphids, bacteria, and spiders
b. bacteria and ants
c. clover, bacteria, and grasshoppers
d. beetles, fungi, and bacteria
e. spiders, grasshoppers, and grass

5. After clear-cutting, timber companies cannot afford to wait for the long process of _____ to occur naturally; they usually replant trees right away.
a. mutualism
b. succession
c. coevolution
d. decomposition
e. herbivory

6. In an ecosystem the _____ is always greater than the _____ .
   a. number of producers . . . number of primary consumers
   b. biomass of secondary consumers . . . biomass of producers
   c. energy used by primary consumers . . . energy used by secondary consumers
   d. biomass of producers . . . biomass of primary consumers
   e. energy used by primary consumers . . . energy used by producers

7. Under which of the following circumstances would interspecific competition be most obvious?
   a. when resources are most abundant
   b. in a stable, long-established ecosystem
   c. when organisms have quite different ecological niches
   d. among species whose trophic levels are different
   e. when a foreign organism is first introduced into a community

8. Milkweed plants produce bad-tasting and poisonous compounds that deter most plant-eaters. But the caterpillars of monarch butterflies are able to eat milkweed leaves without being harmed. In fact, the chemicals obtained from milkweed actually protect the monarch from insect-eating birds. This example illustrates
   a. coevolution.
   b. competitive exclusion.
   c. mutualism.
   d. the effect of a keystone predator.
   e. succession.

9. Two species of birds called cuckoo doves live in a group of islands off the coast of New Guinea. Out of 33 islands, 14 have one species, 6 have the other, 13 have neither, and none has both. What might best explain this? The two species of birds could
   a. be on different trophic levels.
   b. have similar niches.
   c. be detritivores.
   d. have different niches.
   e. be keystone predators.

10. If you wanted to determine whether succession is occurring in a certain area, it might be useful to
    a. determine how many trophic levels are represented.
    b. diagram its food web.
    c. look at its species diversity.
    d. look at old pictures of the area.
    e. measure its productivity.

11. Which of the following are also called detritivores?
    a. mushrooms
    b. scavengers
    c. decomposers
    d. bacteria
    e. tertiary consumers

12. An ecosystem is unlikely to be limited by the supply of ____ because it is obtained from the air.
    a. water
    b. carbon
    c. phosphorus
    d. calcium
    e. nitrogen

13. A tropical butterfly is bright red and black. It usually sits out in the open, unmolested by predators. Based on your reading, what would seem like the best hypothesis to explain this?
    a. The butterfly is an introduced species, and predators are not used to it.
    b. Its color pattern allows it to hide in brightly colored flowers.
    c. Predators do not see the butterfly in the same way we do.
    d. It is the top consumer in its community.
    e. Its color pattern is probably a warning that it tastes bad.

Essay

1. A researcher noted that in many small ponds, fish species A preyed on several smaller species, B, C, D, E, and F. He suspects that A may be a keystone predator in the pond communities. What kind of experiment could you suggest to test whether this hypothesis is valid? If A is a keystone predator, how would you expect the experiment to turn out?

2. When Gause cultured *Paramecium aurelia* and *Paramecium caudatum* together, *P. caudatum* was invariably driven to extinction (Module 37.3). When *P. aurelia* and *P. bursaria* were cultured together in another experiment, however, both were able to coexist for months. What kinds of similarities and differences must there be among the three species? What do you think would happen if *P. caudatum* and *P. bursaria* were cultured together? Why?

3. Elmer Q. Retchfickle, a civil-defense expert, has developed a design for a fallout shelter to be used during a nuclear war. Elmer proposes to have the inhabitants of the shelter eat mushrooms, which would be grown in the dark on the body wastes of the residents of the shelter. He thinks that this system would allow shelter residents to remain underground indefinitely. Aside from the questionable nutritional balance of an all-mushroom diet, do you see any problems with this plan from an energy standpoint? Explain.

4. An experimental fish farm consists of algae, crustaceans that eat algae, and fish that eat crustaceans. If the algae grow at a rate of 20 kilograms of algae per square meter of pond per year, how big will the pond have to be to harvest 1,000 kilograms of fish per year? (Assume that energy content is proportional to weight.)

5. In the mountains of California, Stanford ecologist Craig Heller found that in most areas the least chipmunk lived in sagebrush areas and the yellow pine chipmunk lived in higher areas of mixed sagebrush and piñon pines. If the yellow pine chipmunk was absent from an area, the least chipmunk lived in both the sagebrush and sagebrush-piñon pine zones. If the least chipmunk was absent, the yellow pine chipmunk distribution was unchanged. Explain this difference in terms of the concepts discussed in this chapter.

6. The push to develop and settle the Amazon Basin has resulted in the first extensive contact between many groups of indigenous peoples and the outside world. Many of these native people have died from diseases that are seldom fatal in the developed countries, such as measles, chicken pox, and mumps. What kind of coevolution must have occurred between the viruses causing these diseases and their hosts in the developed countries? What will probably occur in the Amazon?

Put Words to Work

Correctly use as many of the following words as possible when reading, talking, and writing about biology:

abiotic reservoir, biogeochemical cycle, biological control, biomass, chemical cycling, coevolution, community, decomposer, decomposition, detritivore, detritus, disturbance, ecological niche, ecological succession, ecosystem, energy flow, food chain, food web, herbivory, interspecific interactions, interspecific competition, invasive species, keystone species, mutualism, nitrogen fixation, predation, primary consumer, primary production, primary succession, producer, quaternary consumer, secondary consumer, secondary succession, species diversity, tertiary consumer

Use the Web

For further review on the topics in this chapter, see the activities on the Web at www.mybiology.com.

Conservation and Restoration Biology 38

Focus on the Concepts

We are in the midst of a biodiversity crisis, a rapid decrease in the diversity of living things. This chapter outlines the dimensions of the crisis and what we can do to conserve and restore species and ecosystems. Focus on these concepts:

- The current rate of species loss is much greater than in the past; we may be pushing more species toward extinction than the disaster that wiped out the dinosaurs. At the current rate of loss, more than half of all plant and animal species will be gone by the end of this century.

- There are many reasons to care about the loss of biodiversity: There are ethical and esthetic reasons to preserve life. We depend on other species for products such as food, clothing, and medicine. We depend on nature for oxygen, stable climate, soil fertility, flood control, purification of air and water, recycling of wastes, pollination of crops, pest control, and protection from UV rays, and other "services."

- Biodiversity has three levels: Genetic diversity among populations makes adaptation possible. Species diversity is the variety of species in an ecosystem or the biosphere. Endangered species are in danger of extinction. Threatened species are likely to become endangered. Ecosystem diversity arises from the pattern of interactions in a community, and it can be affected by habitat degradation and loss of keystone species.

- Habitat destruction is the biggest threat to biodiversity; humans have altered about half of the Earth's land surface. Invasive species disrupt communities by competing with, preying on, or parasitizing native species. Overexploitation by hunting or fishing often exceeds the ability of populations to rebound. Pollution and nutrient overfertilization can overwhelm natural systems. Global warming is changing the climate and challenges ecosystems' and species' abilities to adapt.

- Greenhouse gases—CO_2, CH_4, and N_2O—produced by human activities such as burning fossil fuels trap heat in the atmosphere and are warming global climate. Effects are greater on land and in northern regions. Glaciers and sea ice are melting and seas are rising. Warmer weather comes earlier each year, and heat waves are increasing. Precipitation patterns are changing and hurricanes are stronger.

- Climate change affects life on many levels: The tundra is thawing, forests are drying and more prone to fire, deserts are expanding, and warming of the oceans causes coral bleaching. Spring warming disrupts blooming and mating. Pests and invasive species spread. Phenotypic plasticity and natural selection allow some organisms to adapt. Others move north or to higher elevations, but for some there is no place to go.

- Conservation biologists focus on protecting populations of endangered and threatened species. Small populations can spiral to extinction due to loss of genetic variation. Conservation focuses on understanding species' needs and preserving habitat to support viable populations. Reserves and movement corridors reduce fragmentation.

- Protection often focuses on biodiversity hot spots, many in the tropics, with high species diversity and large numbers of endemic species that are vulnerable to extinction. Zoned reserves surround unaltered ecosystems with areas of regulated human use, contributing to sustainable development. The Yellowstone to Yukon Conservation Initiative aims to protect the ecosystems and organisms of the Rocky Mountains. The Kissimmee River project attempts to bring back a degraded wetland via restoration ecology.

- Humans are a successful species, but we find ourselves in a precarious position, facing ecosystem degradation, destruction of habitats, and loss of biodiversity. The goal of the Sustainable Biosphere Initiative is to acquire the knowledge needed to develop, manage, and conserve Earth's resources for the long-term benefit of humans and the ecosystems that support them. This may be the biggest challenge we have ever faced.

Review the Concepts

Work through the following exercises to review the concepts in this chapter. For additional review, see the activities on the Web at www.mybiology.com. The website offers a pre-test that will help you plan your studies.

Exercise 1   (Modules 38.1–38.8)

The first half of Chapter 38 outlines the extent and causes of the global biodiversity crisis. There are lots of facts and figures meant to illustrate what is happening. As you read the modules, summarize the major points below.

A. List nine general categories of benefits, products, or services that humans derive from the biosphere:

1.

2.

3.

4

5.

6.

7.

8.

9.

B. List five major ways in which human activities are responsible for the decrease in global biodiversity, and a species or habitat endangered by each.

1.

2.

3.

4.

5.

Exercise 2   (Introduction, Modules 38.1–38.4, 38.7–38.8)

Review some of the examples covered in this discussion of the biodiversity crisis by matching each of the species on the right with its part in the biodiversity story.

_____ 1. Introduced predator that endangers birds on Guam	A. Herring gull
_____ 2. Endangered by warming and drying of Costa Rican forests	B. Rhinoceros
_____ 3. Found to be breeding earlier due to warming in the Yukon	C. Fish
_____ 4. Invasive plant that may benefit from global warming	D. Rosy periwinkle
_____ 5. Eggs contaminated by biomagnification of PCBs in the Great Lakes	E. Hot springs prokaryotes
_____ 6. Illegally hunted for its gallbladder	F. Primates
_____ 7. Endangered by logging and mining in Myanmar, but now protected	G. Phytoplankton
_____ 8. Sought by "bioprospectors" in Yellowstone Park	H. Grizzly bear
_____ 9. Starving as global warming melts Arctic sea ice	I. Kudzu
_____ 10. Endangered by bleaching as water warms	J. Polar bear
_____ 11. Populations reduced by the bush meat trade	K. *Homo sapiens*
_____ 12. A tropical plant valuable in cancer treatment	L. Red squirrel
_____ 13. May be damaged by UV exposure due to ozone "hole"	M. Frogs
_____ 14. Large mammal poached for its horn	N. Coral
_____ 15. Food species may accumulate mercury	O. Tiger
_____ 16. Responsible for the current biodiversity crisis	P. Brown tree snake

Exercise 3   (Modules 38.5–38.8)

Global climate change is one of the most serious threats to biodiversity. Double-check your understanding of greenhouse gases, the greenhouse effect, global warming, and their impact on the biosphere by completing each of the following statements.

1. Scientific debate over global warming is _____.
2. Greenhouse gases such as _____, methane, and nitrous oxide are changing the climate.
3. Temperatures have risen _____ degree C in the last 100 years, most of that in the last three decades.
4. Further increases of _____ degrees C are likely by the end of the 21st century.
5. _____ temperatures are rising too.
6. Warming is greater over _____ than sea.
7. The largest increases are in the _____ portion of the Northern Hemisphere
8. Glaciers are melting and sea ice is _____.
9. Melting ice may cause catastrophic coastal _____.
10. Warm temperatures arrive _____ each year.
11. _____ patterns are changing too.
12. Intensity of _____ is increasing.
13. For millennia, the amount of CO_2 in the air did not exceed 300 ppm; now it is _____ ppm and rising.
14. CO_2 and N_2O are released when _____ are burned.
15. Much _____ comes from landfills and feedlots.
16. Scientists have concluded that the increase in greenhouse gases is due to _____ activities.
17. CO_2 in the _____ is a major reservoir for carbon.
18. Plants use CO_2 in the process of _____, storing carbon in organic molecules.
19. The biomass in living things constitutes a large biotic _____ of carbon.
20. Cellular respiration and _____ release carbon back into the air.
21. _____ are biomass that was buried under sediments long ago.
22. Burning fossil fuels floods the atmosphere with _____.
23. _____ might absorb excess CO_2, but not with deforestation accelerating.
24. The _____ absorbs CO_2, but it lowers the pH of seawater and threatens marine life.
25. As northern areas become warmer, _____ may be replaced by shrubs and trees.
26. Increasing drought will cause _____ to expand.
27. Forests are drying out sooner in the season, setting the stage for disastrous _____.
28. Mismatch between temperature and day length causes flowers to _____ at the wrong time.
29. Warming seas may cause corals to expel their _____, a phenomenon known as bleaching.
30. In North America, the distributions of many species have moved _____.
31. _____ face starvation as sea ice melts away.
32. Species in _____ habitats, such as Costa Rican frogs, may have no place to go.
33. As mosquitoes move to higher elevations, they carry deadly _____ with them.
34. Pest species such as bark beetles can complete their _____ in a shorter period.
35. Phenotypic _____ may allow some species to adjust to climate change.
36. Birds called great tits lay their eggs _____, when caterpillars are available.
37. Red squirrels seem to be showing some evolutionary _____, breeding earlier each year.
38. Animals with high genetic _____ and short life spans may be able to adapt.
39. Evolutionary change is unlikely to save larger animals with _____ life spans.
40. As many as _____ % of Earth's plants and animals may face extinction due to climate change.

What can we do to defuse the biodiversity crisis and protect the biosphere? The rest of the chapter offers some potential strategies for protecting organisms and ecosystems. Review some of these conservation concepts by completing this crossword puzzle.

Across

2. _____ hot spots make up only 1.5% of the Earth's surface, but are home to one-third of all plants and vertebrates.

6. Conservation biologists often focus on protecting threatened or _____ species.

8. Habitat loss can cause _____, the splitting and isolation of populations.

10. To avoid _____, conservationists focus on maintaining enough habitat for a minimum viable population size.

11. Governments have set aside about 7% of the world's land in various kinds of _____.

14. In small populations, _____ and genetic drift may reduce genetic diversity.

15. Biodiversity hot spots are small areas with large numbers of _____ or endangered species

17. _____ reserves consist of undisturbed natural areas and buffer areas with compatible human uses.

19. Loss of _____ may result in the loss of half the species in biodiversity hot spots in the next 10–15 years.

20. Identifying and protecting _____ species may help preserve whole communities.

Down

1. _____ habitats often serve as corridors, and government policy often protects them.

3. A reduced population can enter a downward _____ of smaller and smaller population size.

4. A small population often suffers from reduced genetic _____.

5. Human activities often create many _____ between communities, reducing diversity.

7. Conservation biology aims to sustain the diversity of entire ecosystems and _____.

9. Where habitat is fragmented, narrow strips or patches may serve as a movement _____.

12. _____ development is long-term prosperity of human societies and ecosystems that sustain them.

13. Corridors are especially important for species that _____.

16. Because _____ species are, by definition, limited to specific areas, they are very vulnerable.

18. Most existing parks and ecological reserves are too _____.

Exercise 5   (Module 38.11)

Some locations harbor much more biodiversity than others. Biologists have identified several trends and patterns associated with the distribution of organisms and have pinpointed biodiversity hot spots—small areas with large concentrations of species. A biodiversity hot spot would most likely be characterized by which of the descriptions in each pair below?

1. Closer to the equator
2. Sensitive to habitat degradation
3. Few endangered species
4. High plant diversity, but few animals
5. Likely to be grassland or coniferous forest
6. Many endemic species
7. High potential extinction rate

Farther from the equator
Resistant to habitat degradation
Many endangered species
High plant and animal diversity
Likely to be tropical forest
Few endemic species
Low potential extinction rate

Exercise 6   (Module 38.13)

Biologists and conservationists increasingly try to see the bigger picture and to preserve and restore whole landscapes and ecosystems. They seek to identify biodiversity hot spots, link protected areas (the Yellowstone to Yukon Conservation Initiative), and to rehabilitate damaged ecosystems (the Kissimmee River) when possible. Review managing ecosystems by answering the following questions about wolves and their role in the Yellowstone to Yukon Conservation Initiative.

1. Is the area of the Yellowstone to Yukon Conservation Initiative a biodiversity hot spot? Why start there?

2. How does this project relate to the subject of landscape ecology?

3. In what way does the project use movement corridors?

4. Why focus on the gray wolf?

5. Why can't species such as wolves thrive in the current system of parks and reserves?

6. How did the decline of the wolf population affect the ecosystems where they used to live?

7. Has reintroduction of wolves presented any problems or possible problems? Describe.

8. Has reintroduction of wolves had an effect on other species in Yellowstone? Explain.

The Yellowstone to Yukon Conservation Initiative and Kissimmee River project illustrate the application of the principles of conservation ecology to restoring, managing, and sustaining ecosystems. Match each description with a principle or method.

_____ 1. Promoting prosperity of both societies and ecosystems

_____ 2. Returning damaged ecosystems to their natural state

_____ 3. The strategy of the Kissammee River Project

_____ 4. Using living organisms to detoxify polluted ecosystems

_____ 5. Surrounding undisturbed areas with areas devoted to low-impact human uses

_____ 6. Pioneered by Costa Rica

_____ 7. A project to acquire the information necessary for the development and conservation of Earth's resources

_____ 8. Application of ecological principles to the study of a collection of ecosystems

_____ 9. The overall strategy of using ecological principles to heal degraded ecosystems

_____ 10. Requires us to connect the life sciences with the social sciences, economics, and humanities

A. Restoration ecology
B. Bioremediation
C. Sustainable Biosphere Initiative
D. Establishing zoned reserves
E. Sustainable development
F. Landscape ecology

Test Your Knowledge

Multiple Choice

1. The greatest threat to global biodiversity is
   a. natural disasters such as storms.
   b. pollution.
   c. overexploitation of natural resources.
   d. competition of invasive species with native species.
   e. human alteration of habitats.

2. Which of the following are the main greenhouse gases contributing to global warming?
   a. PCB, CO_2, N_2O
   b. CH_4, CO_2, N_2O
   c. Y2Y, CH_4, CO_2
   d. N_2O, PCB, CH_4
   e. CH_4, CFC, PCB

3. Chemical pesticides and PCBs do the greatest harm to top predators such as gulls and ospreys because
   a. these animals are often exposed to chemical spraying.
   b. their high metabolism is particularly sensitive to very low concentrations of toxics.
   c. these animals are usually exposed to more than one industrial chemical at a time.

   d. concentrations of toxics become magnified as these substances move through food chains.
   e. these animals tend to seek their prey in areas near human activity.

4. Which of the following is thought to be a consequence of global warming?
   a. coastal flooding
   b. wildfires
   c. decrease in Arctic sea ice
   d. changes in distribution of species
   e. all of the above

5. Humans have altered _____ of Earth's land surface.
   a. very little
   b. about 25%
   c. about 50%
   d. about 80%
   e. virtually all

6. Madagascar is known for its
   a. large area of undisturbed tropical forest habitat.
   b. extensive wetland areas.
   c. melting glaciers.
   d. large number of endemic species.
   e. coral reefs.

7. Human activities such as clear-cutting and road building often create edges between areas of different habitat. Edges created by human activities
   a. are usually more abrupt than natural boundaries between habitats.
   b. may have their own species that are not present in adjacent habitats.
   c. may be dominated by relatively few species.
   d. can decrease biodiversity.
   e. all of the above.

8. Some of the most successful ecosystem restoration projects have been carried out in
   a. desert areas.
   b. wetlands.
   c. old-growth forests.
   d. tropical forests.
   e. coral reefs.

9. The three components of biodiversity are
   a. producer, consumer, and decomposer.
   b. ecosystem, species, and genetic.
   c. community, ecosystem, and biosphere.
   d. plant, animal, and microbe.
   e. polar, temperate, and tropical.

10. Invasive species
    a. are often endangered.
    b. usually increase biodiversity.
    c. often enhance the habitat for native species.
    d. usually reduce biodiversity.
    e. are especially numerous in biodiversity hot spots.

11. We do not know the current rate of species loss, mainly because
    a. scientists disagree on whether humans are having an impact on the biosphere.
    b. there is no way of knowing how much habitat has been destroyed.
    c. we don't know how many species exist.
    d. new species are evolving at a rapid rate.
    e. we are uncertain how fast global climate change is occurring.

12. So far, global climate change is having its greatest effect on
    a. tropical forests
    b. arctic regions
    c. the deep ocean
    d. deserts
    e. temperate forests

13. The Yellowstone to Yukon Conservation Initiative has focused on the wolf because
    a. wolves were nearly extinct at one time
    b. they roam over huge areas
    c. wolves are a keystone species
    d. habitats without wolves had become degraded
    e. all of the above

14. Very small populations often spiral to extinction because
    a. their genetic variation is reduced
    b. mates can't find each other
    c. small numbers have difficulty finding prey
    d. they are no longer able to migrate
    e. they run out of important resources

Essay

Imagine that you have been asked to speak to a high school biology class on the topic of the global biodiversity crisis. At the end of your talk, the students have the following questions. Give short, succinct answers to each question.

1. Whooping cranes, condors, and other endangered animals have been raised in captivity and then reintroduced to the wild. Are there any problems doing that with other endangered species?

2. I heard that destroying tropical rain forests is a more serious problem than cutting down forests in the United States? Is that true? Why?

3. Why should I care what happens in forests or wetlands on the other side of the world? How can it affect me?

4. I have heard that there is a lot of disagreement over whether global climate change is really happening. Wouldn't it be better to wait until we are sure before trying to do something about it?

5. A friend told me that burning fossil fuels is destroying the ozone layer. Is this true?

6. People have set aside millions of acres of parks and reserves around the world. Isn't that enough to protect most species?

7. Why are we so worried that plants and animals are going extinct? Hasn't that happened many times—like when the dinosaurs disappeared?

Apply the Concepts

Multiple Choice

1. A conservation group is trying to save Puerto Rico's endemic species. "Endemic" means
   a. they live nowhere else.
   b. predatory.
   c. they are endangered.
   d. little known.
   e. they migrate from place to place.

2. If we fail to do anything about the effects of global warming, species will have to fall back on
   a. genetic drift.
   b. restoration ecology.
   c. biological magnification.
   d. phenotypic plasticity.
   e. biophilia.

3. In assessing whether a bird species is endangered, biologists might look at
   a. reproductive rate.
   b. population size.
   c. habitat loss.
   d. population fragmentation.
   e. all of the above.

4. A newspaper headline shouted, "Ecologists Warn that Dead Zone is Growing." This refers to
   a. tropical deforestation.
   b. an increase in ozone depletion.
   c. a biodiversity hot spot.
   d. global warming's effect on vegetation.
   e. an oxygen-depleted area in the ocean.

5. A hybrid car produces less CO_2, which might reduce
   a. global warming.
   b. acid rain.
   c. ozone depletion.
   d. bioremediation.
   e. all of the above.

6. Imagine that you have been given the job of setting aside habitat to preserve the diversity of birds in a swath of lowland rainforest. Three plans have been proposed, each of them totaling 200 km². Plan A consists of 20 small scattered reserves. Plan B is 1 large reserve. Plan C is 5 medium-sized reserves connected by corridors. Which of the following do you think ranks the reserves in order of potential biodiversity, highest to lowest?
   a. A C B
   b. A B C
   c. C B A
   d. B C A
   e. B A C

7. The introduction to Chapter 32 described how various plants can be used to "soak up" and remove toxic chemicals from the soil. This is an example of
   a. bioremediation.
   b. biological magnification.
   c. bioprospecting.
   d. biological augmentation.
   e. sustainable development.

8. Which of these countries do you suppose produces the most carbon dioxide per person?
   a. United States
   b. Mexico
   c. Indonesia
   d. India
   e. Tanzania

9. In an effort to save an endangered population, wildlife managers might try to enhance _____ to help reduce _____ .
   a. population fragmentation . . . biodiversity hot spots
   b. biological magnification . . . population fragmentation
   c. movement corridors . . . population fragmentation
   d. biological magnification . . . the number of endemic species
   e. sustainable development . . . biodiversity hot spots

10. It is important for us to
    a. protect and preserve diverse species and ecosystems.
    b. appreciate and understand nature.
    c. encourage sustainable development.
    d. improve the human condition.
    e. all of the above.

Essay

1. The top predators in an ecosystem are often regarded as "indicator species"—indicators of the health of the entire system. For this reason, researchers monitor jaguars in the rain forest of Central America and spotted owls in the coniferous forest of the Pacific Northwest. Why do top predators reflect the integrity of their ecosystems?

2. Many factors—both natural and caused by humans—contribute to global climate change. State whether you think each of the following would tend to increase or decrease the possibility of global warming, and why.
   a. Conversion from coal-fired to solar and hydroelectric power plants
   b. Large-scale tree-planting programs
   c. Deforestation
   d. Increasing the average efficiency (gas mileage) of automobiles
   e. Slowing world population growth
   f. Warming of climate, speeding up decomposition of organic matter

3. Marine algae need iron in trace amounts, and their growth seems to be limited, especially in the Antarctic Ocean, by the lack of iron in seawater. Some biologists have suggested that fertilizing the sea with iron compounds might help reduce global warming. How would this work? Do you think we ought to try it? Why or why not?

4. Explain how the following might be employed in restoring an area of patchy clear-cuts into habitat suitable for species that live in large stands of mature forest, such as the spotted owl or red-cockaded woodpecker: bioremediation, restoration ecology, corridors, establishing zoned reserves.

5. Thousands of acres of tropical forest in Central and South America have been cut, burned, and converted to pasture for cattle. Much of the beef raised there is exported, bringing valuable foreign exchange into poor countries. A lot of the beef ends up in the United States in luncheon meats, hamburger, baby foods, and pet foods. (American fast-food chains used to be major buyers of Latin American beef, but pressure from environmentalists has reduced their reliance on this source.) The land cannot be grazed for very long; heavy rains leach away nutrients, and the rains and overgrazing cause soil erosion. Soon pasture is covered by scrub, and it is uncertain when, if ever, the forest will be able to regrow. But surely a few hamburgers can't affect the forest very much? Or can they?

Using the information that follows, see if you can calculate how much tropical forest is destroyed to produce a four-ounce hamburger: An acre of rain forest consists of about 800,000 pounds of living things. After forest is cut down and grass is planted, there is enough food produced to grow 50 pounds of cattle per acre per year. The pasture land can be used for about 8 years before the soil is exhausted and the land is no longer usable. Only about half a steer can be made into hamburger; the other half is bone, skin, etc. One hamburger contains 4 ounces of beef. One pound is 16 ounces, and one acre is 43,793 square feet. (Based on "Our Steak in the Jungle," by C. Uhl and G. Parker, *BioScience*, Vol. 36, No. 10, p. 642.)

Questions: How many pounds of forest life are destroyed for each hamburger? How many square feet of forest are destroyed per hamburger? How many hamburgers do you eat in a year? How much forest could you destroy in a year?

Put Words to Work

Correctly use as many of the following words as possible when reading, talking, and writing about biology:

biodiversity, biodiversity crisis, biodiversity hot spot, biological magnification, conservation biology, endangered species, endemic species, landscape ecology, movement corridor, ozone layer, phenotypic plasticity, restoration ecology, sustainable development, threatened species, zoned reserve

Use the Web

For further review on conservation and restoration biology, see the activities and questions on the Web at www.mybiology.com.

Answers

Chapter 1

Review the Concepts

Exercise 1: 1. Biosphere   2. All of the organisms of a particular area, plus the nonliving environment that affects them   3. Community   4. All the organisms of one species in a given area   5. Organism   6. Organ system   7. A body structure consisting of several tissues, performing particular functions   8. Tissue   9. Cell   10. A functional component of a cell   11. Molecule

Exercise 2:

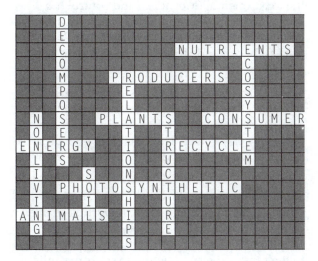

Exercise 3: The structure and function of a college student's *cells*, and all her other characteristics—from facial features to the functioning of her brain—are determined by the *genetic information in DNA molecules*, which shaped these features during her *growth and development*. The student *processes energy* from food to do work—walking, talking, and studying. In performing these actions, she *responds to stimuli* from the environment, and various mechanisms *regulate* her body's activities to adjust to environmental changes. Like her other characteristics, her ability to *reproduce* results from the precise *order* of her molecules, cells, and organ systems. All of these human characteristics have been shaped by natural selection over time—the process of *evolutionary adaptation*.

Exercise 4: 1. Eukarya, Plantae   2. Archaea or Bacteria   3. Archaea or Bacteria   4. Eukarya, Animalia   5. Eukarya, Fungi   6. Eukarya, protists   7. Archaea or Bacteria   8. Eukarya, Animalia   9. Eukarya, Plantae   10. Eukarya, protists

Exercise 5: 1. related   2. Charles Darwin   3. ancestral   4. population   5. varied   6. heritable   7. predators   8. reproduce   9. offspring   10. favorable   11. unequal   12. natural selection   13. variations   14. environmental   15. reproductive   16. species   17. evolution   18. time   19. theme   20. unity   21. diversity

Exercise 6: 1. inductive reasoning   2. discovery   3. hypothesis   4. scientific method   5. deductive reasoning   6. observation   7. question   8. hypothesis   9. experiment   10. experimental   11. control   12. factor   13. prediction   14. deductive reasoning

Exercise 7: See Modules 1.9 and 1.10 for ideas. Recall that science is seeking natural causes for natural phenomena through observation and experiment. Technology is the application of scientific knowledge for a specific purpose.

Test Your Knowledge

Multiple Choice: 1. e   2. c   3. b   4. d   5. c   6. c   7. e   8. b   9. e   10. c

Essay: 1. Prokaryotes are simple, single-celled organisms whose cells lack a nucleus. The two kinds of prokaryotes are classified in Domain Bacteria and Domain Archaea, two quite distinct groups distinguished mainly on the basis of molecular differences. Organisms in Domain Eukarya—plants, animals, fungi, and several kingdoms of protists—all have more complex eukaryotic cells containing a nucleus and other organelles.

2. There are several kingdoms of protists—simple organisms such as algae and protozoa. Like plants, animals, and fungi, protists have eukaryotic cells. But most protists are small and single-celled. Plants, animals, and fungi are all multicellular. Plants, such as pine trees and rosebushes (Kingdom Plantae), differ from the others in that they are photosynthetic and have cells with rigid walls made of cellulose. The fungi of Kingdom Fungi, such as mushrooms and molds, decompose

the remains of dead organisms and absorb the nutrients. Animals (Kingdom Animalia), from worms to human beings, have cells without rigid walls and eat other organisms, and most move actively in search of food.

3. Science deals with questions about natural phenomena that have natural causes. Some examples of questions that potentially could be answered by science are, What particles make up an atom? What makes a cell divide? How do birds find their way when they migrate? Science does not deal with questions that are not subject to experimental test or that are not concerned with the natural world and its laws: What is the purpose of human existence? Why is there a universe? Is disease caused by evil spirits?

4. Chemical structure—the arrangement of atoms and molecules—is the fundamental level of order that underlies all the properties of life. The specific properties of a cell, organ, or organism arise from the combinations and arrangements of the molecules of which the organism is composed. The structure of DNA—most notably the particular sequence of nucleotides that make up the DNA molecule—provides a blueprint for growth and development and information that cells need to make the molecules involved in their structure, energy use, and response to stimuli. Replication, or copying, of DNA underlies the organism's ability to reproduce. The evolution of the diverse forms of life stems from changes in DNA sequences over time.

Apply the Concepts

Multiple Choice: 1. e   2. c   3. d   4. b   5. c   6. c   7. a   8. c   9. e

Essay: 1. The squirrel depends on an oak tree for shelter, an escape route, and food. It may live in a cavity hollowed out by a woodpecker. It buries nuts, adding to the soil organic matter that might be consumed by insects and worms. Sometimes the buried acorns sprout and grow into trees. Hawks and owls prey on squirrels, and various parasites depend on them. Chipmunks sometimes compete with squirrels for food.

2. The ancestors of beavers may have been ratlike animals that occasionally ventured into the water to escape from predators or obtain food. There may have been heritable variation among the rodents in the population with regard to the shapes of their tails and feet. Those rodents with more flattened tails and more webbed feet were able to swim more efficiently, thus escaping predation and

gathering food more effectively, making them more likely to leave offspring than more ratlike individuals. Their offspring inherited their flattened tails and webbed feet, and the proportion of individuals with flattened tails and webbed feet increased over time, until eventually all the individuals in the population looked like beavers. Natural selection occurred as heritable variations were exposed to environmental factors that favored the reproductive success of some individuals over others.

3. Jason's test of MegaGro was not scientifically valid because it was not a controlled experiment. There are any number of factors that might have caused this year's crop of tomatoes to exceed last year's—rainfall, seed type, sunny weather, location of plants, and so on. Jason needs to plant two groups of tomato plants side by side at the same time. Only one factor—the amount of MegaGro—should be allowed to differ between the experimental (MegaGro) and control (no MegaGro) plants. Then if one group of plants produces more tomatoes than the other, Jason can safely conclude that the fertilizer made the difference.

4. Set up an obstacle course for oilbirds to fly through. Divide the birds to be tested into two groups. Plug the ears of the birds in one group so that they cannot hear. Then compare how these experimental birds navigate the obstacle course with the performance of control birds that can hear.

5. The long, narrow, flexible fingers and sensitive nerve endings in the fingertips enable us to grasp and manipulate objects. The flat shape of a leaf and its orientation perpendicular to the rays of the sun allow it to capture sunlight efficiently for the process of photosynthesis. The hawk's strong, curved, sharp beak is used for tearing meat. A frog's long, muscular legs, with their webbed feet, are used in jumping and swimming.

6. You are an organism composed of organ systems such as the circulatory and skeletal systems. Each system is made up of organs, such as the heart or an artery. An organ is made of tissues, and a tissue is made of cells, the smallest living units. Each cell is made up of molecules, which are made of atoms. A cell is alive; it uses energy, responds to stimuli, grows, and reproduces. The molecules that make up a cell are not alive, and they do not have all these properties.

7. Living things are composed of cells. They are complex and highly ordered, and carry out

chemical processes, according to the genetic information in their DNA. They take in energy and use it to perform work, and they are able to respond to stimuli and regulate their internal environment. They also grow, develop, reproduce, and evolve. Some nonliving things possess some of these properties, but only living things are characterized by all of them.

8. Human, chimpanzee, and goldfish DNA are all made of the same chemical subunits; they all have the same fundamental double-helix chemical structure; and they all use the same alphabet to encode hereditary information. The four nucleotides in the DNA of chimps and humans are arranged in similar sequences and spell out similar genetic messages. The nucleotides of human DNA and goldfish DNA are less similar in sequence and spell out rather different messages.

Chapter 2

Review the Concepts

Exercise 1: 1. Mg M   2. O L   3. Zn T   4. H L
5. Cu T   6. I T   7. C L   8. Ca M   9. P M   10. N L
11. Na M   12. Fe T

Exercise 2: 1. C   2. E   3. C   4. C   5. E   6. E
7. C   8. C   9. E   10. C

Exercise 3: 2. N 7 14 7 7 7   3. Cl 17 35 17 18 17
4. O 8 16 8 8 8   5. O 8 17 8 9 8

Exercise 4: 1. The calcium atom gives away 2 electrons, one to each of 2 Cl atoms, producing a Ca^{++} ion and 2 Cl^- ions. These ions form $CaCl_2$.
2. The nitrogen atom shares its electrons with 3 hydrogen atoms, forming ammonia, NH_3.

Exercise 5: 1. chemical   2. cohesion   3. surface tension   4. oxygen   5. hydrogen   6. share
7. covalent   8. nonpolar   9. electronegative
10. negative   11. positive   12. polar   13. neutral
14. negatively   15. positively   16. hydrogen
17. four   18. DNA   19. protein   20. solid
21. liquid   22. gas   23. solvent   24. aqueous
25. solutes   26. heat   27. temperature   28. cup
29. bathtub   30. hydrogen bonds   31. less
32. releases   33. absorbs   34. releases   35. fast
36. evaporate   37. boiling point   38. evaporative
39. less   40. floats

Exercise 6: 1. 4   2. 7   3. 4   4. 7   5. 3   6. 11   7. 2
8. 3   9. 13   10. 8   11. 6   12. 2

Exercise 7: 2 H_2O_2 (reactants) → 2 H_2O + O_2 (products)

Exercise 8:

Test Your Knowledge

Multiple Choice: 1. b   2. c   3. e   4. d   5. b   6. b
7. e   8. a   9. c   10. c   11. e   12. b   13. d

Essay: 1. See Table 2.2 in the text.

2. Iron is an element, and there are atoms of iron. Water is a compound, containing two elements. The smallest particle of water is a molecule, composed of two hydrogen atoms and one oxygen atom. There are no atoms of water; if a water molecule is broken up into hydrogen and oxygen atoms, it is no longer water.

3. Temperature is the average speed of molecules in a body of matter. Heat is the total amount of energy due to movement of molecules. Water molecules in a teakettle move faster, but there are many more of them in a swimming pool, so even though the molecules in a swimming pool are moving slower, their total amount of energy is greater.

4. Acid precipitation forms when air pollutants from burning fossil fuels—sulfur and nitrogen oxides—combine with water vapor in the air to form sulfuric and nitric acids. Acids can break down chemicals in animals, and low pH can interfere with chemical processes, especially in the eggs and young of aquatic organisms. Acid in the soil can lead to mineral imbalances that affect plants.

5. The shared electrons in a water molecule are pulled more strongly toward the oxygen atom than the hydrogen atoms. This makes the oxygen atom partially negative and the hydrogen atoms partially positive, causing the water molecule to be

polar. The + and − charged regions on each water molecule are attracted to oppositely charged regions on adjacent molecules; these attractions are called hydrogen bonds. When water is heated, the heat energy first disrupts the hydrogen bonds and then makes the molecules move faster. Because heat is absorbed as bonds break, water can absorb and store much heat while warming only a small amount.

6. Water evaporates when the fastest molecules move so rapidly that they escape from the liquid and form water vapor. This leaves the molecules still in the liquid with a lower average speed, and as a result, the remaining liquid has a lower temperature.

Apply the Concepts

Multiple Choice: 1. e   2. d   3. a   4. c   5. e   6. d
7. b   8. a   9. c   10. a

Essay: 1. It will form two covalent bonds, because it needs to share 2 electrons with other atoms to have a complete outer shell of 8 electrons. H_2S.

2. Give a plant water containing a radioactive isotope of oxygen, H_2O^*, and look to see whether the sugar made contains the radioactive oxygen $C_6H_{12}O_6^*$. Carry out a similar test with CO_2^*.

3. It takes a large amount of heat to break the hydrogen bonds in water before the water warms up much, so water can absorb much heat with a small rise in temperature. As the rock loses heat it cools off a lot, but this same amount of heat absorbed by the water does not raise the temperature of the water much, because most of the heat energy is used to break hydrogen bonds.

4. Water molecules can form hydrogen bonds with the positively and negatively charged atoms in a sugar molecule, so the water molecules can surround individual sugar molecules. Water molecules are not attracted to oil molecules; the water molecules stick to each other and do not surround the oil molecules.

5. Each atom is characterized by a certain number of electrons, arranged in energy shells. For most atoms, the outermost shell can hold up to 8 electrons. Carbon has only 4 electrons in its outer shell. It tends to share electrons—form bonds—with four other atoms, filling its outer shell. Thus it tends to form four bonds, and methane is CH_4, not CH_6.

6. Seattle is much closer to the ocean, which stores heat and stabilizes climate. During the summer, the water absorbs heat, cooling Seattle, and during the

winter the water releases stored heat to the air, warming the city.

7. Water molecules are more polar than those of most other substances. They are attracted to one another, bonded by hydrogen bonds. These bonds are hard to break, so water must absorb more heat to vaporize than most other substances. The easier-to-vaporize alcohol molecules tend to evaporate first.

Chapter 3

Review the Concepts

Exercise 1:

1.

```
     H   H   H   H   H
     |   |   |   |   |
 H — C — C — C — C — C — H
     |   |   |   |   |
     H   H   H   H   H
```

Carbon has 4 electrons in an outer shell that holds 8. It completes its outer shell by sharing with four other atoms.

2.

```
     H   H   H   H              C₄H₁₀
     |   |   |   |
 H — C — C — C — C — H
     |   |   |   |
     H   H   H   H
```

C_4H_{10}

3.

```
     H   H   H   H   H          C₅H₁₀
     |   |   |   |   |
 H — C — C = C — C — C — H
     |           |   |
     H           H   H
```

C_5H_{10}

or

```
 H   H   H   H   H
 |   |   |   |   |
 H — C = C — C — C — C — H
         |   |   |
         H   H   H
```

4.

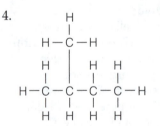

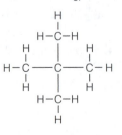

C_5H_{12}. They are isomers.

5.

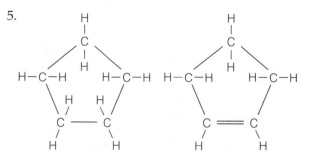

Without double bond: 10 Hs. With double bond: 8 Hs.

Exercise 2: 4 hydroxyls, 6 carbonyls, 2 carboxyls, 5 aminos, 1 phosphate, 1 methyl

Exercise 3: 1. D 2. D 3. H 4. H 5. D 6. D 7. H 8. D 9. H 10. H

Exercise 4: 1. monosaccharides 2. polysaccharides 3. proteins 4. fats 5. starch 6. glucose 7. dehydration 8. water 9. hydrolysis 10. disaccharide 11. ring 12. six 13. cellulose 14. linked 15. digesting 16. fiber 17. disaccharide 18. glucose 19. fructose 20. monosaccharide 21. starch 22. glucose 23. obesity 24. type 2 diabetes 25. hydrophilic 26. isomers 27. $C_6H_{12}O_6$ 28. dehydration 29. glycogen 30. starch 31. glucose

Exercise 5:

Exercise 6: 1. transport 2. contractile 3. defensive 4. structural 5. signal 6. receptor 7. storage 8. enzyme

Exercise 7: See Figures 3.12 A, B, and C in the text.

Exercise 8: 1. quaternary DF 2. tertiary AIJK 3. secondary EH 4. primary BCG

Exercise 9: 1. H 2. G 3. J 4. D 5. F 6. D 7. C

8. K 9. J 10. I 11. A 12. E 13. B 14. K or D

Exercise 10: 1. fats 2. steroids 3. energy 4. glycerol 5. unsaturated 6. plant oils 7. cardiovascular disease 8. butter 9. corn oil 10. cell membranes

Test Your Knowledge

Multiple Choice: 1. c 2. a 3. c 4. e 5. d 6. c 7. e 8. c 9. d 10. d 11. b 12. e 13. d

Essay: 1. Structural proteins form body parts such as fibers in tendons and ligaments. Contractile proteins in muscles are responsible for movement. Storage proteins such as albumin store amino acids. Defensive proteins called antibodies fight infection. Transport proteins such as hemoglobin carry substances. Some hormones are signal proteins, which convey messages from cell to cell. Receptor proteins respond to signals. An enzyme is a protein that catalyzes chemical changes.

2. Animal fats are saturated; their fatty acids contain no double bonds and can pack together tightly and solidify at room temperature. Plant oils are unsaturated; their fatty acids contain double bonds, which cause kinks that prevent them from packing together and solidifying.

3. The secondary and tertiary levels of protein structure are maintained mostly by relatively weak hydrogen bonds and ionic bonds. These weak bonds are easily disrupted by changes in the protein's environment.

4. Glucose: one circle. Maltose: two joined circles. Starch: a chain of linked circles. See Figures 3.4, 3.5, and 3.7 in the text. Maltose is a disaccharide. Starch is a polysaccharide.

5. See Figures 3.14A–3.14D in the text.

Apply the Concepts

Multiple Choice: 1. e 2. d 3. d 4. d 5. d 6. b 7. e 8. a 9. a 10. e 11. a 12. a 13. a

Essay: 1. Every starch molecule is a polymer of glucose monomers, which are all the same, so starch molecules differ only in length. A protein molecule can be composed of 20 kinds of amino acid monomers. These can be arranged in chains of an almost infinite number of different sequences and lengths, producing many different kinds of proteins.

2. A protein is folded in a specific three-dimensional shape, and the function of a protein such as an enzyme usually depends on its ability to bind to some other molecule because of its specific shape. The shape of these enzyme molecules allows them to fit a starch molecule. The glucose monomers of cellulose are linked together in a different orientation,

forming a rod instead of a coil. Apparently the enzymes do not fit this structure.

3. The hydrogen and ionic bonds that maintain the three-dimensional shape of a protein are fairly weak. A slight change in the environment can alter these bonds, change the shape of the protein, and impair its function, even if the covalent bonds that hold the amino acids together in the protein are unaffected. The main function of starch is to store glucose; the three-dimensional shape of the molecule is relatively unimportant. A much more drastic change is needed to break the covalent bonds in the starch molecule, separate the glucose subunits, and alter its function.

4. See Figure 3.1 in the text. Make sure every C forms four bonds and every H forms one bond. Isomers have the same molecular formulas, but their structures differ.

5. The body stores excess glucose in the form of glycogen, a polysaccharide. We can draw on this store of sugar between meals. Since Fred cannot produce glycogen, he cannot store as much sugar and must eat more often to provide his cells with glucose "fuel."

6. The number of possible tripeptides is $20 \times 20 \times 20 = 8,000$. The number of possible polypeptides 100 amino acids long is 20^{100}—a very large number indeed!

Chapter 4

Review the Concepts

Exercise 1: 1. visible light   2. visible light
3. electrons   4. eye cornea and lens   3. glass lenses   6. electromagnets   7. none   8. $1,000\times$
9. $100,000\times$ +   10. dust particle   11. smallest bacterium   12. small molecules   13. 0.1 millimeter
14. 0.2 micrometer   15. 2 nanometers
16. no magnification   17. limited magnification and resolving power   18. cannot be used on living specimens

Exercise 2: *Cell 1:* Surface area = 96 mm². Volume = 64 mm³. Surface/volume = 1.5.
Cell 2: Surface area = 384 mm². Volume = 512 mm³. Surface/volume = 0.75.
1. Cell 2.   2. Cell 2.   3. Cell 1.   4. Cell 1.
5. Divide the larger cell into eight cells the size of the smaller one.

Exercise 3: For labels, see Figure 4.3 in the text.
Capsule: Sticky coat protects cell
Cell wall: Protects and maintains cell shape
Plasma membrane: Surrounds cell and separates it from environment   Nucleoid: location of bacterial chromosome   Bacterial chromosome: Hereditary material (DNA)   Ribosome: Assembles proteins

Prokaryotic flagella: Propel cell   Pili: Attachment to surfaces

Exercise 4: *Prokaryotic cell:* prokaryotic, small, plasma membrane, cell wall, cytoplasm, ribosomes, bacterial flagellum. *Plant cell:* eukaryotic, large, membranous organelles, plasma membrane, cell wall, cytoplasm, ribosomes, nucleus, nuclear envelope, rough ER, smooth ER, Golgi, peroxisome, central vacuole, mitochondrion, chloroplast, cytoskeleton, plasmodesmata. *Animal cell:* eukaryotic, large, membranous organelles, plasma membrane, cytoplasm, ribosomes, nucleus, nuclear envelope, rough ER, smooth ER, Golgi, lysosome, peroxisome, mitochondrion, cytoskeleton, flagellum, centriole, extracellular matrix, cell junctions.

Exercise 5: 1. D   2. I   3. A   4. E   5. B   6. D
7. G   8. A   9. G   10. F   11. C   12. B   13. E
14. G   15. H   16. I   17. F   18. J   19. A   20. B
21. D   22. F   23. F   24. E   25. J   26. D   27. E

Exercise 6: See Figure 4.13 in the text.   1. rough ER—vesicle—Golgi—vesicle—plasma membrane—outside   2. rough ER—vesicle—Golgi—lysosome   3. rough ER—vesicle—Golgi—vesicle—plasma membrane   4. smooth ER—vesicle—Golgi—vesicle—plasma membrane—outside

Exercise 7: *Chloroplast:* Found in plants and some protists; carries out photosynthesis; converts energy of sunlight into chemical energy in sugar. *Mitochondrion:* Found in all eukaryotes; carries out cellular respiration; converts chemical energy of foods into chemical energy in ATP. Both evolved by endosymbiosis from free-living prokaryotes.

Exercise 8: *Microfilaments:* solid rods, actin, work with myosin to change shape, muscle contraction. *Intermediate filaments:* Ropelike structure, fibrous proteins, reinforce cell shape and anchor organelles. *Microtubules:* hollow tubes, tubulin, help change shape, move chromosomes, act as tracks, shape and support cell, in cilia, in flagella, in centrioles, 9 + 2 pattern, dynein arms cause bending, extend from basal body, don't function in PCD

Exercise 9: 1. E   2. D   3. A   4. B   5. C   6. F
7. F   8. D   9. B   10. C   11. D   12. F

Exercise 10: See Figure 4.4 and Table 4.23 in the text.

Test Your Knowledge

Multiple Choice: 1. e   2. b   3. b   4. e   5. a   6. b
7. e   8. c   9. b   10. b   11. a

Essay: *1.* An electron microscope has greater resolving power and is capable of greater magnification than a light microscope, enabling it to reveal finer details than a light microscope. An electron microscope cannot be used to view living specimens, because they must be held in a vacuum and all air and liquids must be removed. A light microscope is preferable for viewing living specimens.

2. Prokaryotic cells are smaller and much simpler than eukaryotic cells. Prokaryotes lack a membrane-enclosed nucleus and other membranous organelles that compartmentalize eukaryotic cells.

3. There are three major structures present in plant cells but lacking in animal cells. Rigid cell walls containing cellulose surround and support plant cells. Chloroplasts in plant cells perform photosynthesis. Plant cells usually contain central vacuoles that may help carry out metabolic processes or store water, food, or other chemicals.

4. Membranes separate metabolic processes and allow processes that require different conditions to be carried out simultaneously. They also increase the total surface area of membranes, where many metabolic processes occur.

5. Chloroplasts carry out photosynthesis, capturing the energy of sunlight and storing it in sugar molecules. Mitochondria carry out cellular respiration, converting the chemical energy in sugar and other foods into the chemical energy of ATP, which serves as a cellular fuel.

Apply the Concepts

Multiple Choice: 1. d   2. d   3. c   4. a   5. a   6. b   7. b   8. c   9. a   10. e

Essay: 1. The TEM would be best for examining fine details within cell organelles because it is capable of greater magnification and resolving power and uses a beam of electrons that passes through thinly sliced sections. The SEM would be best for studying bumps on the cell surface because its electron beam scans the surface and forms an image of the outside of the cell. Electron microscopes are not suitable for observing living specimens because they must be placed in a vacuum and the air and liquids must be removed, so a light microscope would be best for observing changes in the nucleus as the cell prepares to divide and in cell shape as the cell moves.

2. Surface area = 150 μm^2. Volume = 125 μm^3. Surface/volume = 1.2. Surface area = 600 μm^2. Volume = 1000 μm^3. Surface/volume = 0.6. The smaller cell has the greater surface-to-volume ratio, which would enable it to exchange materials more efficiently with its environment via its membrane. The surface-to-volume ratio of the larger cell would equal that of the smaller cell if it were divided up into eight smaller cubes, 5 μm on each side.

3. The amylase molecules are produced by ribosomes attached to the rough ER and deposited inside the ER. Transport vesicles filled with the protein bud from the ER and fuse with the Golgi apparatus, where the proteins may be modified. Transport vesicles then bud from the Golgi apparatus and fuse with the plasma membrane, dumping amylase outside the cell.

4. The mitochondrion is a sort of cellular "power plant," converting the energy in food molecules into the energy of ATP, which in turn provides energy for many cellular activities. Interfering with the mitochondria could reduce the supply of ATP fuel and thus impair many cellular processes.

5. As a muscle cell grows, ribosomes, ER, and Golgi would work together to produce more protein and cell membrane. Microfilaments, made of the protein actin, are responsible for the contractile activity of muscle and would increase in number in a growing muscle cell. Smooth ER stores calcium that is involved in the contraction process, and it too would grow. Mitochondria provide energy for muscle contraction and other cellular activities and would perhaps increase in number.

Chapter 5

Focus on the Concepts

Exercise 1: 1. phospholipid   2. glycoprotein   3. carbohydrate I.D. tags   4. extracellular fluid   5. glycolipid   6. transport protein   7. fiber of extracellular matrix   8. cholesterol   9. receptor protein   10. enzyme   11. cytoplasm

Exercise 2: 1. E   2. B   3. A   4. F   5. DGH   6. G   7. C   8. B   9. H   10. E   11. CF   12. D   13. I   14. D   15. C

Exercise 3: 1. Out   2. Out   3. In   4. None   5. In   6. Out   7. In   8. In   9. None   10. Out

Exercise 4: 1. phospholipid   2. nonpolar (hydrophobic)   3. phospholipid   4. protein   5. fluid   6. glycolipids   7. glycoproteins   8. identity   9. receptor   10. diffusion   11. passive   12. concentration   13. more   14. less   15. isotonic   16. hypertonic   17. out of   18. selectively   19. osmosis   20. transport   21. energy   22. facilitated   23. active   24. ATP   25. endocytosis   26. phagocytosis   27. vesicle   28. lysosome   29. exocytosis

Exercise 5: 1. work   2. kinetic   3. thermodynamics   4. first   5. transformed   6. potential   7. heat   8. second   9. entropy   10. more   11. endergonic   12. exergonic   13. releases   14. ATP   15. ADP   16. phosphate   17. potential   18. hydrolysis   19. phosphate   20. ADP   21. work   22. mechanical   23. transport   24. chemical   25. phosphorylation   26. heat   27. membranes   28. endergonic   29. coupling   30. activation   31. barrier   32. protein

33. catalyst 34. increases 35. changed 36. lower
37. metabolism

Exercise 6: 1. Kinetic energy is energy that is doing work, and potential energy is stored energy.
2. Exergonic reactions release energy, and endergonic reactions consume energy.   3. In a chemical reaction, reactants are changed into products.
4. ADP is phosphorylated (a phosphate group is added), producing ATP. When ATP is hydrolyzed, a phosphate group is removed from ATP, and ADP is produced.   5. Enzymes lower the energy of activation for a reaction, so the reaction with an enzyme requires less energy to get it started than the reaction without the enzyme.   6. In photosynthesis, the energy of sunlight is used to make glucose from carbon dioxide and water. In cellular respiration, oxygen and glucose combine, and the energy in glucose is used to make ATP.   7. First law: Energy can be transferred or transformed but not created or destroyed. Second law: Energy changes are accompanied by an increase in disorder, or entropy.
8. Mechanical work is movement of the cell or body. Transport work is moving substances through cell membranes. Chemical work is building molecules.

Exercise 7: 1. See Figures 5.15 and 5.16 in the text. Substrate enters enzyme active site and slightly changes site to fit. Products leave and enzyme recycles.   2. Shape of enzyme altered by heat or pH, deforming active site so substrate will not fit.   3. See Figure 5.16 in text. Competitive inhibitor is shaped like substrate and blocks active site. Noncompetitive inhibitor can be any shape; attaches to enzyme and alters shape of active site so substrate will not fit.

Test Your Knowledge

Multiple Choice: 1. a   2. c   3. b   4. b   5. d   6. c
7. b   8. a   9. e   10. b   11. d   12. b

Essay: 1. Large and polar molecules cannot easily diffuse through cell membranes. Transport proteins provide pores by which these substances can pass through membranes.

2. The enzyme is shaped in such a way that its active site forms a "pocket" that holds the substrate.

3. Molecules slow down at lower temperatures, slowing the rate of contact between reactant molecules and the enzyme's active site. Higher temperatures denature the enzyme, altering its shape and impairing its specific function.

4. Larger molecules contain more covalent bonds, so they have more potential energy stored in their complex structure. It takes energy to form these bonds, so the reactions by which amino acids join to form proteins are endergonic. The cell obtains energy for endergonic reactions by coupling them to exergonic reactions, the most important of which is the hydrolysis of ATP.

5. A cell will gain water by osmosis if the solution outside the cell is hypotonic, that is, if the outside solution has a lower total concentration of solutes than the cytoplasm. It will lose water if the solution outside is hypertonic, that is, if it has a higher total concentration of solutes than the cytoplasm. Water gain by a plant cell is least serious; the cell wall prevents bursting.

Apply the Concepts

Multiple Choice: 1. b   2. e   3. c   4. b   5. a   6. d
7. e   8. b   9. d   10. d   11. e   12. e   13. b

Essay: 1. Paper does not burn spontaneously (at room temperature) because energy is required to get this reaction started—energy of activation. This creates an energy barrier that prevents the reaction from happening spontaneously. A flame supplies energy of activation.

2. The hypertonic solution would cause water to move out of the stomach cells by osmosis, causing dehydration.

3. You can't win: The first law of thermodynamics says that energy cannot be created or destroyed, so you can't come out ahead in the energy game. You can't break even: The second law of thermodynamics says that energy changes are accompanied by an increase in disorder, or entropy. Heat, a form of disorder, is created during energy conversions; some energy is always lost to the surroundings as heat, so you can't break even.

4. The insecticide molecule is the same shape as the substrate molecule. It fits into the active site of the enzyme, preventing the enzyme from acting on its normal substrate, and thus interfering with nervous system function.

5. The lecithin molecules coat the droplet with their nonpolar tails embedded in the fat and the polar heads projecting into the surrounding water.

Chapter 6

Review the Concepts

Exercise 1: 1. respiration 2. oxygen 3. carbon dioxide 4. cellular respiration 5. glucose
6. oxygen 7. carbon dioxide 8. water 9. ATP
10. 2,200 11. sun 12. photosynthesis 13. chemical 14. CO_2 15. H_2O 16. glucose 17. food
18. air 19. aerobically 20. 40 21. anaerobically
22. 2 23. Slow 24. aerobic 25. thin 26. blood
27. mitochondria 28. myoglobin 29. dark
30. fast 31. thicker 32. fewer 33. less 34. glucose 35. lactic acid 36. ache 37. fast 38. slow
39. fast

Exercise 2:

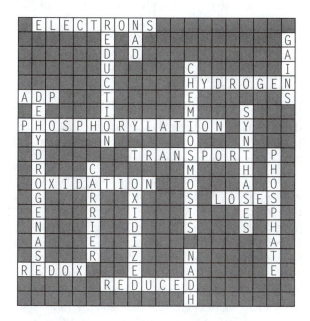

Exercise 3: 1. electrons carried by NADH   2. electrons carried by NADH and $FADH_2$ 3. glycolysis
4. glucose   5. pyruvate   6. Citric acid cycle
7. oxidative phosphorylation   8. cytoplasm
9. CO_2 10. CO_2 11. mitochondrion   12. ATP
13. ATP   14. ATP

Exercise 4: 1. I   2. K   3. E   4. C   5. C   6. D   7. B
8. J   9. F   10. A   11. L   12. G   13. H

Exercise 5: 1. NAD^+ 2. NADH   3. pyruvic acid
4. acetyl CoA   5. CO_2 6. coenzyme A   7. CO_2
8. NADH   9. ATP   10. $FADH_2$

Exercise 6: 1. final   2. ATP   3. oxidative phosphorylation   4. electron   5. inner   6. NADH   7. the
citric acid cycle   8. O_2 9. H_2O 10. H^+ ions
11. active transport   12. potential   13. ATP
14. inner compartment of the mitochondrion
15. ATP synthases   16. ADP   17. ATP

Exercise 7: See Figures 6.10 and 6.11 in the text.

Exercise 8: 1. OP   2. G   3. G   4. G   5. OP   6. CA
7. OP   8. OP   9. CA   10. OP   11. OP   12. CA

Exercise 9: 1. facultative   2. yeasts   3. pyruvate
4. muscle   5. oxygen   6. NADH   7. ethanol
8. lactic   9. ATP   10. glycolysis   11. CO_2
12. anaerobes   13. anaerobic   14. less
15. glycolysis   16. lactate

Exercise 10: 1. carbohydrates   2. fats   3. proteins
4. sugars   5. glycerol   6. fatty acids
7. amino acids   8. glucose

Exercise 11: See Figure 6.16 in the text.

Test Your Knowledge

Multiple Choice: 1. b   2. e   3. c   4. b   5. d
6. c   7. c   8. a   9. b   10. e   11. e   12. a

Essay: 1. Breathing allows our lungs to supply
cells with the oxygen used in cellular respiration
and expel the carbon dioxide produced by cellular
respiration.

2. Aerobic cellular respiration is more efficient
(it produces more ATP per molecule of glucose
consumed), but it must take place in an oxygen-
containing (aerobic) environment. Fermentation
requires no oxygen, but it is much less efficient
(it produces much less ATP per glucose).

3. Oxidative phosphorylation occurs in the third
(electron transport–chemiosmosis) stage of cellular respiration. Substrate-level phosphorylation
occurs in glycolysis and the citric acid cycle. In oxidative phosphorylation, ATP synthase harnesses
the flow of H^+ ions down their concentration gradient to make ATP. In substrate-level phosphorylation, an enzyme transfers a phosphate group
from an organic molecule to ADP to make ATP.
Oxidative phosphorylation makes the most ATP
under aerobic conditions. Substrate-level
phosphorylation makes the most ATP under
anaerobic conditions.

4. Rotenone prevents electrons from passing from
an electron carrier molecule to the next. Cyanide
and carbon dioxide block the passage of electrons
to oxygen. In either case, the energy of moving
electrons cannot be captured. Oligomycin blocks
the passage of H^+ ions through ATP synthase, so
ATP cannot be made. DNP makes the membrane of
the mitochondrion leaky to H^+ ions, thus abolish-
ing the H^+ concentration gradient whose energy is
harnessed to make ATP.

5. These processes provide small organic molecules
that the cell can use in the synthesis of organic molecules not obtained in food.

Apply the Concepts

Multiple Choice: 1. b   2. a   3. c   4. c   5. d   6. d
7. c   8. d   9. b   10. d   11. d   12. a   13. c

Essay: 1. Organisms that rely on aerobic cellular
respiration are limited to environments where oxygen is available. Fermenters are able to live in
anaerobic environments such as mud, manure
piles, and sealed wine vats, free from competition
with aerobic organisms.

2. Oxygen picks up electrons at the end of the electron transport chain. When oxygen runs out, there is
no place for electrons to go. They "pile up," and electron transport stops. When electron transport stops,
NADH and $FADH_2$ pile up, and NAD^+ and FAD are
not regenerated. Without NAD^+ and FAD to pick up
electrons and hydrogens, the Krebs cycle must stop.

3. Because FAD and NAD^+ are constantly recycled,
the cell requires only a small supply of them.

Without NAD^+ and FAD to pick up electrons and hydrogens from glycolysis and the citric acid cycle and deliver them to the electron transport chain, cellular respiration would stop.

4. If the victim had suffocated, the lack of oxygen would have caused electrons to pile up, and all the electron carriers would have held electrons. They would have been in the reduced state. Because the carriers were in the oxidized state—lacking electrons—Zyme knew that oxygen had been available in the cells to pick them up, so the victim did not suffocate.

5. At first the yeasts in the flask were able to produce sufficient ATP via aerobic cellular respiration. As the yeast population grew, the oxygen in the flask was used up, and they turned to fermentation to produce ATP. Because fermentation is less efficient than aerobic respiration, the yeasts had to consume glucose at a faster rate to maintain their supply of ATP. Ethanol is a waste product of fermentation.

Chapter 7

Review the Concepts

Exercise 1: pine tree, green bacterium, rosebush, seaweed, moss, alga, grass

Exercise 2:

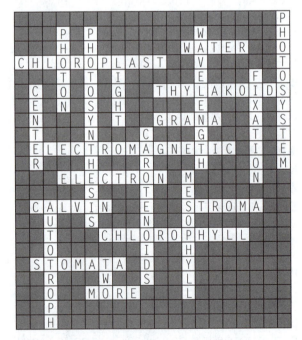

Exercise 3: $6 CO_2 + 12 H_2O \rightarrow C_6H_{12}O_6 + 6 H_2O + 6 O_2$. The C and O in $C_6H_{12}O_6$ come from CO_2. The H in $C_6H_{12}O_6$ comes from H_2O. The O in O_2 comes from H_2O. The H in H_2O comes from H_2O and the O comes from CO_2. H_2O is oxidized, and CO_2 is reduced.

Exercise 4: 1. light 2. H_2O 3. outer membrane of chloroplast 4. CO_2 5. light reactions 6. $NADP^+$ 7. Calvin cycle 8. ADP + P 9. stroma 10. granum 11. thylakoids 12. electrons 13. ATP 14. NADPH 15. O_2 16. sugar

Exercise 5: 1. B 2. A 3. H 4. C 5. B 6. A 7. D 8. H 9. G 10. F 11. C 12. E 13. I 14. D 15. A 16. A 17. H 18. G

Exercise 6: a. 5 b. 9 c. 2 d. 6 e. 3 f. 7 g. 8 h. 4 i. 1 Gamma waves have the most energy. Blue and red light are used in photosynthesis.

Exercise 7: 1. photosynthesis 2. chloroplasts 3. mesophyll 4. light 5. chlorophyll 6. thylakoids 7. Calvin 8. stroma 9. sugar (or glucose or G3P) 10. photons 11. blue 12. red 13. green 14. carotenoids 15. reaction center complex 16. electron 17. H^+ 18. stroma 19. thylakoids 20. gradient 21. ATP synthase 22. ATP 23. photophosphorylation 24. water 25. water 26. oxygen 27. oxygen 28. light 29. chlorophyll 30. acceptor 31. NADPH 32. Calvin 33. photosynthesis 34. stroma 35. carbon dioxide 36. ATP 37. NADPH 38. G3P 39. glucose 40. fixation 41. respiration 42. ATP

Exercise 8: 1. C_3 2. C_4 3. C_4 4. CAM 5. C_3 6. C_3 7. C_4 or CAM 8. C_4 or CAM 9. CAM 10. C_3

Exercise 9: 1. warm 2. cool 3. cool 4. cool 5. warm 6. cool 7. cool 8. warm 9. cool 10. cool 11. warm 12. warm

Exercise 10: For example, "CFCs, used as refrigerants and propellants, are changed by the sun into reactive free radicals, which destroy the ozone that protects life from UV radiation."

Test Your Knowledge

Multiple Choice: 1. b 2. a 3. a 4. c 5. e 6. b 7. b 8. d 9. b 10. e 11. b 12. a 13. d

Essay: 1. If a plant is given carbon dioxide containing radioactive ^{18}O, no labeled oxygen atoms appear in the O_2 produced by the plants, showing that the oxygen in O_2 does not come from CO_2. If the plants are given water containing ^{18}O, the O_2 gas they produce does contain labeled O, showing that the oxygen in O_2 comes from H_2O.

2. See Figure 7.5 in the text.

3. Photosynthesis allows producers to manufacture organic food molecules from inorganic reactants. This process produces virtually all the organic matter required as food by plants and all other living things on Earth.

4. Increased burning of fossil fuels (such as oil and gas) releases CO_2 into the atmosphere. Deforestation reduces global photosynthesis,

reducing removal of CO_2 from the atmosphere by plants. Both of these activities contribute to an increased greenhouse effect, because more CO_2 in the atmosphere could trap more heat. A temperature increase of just a few degrees could melt polar ice, flooding coastal regions. Storms could intensify and droughts become more common. We could reduce our contribution to the greenhouse effect by using fossil fuels more efficiently, recycling forest products, and replanting trees on deforested land.

Apply the Concepts

Multiple Choice: 1. a   2. d   3. e   4. c   5. a   6. c   7. c   8. a   9. c   10. e   11. d

Essay: 1. Plants would grow most slowly under green lights, because green light is not absorbed well by photosynthetic pigments. Green is reflected, which is why plants are green.

2. Carotenoids absorb all colors except yellow and orange. Because carotenoids absorb green light, they are able to capture some light energy not absorbed by chlorophyll and channel this energy to chlorophyll for use in photosynthesis.

3. Energy from the movement of excited electrons along a chain of electron carriers is used to pump H^+ through the membrane into the thylakoid space. This buildup of H^+ lowers the pH of the solution in the thylakoid space.

4. In photosynthesis, high-energy electrons come from chlorophyll whose electrons are excited by light. These electrons are replaced by splitting water. In cellular respiration, glucose is the source of high-energy electrons. In both processes, movement of the electrons through electron carriers is harnessed to move H^+ through a membrane; this energy is ultimately used by ATP synthase to make ATP. In photosynthesis, the high-energy electrons are also used to make glucose. In cellular respiration, the electrons combine with O_2 and H^+ to form water. Photophosphorylation takes place in the thylakoid membranes of the chloroplast; oxidative phosphorylation takes place along the inner membrane of the mitochondrion. In both processes, ATP synthase harnesses flow of H^+ through the membrane to make ATP.

5. In the light reactions of photosynthesis, sunlight excites chlorophyll in a leaf, causing it to give up excited electrons. High-energy electrons move through electron carriers, which generate an H^+ gradient, which in turn is harnessed to make ATP. The electrons and hydrogen are picked up by $NADP^+$, forming NADPH. In the Calvin cycle, energy from the ATP and hydrogen and electrons from NADPH are used to reduce CO_2, forming

sugar. When you consume the plant, the sugar is "burned" in cellular respiration. The glucose is oxidized, and its hydrogen and electrons are transferred to NAD^+ and FAD, forming NADH and $FADH_2$. These molecules deliver the high-energy electrons to the electron transport system in the mitochondrion. The electrons move through electron carriers, which generate an H^+ gradient, which in turn is harnessed by ATP synthase to make ATP. The ATP is then hydrolyzed to ADP and P to power muscle movement.

6. Van Helmont was correct in concluding that the soil did not contribute much material to the growth of the tree, but he was wrong in thinking that the weight gain must have come from the water. What van Helmont couldn't know (but we do) is that water supplies only the hydrogen atoms used in photosynthesis to build the carbohydrates that make up most of a plant. The bulk of the atoms in carbohydrate molecules—all the carbon and oxygen— come from CO_2 that the plant gets from the air!

Chapter 8

Review the Concepts

Exercise 1: 1. like   2. asexual   3. DNA   4. DNA   5. chromosomes   6. identical   7. mother   8. binary fission   9. circular   10. smaller   11. duplicates or replicates   12. plasma membrane   13. wall   14. genomes   15. egg   16. sperm   17. fertilization   18. variation

Exercise 2:

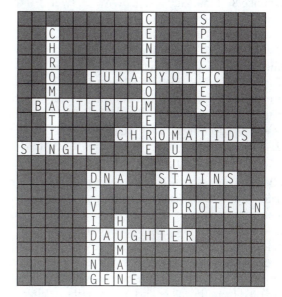

Exercise 3: 1. interphase, activity between divisions   2. G_1, cell growth following division   3. S, growth and DNA synthesis   4. G_2, growth and activity between DNA replication and division

5. mitotic phase, mitosis plus cytokinesis   6. mitosis, division of nucleus and chromosomes
7. cytokinesis, division of cytoplasm

Exercise 4: *Nucleus:* I—envelope intact, chromatin dispersed; P—envelope intact, chromosomes condense; PM—envelope breaks down; M—none; A—none; T—daughter nuclei and envelopes form. *Spindle:* I—none; P—spindle begins to form from microtubules; PM—spindle microtubules attach to chromosomes and move them toward center of cell; M—fully formed; A—microtubules guide chromosomes to poles; T—breaks down. *Chromosomes:* I—duplicated but dispersed as chromatin; P—chromatin coils to form shorter, thicker pairs of chromatids joined at centromeres; PM—spindle microtubules attach to kinetochores, move chromosomes to center of cell; M—line up on metaphase plate; A—sister chromatids separate move toward poles; T—chromosomes uncoil to form chromatin. *Cell size and shape:* I—rounded, doubles in volume; P—rounded; PM—rounded; M—rounded; A—spindle elongates cell; T—elongation continues; cytokinesis splits cell into two smaller rounded cells. *Sketches:* See Module 8.6 in the text.

Exercise 5: 1. interphase   2. telophase
3. metaphase   4. prophase   5. prometaphase
6. anaphase

Exercise 6: 1. F   2. A   3. C   4. B   5. D   6. E

Exercise 7: For example: In animals, microfilaments produce a cleavage furrow that pinches the cell apart, while in plants, vesicles align in a cell plate, where a cell wall grows to split the cell.

Exercise 8: 1. asexual   2. growth   3. liver   4. repairs   5. skin   6. digestive tract   7. replaced
8. numbers   9. types   10. mitosis   11. cultures
12. surface   13. anchorage   14. density-dependent inhibition   15. growth factors   16. control
17. stops   18. G_1 19. S   20. receptor   21. nerve or muscle   22. cancer   23. inhibiting   24. 20–50
25. tumor   26. benign   27. malignant   28. metastasis   29. Sarcomas   30. Leukemias   31. radiation
32. chemotherapy   33. cell division   34. spindle

Exercise 9: 1. Homologous pairs of chromosomes called autosomes are found in the somatic cells of both males and females. Sex chromosomes are an additional pair of chromosomes—homologous in females (who have two X chromosomes) but not homologous in males (who have X and Y chromosomes).   2. The two chromosomes of a homologous pair carry genes for the same inherited traits at the same place, or locus. One is inherited from the father and one from the mother.
3. The two chromatids of a single chromosome are exactly identical, having been produced via the process of DNA replication during the S phase of interphase.
4. A diploid cell contains two homologous sets of chromosomes. A haploid cell contains one chromosome

set. Diploid cells divide to form haploid cells in meiosis.   5. Gametes are haploid reproductive cells—sperm and eggs. Somatic cells are body cells other than gametes. In humans, somatic cells are diploid.   6. An egg is haploid. A zygote, formed by fusion of a sperm and egg, is diploid.   7. In meiosis, a diploid cell divides to form haploid cells. In fertilization, haploid cells (egg and sperm) fuse to form a diploid cell (zygote).   8. Mitosis produces two cells that are genetically identical to the mother cell. Meiosis is the division of a diploid cell to produce four haploid cells that are not genetically like the mother cell or each other.
9. These are the sex chromosomes. They are mostly not homologous. A female mammal has two X chromosomes; a male has an X and a Y.   10. A gene that determines a particular characteristic is located at a particular place (called its locus) on a chromosome.

Exercise 10: See Module 8.14 in the text for appearance of chromosomes, homologous chromosomes, sister chromatids, and crossing over.   1. interphase I   2. prophase I   3. metaphase I   4. anaphase I
5. telophase I and cytokinesis   6. prophase II
7. metaphase II   8. anaphase II   9. telophase II and cytokinesis   10. meiosis I   11. meiosis II

Exercise 11: 1. produces daughter cells identical to parent cell   2. involves two cell divisions   3. produces four daughter cells   4. sister chromatids of each chromosome separate   5. homologous pairs (tetrads) line up at metaphase plate   6. crossing over occurs between homologous chromosomes
7. provides for asexual reproduction, growth, replacement, repair

Exercise 12:

1.

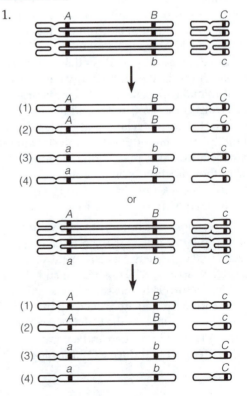

2.

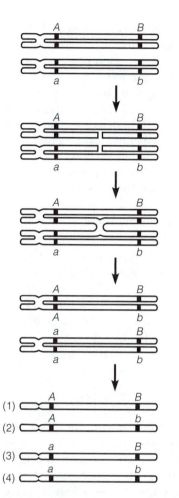

3.

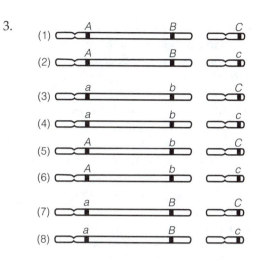

8 combinations are possible

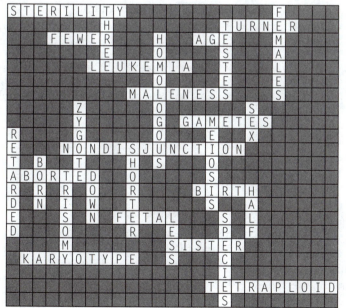

Exercise 13:

Exercise 14: 1. C, X   2. D, W   3. B, Z   4. A, Y

Test Your Knowledge

Multiple Choice: 1. d   2. a   3. d   4. b   5. a   6. c
7. d   8. b   9. a   10. c   11. c   12. b   13. e   14. d
15. b

Essay: 1. In asexual reproduction, the cells or buds that give rise to offspring are produced by mitosis. Because in this process genetic information is copied and passed on to identical daughter cells, the offspring produced by asexual reproduction are exactly like the single parent and each other. In sexual reproduction, each offspring inherits a unique combination of chromosomes and genes from two parents. Because of this genetic recombination, the offspring produced by sexual reproduction are not exactly like either parent or each other.

2. In mitosis, the nucleus and duplicated chromosomes divide and are equally distributed to daughter cells. In cytokinesis, the cytoplasm divides in two and is distributed to the daughter cells.

3. Cancer cells escape from the control mechanisms that regulate division in normal cells. Cancer cells do not need to be anchored to a solid surface to grow. They seem unaffected by density-dependent inhibition and will pile up on one another in cell culture. They either don't require growth factors or manufacture their own growth factors. Unlike normal cells, cancer cells can divide indefinitely. If they stop dividing, they stop at random points in the cell cycle, not at the restriction point.

4. Mitosis produces two cells that are genetically identical to the mother cell. Meiosis is the division of a diploid cell to produce four haploid cells that are not genetically like the mother cell or each other. In humans, mitosis is responsible for producing diploid somatic cells for growth, replacement of worn-out cells, or body repair. Meiosis produces haploid gametes (eggs and sperm) for sexual reproduction. Many (but not all) somatic cells undergo mitosis; special cells in ovaries and testes undergo meiosis.

5. Because sister chromatids are produced when DNA is duplicated, the genes of sister chromatids are identical. Homologous chromatids carry genes for the same traits (eye color, for example) at corresponding places, or loci, but they might carry different versions of the gene (blue or brown) at these loci.

6. The two chromosomes—maternal and paternal—of a homologous pair carry different genetic information. Because the two chromosomes of each pair line up independently of other pairs at metaphase I of meiosis, many different combinations of chromosomes are possible in the gametes produced by meiosis. Also in meiosis, homologous chromosomes exchange corresponding segments in the process of crossing over, creating chromosomes with a variety of gene combinations. Finally, the tremendous variety of eggs and sperm create many possible combinations at fertilization, further increasing genetic variation.

7. Sometimes a pair of homologous chromosomes fails to separate during meiosis I. One of the cells resulting from this division ends up with 24 chromosomes, the other with 22 chromosomes. All of the gametes produced by meiosis II have abnormal numbers of chromosomes. Some have 24 chromosomes—one of each kind plus an extra of one kind. Others have 22 chromosomes—one of each kind but one missing. Sometimes meiosis I is normal, but a pair of sister chromatids stays together during meiosis II. In this case, some gametes have the normal number of chromosomes, but some have one extra and some are one short. Fertilization of these abnormal gametes by a normal egg or sperm (containing the normal 23 chromosomes) produces a zygote with 47 or 45 chromosomes. Such genetic abnormalities have serious consequences.

8. An extra copy of chromosome 21 is usually not fatal, but it does produce Down syndrome. In most other situations in which an abnormal number of chromosomes occurs, the offspring is spontaneously aborted, so these abnormalities must be so serious as to be fatal.

Apply the Concepts

Multiple Choice: 1. d 2. b 3. c 4. c 5. a 6. b 7. a 8. d 9. d 10. d 11. d 12. b 13. b 14. d

Essay: 1. Cancer is uncontrolled cell division. Chemotherapy (and other cancer therapies such as radiation) slows tumor growth by interfering with cell division. Cell division is most rapid in the skin and digestive tract lining, where it replaces lost cells, so the side effects of chemotherapy are greatest on skin and intestinal lining.

2. Cells with between 100 and 200 units were in S phase, in the process of replicating their DNA.

3. These data suggest that prophase is the longest stage of mitosis. Prophase and prometaphase make up about half the process of mitosis. Telophase is also fairly long, lasting about a third of mitosis. Metaphase and anaphase take a relatively short time to occur.

4. There would be 39 different kinds of chromosome—39 homologous pairs. (In female mammals, the two X chromosomes, as well as each pair of autosomes, look alike.)

5. Possible alignments: *ABC* and *abc*, *ABc* and *abC*, *Abc* and *aBC*, and *AbC* and *aBc*. These same eight combinations are possible in eggs or sperm. There are 8 × 8 = 64 combinations possible in the zygotes.

6. Parental-type gametes: *QR* and *qr*. Recombinant-type gametes: *Qr* and *qR*.

7. XXY, XYYY, and XXYY would produce male phenotypes, XXXX and X female phenotypes. A single Y chromosome produces "maleness"; the absence of a Y chromosome produces "femaleness."

Chapter 9

Review the Concepts

Exercise 1: 1. S 2. G 3. C 4. Q 5. X 6. K
7. W 8. L 9. A 10. T 11. H 12. U 13. D
14. E 15. V 16. B 17. Y 18. M 19. N 20. F
21. I 22. Z 23. J 24. R 25. P 26. O

Exercise 2: 1. a. green, yellow b. *GG*, *gg* c. all *Gg*
2. a. 1 *GG*:2 *Gg*:1 *gg* b. 3 green:1 yellow 3. a. both
Bb b. white *bb*

Exercise 3: 1. *BBSS* 2. *bbss* 3. *BS* 4. *bs* 5. *BbSs*
6. *BS* 7. *Bs* 8. *bS* 9. *bs* 10. *BS* 11. *Bs* 12. *bS*
13. *bs* 14. *BBSS* 15. *BBSs* 16. *BBSs* 17. *BbSS*
18. *BBss* 19. *BbSS* 20. *BbSs* 21. *BbSs* 22. *BbSs*
23. *BbSs* 24. *Bbss* 25. *bbSS* 26. *Bbss* 27. *bbSs*
28. *bbSs* 29. *bbss* 30. 9/16 31. brown, short-
haired 32. 3/16 33. brown, long-haired
34. 3/16 35. white, short-haired 36. 1/16
37. white, long-haired

Exercise 4: 1. a. white b. brown 2. a. *bb* b. *Bb*
or *BB* 3. white 4. *Bb* 5. all brown, 0 white
6. heterozygous 7. homozygous

Exercise 5: 1. 0 2. 1 3. 1 4. 1/6 5. 5/6 6. independent 7. product 8. 1/6 9. 1/6 10. 1/36
11. multiplication 12. 1/2 13. 1/2 14. homozygous 15. multiplication 16. 1/2 17. 1/2
18. 1/4 19. addition 20. sum 21. 1/36 22. 1/36
23. 2/36 (1/18) 24. 1/2 25. 1/2 26. 1/2 27. 1/2
28. 1/4 29. 1/4 30. 1/4 31. 1/4 32. 1/2

Exercise 6: 1. none 2. *Ww* and *Ww* 3. 1/4

Exercise 7: 1. *ss* 2. ? 3. ? 4. ? 5. *Ss* 6. *Ss*
7. *Ss* 8. *ss* 9. ? 10. *ss* 11. *Ss* 12. ? 13. *Ss*
14. *SS* 15. ? 16. ? 17. half colored 18. half
colored 19. ? 20. half colored 21. fully colored
22. ? 23. ? 24. half colored 25. half colored
26. blank 27. fully colored 28. fully colored

Exercise 8: 1. T 2. F (one-fourth) 3. F (thousands)
4. F (normal parents who are heterozygous) 5. T
6. F (Caucasians) 7. F (not evenly distributed)
8. T 9. T 10. F (not necessarily more common)
11. T 12. F (less common) 13. T 14. F (There is debate among geneticists.) 15. T 16. F (cystic fibrosis)
17. T 18. T 19. F (There is now a test) 20. T

Exercise 9: 1. testing (screening) 2. carrier
3. fetal 4. AFP 5. Tay-Sachs 6. one-fourth
7. phenylketonuria 8. amino acid 9. diet
10. environment 11. amniocentesis 12. amniotic
13. cells 14. karyotype 15. Down 16. chorionic
villus 17. ultrasound 18. 35

Exercise 10: 1. B 2. A 3. D 4. B 5. C 6. C
7. E 8. D 9. E 10. A

Exercise 11: 1. *B* 2. *b* 3. *S* 4. *s* 5. *B* 6. *b* 7. *S*
8. *s* 9. *B* 10. *S* 11. *b* 12. *s* 13. *S* 14. *B* 15. *S*

16. *B* 17. *s* 18. *b* 19. *s* 20. *b* 21. *b* 22. *B*
23. *S* 24. *s* 25. *b* 26. *B* 27. *S* 28. *s* 29. *b*
30. *S* 31. *B* 32. *s* 33. *S* 34. *b* 35. *S* 36. *b*
37. *s* 38. *B* 39. *s* 40. *B*

Exercise 12: 1. B 2. C 3. E 4. A 5. D

Exercise 13: 1. E 2. D 3. D 4. A 5. C 6. B
7. A

Exercise 14: 1. X 2. dominant 3. X^CY 4. X^cY
5. daughters 6. sons 7. X 8. Y 9. carriers
10. X^CX^c 11. half 12. half 13. half 14. all
15. father 16. carriers 17. mother 18. X 19. *c*
20. *SRY* 21. X 22. ancestry

Test Your Knowledge

Multiple Choice: 1. c 2. e 3. d 4. b 5. e
6. b 7. b 8. a 9. d 10. e 11. c 12. a 13. b
14. b 15. e 16. a 17. c

Essay: 1. Mendel studied peas, which are easy to
grow and come in many readily distinguishable
varieties. He studied seven characteristics that each
occur in two distinct forms. He first made sure that
he had true-breeding varieties. Mendel carefully
controlled which plant mated with which and kept
scrupulous records of the matings and their results.
Finally, he analyzed and interpreted his results
mathematically, devising hypotheses that he tested
and retested experimentally.

2. Each flip is an independent event, with a probability of 1/2. The probability of getting two heads
is the product of the individual probabilities, 1/2 ×
1/2 = 1/4. The probability of coin 1 coming up
heads and coin 2 coming up tails is also the
product of the individual probabilities, 1/2 × 1/2
= 1/4. But there is another way that you can get
one head and one tail—coin 1 could come up tails
and coin 2 could come up heads. The combined
probability of this occurring is also 1/2 × 1/2
= 1/4. There are two different ways to get one
head and one tail. The probability of an event that
can occur two different ways is the sum of the separate probabilities of the different ways, in this case
1/4 + 1/4 = 1/2.

3. Peas and fruit flies are easy to raise in a small
space at little expense. They have many easy-
to-spot characteristics, and it is easy to control
matings between different varieties. They both
produce many offspring in a relatively short
time.

4. Humans who inherit X chromosomes from both
mother and father (XX) are female. Humans who
inherit an X chromosome from the mother and a Y
chromosome from the father (XY) are male. The
SRY gene on the Y chromosome appears to trigger
testis development. In the absence of this gene, an

individual develops ovaries and is female. In grasshoppers and crickets, an individual with one sex chromosome (XO) is male, and one with two sex chromosomes (XX) is female. Sperm cells with or without sex chromosomes determine the sex of the offspring. In certain fishes, butterflies, and birds, a ZZ individual is male, a ZW individual is female, and eggs determine the sex of offspring. In ants and bees, diploid individuals are female and haploid individuals are male.

5. Many genetic tests detect conditions (Huntington's disease) that are not yet treatable; does every afflicted individual need or want to know? Then what? Tests should always be accompanied by counseling; is this always the case? Are there enough counselors? What about "do-it-yourself" tests, from mail-order or online labs? A "bad" result might stigmatize an individual; might she lose her job or be denied health or life insurance? Will genetic testing only be available to those who can afford it?

Apply the Concepts

Multiple Choice: 1. c 2. d 3. c 4. b 5. a 6. d 7. c 8. c 9. c 10. c 11. d 12. b

Essay: 1. The parents are both heterozygous, *Aa*. The daughter is homozygous recessive, *aa*. The probability is 1/4.

2. 9/16 purple green, 3/16 purple yellow, 3/16 white green, 1/16 white yellow

3. The man is heterozygous, *Ff*. His wife is homozygous recessive, *ff*. The children with freckles are *Ff*; those without freckles are *ff*.

4. Pink × pink yields 1/4 red, 1/2 pink, 1/4 white. Pink × white yields 1/2 pink, 1/2 white.

5. The man is $I^A i$, the woman $I^B i$. The probability of a type B child is 1/4.

6. All four individuals have about the same skin color. Couple 2 could have children with the widest range of skin colors, ranging from *AABBCC* to *aabbcc*. (Couple 1 could have children ranging only from *aaBBCC* to *aabbCC*.)

7. Prediction: 1/4 red eyes and straight wings, *RrSs*; 1/4 red eyes and curled wings, *Rrss*; 1/4 pink eyes and straight wings, *rrSs*; and 1/4 pink eyes and curled wings, *rrss*. The actual results of the cross indicate that the genes for eye color and wing shape are linked. In the heterozygous parent, *R* and *S* are on one chromosome, and *r* and *s* on the homologous chromosome. Most gametes were *RS* and *rs*, but crossing over resulted in a small percentage of *Rs* and *rS* gametes being produced.

8. Genes *s* and *b* are farther apart than *h* and *b*. The order of the genes on the chromosome is *shb* (or *bhs*).

9. The genes for sex-linked disorders are on the X chromosome. If both parents are normal but have a child with a sex-linked disorder, the mother must carry the allele for the disorder on one of her X chromosomes, and she will pass it on to half of her children. Half of her daughters will carry the allele for coloboma iridis, but they will be of normal phenotype because they will inherit the normal allele from their father. Half of her sons will also inherit the coloboma iridis allele. Because they inherit Y chromosomes from their father (and no normal allele that could mask the coloboma iridis allele), half the sons will exhibit coloboma iridis. Thus, one-fourth of their children—all boys—will have coloboma iridis.

10. If Diane's brother has hemophilia (and presumably her father does not), her mother must be a carrier of the hemophilia allele, and she would pass this allele on to half her daughters. The probability that Diane is a carrier is 1/2. If Diane is a carrier, there is a probability of 1/4 that she and Craig will have a child—a son—with the disease. The probability both that Diane is a carrier and that Diane and Craig will have a child with hemophilia is $1/2 \times 1/4 = 1/8$.

Chapter 10

Review the Concepts

Exercise 1: 1. N 2. D 3. DR 4. SPB 5. B 6. I 7. D 8. SP 9. ACGT 10. A 11. K 12. L 13. S 14. D 15. W 16. C 17. E 18. B (or N) 19. L 20. F 21. J 22. O 23. H 24. ACGU 25. U 26. Q (or D or R) 27. V

Exercise 2: 1. sugar-phosphate backbone 2. phosphate group 3. sugar (deoxyribose) 4. double helix 5. complementary base pair 6. adenine (A) 7. guanine (G) 8. thymine (T) 9. hydrogen bond 10. cytosine (C) 11. nucleotide 12. polynucleotide 13. pyrimidine bases 14. purine bases

Exercise 3: See Figure 10.4A in the text.

Exercise 4: 1. D 2. D 3. A 4. F 5. E 6. B 7. C

Exercise 5:

Exercise 6: 1. nucleus   2. RNA polymerase
3. promoter   4. nucleotides   5. terminator
6. messenger RNA   7. processed   8. introns
9. exons   10. cytoplasm   11. amino acid
12. transfer RNA   13. amino acid   14. codon
15. enzyme   16. anticodon   17. Ribosomes
18. ribosomal RNA   19. polypeptide   20. initiator
21. start codon   22. polypeptide   23. amino acid
24. amino acids   25. stop codon   26. ribosome
27. ribosome   28. protein

Exercise 7: 1. transcription   2. translation
3. DNA   4. mRNA   5. RNA polymerase
6. amino acid   7. tRNA   8. anticodon   9. large
ribosomal subunit   10. initiator tRNA   11. initiation   12. small ribosomal subunit   13. mRNA
14. start codon   15. polypeptide   16. peptide bond
17. elongation   18. codons   19. termination
20. polypeptide   21. stop codon

Exercise 8: 1. Met-Pro-Asp-Asn-Ile-Lys   2. Met-Pro-Asp-His-Ile-Lys   3. one base changed   4. substitution (tenth base changed from A to C)   5. one amino acid changed (Asn to His)   6. Met-Pro-Asp-Glu-Tyr   7. one base changed   8. Insertion (tenth base, G, inserted between C and A)   9. two amino acids changed and chain shortened by one amino acid, due to change to stop codon   10. effect of mutation in mRNA 3 greatest, due to reading frame shift and alteration of all codons following insertion. (Deletion of a base would have a similar effect.)

Exercise 9: 1. E   2. G   3. B   4. O   5. B   6. R
7. U   8. A   9. S   10. D   11. V   12. U   13. T
14. I   15. F   16. L   17. J   18. Z   19. Y   20. P
21. N   22. L   23. C   24. Q   25. X   26. K   27. W
28. M   29. H   30. U

Exercise 10: 1. B   2. D   3. B   4. D   5. C   6. A
7. B   8. A   9. A   10. D   11. C   12. A   13. A   14. A

Test Your Knowledge

Multiple Choice: 1. a   2. c   3. c   4. b   5. c   6. e
7. a   8. b   9. a   10. d   11. c   12. c   13. b   14. e
15. d   16. a

Essay: 1. See Figure 10.3D in the text.

2. A single-ringed pyrimidine, such as T, must pair with a double-ringed purine, such as A. Two pyrimidines—A and C—would not be large enough to reach across the double helix. T specifically pairs with A and not G because T and A have complementary chemical side groups that form hydrogen bonds.

3. Twenty amino acids are used in building proteins. There are only four nucleotides in DNA, so a one-base code could specify only 4 amino acids. A two-base code could specify only 4^2, or 16 amino acids. In a triplet code, 4^3, or 64, combinations of bases are possible, more than enough to code for 20 amino acids.

4. A mutation is a change in the nucleotide sequence of DNA. Some mutations are spontaneous, but many are caused by X-rays, ultra-violet light, and chemicals. Most mutations are harmful because they alter protein amino acid sequences and impair the function of proteins. Some mutations do not alter amino acid sequence; others lead to an improved protein or one with new capabilities that enhance an organism's success. Mutations create genetic diversity that makes evolution by natural selection possible.

5. A base substitution can result in no change to a protein or, most often, a change in one amino acid. Effects on the organism may be minimal. A base deletion shifts the sequence of triplet groupings in a gene, causing a drastic change in the amino acid sequence downstream from the deletion. This can have a profound effect on the protein and the organism.

6. Messenger RNA (mRNA) is transcribed from a gene and carries the instructions for making a particular polypeptide. A ribosome is the site of translation, the process of making a polypeptide according to the mRNA message. Transfer RNAs act as translators, each kind matching a particular amino acid with a particular codon in the mRNA. The ribosome "reads" the mRNA one codon at a time, and tRNAs deliver their amino acids, which are added to the polypeptide chain one at a time.

Apply the Concepts

Multiple Choice: 1. a   2. a   3. c   4. e   5. c   6. d
7. b   8. a   9. a   10. c   11. c   12. d

Essay: 1. Extract DNA from dead type A *E. coli* and mix it with live type V *E. coli*. If the type V *E. coli* absorb the DNA and genes are made of DNA, the type V bacteria will be able to grow on the simple medium without the vitamin. For comparison, extract protein from dead type A *E. coli* and mix it with live type V *E. coli*. If the type V *E. coli* absorb the protein and genes are made of protein, the type V bacteria will be able to grow on the simple medium without the vitamin. Prediction: DNA will transform the type V bacteria, but protein will not.

2. In the DNA double helix, T and A bases pair up; wherever there is a T nucleotide in one strand, there is a complementary A nucleotide in the other. Thus the amount of T is equal to the amount of A. The specific nucleotide sequences of goldfish and human DNA are different, so the number of A-T pairs is different in the two species.

3. They are both right. Eric is talking about phenotype and Renee about genotype. The nucleotide sequences of your genes—genotype—determine the amino acid sequences of proteins that cause your hair to be curly or straight—your phenotype.

4. Met-Glu-Leu-Ser-Ile-Asp. First you have to transcribe DNA base sequence into mRNA base sequence, then translate into amino acid sequence. Remember that translation always starts at the AUG start codon.

5. Because a three-base codon specifies each amino acid, the message coding for the polypeptide would consist of 233×3, or 699 nucleotides (actually 702, if the stop codon is included). The rest of the nucleotides are extras. Translation of protein does not have to start and stop at the ends of the mRNA. It starts at a start codon (AUG) and stops at a stop codon (UAA, UAG, or UGA). The nucleotides from the start codon to the stop codon are the only ones that spell out the amino acid sequence of the polypeptide.

6. The virus may be able to insert its genes into a nerve cell's DNA and remain latent within the cell as a provirus. From time to time the provirus may begin reproducing complete viruses, causing disease symptoms.

7. In a eukaryotic cell, the RNA transcript is processed before translation. Noncoding introns are removed and exons are spliced together to produce the mRNA. A prokaryote does not process its RNA before translation, so the introns from the eukaryotic gene are not removed. When this unedited RNA is translated, the wrong polypeptide is produced.

8. Some bacteria are combining their DNA with the DNA from the other strain, producing bacteria with the characteristics of both strains. It could be that some bacteria are dying, and their DNA is being taken up from the medium by living bacteria—a process called transformation. Bacterial genes could be transferred from bacterium to bacterium by a bacteriophage—transduction. Or the bacteria could be undergoing conjugation—bacterial mating—in which DNA is transferred from one bacterium to another.

9. The hybrid virus would only be able to attack lung cells because its coat contains structures (like the glycoprotein spikes of the mumps virus or HIV) that "dock" with lung cells, not nerve cells. Once a cell is infected, the polio virus RNA replicates and directs the manufacture of polio virus proteins. Therefore the viruses that fill the infected cell would contain only polio virus RNA and protein, because only polio RNA instructions got in.

Chapter 11

Review the Concepts

Exercise 1: *lac* operon: 1. E   2. G   3. A   4. B   5. F
6. C   7. H   8. D;   *trp* operon: 1. E   2. A   3. G
4. B   5. F   6. C   7. H   8. D

Exercise 2: *Stomach gland cell:* citric acid cycle enzyme gene, digestive enzyme gene. *Hair follicle cell:* keratin gene, citric acid cycle enzyme gene. *Stem cell:* hemoglobin gene, citric acid cycle enzyme gene.

Exercise 3:

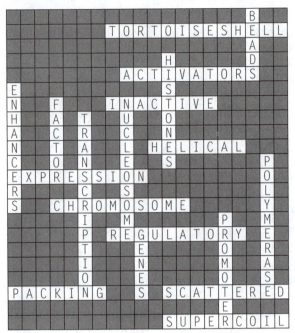

Exercise 4: 1. E   2. F   3. D   4. H   5. B   6. C
7. A   8. G

Exercise 5: 1. DNA unpacking and changes
2. transcription   3. addition of cap and tail
4. splicing   5. flow through nuclear envelope
6. mRNA breakdown   7. translation   8. cleavage/modification/activation   9. protein breakdown

Exercise 6: 1. head   2. tail   3. genes   4. receptor
5. transcription   6. translated   7. activates
8. mRNA   9. head   10. tail   11. mitoses (cell divisions)   12. proteins   13. segments   14. genes
15. homeotic   16. antennae   17. wings
18. mutant   19. microarray   20. single
21. reverse   22. cDNAs   23. complementary
24. transcribed   25. leukemia

Exercise 7: 1. H, R   2. D,W   3. B,S   4. E,V
5. C,Q   6. A,U   7. F,P   8. G,T

Exercise 8: 1. F   2. A   3. J   4. D   5. G   6. C   7. E
8. I   9. H   10. B

Exercise 9: 1. control   2. division   3. mutations
4. stimulate   5. inhibit   6. carcinogens   7. mutagens   8. oncogenes   9. proto-oncogenes
10. growth   11. division   12. mutation   13. excess
14. tumor-suppressor   15. repair   16. one
17. several   18. age   19. signal   20. protein
21. growth   22. cell division   23. absence
24. transcription   25. increased   26. exams
27. smoking   28. sun   29. high   30. fat

Exercise 10: 1. C   2. F   3. D   4. B   5. C   6. C
7. G   8. A   9. E

Test Your Knowledge

Multiple Choice: 1. d   2. a   3. c   4. a   5. e   6. e
7. d   8. a   9. b   10. e   11. e   12. b   13. e   14. e
15. c

Essay: 1. Differentiated fern cells can be made to dedifferentiate and regrow the entire organism. Differentiation apparently does not irreversibly change the DNA.

2. Most genes in plants and animals consist of regions that code for polypeptides, called exons, interrupted by long noncoding segments, called introns. Both introns and exons are transcribed from DNA into RNA, then the introns are removed and the remaining exons linked together—a process called RNA splicing. Introns may contain nucleotide sequences that regulate gene activity. The splicing process may help control flow of mRNA from nucleus to cytoplasm. In some cases splicing can occur in more than one way, producing different mRNA molecules from the same transcript.

3. A homeotic gene is a master control gene that functions during development. It regulates a battery of other genes, switching them on or off and thus shaping large-scale aspects of body plan such as development of appendages.

4. A gene in the egg cell codes for a protein that signals surrounding follicle cells. The protein binds to the membrane of a follicle cell and through a series of relay proteins activates transcription factors in the target cell. This triggers transcription and translation of specific genes into proteins. These follicle cell proteins in turn act on the egg cell, causing it to localize a kind of mRNA at the end of the egg that will later become the head. After fertilization, the "head" mRNA is translated into a regulatory protein that acts on other genes, which trigger the pattern of gene expression that divides the embryo into segments from head to tail.

5. Cancer cells escape from normal controls that regulate division and growth, and multiply excessively. This is usually triggered by several mutations. A proto-oncogene is a gene that normally makes a protein that helps trigger cell division. A mutation can change this gene into an oncogene that codes for a hyperactive protein or an excess of protein, stimulating cell division more than normal. A second kind of mutation occurs in a tumor-suppressor gene. This mutation keeps a protein that normally blocks cell division from being made. The combined effect of these mutations is the uncontrolled cell division characteristic of cancer.

6. Animal cloning is achieved by replacing the nucleus of an egg or zygote with the nucleus of a somatic cell. The cell formed then divides repeatedly, forming a ball of cells called a blastocyst. In the right environment (i.e., the uterus of a surrogate mother), this blastocyst may grow into a new individual, genetically identical to the nuclear donor. This is called reproductive cloning. Alternatively, embryonic stem cells can be harvested from the blastula. These cells can be grown indefinitely in the laboratory, and they have the potential for differentiating into virtually any kind of somatic cell. Such embryonic stem cells may be used to repair or replace injured or diseased organs—a procedure called therapeutic cloning.

7. Embryonic stem cells are capable of developing into a wider variety of cell types than adult stem cells, but there are ethical concerns about obtaining cells from embryos. Adult stem cells are easier to obtain, but their developmental potential is limited to fewer cell types.

8. To avoid cancer a person should avoid carcinogens—UV radiation, unnecessary X rays, tobacco, alcohol, and so on. A high-fiber, low-fat diet, with foods high in vitamin C, vitamin E, and substances related to vitamin A, also reduces the risk of cancer.

Apply the Concepts

Multiple Choice: 1. c   2. b   3. c   4. a   5. e   6. c
7. a   8. c   9. c   10. b

Essay: 1. Normally, when lactose is absent, the repressor binds to the operator site, blocking gene transcription, and no enzymes for using lactose are made. When lactose is present, it binds to the repressor and changes the shape of the repressor in such a way that it can no longer bind to the operator. Genes are then transcribed and enzymes for using lactose are made. If the mutation altered the repressor in such a way that it could no longer bind to the operator, the genes would be transcribed and enzymes for using lactose would be made all the time, whether or not lactose was present.

2. Many answers are possible, but the following is an example: Liver cells are small and metabolically active, making and breaking down many substances. Muscle cells are long and thin and have the ability to contract or shorten. Salivary gland cells form saclike clusters and secrete saliva. Liver cells make fibrinogen, a blood-clotting protein. Genes for making fibrinogen would be active in

liver cells but not in the other cells. The genes for building contractile proteins would be active in muscle cells, and the gene for making amylase, a digestive enzyme, would be active in salivary gland cells. All the cells perform glycolysis, the first process in breaking down sugar, so genes that code for glycolysis enzymes would be active in all the cells. The gene that codes for the blood protein hemoglobin would not be active in any of the cells.

3. The minimum number of nucleotides needed to code for the protein is $258 \times 3 = 774$ (or 777, if you wish to include the 3 nucleotides of the stop codon). There are numerous nucleotides upstream from the AUG start codon and downstream from the stop codon. Only the 777 nucleotides from start to stop actually code for the protein. The mRNA is much shorter than the gene because noncoding introns in the gene are cut out, and the remaining exons are joined together to form the final mRNA in the RNA splicing process that occurs in the nucleus.

4. A gene might code for a repressor or activator protein that binds to the DNA of other chromosomes and turns off or enhances transcription of genes at those sites. A gene may code for production of a signal molecule that travels through the bloodstream to a distant site, attaches to a receptor and activates a signal transduction pathway that similarly affects gene transcription.

5. Different carcinogens can cause the same kind of cancer because they all might cause mutations in bone marrow cells that change proto-oncogenes into oncogenes and/or inactivating tumor-suppressor genes. The oncogenes and damaged tumor-suppressor genes, no matter what changes them, have the same effect—to stimulate the uncontrolled cell division characteristic of this kind of cancer.

6. Cloning a woolly mammoth is not as far-fetched as it sounds (and a lot easier than cloning a dinosaur!). You would have to replace the nucleus of an elephant egg or zygote with a nucleus obtained from a frozen mammoth somatic cell, for example a skin cell. The resulting "zygote" would then be injected into the uterus of a surrogate mother elephant. It is possible that the two species are closely enough related that the mammoth nucleus and the elephant egg cytoplasm would be compatible, and if the mammoth nucleus was undamaged, the zygote might be able to develop into a baby mammoth. Wow!

Chapter 12

Review the Concepts

Exercise 1: 1. F 2. H 3. B 4. L 5. P 6. G 7. J 8. D 9. Q 10. A 11. O 12. N 13. C 14. K 15. E 16. I 17. M

Exercise 2: 1. vectors 2. gene 3. restriction

4. sequences 5. restriction 6. sticky 7. plasmids 8. complementary 9. ligase 10. replication 11. recombinant 12. artificial 13. viruses 14. transformation 15. cloning 16. library 17. introns 18. reverse 19. exons 20. retrovirus 21. bacteria 22. genomic 23. expressed 24. protein 25. probe 26. UUCACAUC 27. radioactive 28. secrete 29. hormones 30. cancer 31. vaccines 32. viruses 33. Yeast 34. integrate 35. sugars 36. milk

Exercise 3: 1. isolating a plasmid from *E. coli* 2. obtaining copies of a gene and protein from cloned bacteria 3. cloning recombinant DNA 4. using a nucleic acid probe to find a nucleic acid sequence 5. cutting DNA with a restriction enzyme 6. extracting DNA from a eukaryotic cell 7. joining a plasmid and a DNA fragment using DNA ligase 8. inserting a plasmid into a bacterium via transformation 9. using reverse transcriptase to make cDNA 10. mixing plasmids and DNA fragments with sticky ends

Exercise 4: 1. E 2. A or G 3. A or G 4. F 5. D 6. C 7. B

Exercise 5:

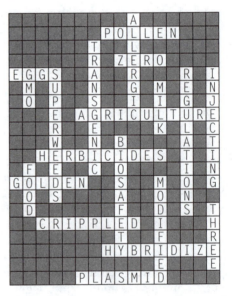

Exercise 6: 1. Technical issues: Using the right virus vector to get the corrected gene into the cell, inserting the gene into the correct part of the genome, building control mechanisms so the gene will produce the proper amount of gene product at the right time in the right places, and inserting genes so they do not harm other life functions 2. Ethical issues: Who will have access to such an expensive and difficult procedure? Should gene therapy be reserved for treating serious diseases, or could we use it to "enhance" a healthy person? How much risk (the possibility of causing cancer, etc.) is acceptable? Should we "tamper with evolution" by eliminating genetic defects in eggs, sperm,

and zygotes, thus altering the genetic diversity of future generations?

Exercise 7: 1. DNA 2. Heat 3. Cool 4. DNA polymerase 5. Heat

Exercise 8: 1. Joe should be held and Sam released. Joe's profile matches the DNA from the crime scene. 2. The smallest DNA fragments move the fastest, winning the "race" to the bottom of the gel. 3. Diagram A depicts Sam's DNA, and diagram B Joe's DNA. The length of the longest STR segments are the same for Sam and Joe; this DNA was slowest and is represented by the line nearest to the top of the electrophoresis gel. But Joe had the smallest STP sequence, closest to the bottom of the gel, and the second largest, closer to the top.

Exercise 9: 1. Indicate on diagram, within each CCGG sequence 2. Father—two places; Jake—one place; Jennifer—two places 3. Father—three fragments; Jake—two fragments; Jennifer—three fragments 4. Father—large, medium, and small fragments; Jake—two fairly large fragments; Jennifer—large, medium, and small fragments 5. Father and Jennifer: A large fragment near the top, a medium fragment in the middle, and a small fragment at the bottom; Jake: Two fairly large fragments near the top. Like this:

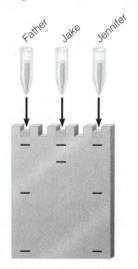

6. Jennifer carries the Huntington's allele, like her father. Jake does not carry the allele. (Looking at the DNA base sequences, your first impulse might have been to think that Jake has Huntington's. The trick here is that the father's DNA and Jennifer's match *exactly*, but Jennifer's DNA sequence is shown *upside down* relative to her father's. There is no upside down to real DNA or the enzyme, which cuts both into three pieces, showing that Jennifer has inherited the Huntington's allele from her father.

Exercise 10: 1. Q 2. H 3. T 4. R 5. G 6. L 7. X 8. K 9. J 10. U 11. N 12. S 13. V 14. Z 15. P 16. W 17. B 18. O 19. D 20. Y 21. F 22. A 23. E 24. C 25. M 26. I

Test Your Knowledge

Multiple Choice: 1. b 2. e 3. d 4. a 5. c 6. a 7. d 8. e 9. c 10. d 11. a

Essay: 1. A restriction enzyme is used to cut DNA at a particular nucleotide sequence, called a recognition sequence. It cuts the two strands of DNA unevenly, producing DNA fragments with single-stranded ends, called "sticky ends." The sticky ends produced by a restriction enzyme are complementary, and complementary fragments produced by the enzyme can hydrogen bond to one another by their single-stranded ends. An enzyme called DNA ligase can join these loosely linked pieces of DNA by catalyzing the formation of covalent bonds between adjacent nucleotides. A fragment of DNA can be spliced into a plasmid, a small, circular bacterial DNA molecule. A bacterium can take up the plasmid from solution by transformation. When the bacterium replicates its larger chromosome, it also replicates the plasmid. It divides repeatedly, producing a clone of cells containing the recombinant plasmid.

2. Bacteria and yeasts have plasmids that can be used as gene vectors, and they can be rapidly and cheaply grown in large quantities. They will secrete protein products into their growth medium, where they can be collected and then purified.

3. Recombinant DNA technology can be used to produce genetically modified (GM) crop plants that resist herbicides, insects, and disease and can grow in poor soil. DNA technology also produces drugs, vaccines, and hormones and has been used to produce transgenic animals.

4. Several human proteins, such as insulin and growth hormone, are produced by recombinant bacteria and yeasts and used to treat diseases. Genetically engineered viruses and bacteria are also used in the production of vaccines and drugs used against various diseases. Diagnostic tests look for genetic diseases, cancer genes, or DNA of infectious organisms. Recombinant DNA technology is beginning to be used in gene therapy—alteration of genes in human cells to remedy genetic diseases.

5. Genetic engineering could create hazardous pathogens that could infect humans, animals, or plants. New organisms could also transfer genes to other species, harming them or making them harmful. Recombinant organisms could also compete with and displace wild species, with unpredictable consequences. Plants with herbicide genes might spread out of control, or those with genes that kill insects kill useful or harmless species. Humans could be allergic to proteins in recombinant foods. Strict laboratory procedures prevent accidental infection and escape of genetically engineered organisms. Usually, microorganisms are genetically crippled,

so they cannot survive outside the laboratory. Some dangerous experiments are banned. Genetically modified crop plants and animals are tested under controlled conditions before being released.

6. The potential dangers of recombinant DNA technology present some ethical questions: Are possible benefits worth the risks? How do we weigh the possible benefits against the risks? How do we feel about altering species? About adding new species to an already precarious environment? Do we have a right to forgo recombinant DNA technology if it can solve environmental and medical problems? Should we alter our own genes? What uses will be made of genetic information? One of the biggest ethical questions concerning human gene therapy is whether we should alter evolution by eliminating genetic "defects" in our descendants.

7. Restriction enzymes cut DNA into fragments at specific sequences called restriction sites. The cut-up DNA is placed on a gel and pulled through the gel by the attraction of the negatively charged phosphate groups on the DNA to a positive electrode. This method is called electrophoresis. Smaller fragments move through the gel faster, and this separates fragments of different sizes. Fragments of different sizes accumulate in bands in the gel, and the resulting DNA fingerprint looks a bit like a supermarket bar code. Each individual has a unique DNA fingerprint, because the nucleotide sequence of each person's DNA is unique. This means that restriction sites differ among different individuals, and restriction enzymes cut the DNA into different-sized fragments, which produce a different restriction fragment band pattern.

8. Cross-species comparisons can help us deduce evolutionary relationships. For example, comparisons of genomes confirms that living things fall into three large domains, and that chimps are our closest relatives. Disease genes in humans, such as a gene for Parkinson's disease, occur in other species, such as rats and finches. These animals might offer clues to how the gene causes disease in humans. Comparing the genomes of humans and other species also reveals those genes that seem to be evolving faster in humans than in other animals, such as the *FOXP2* gene that functions in speech.

Apply the Concepts

Multiple Choice: 1. b 2. b 3. c 4. c 5. b 6. a 7. c 8. e 9. d 10. a 11. e 12. a

Essay: 1. Knowing the amino acid sequence of EGF, you could use the genetic code chart to figure out the corresponding DNA nucleotide sequence and then program the machine to synthesize a gene coding for the protein. Use a restriction enzyme to cut the gene and plasmid, producing complementary sticky ends. Combine the gene and plasmid, and use ligase to bond them. Introduce the plasmid into *E. coli* via transformation. The bacteria will replicate their DNA and divide in the growth medium, producing a clone of bacteria with the EGF gene. When you have grown a sufficient quantity of the bacteria, extract the EGF.

2. STRs are short nucleotide sequences (such as AGAT) repeated many times in a row in the DNA. There are 13 STR sites used in forensic analysis of DNA samples where the number of repeats varies between different individuals. To create a profile, a technician uses PCR to amplify the regions of DNA that include the STR sites. This can be done by using primers known to flank the sites. The resulting fragments are compared using gel electrophoresis. Different locations of the bands in the gel show the different sizes of the DNA fragments from the STR sites. The sizes can vary quite a bit—in one person AGAT might be repeated 10 times and in another it might be repeated 32 times. There are as many as 80 known variations for some STR sites, so with 13 sites and all that variation, each STR profile is unique.

3. She could use a nucleic acid probe—an RNA or single-stranded DNA with the complementary base sequence. Such a probe might be an RNA fragment with the sequence UACCGAUAG, for example. This probe contains nucleotides labeled with radioactive isotope or dye. She could mix it with each of the clones. It bonds to the desired complementary base sequence, and the radioactivity or color tags the clone (or the DNA fragments in a gel) with the sequence she is trying to find.

4. This bacterium could grow only under laboratory conditions. If it accidentally escaped from the lab, it would not be able to survive and thus would not cause harm in the outside environment.

5. The remains are probably those of Manuel, because the DNA fingerprint of the remains contains bands matching some of those in the father's DNA fingerprint and some of those in the mother's DNA fingerprint, but no bands that are found in neither of the parents' DNA fingerprints.

6. The lack of blood enzyme is a better candidate for gene therapy because bone marrow cells can be removed from the patient, their genes can be altered, and the cells can be reintroduced to the patient. The reintroduced cells can continue to proliferate through life, curing the disease. At this time it is not possible to genetically engineer specialized cells such as nerve cells that do not continue to proliferate, because it is not possible to get the corrected gene into the cells and have them function properly.

Chapter 13

Review the Concepts

Exercise 1: 1. E 2. D 3. G 4. F 5. B 6. H 7. A 8. C

Exercise 2: 1. Charles Darwin 2. evolution
3. vary 4. heritable 5. environment 6. survive
7. reproduce 8. beneficial 9. natural selection
10. desirable 11. beaks 12. pesticide 13. population 14.alleles 15. environment 16. goal
17. population 18. adapt

Exercise 3:

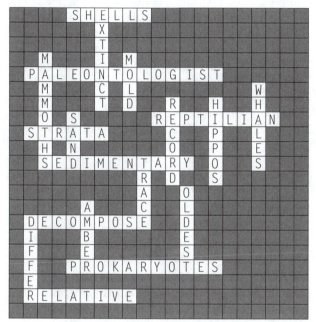

Exercise 4: 1. comparative anatomy 2. molecular biology 3. molecular biology 4. fossil record
5. comparative anatomy 6. natural selection
7. artificial selection 8. biogeography 9. molecular biology 10. biogeography 11. comparative anatomy 12. comparative anatomy 13. fossil record 14. molecular biology 15. comparative anatomy 16. artificial selection

Exercise 5: The following illustrate homologies: 1, 2, 3, 5, 9, 11, 12, 14, 15

Exercise 6: 1. population 2. individual 3. population 4. gene pool 5. microevolution 6. mutation, sexual reproduction 7. genetic 8. mutation
9. body (somatic) cells 10. mutations in gametes
11. harmful 12. duplication mutation
13. prokaryotes 14. 1/100,000 15. sexual reproduction 16. crossing over, independent assortment, random fertilization

Exercise 7: B. *RR, Rr, rr* C. 36, 48, 16 D. 0.36, 0.48, 0.16 E. 72, 48, 0 F. 0, 48, 32 G. $p = 0.6, q = 0.4$ H. 0.6, 0.4 J. *RR, Rr, rr* K. 0.36, 0.48, 0.16 L. 72, 48, 0 M. 0, 48, 32 N. $p = 0.6, q = 0.4$

O. The genotype and allele frequencies stayed the same in the second generation and will continue to do so in succeeding generations. This illustrates that the frequency of each allele in the gene pool will remain constant unless acted on by other agents. Sexual reproduction alone does not lead to microevolution—the so-called Hardy-Weinberg principle.   P. Whether an allele is dominant or recessive has nothing to do with its probability of being inherited, so the frequencies of the *R* and *r* alleles remain constant from generation to generation.

Exercise 8: 1. No gene flow. The population is not isolated; there is gene flow—migration of individuals with yellow alleles into the population.   2. No mutation. A mutation alters the frequency of red and yellow alleles in the gene pool.   3. Very large population. Because the flower population is small, a chance event such as a landslide can significantly alter the gene pool. This is called genetic drift.
4. No natural selection. Individuals differ in reproductive success, changing the relative frequencies of red and yellow alleles in the gene pool. This is natural selection, and it is the only situation in this exercise that results in adaptation to the environment.
5. Random mating. Mating is not random because a yellow plant does not stand an equal chance of mating with every other plant in the population. It is more likely to mate with a yellow plant.

Exercise 9: 1. struggle 2. fittest 3. fitness
4. phenotype 5. phenotype 6. genotype 7. gene pool 8. population 9. allele 10. diploid
11. variation 12. environment 13. balancing
14. heterozygote 15. frequency 16. neutral
17. dimorphism 18. secondary 19. intrasexual
20. intersexual 21. mate 22. genes 23. bell
24. stabilizing 25. directional 26. change
27. diversifying 28. intermediate 29. Natural selection 30. compromises

Test Your Knowledge

Multiple Choice: 1. e 2. b 3. d 4. d 5. c 6. a 7. c 8. e 9. b 10. b 11. e 12. d 13. b

Essay: 1. Because organisms produce more offspring than the resources in their environment can support, only a fraction of these offspring survive and reproduce themselves. Individuals in a population vary in their characteristics, and many of these varying traits are inherited. Those individuals with traits that fit them best to their environment are most likely to survive and reproduce, and thus these individuals leave more offspring than less fit individuals. Over time, the fitter types make up a larger fraction of the population, and through this process of natural selection the population adapts to its environment.

2. There are many well-researched examples of adaptation through natural selection—Galapagos finches, pesticide resistance in insects, antibiotic resistance in bacteria. Biogeography, the geographical distribution of species, suggests that species change when they move into new habitats. Animals on islands resemble those on the nearest land mass more than those in more distant localities. The fossil record shows sequential changes in living things over time. Comparative anatomy shows the relationships among organisms descended from a common ancestor. Molecular biology shows that organisms thought to be more closely related on the basis of other criteria also have a greater proportion of their DNA and proteins in common.

3. A population is a group of individuals of a particular species living in a particular place at a particular time. A species is all the populations whose individuals can potentially interbreed. The white-tailed deer on an island constitute a population; all white-tailed deer are members of a particular species.

4. Homologous structures—anatomical or molecular—are features that often have different functions but are structurally similar because of common ancestry. The forelimbs of humans, cats, bats, and whales are similar in structure but differ in shape and function and show descent of these animals from a common land mammal ancestor. The genes of these animals are more similar to each other than any of their genes are to the genes of fish, also showing common ancestry. Homologous structures enable us to reconstruct the branching sequence of the evolutionary "tree," which species have common ancestors and are on the same branch and how the branches connect.

5. Darwin would have said that both horses and zebras are descended from a common ancestral species that lived in the past. As the descendants of this species spread into different environments, they adapted in different ways to different conditions. Horses and zebras eventually became distinct species, but their homologous structures show that they are related—descended from the same ancestor.

6. In a small population, a chance event might alter the frequency of alleles in the gene pool so as to increase the frequency of an allele that would otherwise be disadvantageous. In a diploid population, a recessive allele affects the phenotype only when two copies of it are present in a homozygous individual. Many harmful recessive alleles in a population might be "hidden" in heterozygous individuals, not exposed to the effects of natural selection. Sometimes natural selection actually favors heterozygotes—a phenomenon known as heterozygote advantage. In this case, heterozygotes might have greater reproductive success than either homozygote, further increasing the frequency of an allele that is harmful in the homozygous recessive state.

7. All these phenomena alter allele and genotype frequencies in the gene pool, but in different ways. In a small population, a chance event could kill some individuals and this random "sampling error" could significantly alter the gene pool. If individuals prefer some types of potential mates to others—that is, if mating is nonrandom—certain alleles are more likely to be passed on than others, and the gene pool can change. Migration of individuals in or out of the population can add new alleles or remove alleles from the gene pool. Mutation creates a new allele, altering allele frequencies. If a certain allele enables individuals with that allele to better survive and reproduce, that allele is more likely to be passed on to the next generation than others, and its frequency in the gene pool will increase. This is natural selection, and it is the only circumstance that leads to adaptation.

Apply the Concepts

Multiple Choice: 1. e 2. a 3. a 4. c 5. b 6. d 7. c 8. b 9. e 10. d 11. d 12. b

Essay: 1. Lamarck would have said that the ancestral butterfly had a shorter proboscis. It stretched the proboscis to reach the nectar deep in the flower, and this change was passed on to its offspring. In this way, the proboscis got longer and longer. Darwin would have said that the ancestral butterflies varied—some had longer proboscises than others—and that proboscis length is inherited. Those butterflies with the longest proboscises were able to get the most food from the flowers and were therefore better able to survive and reproduce. A greater proportion of butterflies in the next generation inherited longer proboscises, and over time, this survival of the fittest resulted in an increase in the average length of the proboscis.

2. Darwin saw that by selecting individuals with desired characteristics as breeding stock, humans play the role of the environment and bring about differential reproduction. If this artificial selection can bring about great change in pigeons in a few hundred years, natural selection must be able to cause even greater changes over millions of years.

3. An individual cannot adapt in an evolutionary sense. Each plant has particular characteristics; it can either stand heat or it cannot. Lindsay recognized that individual plants probably varied in their ability to withstand heat. Perhaps plants with thinner leaves or shallower roots are more likely to

wilt and die. Those with thicker leaves or deeper roots might withstand heat and drought better and are better able to survive and reproduce, passing their genes for the characteristics that enable them to withstand the heat on to the next generation. In the next generation of wildflowers, a greater proportion of individuals will be of the more fit type. Thus individual plants do not adapt, but the population does.

4. Use the formulas $p + q = 1$ and $p^2 + 2pq + q^2 = 1$ to answer these questions. Nine percent (0.09) of the individuals in the population are homozygous recessive (white). If $q^2 = .09$, then $q = 0.3$. Since $p = 1 - q$, $p = 0.7$. Frequencies of homozygous dominant, heterozygous, and homozygous recessive individuals are 0.49, 0.42, and 0.09, respectively. Frequency of the green allele is 0.7. Frequency of the white allele is 0.3. The genotype and gene frequencies will stay the same in the next generation if the population is at Hardy-Weinberg equilibrium.

5. The population could be made to deviate from Hardy-Weinberg equilibrium in several ways. One could randomly scoop out some fish. In such a small population, this could probably alter allele and genotype frequencies by chance—genetic drift. One could cause gene flow by adding white fish to the aquarium. Subjecting the fish to X-rays might cause mutations that would alter the makeup of the gene pool. One could segregate the fish so that mating is nonrandom—only allowing fish of the same color to mate with each other, for example. Putting in a predator that can see and catch the white fish more easily than the green ones would result in natural selection in favor of the green fish and adaptation of the fish population to their environment.

6. The data indicate that the owls are catching a larger proportion of the dark mice than of the light mice. Perhaps the dark mice are easier to see against the white sand. If coloration is inherited, the light mice are more likely to pass their genes on to future generations, so a larger proportion of the mice will be light colored. Directional selection is occurring, and the mouse population is adapting to its environment.

Chapter 14

Review the Concepts

Exercise 1: 1. C 2. D 3. A 4. B

Exercise 2: 1. Because both are extinct, we must rely on appearance rather than the biological species concept. 2. Do tigers and lions interbreed under natural conditions? Is a tiglon fertile?

3. Appearance can be deceptive; they can interbreed. But would a Chihuahua really be able to mate with a Saint Bernard? 4. The criterion of interbreeding is useless for asexual organisms; we have to classify them by appearance and biochemical features. 5. A and C are becoming separate species, but there is still some gene flow between them via B. 6. If it is difficult to distinguish species on the basis of appearance, it might be difficult to know whether isolated populations are of the same or different species. 7. Are the birds the same species if they look alike? If the populations are separated, it might not be possible to determine whether they could interbreed in nature.

Exercise 3: 1. prezygotic, temporal isolation 2. postzygotic, hybrid breakdown 3. postzygotic, reduced hybrid fertility 4. prezygotic, mechanical isolation 5. prezygotic, behavioral isolation 6. prezygotic, gametic isolation 7. prezygotic, habitat isolation 8. postzygotic, reduced hybrid viability

Exercise 4: There are many possible answers for each of these questions, for example: 1. ocean, mountain range 2. mountain range, river 3. separate river basins, dry land between lakes 4. mountain range, ocean 5. ocean, mountain range 6. land bridge, deep water between coral reefs

Exercise 5: 1. yes, yes, 36 2. no, no, 33 3. no, no, 35 4. yes, yes, 36 5. yes, yes, 28 6. no, no, 21

Exercise 6: 1. population 2. species 3. geographical barrier 4. genetic drift 5. natural selection 6. interbreed 7. pre 8. behavioral 9. post 10. reduced hybrid viability 11. reinforce 12. fusion 13. hybrid 14. gene 15. species 16. biological 17. isolating 18. allopatric 19. punctuated 20. polyploid 21. tetraploid 22. meiosis 23. isolated 24. species 25. sympatric

Exercise 7: There are many possible examples, such as the following: 1. As a large lake gradually dries up, fish or frogs could be isolated in smaller ponds or springs. 2. Wildflowers or beetles could evolve on isolated mountains surrounded by lowlands. 3. Insects could evolve in isolated caves. 4. Unique plants could evolve on rock outcrops that have different minerals from the surrounding soil.

Exercise 8: 1. H 2. I 3. E 4. F 5. A 6. B 7. D 8. C 9. G

Test Your Knowledge

Multiple Choice: 1. b 2. b 3. e 4. c 5. a 6. d 7. d 8. a 9. e 10. b

Essay: 1. It is not possible to see whether fossil species could have interbred. Isolated populations may not naturally have the opportunity to interbreed. The interbreeding criterion does not apply to asexual organisms.

2. A splinter population is cut off from the parent population by a geographical barrier such as a mountain range, a canyon, or a stretch of ocean. Changes in allele frequencies occur in the splinter population, caused by mutation, natural selection, and genetic drift. These changes are undiluted by gene flow from the parent population. Finally, the splinter population has diverged so much from the parent population that individuals from the two populations can no longer interbreed. The presence of prezygotic or postzygotic barriers tells us that they have become separate species.

3. Imagine two species of flowers: species A, with a diploid number of 10 ($2n = 10$), and species B, with a diploid number of 12 ($2n = 12$). Two things have to happen to produce a fertile hybrid—hybridization and then an error in cell division. Pollen from species A ($n = 5$) fertilizes an egg from species B ($n = 6$). This produces a zygote with 13 chromosomes. This zygote can develop into a hybrid plant, but the hybrid would be sterile because its chromosomes cannot pair during meiosis. However, an error during cell division could duplicate chromosomes, producing a bud or a plant with 26 chromosomes. Each chromosome would be able to pair in meiosis, producing gametes with 13 chromosomes. Self-fertilization among the gametes would produce a zygote with 26 chromosomes ($2n = 26$). This hybrid would be fertile, because all the chromosomes would be able to pair in meiosis.

4. A small, isolated population is more likely to undergo speciation, because it is more likely to be affected by natural selection and genetic drift in a local area and less affected by gene flow that would make it more like other populations.

5. According to the gradual model, isolated populations gradually evolve differences as they adapt to local environments. There is a smooth, gradual process of change from one species to the next, through an incremental accumulation of microevolutionary differences. The punctuated equilibrium model suggests that evolution occurs in spurts. New species rapidly diverge from the parent stock and then remain unchanged for long periods. The fossil record often suggests the sudden appearance, persistence, and sudden disappearance of species predicted by the punctuated equilibrium model, rather than the gradual transitions predicted by the gradual model (although not all changes show up in fossils).

Apply the Concepts

Multiple Choice: 1. a   2. b   3. e   4. d   5. a   6. e
7. b   8. c   9. b   10. c

Essay: 1. The west has much more rugged terrain than the east. There are many mountain ranges and canyons that would be formidable geographical barriers, isolating populations and promoting allopatric speciation.

2. A cell from a mule would contain 63 chromosomes, 32 from the horse gamete and 31 from the donkey gamete. A mule is sterile because, in meiosis, its chromosomes are unable to pair up and then separate to form haploid cells.

3. This is a gray area, because dogs and coyotes probably would not interbreed under natural conditions in the wild. If dogs and coyotes are different species, this shows us that they are closely related, having recently evolved from a common ancestral species.

4. Dry stretches of desert between streams constitute geographical barriers that isolate fish populations in desert streams, much as stretches of ocean separate land animals on islands. Isolated fish populations might undergo different mutations and genetic drift and be subjected to different kinds of natural selection. They might evolve into different species, like animals isolated on oceanic islands.

5. Flies in newer populations might be less picky because they have not been separated from other populations as long, so they still share some characteristics (courtship signals, and so on) and can be "fooled" by flies of other species. Perhaps females are pickier than males because females only mate once and males can mate repeatedly. Females who make mistakes are not as successful at reproducing, so there is strong selection in favor of reproductive barriers—females who can tell the difference between their own species and other species.

6. According to the punctuated equilibrium model, species evolve rapidly and then remain unchanged for long periods. If the banana-eating moths are native to Hawaii, and bananas have only been there for 1,000 years, this suggests that a new species of moth evolved there in 1,000 years or less. This is a short time by evolutionary standards and is a bit of evidence in favor of the punctuated equilibrium model.

7. This criticism is not valid. Formation of a new species by allopatric speciation might take thousands of years. We have not been able to study species for long enough to see the whole process happen, but there is a lot of circumstantial evidence—Darwin's finches, cichlids in Lake Victoria, and so on. More important, we have seen many species of plants appear virtually overnight, via sympatric speciation—hybridization and polyploidy.

Chapter 15

Review the Concepts

Exercise 1: 1. M 2. C 3. B 4. L 5. E 6. K 7. A 8. I 9. G 10. D 11. N 12. J 13. H 14. F

Exercise 2: 1. simple molecules 2. energy from lightning, UV 3. small organic monomers 4. polymerization 5. polypeptides and lipids 6. simple RNA replication 7. lipid/polypeptide protobionts 8. self-replicating RNA molecules. 9. enclosure of RNA by "membrane" 10. natural selection 11. the first "cells"

Exercise 3: A. 11, Phanerozoic, Mesozoic B. 14, Phanerozoic, Cenozoic C. 5, Proterozoic D. 2 Archaean E. 8, Phanerozoic, Paleozoic F. 3, Proterozoic G. 13, Phanerozoic, Cenozoic H. 7, Phanerozoic, Paleozoic I. 9, Phanerozoic, Paleozoic J. 4, Proterozoic K. 1, Archaean L. 10, Phanerozoic, Paleozoic M. 12, Phanerozoic, Mesozoic N. 6, Phanerozoic, Paleozoic

Exercise 4: 1. The mammoth is approximately 17,190 years old. (After 5,730 years, the ^{14}C-to-^{12}C ratio would be half as great. After 11,460 years it would be 1/4 as great, and after 17,190 years it would be 1/8 as great.) The minimum and maximum ages of the fossil are 15,471 to 18,909 years.

2. Half the potassium-40 would disappear in the first 1.3 billion years, leaving 0.5 gram. Half of this would disappear by 2.6 billion years, leaving 0.25 gram.

Exercise 5: 1. S 2. O 3. G 4. J 5. N 6. Z 7. A 8. L 9. R 10. U 11. E 12. F 13. C 14. K 15. I 16. B 17. V 18. W 19. X 20. Y 21. T 22. D 23. M 24. P 25. Q 26. H

Exercise 6: 1. G 2. B 3. D 4. C 5. E 6. A 7. F

Exercise 7: 1. phylogeny 2. fossil 3. homologies 4. convergent 5. analogy 6. Systematics 7. taxonomists 8. binomial 9. genus 10. species 11. species 12. domain 13. phylogenetic 14. cladistics 15. clades 16. monophyletic 17. closely 18. ancestral 19. derived 20. out-group 21. parsimony 22. birds 23. derived 24. systematics 25. base 26. closely 27. mitochondria 28. ribosomal 29. clock 30. 550 31. hypotheses 32. 99% 33. yeasts 34. duplication 35. plants 36. animals 37. domain 38. Bacteria 39. Archaea 40. Eukarya 41. animals 42. kingdoms 43. horizontal 44. viral 45. bacteria

Exercise 8: 1. species 2. genus 3. family 4. order 5. class 6. phylum 7. kingdom 8. domain

Exercise 9: 1. mosses, ferns, conifers, flowering plants 2. green algae 3. homologies 4. shared derived characters; shared derived characters 5. chlorophyll *b* 6. protected embryos 7. chlorophyll *b*, or protected embryos, or vascular tissues 8. seeds 9. shared derived characters 10. out-group: green algae, mosses; in-group: ferns, conifers, flowering plants 11. clade or monophyletic taxon 12. conifers and flowering plants 13. green algae, mosses, ferns, conifers, flowering plants, plants with seeds, plants with vascular tissues, plants, organisms with chlorophyll *b*

Exercise 10: 1. B or F 2. F or B 3. D 4. A or C 5. C or A 6. E

Test Your Knowledge

Multiple Choice: 1. d 2. a 3. c 4. e 5. a 6. b 7. d 8. a 9. e 10. a 11. b 12. c 13. b

Essay: 1. Miller's experiment showed how simple organic compounds could have formed from inorganic chemicals on the primitive Earth. A flask containing warm water simulated the sea, while CH_4, H_2, and NH_3 simulated the early atmosphere. Electrical sparks introduced into the flask simulated lightning. A condenser cooled water vapor and caused "rain" that washed dissolved compounds into the "sea." As material circulated through the apparatus, the sparks triggered chemical reactions that formed a variety of organic compounds, including a mixture of amino acids. Scientists now think that the early atmosphere was made mostly of N_2 and CO_2. Experiments with this mixture have not produced organic molecules.

2. When the land masses fused, the total amount of shoreline and shallow coastal areas was reduced, and many marine species that inhabit shallow waters may have died as their habitats shrank. Ocean current would have shifted, altering climate. Less land would have been near the moderate coastal climate, so the climate in the interior may have become more harsh, leading to extinction of land species. There were also massive volcanic eruptions in what is now Siberia, which may have affected climate. Mass extinction of ancient species may have opened up opportunities for the adaptive radiation of the survivors into new species. Gymnosperms (cone-bearing plants) and reptiles diversified in the colder, drier climate on land.

3. These events occurred in the following order: First prokaryotes, origin of eukaryotes, origin of animals, appearance of first vertebrates, movement of plants and animals onto land, dominance of dinosaurs and cone-bearing plants, origin of flowering plants, diversification of mammals, appearance of humans.

4. When plates collide, their edges may crumple and form mountain ranges. The Himalayas formed this way when India collided with Eurasia. Volcanoes often arise where plates grind past each other or one plate overrides another.

5. Domain Eukarya, Kingdom Animalia, Phylum Chordata, Subphylum Vertebrata, Class Mammalia, Order Primates, Family Hominidae, Genus *Homo*, Species *sapiens*.

6. Cladistics is a method that groups organisms into clades according to their evolutionary descent. A clade is an ancestral species and all its descendants. (In other words, the clade is monophyletic.) The organisms of a clade have shared derived characters unique to the clade. Ideally, each genus, order, or phylum in the taxonomic hierarchy is a clade. Cladistics has shaken the tree of life a bit. It suggests, for example, that eukaryotes are more closely related to archaea than they are to bacteria, even though archaea and bacteria are both prokaryotes.

Apply the Concepts

Multiple Choice: 1. e   2. c   3. b   4. c   5. d   6. c
7. b   8. a   9. b   10. c   11. a   12. c   13. e   14. b

Essay: 1. There would be 200 mg after 5,730 years, and 100 mg after 11,460 years. The tree was buried about 11,460 years ago.

2. The fact that UV light breaks chemical bonds poses some problems for the hypothesized role of UV in the origin of organic compounds. If UV light provided energy for the reactions that caused inorganic compounds to combine and form simple organic compounds, why did it not then cause these same organic compounds to break apart?

3. Sediments from the end of the Cretaceous contain a layer of iridium, an element rare on Earth but common in meteorites. The iridium layer may be fallout from a huge cloud of dust produced when a large meteorite or asteroid hit the Earth, probably near Mexico, where a large crater has been discovered. The dust could have blocked light and disrupted the climate for an extended period, killing off many plants and the dinosaurs that depended on them. At the same time, continental drift may have caused climatic changes. Massive volcanic eruptions in what is now India released particles into the atmosphere, perhaps contributing to the cooling. Finally, it is possible that a combination of factors is responsible for the demise of the dinosaurs. Many forms of marine life also perished; any of these hypotheses could account for the extinction of both dinosaurs and marine species.

4. One way to determine how closely related the birds are would be to compare their DNA base sequences or protein amino acid sequences. The greater the similarity between base sequences or amino acid sequences, the closer the relationship.

5. Apparently, the similarities between the two groups of vultures are analogous, not homologous.

The birds probably have similar lifestyles and live in similar environments. They have apparently been similarly shaped by natural selection, even though they are not closely related. This is called convergent evolution.

6. If the "vultures" of America and Africa are not descended from a vulture ancestor, they do not form a clade. If the vultures do not share derived "vulture" characters that they inherited from a common ancestor, then "vultures" is not a monophyletic taxon.

7. Complex structures such as the eye can evolve from much simpler structures having the same basic function. In fact, several stages of eye evolution are seen among living mollusks—from a simple patch of light-detecting cells, to a cup-light structure that can locate a light source, to a complex camera-like eye with an iris and lens that can form sharp images. Natural selection could have shaped a complex eye from a simple light detector in many gradual steps.

Chapter 16

Review the Concepts

Exercise 1: 1. B   2. A   3. B   4. A   5. B   6. A
7. A   8. A   9. B   10. B   11. A

Exercise 2:

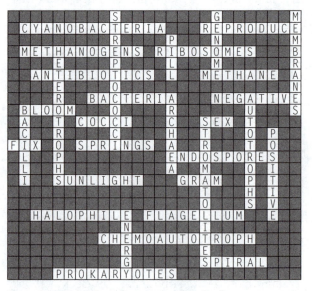

Exercise 3: 1. photoheterotroph   2. chemoautotroph   3. chemoheterotroph   4. photoautotroph   5. photoautotroph   6. chemoheterotroph

Exercise 4: 1. S, A   2. Q, B   3. O, B   4. T, B   5. O, B   6. M, A   7. N, B   8. Q, B   9. R, B   10. N, B   11. P, A   12. M, A

Exercise 5: 1. prokaryotes   2. eukaryotes   3. pathogens   4. poisons (toxins)   5. exotoxins   6. diarrhea   7. beef   8. botulinum   9. endotoxin   10. typhoid   11. sexually   12. endospores

13. weaponized   14. water-treatment   15. sewage
16. antibiotics   17. nutrients 18. organic
19. wastes   20. dead organisms   21. nitrogen
22. treatment   23. Anaerobic   24. sludge
25. aerobic   26. biofilm   27. copper   28. gold
29. bioremediation   30. tick

Exercise 6: 1. A, F, O   2. B   3. D, P   4. E   5. Q
6. L   7. H   8. C   9. G, N   10. E, I, J, K, M, Q

Exercise 7: 1. O, R, T   2. C   3. G, H, I   4. D
5. M, N   6. I, Q   7. A   8. I   9. F   10. H   11. P, Q
12. L   13. F   14. J, K   15. N   16. D, E, F   17. G
18. H, L, M   19. C   20. J   21. D   22. Q   23. A
24. E   25. I

Exercise 8: 1. H   2. F   3. E   4. A   5. C   6. D
7. G   8. B

Test Your Knowledge

Multiple Choice: 1. a   2. a   3. c   4. e   5. e   6. c
7. c   8. a   9. b   10. d   11. d

Essay: 1. Bacteria cause tuberculosis, cholera, bubonic plague, anthrax, and typhoid fever. Most bacteria cause illness via poisons—exotoxins secreted from cells or endotoxins that are components of cell walls.

2. Prokaryotes are important in the cycling of chemical elements between living things and the nonliving environment. Cyanobacteria restore oxygen to the atmosphere. Some bacteria take nitrogen from the air and convert it to forms in the soil that can be used by plants to make proteins and nucleic acids. Prokaryotes also act as decomposers, breaking down organic matter in dead organisms and body wastes and converting them to inorganic forms that are available to organisms for reuse. Some prokaryotes might be useful in similarly breaking down spilled petroleum and toxic wastes.

3. Primitive cyanobacteria, such as those that formed fossil stromatolites, were among the prokaryotes that dominated Earth from about 3 billion to about 1 billion years ago. Their photosynthesis released enough oxygen (O_2) gas to make the atmosphere aerobic, setting the stage for the evolution of aerobic organisms such as animals.

4. The endosymbiosis hypothesis suggests that eukaryotic cells may have started out as cooperative communities of prokaryotic cells. An ancestral prokaryote may have ingested small heterotrophic bacteria that were able to use O_2 to release energy from organic molecules via cellular respiration. These cells remained alive inside the host cell and continued to perform respiration. They eventually became mitochondria. Similarly, small photosynthetic prokaryotes may have taken up residence inside host cells and become chloroplasts. This is called primary endosymbiosis. Later, certain red and green algae produced via primary endosymbiosis were taken up by other heterotrophic cells, giving rise to the ancestors of dinoflagellates, euglenozoans, etc. This is called secondary endosymbiosis.

5. Multicellularity probably arose several times, with today's seaweeds, plants, animals, and fungi evolving from different protist ancestors. Unicellular protists may have formed colonies of interconnected cells. Such a colony might form when a cell divided and its offspring remained connected. The cells gradually became more specialized and interdependent, with specialized cells coming to perform specific tasks, such as locomotion, feeding, gas exchange, and reproduction.

Apply the Concepts

Multiple Choice: 1. b   2. c   3. c   4. a   5. e   6. a
7. d   8. e   9. a   10. c   11. b   12. c

Essay: 1. If the diphtheria bacterium is like most bacteria, it causes illness by producing a toxin (in this case an exotoxin secreted by the bacteria) that can cause damage or interfere with body functions at places removed from the site of infection.

2. The number of bacteria would double every 20 minutes, so they would double 18 times in 6 hours. After 20 minutes, there would be 200, after 40 minutes 400, after an hour 800, and so on. There would be more than 26 million bacteria at the end of 6 hours!

3. The chloroplasts of green algae are thought to be the products of primary endosymbiosis, in which a photosynthetic prokaryote took up residence inside a heterotrophic eukaryote. The prokaryote evolved into a chloroplast with two membranes—an inner membrane that was the plasma membrane of the original prokaryote, and an outer membrane derived from the plasma membrane of the host cell (essentially the membrane of a "food" vacuole). The chloroplasts of euglenoflagellates are thought to be the products of *secondary* endosymbiosis, in which a green alga took up residence inside a heterotrophic eukaryote. The alga evolved into a chloroplast with the two membranes of the green alga chloroplast, then the alga plasma membrane, and an outer membrane derived from the plasma membrane of the host cell—four membranes in all.

4. Biologists used to think that these nutritional modes indicated evolutionary relationships, so it made sense to put all algae in one group, for example. It seemed that if all algae did photosynthesis, they must be related. More recent studies of genes, cell structure, and biochemistry suggest that these different modes of nutrition evolved and were lost multiple times on various branches of the protist

phylogenetic tree, so they do not reflect ancestry. For example, diatoms, brown algae, and water molds are now all thought to be related. Diatoms and brown algae are photosynthetic, but water molds apparently lost their chloroplast and are more fungus-like. Thus they are related, but they don't share the same nutritional mode. But that doesn't mean that water molds are closely related to slime molds, which are related to amoebas! Euglenozoans and green algae are both green, but euglenozoans appear to be related not to green algae, but to the very simplest protists, which are not photosynthetic. Euglenozoans are green because their chloroplasts were green algae taken up by ancestors of the euglenozoans in a process of secondary endosymbiosis. Protist genealogy can be confusing!

Chapter 17

Review the Concepts

Exercise 1: 1. On dry land, plants could dry out. They are covered by a waxy cuticle that prevents water loss. 2. A Cuticle prevents drying, but can interfere with gas exchange. Gasses diffuse through pores called stomata. 3. On land, light and gases are above ground, but water and minerals in the soil. Specialized leaves obtain light from the sun and carbon dioxide from the air, while underground roots absorb water and minerals. A vascular tissue transport system connect roots and leaves via the stem. Plants also have specialized growth regions called apical meristems at the tips of stems and roots. 4. Because a plant is not buoyant in air, it must be able to support itself against the pull of gravity. Cell walls of some plant tissues are thickened and reinforced with lignin for rigidity. 5. Plant gametes are protected in male and female *gametangia. 6. Some plants have swimming sperm, but in most, sperm-producing cells are carried through the air by pollen grains. 7. In plants, eggs are fertilized and embryos *develop in and depend on protective gametangia. 8. All plant life cycles involve *alternation of a haploid generation that produces gametes with a diploid generation that produces *spores, which sometime disperse to new locations. Pines and roses have spores, but their embryos are dispersed in seeds.

Exercise 2: A. 3, gymnosperms, angiosperms B. 1, bryophytes C. 4, angiosperms D. 2, bryophytes, seedless vascular plants (club mosses and ferns)

Exercise 3: 1. sporophytes 2. haploid 3. diploid 4. gametophyte 5. gametes 6. gametangia 7. sperm 8. flagella 9. fertilization 10. zygote 11. sporangium 12. spores 13. meiosis

Exercise 4: 1. meiosis 2. spores 3. mitosis and development 4. gametophyte 5. sperm 6. egg 7. fertilization 8. zygote 9. mitosis and development 10. sporophyte 11. sporangia 12. haploid phase 13. diploid phase

Exercise 5: 1. pine tree 2. pollen cones, ovulate cones 3. pollen cones, pollen grains 4. ovulate cones, ovules 5. ovule 6. ovule 7. eggs, sperm 8. pollen grain, sperm, egg 9. zygote, embryo, seed 10. seed, embryo 11. pine tree

Exercise 6: 1. D 2. E 3. I 4. A 5. F 6. C 7. G 8. H 9. B

Exercise 7: 1. meiosis 2. pollen grain (male gametophyte) 3. female gametophyte 4. egg 5. pollination 6. stigma 7. pollen tube 8. sperm 9. fertilization 10. zygote 11. seed 12. embryo 13. seed coat 14. food supply 15. fruit 16. sporophyte 17. ovary 18. ovule 19. anther 20. diploid phase 21. haploid phase

Exercise 8: 1. cones 2. flowers 3. wind 4. insects, and so on 5. a year or more 6. 12 hours or so 7. more than a year 8. days or weeks 9. seeds unprotected 10. seeds protected and dispersed by fruit

Exercise 9: 1. A 2. G, A 3. F 4. F 5. G 6. M 7. A 8. M 9. M, F 10. G 11. A 12. F 13. A 14. G, A 15. F 16. G 17. M, F, G, A 18. G 19. A 20. M, F 21. A 22. A 23. M, F, G, A 24. M

Exercise 10: 1. 50,000 2. 50 3. 20 4. 40 5. 25 6. 5,000, 290,000 7. 25

Exercise 11: 1. J 2. E 3. P 4. Q 5. M 6. C 7. T 8. W 9. Z 10. U 11. S 12. F 13. D 14. K 15. V 16. Y 17. O 18. X 19. I 20. B 21. R 22. A 23. N 24. L 25. G 26. H

Exercise 12: 1. three 2. spores 3. hyphae 4. mycelium 5. mating types 6. nuclei 7. heterokaryotic 8. fruiting body 9. diploid 10. meiosis

Exercise 13:

Test Your Knowledge

Multiple Choice: 1. a   2. d   3. b   4. d   5. a   6. b
7. c   8. e   9. b   10. e   11. d   12. e   13. d   14. c
15. c

Essay: 1. Algae are supported by water, and they lack supporting tissues. The leaves and stems of plants contain rigid supporting elements. The whole body of an alga is exposed to water, light, and minerals, while only aboveground parts of the plant are exposed to light and only the roots are in contact with a reliable supply of nutrients and water. Plants have discrete organs—roots, stems, and leaves, and specialized growing regions at the tips of stems and roots—while algae do not have true organs with specialized functions. Plants have mycorrhizae, but algae do not. Plants are covered by a cuticle that prevents drying, pierced by stomata for gas exchange, but algae have no need for such a covering or stomata. Plants have vascular tissues that transport nutrients and water, but algae do not. In plants, zygotes and embryos develop within the female gametangium.

2. Many people think all are plants because they are rigid, "rooted" in one spot, and cannot move. Seaweeds lack many of the specialized interdependent tissues and organs seen in plants, such as leaves, stems, roots, and vascular tissues, and the gametes and embryos of seaweeds are unprotected. Fungi are heterotrophic, while plants and seaweeds are autotrophic—capable of making their own food via photosynthesis.

3. Gymnosperms and angiosperms—the cone-bearing and flowering plants—both produce seeds. Seed plants do not require water for fertilization. Instead of sperm that swim to the eggs, seed plants produce pollen grains that protect male gametophytes and carry them through the air to the female parts of the plant. The seed is a survival packet for the embryo—a protective coat and a food supply that enable the embryo to get a good start.

4. Gymnosperms—conifers such as pines and firs—supply most lumber for construction, wood pulp for making paper, and most other forest products. Forests are also important for wildlife habitat, watersheds, and recreation. We can reduce the cutting of forests by recycling lumber and paper, and by cutting trees no faster than they can be replanted.

5. See Figure 17.8B in the text.

6. Animals assist angiosperms by carrying pollen from anthers to carpels and by dispersing seeds. Flowers produce nectar and pollen, which are attractive to animals, and advertise their presence with showy shapes, colors, and scents. Fruits sometimes cling to the fur of animals or attract animals to eat them and thereby disperse seeds in their droppings.

7. The threadlike hyphae of fungi grow through their food and secrete enzymes that digest the food outside their bodies. The fungal cells then absorb the digested nutrient molecules through their cell membranes.

8. A lichen consists of a fungus growing in a mutualistic relationship with green algae or cyanobacteria. Their exact roles are not completely understood, but it is known that the fungus receives food from the algae or cyanobacteria, and the photosynthetic cells receive protection, water, and certain minerals from the fungus.

Apply the Concepts

Multiple Choice: 1. e   2. d   3. c   4. a   5. b   6. e
7. b   8. b   9. c   10. d

Essay: 1. Like seed plants, ferns have true roots, stems, and leaves; vascular tissues that transport water and nutrients; and a life cycle in which the sporophyte stage predominates. Like mosses, ferns have flagellated sperm that swim to the egg.

2. Mosses and ferns are dispersed in the form of spores. Gymnosperms and angiosperms are dispersed in the form of seeds. Spores are very small and can be carried by the wind for long distances, but they are relatively vulnerable and short-lived. Seeds are bigger and cannot travel as far as fast (at least without help from animals), but they are protected by a seed coat and carry their own food supply that the embryo can draw on for a good start.

3. What probably happened is that the farmer killed beneficial mycorrhizal fungi along with the harmful parasites. Without the help of the mycorrhizae, the trees are not able to absorb water and soil minerals effectively.

4. Fungi can actually grow into other cells, where they are protected. They grow very rapidly, which enables them to come back quickly if not completely killed off. (Also, since they are eukaryotes, their metabolism is more like our own cells than is the metabolism of pathogenic bacteria. This makes them harder to kill than bacteria, because chemicals that injure fungi might injure human cells as well.)

5. Naming a lichen is tricky because the lichen consists of two organisms—a fungus and an alga or cyanobacterium. Does the lichen get one name or two?

Chapter 18

Review the Concepts

Exercise 1: The black line goes between prokaryotes and protists; prokaryotes have simple cells lacking nuclei and other organelles, while the other organisms have more complex eukaryotic cells. The blue line goes above protists; they are primarily unicellular, and those above them multicellular. The green line goes between plants and fungi; plants are autotrophs, and fungi and animals are heterotrophs. The red line goes between fungi and animals; fungi digest their food and then absorb the nutrients; animals ingest food and then digest it.

Exercise 2: See Figure 18.2B in the text. A colony of cells may have become a hollow sphere. Cells of the colony then specialized and folded inward, forming two layers of cells performing various functions.

Exercise 3: 1. Flatworms, crustaceans, insects, humans, and so on   2. Parts arranged in a circle around a central axis   3. No dorsal and ventral surfaces   4. Anterior and posterior ends 5. Distinct head   6. Sedentary or passively drifting 7. Head with sense organs contacts environment first

Exercise 4: See Figure 18.3C in the text.

Exercise 5: 1 Porifera   2. bilateral   3. Cnidaria 4. Platyhelminthes   5. pseudocoelom   6. no notochord or backbone   7. Chordata   8. jointed legs 9. no jointed legs   10. Arthropoda

Exercise 6: 1. Porifera   2. Cnidaria   3. sea 4. radial   5. pores   6. two   7. flagella   8. enters 9. exits   10. bacteria   11. skeletal   12. nerves 13. tissues   14. protists   15. jellies   16. digestive 17. nervous   18. tissues   19. two   20. organ systems   21. polyp   22. medusa   23. Jellies 24. hydra   25. tentacles   26. cnidocytes 27. underneath   28. gastrovascular   29. wastes

Exercise 7: 1. Cnidaria; Platyhelminthes; Nematoda   2. jelly, hydra; planarian, tapeworm; *Caenorhabditis, Trichinella*   3. radial; bilateral; bilateral   4. 2; 3; 3   5. medusa or polyp; flattened ribbon or leaf shape; cylinder   6. none; none; pseudocoelom   7. incomplete; incomplete or none; complete   8. fresh water and salt water; fresh water, salt water, soil, and parasites in animals; water, soil, and parasites in animals and plants 9. build coral reefs; human and animal parasites; human, animal, and plant parasites

Exercise 8: 1. E   2. G   3. C   4. G   5. D   6. C 7. B   8. F   9. G   10. E   11. D   12. A   13. C 14. D   15. D   16. E   17. E   18. G   19. C   20. E

Exercise 9: 1. Annelida   2. segments   3. sea 4. fresh water   5. walls   6. nervous   7. excretory 8. closed   9. digestive   10. movement 11. flexibility   12. mobility   13. circular   14. feces (castings)   15. polychaetes   16. ocean   17. tubes 18. appendages   19. feeding   20. leeches 21. blood   22. carnivores   23. fresh water 24. teeth   25. blood clots   26. Arthropods

Exercise 10:

Exercise 11: Circle the following: 1, 4, 6, 8, 9, 10, 12, 13, 14, 15, 16.

Exercise 12: 1. dorsal hollow nerve cord   2. notochord   3. pharyngeal slits   4. muscular post-anal tail. See Figure 18.14B in the text.

Exercise 13: 1. G   2. B   3. D   4. E   5. A   6. A 7. C   8. F

Test Your Knowledge

Multiple Choice: 1. b   2. c   3. a   4. a   5. b   6. e 7. d   8. a   9. c   10. d   11. c   12. e   13. a   14. a

Essay: 1. Animals are multicellular, heterotrophic eukaryotes that obtain nutrients by ingestion. Animal cells lack walls and are held together by unique proteins and junctions. Animals are diploid, except for eggs and sperm. Their embryos go through a blastula and gastrula stage, and some pass through a larval stage before becoming adults. Development is controlled by *Hox* genes. Most animals have three tissue layers, muscle cells, and nerve cells.

2. Some characteristics important in determining an animal's phylum are number of cell layers, body symmetry, presence of tissues and organs, presence and type of body cavity, embryonic development of the body cavity, and presence or absence of segmentation.

3. All animal phyla discussed in this chapter have some kind of body cavity, except for sponges, cnidarians, and flatworms. The body cavity aids in movement, cushions internal organs, and may stiffen the body and help in circulation.

4. The mantle secretes the shell, but it can have other functions. In some bivalves, such as scallops,

sensory structures such as eyes are arranged along the edge of the mantle. It functions in waste disposal in some mollusks. The mantle cavity houses the gills, and it is modified as a kind of lung in land snails. In squids, the mantle cavity can expel water for jet propulsion. Snails have a one-piece coiled shell. The two-part hinged shell of a clam can be closed for protection or opened for feeding. Squids have a small internal shell that stiffens the body for swimming.

5. Insects are arthropods, so they are covered by a tough exoskeleton that protects them from injury and drying out. They have jointed appendages that are modified for various functions—locomotion, sensing the environment, and feeding. Many insects have highly specialized mouthparts. Insects are small and can reproduce rapidly. They are the only invertebrates with wings, and they use their flying ability to find food, escape predators, and disperse to new habitats. Most insects undergo complete metamorphosis; larvae specialized for feeding change into adults specialized for reproduction.

Apply the Concepts

Multiple Choice: 1. d 2. c 3. a 4. e 5. c 6. a 7. d 8. e 9. b 10. c 11. b

Essay: 1. Like all other animals, sponges are eukaryotic, multicellular heterotrophs that lack cell walls, ingest their food, and are mostly diploid.

2. The temporary digestive cavity of *Trichoplax* suggests a hypothesized stage in animal evolution when a relatively unspecialized colony of cells may have folded inward to form a digestive cavity, much as a blastula becomes a gastrula during animal development.

3. Pigs and people are infected when they eat undercooked meat. If meat is thoroughly cooked, the worms will be killed. Pigs can also become infected by ingesting the feces of infected humans or animals. A way to prevent the spread of the worms might be to use good sanitation and keep the pigs away from human feces.

4. Your criteria might include number of species or individuals or how widespread the animals are. By these criteria, nematodes, mollusks, and arthropods are very successful. Discuss particular structural, functional, or behavioral features or adaptations that contribute to their success.

5. These larvae, called trochophore larvae, suggest that mollusks and annelids are related. These two phyla are on adjacent branches of the animal phylogenetic tree. They are both bilaterally symmetrical animals with a coelom that forms within masses of cells in the middle tissue layer.

6. The insect body plan is modular—each embryonic segment is a building block that develops independently of the other segments. Structural genes in different segments or groups of segments might be the same, but different changes in homeotic genes in each area might produce different structures. For example, if a homeotic gene in a posterior segment tells a "leg" gene to turn on a little sooner or stay turned on a little longer than other leg genes, the grasshopper's long hind legs might result. Similarly, differences in gene expression, and not in the genes themselves, might produce the broader wings or tubular mouthparts of a moth. One would expect the genes for building legs, mouthparts, or wings to be similar. Differences in gene expression, controlled by small changes in homeotic genes, make the structures turn out differently.

Chapter 19

Review the Concepts

Exercise 1: 1. Tunicates 2. Lancelets 3. Hagfishes 4. Lampreys 5. Lampreys 6. Sharks and rays 7. Ray-finned fishes, lobe-fins, amphibians, reptiles, and mammals 8. Lobe-fins 9. Amphibians 10. Reptiles and mammals 11. Reptiles and mammals 12. Mammals

Exercise 2: 1. C 2. D, E 3. E 4. C 5. A, B 6. D 7. E 8. A 9. D 10. D 11. E 12. D, E 13. A 14. A 15. D 16. B 17. A, B

Exercise 3: 1. T 2. F, shallow fresh water 3. F, lobe-finned fishes 4. T 5. F, Several important fossils give us information 6. F, several new discoveries show us what bridged the gap 7. T 8. F, limbs that were like paddles 9. T 10. F, fish with necks 11. T 12. F, Modern amphibians, reptiles, and mammals

Exercise 4: 1. A 2. B, M 3. R 4. B, M 5. B 6. M 7. R 8. M 9. A 10. M 11. A 12. M 13. M 14. B, M 15. A 16. M 17. A, R 18. A 19. M 20. B 21. R 22. A 23. B 24. A 25. M 26. B 27. R 28. B 29. R 30. M

Exercise 5: 1. brain 2. head 3. vertebral column 4. jaws 5. lungs or derivatives 6. lobed fins 7. legs 8. amniotic egg 9. milk 10. chordates 11. craniates 12. vertebrates 13. jawed vertebrates 14. tetrapods 15. amniotes

Exercise 6: 1. Sensitive grasping hands with opposable thumbs 2. Eyes positioned for depth perception, limber shoulder joints, and opposable thumbs 3. Eyes positioned for depth perception and sensitive hands with opposable thumbs 4. Limber shoulder and hip joints, depth perception

5. Eyes positioned for depth perception and sensitive hands with opposable thumbs   6. Eyes positioned for depth perception, limber shoulder joints, sensitive hands

Exercise 7: 1. E, F, G, H, I   2. C, D, E, F, G, H, I
3. E, F, G, H, I   4. G   5. C   6. D   7. A, B   8. B
9. A   10. H   11. A   12. C   13. D   14. E   15. F
16. G   17. C   18. D   19. E, F, G, H, I   20. C
21. D   22. E   23. A   24. C, D, E, F, G, H, I   25. C
26. D   27. H   28. K   29. P   30. R   31. O   32. J
33. Q   34. M

Exercise 8: 1. B   2. E   3. C   4. F   5. A   6. D
7. M   8. J   9. G   10. L   11. I   12. H   13. K

Exercise 9: 1. Researchers have found and studied fossils of hundreds of ancient hominids, dating back 7 million years. They have identified dozens of species of ancient hominids. In addition, they can now use DNA comparisons to track our origins. The "missing link" is no longer missing.
2. Humans and chimps are closely related; their genomes are 96% identical. But humans are not the descendants of modern-day chimps. An ape that lived 5–7 million years ago appears to have been the common ancestor of both chimps and humans.
3. The hips, knees, and skulls of *Australopithecus* fossils show that hominids walked upright when their brains were still chimp-sized.   4. The human evolutionary tree is a many-branched bush. At times in the past, several species of hominids coexisted. All but one of the branches died out—the one leading to us—so most fossil hominids were not our ancestors.   5. DNA analysis of fossil Neanderthals shows convincingly that the genes of Neanderthals are different from those of all modern humans. They arrived in Europe long before the ancestors of modern humans, there was no interbreeding between Neanderthals and our ancestors, and Neanderthals went extinct without contributing to the modern human gene pool.   6. It is true that hominids first appeared in Africa and *Homo erectus* spread around the globe. But then *Homo sapiens* evolved in Africa, and in a second wave of migration, displaced earlier hominids all over the world.   7. Humans have an immense impact on virtually all life on the planet. Introduction of nonnative species is just one problem we brought with us as we multiplied and spread around the globe. We also destroy the habitat of other species, pollute water and air, and even alter climate.   8. Skin color is not a significant racial characteristic. Apparently dark skin is an adaptation that protects folate, a vital nutrient, from being broken down by sunlight in tropical latitudes where the sun's UV radiation is most intense. Lighter skin evolved in areas farther from the equator, where sunlight is less intense. There it is important for enough UV to penetrate the skin to catalyze the formation of vitamin D, another important nutrient. Because skin color was a product of natural selection, similar environments produced similar degrees of pigmentation. Widely separated populations may have the same adaptation, no matter how closely they are related, so skin color is not useful in identifying meaningful phylogenetic relationships.   9. Study of a rare language disorder led researchers to discover a gene, called *FOXP2*, that codes for a transcription factor that controls the expression of many genes that shape the brain. Many animals have *FOXP2* genes, but only those with language-learning abilities, such as certain songbirds, have a gene similar to that of humans. So, while it is true that our advanced brains allow us to readily learn and use symbolic language, it is our genes and natural selection that help to shape our brains.

Test Your Knowledge

Multiple Choice: 1. c   2. b   3. c   4. e   5. a   6. c
7. b   8. d   9. e   10. b   11. d   12. e   13. b

Essay: 1. The fins of lobe-finned fishes are different from those of ray-finned fishes like a trout or perch. The fins of lobe-finned fishes are fleshy and contain bones that appear to be homologous with the upper and lower limb bones of tetrapods. Also, some living lobe-fins, the lungfishes, gulp air into primitive lungs. Fossils of *Acanthostega* and *Tiktaalik* suggest that the ancestor of tetrapods was not a fish with lungs that gradually developed legs, but rather a fish with a neck and four leg-like limbs that raised its head above shallow water to breathe air.

2. Birds have feathers, which form a lightweight, streamlined covering and shape the wings. They have many modifications to reduce their weight: no teeth, a reduced tail, honeycombed bones, and air sacs in the body. They have large flight muscles attached to a keeled breastbone. They are endothermic and insulated by feathers, maintaining a high constant temperature and metabolic rate. Birds' circulatory system and lungs are highly efficient in delivering food and oxygen for the rapid metabolism needed for flight. Their senses, large brain, and nervous system are well-developed for complex behaviors and an active lifestyle.

3. Reptile skin is covered by scales made of keratin, a waterproof protein that keeps the body from drying out. Reptiles are also able to lay their amniotic eggs on land. The reptile embryo develops within a hard shell and a protective, fluid-filled sac called the amnion. The young reptile does not go through an aquatic larval stage but rather emerges from the egg as a miniature adult, ready to cope with life on

dry land. Amphibians have thin, moist skin that dries out easily. They must lay their jelly-covered eggs in water, and their tadpole larvae must develop in water.

4. Primates have limber hip and shoulder joints and grasping hands and feet, useful for climbing and manipulating objects. Their hands and feet have an acute sense of touch. Their snouts are short, and their eyes are set close together on the front of their face, which enhances depth perception.

5. Chimpanzees (and bonobos, closely related to chimps) are the living primates most closely related to humans. They are apes and share numerous anatomical similarities with humans. Chimpanzees and humans share behavioral similarities as well. Chimps are very intelligent. They can make and use simple tools; they raid other social groups of their own species; they can learn sign language; and their behavior in front of mirrors indicates that they are, to some degree, self-aware. Biochemical evidence also shows the close relationship between chimps and humans; their genomes are 96% identical.

6. a. Anthropoids have relatively larger brains and depend more on sight and less on sense of smell than prosimians. b. The nostrils of New World monkeys face forward and those of Old World monkeys have nostrils that are narrow and point downward, and their tails are not prehensile. c. Monkeys have tails; apes do not. Monkeys have front limbs that are equal in length to their hind limbs; the front limbs of apes are longer than their hind limbs. d. Most apes walk with their knuckles on the ground; humans are bipedal, walking upright. Humans have shorter jaws and larger brains than other apes. (There are other differences, such as language and use of tools, but chimps share these traits to some degree.)

7. Analysis of mitochondrial DNA of modern humans shows that African lineages are the oldest branches in the human family tree. Studies of mitochondrial DNA (which is passed on maternally) and Y chromosomes (passed on from father to son) in modern human populations show that all living humans inherited their mitochondrial DNA from a woman who lived 160,000–200,000 years ago. The Y chromosome points to an African ancestor. The oldest fossils of *Homo sapiens*—160,000–195,000 years old—have been found in Africa (Ethiopia), so fossils agree with the genetic data.

Apply the Concepts

Multiple Choice: 1. c   2. c   3. a   4. a   5. b   6. e
7. a   8. b   9. c   10. e   11. d   12. e   13. e

Essay: 1. Tunicate—notochord (and three other chordate characteristics). Lancelet—brain.

Hagfish—head. Lamprey—vertebral column. Shark—jaws. Trout—lung or lung derivative. Lungfish—lobe-fin limb structure. Frog—legs. Rattlesnake—amniotic egg. Bear—milk.

2. Birds and mammals are endothermic. Unlike reptiles and amphibians, they maintain a constant high rate of metabolism and body temperature, enabling them to function in a cold environment.

3. Humans must contend with a trade-off between a narrow pelvis adapted for bipedal walking, which limits the width of the hips and birth canal, and our enlarged brain. If fetal development were longer, the baby's head would not pass through the birth canal. This means babies are born helpless and need a long period of parental care, but this long childhood contributes to learning and transmission of culture.

4. Chimpanzees display all of these "human" characteristics. They sometimes strip a blade of grass and use it as a tool to fish for termites. Chimps can learn human sign language, although it is not known how much they use symbolic language in the wild. A chimpanzee looking in a mirror displays behaviors—examining its body, making faces—that suggest it has a concept of self.

5. Hominids are thought to have diverged from other apes 5 to 7 million years ago and started walking upright. A large brain evolved later. A small-brained bipedal hominid 5.0 million years old fits current theory. A large-brained hominid that walked on all fours 2.5 million years ago would be harder to explain.

Chapter 20

Review the Concepts

Exercise 1: 1. The tail is flattened into broad flukes, which propel the dolphin through the water.   2. The hummingbird's long, thin beak enables it to sip nectar from tubular flowers.   3. The sensitive nerve endings and finely controlled muscles of the hand enable you to grasp and manipulate objects.   4. The frog's long legs enable it to leap to safety, and its webbed feet make it a good swimmer.   5. The mosquito's mouthparts allow it to pierce the skin and suck blood.   6. The stomach chambers carry out various steps in breaking down hard-to-digest grass.

Exercise 2: 1. cells   2. tissue   3. epithelial
4. connective   5. nervous   6. organ   7. organ system   8. organism   9. arrangement   10. organism
11. organ systems   12. organs   13. tissues
14. cells

Exercise 3: 1. stratified squamous epithelium 2. multiple layers of flattened cells 3. lining of esophagus, epidermis of skin 4. covers and protects surfaces subject to abrasion 5. simple cuboidal epithelium 6. single layer of cube-shaped cells 7. tubular passageways where urine forms in kidneys 8. simple squamous epithelium 9. linings of lungs and blood vessels 10. exchange of materials by diffusion 11. simple columnar epithelium 12. single layer of elongated, cylindrical cells 13. secretes and absorbs in walls of digestive tract 14. pseudostratified ciliated columnar epithelium 15. single layer of elongated cells with cilia 16. secretes moist mucus; cilia sweep out particles

Exercise 4:

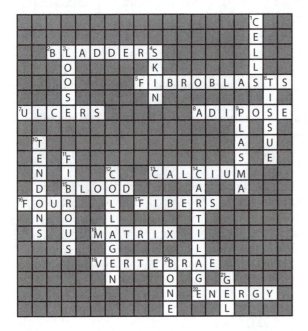

Exercise 5: 1. D 2. C 3. B 4. A 5. B 6. D 7. C 8. B 9. D 10. C 11. D 12. C 13. A 14. A 15. A

Exercise 6: 1. nervous 2. muscular 3. skeletal 4. digestive 5. respiratory 6. circulatory 7. lymphatic 8. integumentary 9. immune 10. excretory 11. endocrine 12. reproductive

Exercise 7: 1. X-rays 2. shadows of hard tissues—bones, cartilage, dense tumors 3. X-ray cross sections combined by computer 4. cross sections and three-dimensional views of normal and abnormal hard and soft tissues and changes in brain, heart, etc. 5. magnets, radio pulses, and computer 6. visualizes soft tissue such as brain; used to plan surgery. 7. radiation from radioactive isotope picked up by detector 8. measuring metabolic activity, especially of brain, tumors

9. combines X-ray sections and 3-D of CT with PET radiation from radioactive isotope 10. detailed anatomy of CT plus metabolic info of PET

Exercise 8: 1. B 2. F 3. E 4. K 5. A 6. I 7. H 8. G 9. D 10. J 11. C 12. L

Exercise 9: See Figure 20.13A in the text.

Exercise 10: 1. increase in temperature 2. eating a meal 3. thermostat in brain 4. pancreas 5. increase in blood temperature 6. increase in blood sugar level 7. nerve impulses 8. hormone—insulin 9. sweat glands and blood vessels in skin 10. cells 11. sweat glands secrete sweat, and surface blood vessels dilate and fill with warm blood, cooling the body 12. cells take up sugar and lower blood sugar level 13. around 37°C 14. 70 to 110 mg of sugar per 100 mL of blood

Test Your Knowledge

Multiple Choice: 1. b 2. d 3. c 4. d 5. b 6. e 7. a 8. d 9. e 10. b

Essay: 1. Epithelial tissue consists of sheets of closely packed cells that cover and line body surfaces. It protects, secretes, and absorbs. Connective tissue consists of sparse cells scattered in a matrix, which usually is a web of fibers embedded in a liquid, jelly, or solid. Most connective tissues form a framework that supports and protects the organs of the body. Muscle tissue consists of bundles of elongated cells called fibers. Muscle cells can contract and are responsible for body movement. Nervous tissue consists of cells called neurons and other supporting cells that form a communication and coordination system within the body. Neurons have elongated extensions that transmit signals to other cells.

2. The digestive system ingests food, breaks it down, and absorbs it into the bloodstream. The respiratory system supplies the blood with oxygen and disposes of carbon dioxide. The circulatory system transports nutrients and oxygen to cells and carries away wastes. The lymphatic system returns fluid to the blood and plays a role in body defense. The immune system protects the body from disease and foreign substances. The excretory system disposes of metabolic wastes. The endocrine system and nervous system control and coordinate body activities via hormones and nerve signals. The male and female reproductive systems produce gametes and support the growth of

the developing embryo. The muscular system moves the body and its parts; the skeletal system supports and protects the body; and the integumentary system covers the body.

3. Computerized tomography uses less powerful (and therefore less potentially damaging) X-rays than conventional X-rays. A CT scan reveals more of the fine detail of soft tissues. It yields cross-sectional views, which can reveal features that might block one another on conventional X-rays, and these cross sections can be combined by computer to give three-dimensional views.

4. In a house, the stimulus is a change in room temperature, and in the body it is a change in the temperature of the blood. The control center is the thermostat in a house, and in the body it is a "thermostat" in the brain. The home thermostat sends a signal to the furnace—the effector— which responds to an increase in room temperature by turning off and allowing the house to cool a bit. When blood temperature goes up, the brain sends a signal to body effectors—sweat glands and blood vessels in the skin. The sweat glands respond by secreting sweat, which increases evaporative cooling. The blood vessels dilate to bring warm blood to the surface, where it loses its heat and cools the body. When the house or body cools, the sensors and control centers signal effectors to warm things up.

5. Small animals have a larger surface-to-volume ratio than larger animals, and most of their cells may be in direct contact with their surroundings, so their body surfaces may be sufficient to exchange materials with the environment. Larger animals have smaller surfaces in relation to overall volume, so they require specialized surfaces for exchange of materials with the environment.

Apply the Concepts

Multiple Choice: 1. b   2. c   3. c   4. e   5. c   6. c   7. e   8. c   9. b   10. e

Essay: 1. A woodpecker displays many features that illustrate the correlation between structure and function. Its keen hearing enables it to detect insects inside a tree. It uses its sharp, chisel-like bill to cut a hole in the tree and its long tongue to probe for insects. Two toes on each foot face forward and two backward, enabling the woodpecker to cling to a vertical tree trunk, and its stout tail is used as a brace as it drills for insects.

2. An anatomist studies the structure of the body and so might be interested in the shape of the fish's fins and body, the structure and arrangement of the penguin's feathers, and how the insect's exoskeleton protects it from its enemies. A physiologist studies body function and so might be interested in how the fish's muscles propel it through the water, how the penguin generates body heat, and how the insect's senses warn it about potential predators.

3. Bone cells are embedded in a matrix of ropelike collagen fibers and hard calcium salts. This combination makes bone hard without being brittle. Simple squamous epithelium is a single layer of thin cells, forming the lining of the lungs and blood vessels. Gases and liquids can easily diffuse through this thin layer, facilitating exchange. Blood cells are small and round and suspended in a liquid matrix, enabling them to be pumped around the body to transport oxygen and defend against infection.

4. The digestive system takes in, breaks down, and absorbs food into the blood. The respiratory system draws in air, and oxygen enters the blood in the lungs. The food and oxygen are transported to the brain by the circulatory system—the heart, blood, and blood vessels. The nervous and endocrine systems control these activities.

5. When blood calcium gets too high, the parathyroid glands slow absorption of calcium by the intestine and speed up excretion of calcium by the kidneys. When blood calcium gets too low, the parathyroids increase absorption by the intestine and slow excretion by the kidneys. This illustrates negative feedback because a change in the concentration of calcium triggers the control mechanisms to counteract further change in the same direction.

Chapter 21

Review the Concepts

Exercise 1: 1. herbivore, bulk feeder   2. omnivore, substrate feeder   3. herbivore, fluid feeder   4. carnivore, suspension feeder   5. omnivore, bulk feeder   6. herbivore, substrate feeder   7. carnivore, bulk feeder   8. carnivore, fluid feeder   9. omnivore, suspension feeder   10. herbivore, bulk feeder

Exercise 2: Label as in Figure 21.3B in the text. Color as follows: *Earthworm:* mouth, pharynx, esophagus, and crop yellow; gizzard red; intestine red and green; anus blue. *Grasshopper:* mouth, esophagus, and crop yellow; gizzard red; stomach red and green; gastric pouches green; intestine green and blue; anus blue. *Bird:* mouth, esophagus, and crop yellow; stomach and gizzard red; intestine red and green; anus blue.

Exercise 3: See Figure 21.4 in the text.

Exercise 4: 1. incisor   2. liver   3. esophageal sphincter   4. anus   5. gallbladder   6. pyloric sphincter   7. pharynx   8. liver   9. molars   10. esophagus   11. duodenum (small intestine)   12. pancreas   13. tongue   14. large intestine   15. canines   16. oral cavity   17. stomach   18. bolus   19. salivary glands   20. small intestine   21. large intestine   22. epiglottis   23. stomach   24. esophagus   25. rectum   26. oral cavity   27. pharynx (trachea is respiratory)   28. gastric gland   29. chyme

Exercise 5: 1. esophagus   2. peristalsis   3. acid chyme   4. gastric glands   5. gastric juice   6. gastrin   7. mitosis   8. hydrochloric acid   9. pepsin   10. proteins   11. small intestine   12. protects   13. Mucous   14. mucus   15. GERD   16. gastric ulcer   17. mucus   18. infection   19. pyloric sphincter   20. small intestine   21. ulcer   22. bile   23. pancreas   24. neutralizes   25. liver   26. fats   27. emulsification   28. lipase   29. proteins   30. amino acids   31. starch

32. monosaccharides   33. absorption   34. villi   35. microvilli   36. capillaries   37. lymph   38. diffuse   39. hepatic portal vein   40. glycogen   41. large intestine   42. water   43. feces   44. anus

Exercise 6: 1. C   2. F   3. B   4. A or E   5. D   6. A or E

Exercise 7: Here's a possible answer, in exactly 25 words: "Every animal must obtain from its diet fuel for metabolism, raw materials for making the animal's own molecules, and substances the animal cannot make itself."

Exercise 8: Answers for this exercise will depend on your body weight. It should take 20–40 minutes for the average person to "burn" the Calories in a cheeseburger, depending on body weight.

Exercise 9:

Exercise 10: 1. I   2. I   3. D   4. D   5. I   6. I   7. D   8. D   9. D   10. I   11. D   12. I   13. I   14. D   15. D   16. D   17. D   18. I   19. D

Test Your Knowledge

Multiple Choice: 1. d   2. a   3. b   4. c   5. c   6. e   7. d   8. c   9. a   10. e   11. b   12. b   13. c   14. d   15. c

Essay: 1. The large polymer molecules in food—proteins, carbohydrates, fats, and nucleic acids—are too large to pass through cell

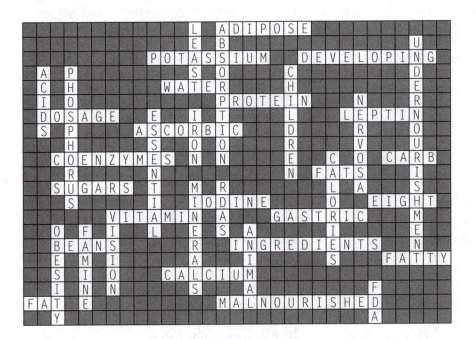

membranes. They cannot be absorbed or taken up by cells unless they are digested first. Also, an animal needs to break down these large molecules to obtain monomers to make its own polymers.

2. Both a grasshopper and a human have alimentary canals with a mouth and an anus. In both, food is cut and chewed by the mouth before it passes through the esophagus. In a grasshopper, chewed food is stored in the crop, ground by the gizzard, digested largely by the stomach, and absorbed in the stomach and gastric pouches. The intestine mainly compacts wastes. Humans have no crop or gizzard. The human stomach begins chemical digestion, but most digestion and absorption occur in the small intestine. The human large intestine has much the same functions as the grasshopper intestine.

3. Saliva contains a slippery glycoprotein that lubricates food. Buffers neutralize food acids and help prevent tooth decay. Antibacterial agents kill food bacteria, and an enzyme in saliva begins the digestion of starch.

4. Herbivores generally have longer, more complex alimentary canals than carnivores, because plants are harder to digest. Horses and elephants house cellulose-digesting microbes in the colon and a large cecum. Some of the nutrients are absorbed by the cecum and colon, but some are lost. Rabbits compensate for this loss by reingesting their fecal pellets, extracting additional nutrients as the already-digested material passes through the alimentary canal a second time. Ruminant mammals such as cattle have four-chambered stomachs. Food is digested by bacteria in two of the compartments, then the softened material (the "cud") is regurgitated and rechewed, then swallowed and passed to another compartment where digestion is completed. The nutrients absorbed by the cow's intestine come from both plant material and the rapidly reproducing bacteria in the stomach.

5. Fad diets—low-carbohydrate diets, low-fat diets, and formula diets—may be inadequate in essential nutrients, may result in loss of body protein, may cause health problems, and may ultimately prove unsatisfying and result in bingeing and regaining lost weight. A balanced diet of 1200 calories or more, combined with moderate exercise, is best for healthy, gradual weight loss.

6. Food labels provide two kinds of information — a list of ingredients and nutrients present in the food. Such information is useful in planning a healthy, balanced diet—for example, obtaining the proper balance of vitamins, minerals, and other nutrients, or avoiding foods high in calories, sugar, saturated fat, and sodium.

Apply the Concepts

Multiple Choice: 1. c 2. b 3. b 4. a 5. e 6. a 7. b 8. b 9. a 10. b

Essay: 1. Muscle is meat, so meat protein will supply James with the right mix of amino acids needed by the body for building muscle. Meat is also high in saturated fats, however, which can contribute to weight gain, heart disease, and cancer. Plant foods are low in saturated fats, and contain fiber, vitamins, and antioxidants that reduce cancer risk. But because no single plant contains all the amino acids needed to build muscle, Michael must be careful to consume a variety of plant foods to get all the amino acids his body needs.

2. It would be most difficult to live without the small intestine, because it is the portion of the digestive system where most food molecules are digested and where nearly all absorption of nutrients takes place.

3. If trypsin and chymotrypsin were manufactured in their active form, they would digest proteins in the cells where they are made and stored. Like pepsin in the stomach, they are produced in an inactive form that does not digest protein, and they are converted to active, protein-digesting enzymes once they enter the intestine.

4. If 4.83 kcal of food energy are liberated for every liter of oxygen consumed, an astronaut would consume 2,500 kcal/4.83 kcal/L = 518 liters of oxygen per day.

5. He could try eliminating all scandium from the diet and looking for ill effects. There are ethical problems with this experiment, because it might harm human experimental subjects. He could experiment with animals, but they might not require scandium as he thinks humans do. If he eliminates scandium from the diet, he might accidentally eliminate other elements as well, and then it would be difficult to determine what effects are due to scandium alone. Finally, if scandium is required, it must be in minute quantities, or this requirement probably would have been discovered already. If this mineral is required in tiny concentrations, how do we make sure none is present in the diet?

6. Calculate by using the conversion factors (kcal "burned" per hour) in Module 21.15 in the text.

Chapter 22

Review the Concepts

Exercise 1: 1. transport 2. breathing 3. breathing 4. transport 5. exchange 6. breathing 7. exchange 8. exchange 9. transport

Exercise 2: 1. C, P 2. A, R 3. D, Q 4. B, S 5. A, R 6. A, R 7. B, S 8. B, S 9. C, P 10. D, Q

Exercise 3: 1. skin 2. oxygen 3. carbon dioxide 4. respiratory 5. water 6. moist 7. large 8. thin 9. gills 10. moist 11. less 12. lamellae 13. blood vessels 14. ventilate 15. opposite 16. blood 17. Countercurrent exchange 18. oxygen 19. more 20. more 21. diffusion 22. air 23. higher 24. easier 25. dry 26. tracheal 27. Tracheal 28. circulatory 29. cells 30. Oxygen 31. carbon dioxide 32. water. 33. lungs 34. water 35. gills 36. frogs 37. mammals 38. Amphibians 39. metabolism 40. chest cavity 41. diaphragm 42. nose 43. filtered 44. nasal 45. pharynx 46. larynx 47. vocal cords 48. tense 49. less tense 50. trachea 51. bronchi 52. bronchioles 53. mucus 54. cilia 55. alveoli 56. millions 57. blood capillaries 58. Carbon dioxide

Exercise 4: See Figure 22.6A in the text.

Exercise 5:

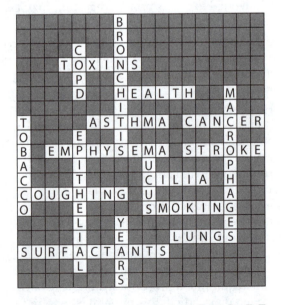

Exercise 6: 1. I 2. I 3. E 4. I 5. I 6. I 7. E 8. E 9. I 10. E 11. E 12. E

Exercise 7: 1. up 2. down 3. up 4. down 5. up 6. up 7. down 8. up 9. up

Exercise 8: 1. circulatory 2. O_2 3. CO_2 4. low 5. high 6. alveoli 7. CO_2 8. drop 9. brain 10. diaphragm 11. ribs 12. CO_2 13. O_2 14. O_2 15. CO_2 16. CO_2 17. higher 18. lower 19. O_2 20. soluble 21. hemoglobin 22. red 23. iron 24. O_2 25. muscles 26. higher 27. higher 28. out of 29. into 30. red 31. hemoglobin 32. water 33. bicarbonate 34. plasma 35. buffering 36. rises 37. CO_2

Exercise 9: A. 5 B. 6 C. 1 D. 11 E. 2 F. 4 G. 9 H. 7 I. 3 J. 8 K. 10

Test Your Knowledge

Multiple Choice: 1. b 2. d 3. a 4. e 5. a 6. c 7. b 8. d 9. a 10. a 11. b

Essay: 1. Air contains a higher concentration of oxygen than water, and it is easier to move across the respiratory surface, but it tends to dry the respiratory surface. There is no danger of the respiratory surface drying out in water, but water contains less oxygen than air and is harder to move.

2. Gills and lungs are both divided into small units that greatly enlarge their surface area for gas exchange. They both have a thin, moist surface through which oxygen passes into numerous blood vessels. They usually have a mechanism for ventilating the respiratory surface. The primary difference between gills and lungs is that gills project outward from the body and lungs are folded inward.

3. There are millions of alveoli in each lung, forming a large respiratory surface. Oxygen dissolves easily in the moisture inside an alveolus and passes easily through the thin wall of the alveolus and into the network of capillaries that cover its surface.

4. The two vocal cords are located in the larynx, or voice box, which is between the pharynx and the trachea. When you exhale, air rushes past the vocal cords. If they are tensed, they vibrate and produce sounds. The more the muscles in the larynx tense the vocal cords, the faster they vibrate, and higher-pitched sounds are produced. When the cords are less tense, they produce lower-pitched sounds.

5. When you inhale, the diaphragm contracts and moves downward, and muscles between the ribs contract and expand the rib cage. This enlarges the chest cavity, and the lungs expand with it. This reduces the pressure in the alveoli to less than that of the atmosphere. Air rushes in from the area of higher pressure outside to the area of lower pressure in the alveoli. Thus, the lungs

expand first, causing air to rush in. This is called negative-pressure breathing.

6. Water passes through the gill in a direction opposite to the flow of blood within each of the many thin lamellae that make up the gill's respiratory surface. As blood picks up more and more O_2, it comes into contact with water that has more and more O_2 available. Thus, a concentration gradient is maintained that favors diffusion of O_2 along the entire length of the lamella. This allows the gill to extract a large percentage of O_2 from the water flowing through it.

Apply the Concepts

Multiple Choice: 1. b   2. c   3. b   4. a   5. b   6. d   7. c   8. d   9. b   10. d   11. a

Essay: 1. When you inhale, air passes through your nose; into the nasal cavity; through the pharynx, larynx, and trachea; and into the bronchi that lead into each lung. The air continues through smaller and smaller bronchioles, which end in tiny alveoli. An O_2 molecule dissolves in the moisture on the surface of an alveolus and diffuses through the thin epithelium of the alveolus and into a capillary. The O_2 molecule dissolves in blood plasma, diffuses through the membrane of a red blood cell, and attaches to one of the iron atoms in a hemoglobin molecule. The blood returns to the heart, which pumps it through a blood vessel to a capillary in the brain. There the O_2 molecule detaches from the hemoglobin molecule and diffuses out of the red blood cell, into the plasma, through the wall of the capillary, and into a brain cell.

2. The respiratory surface should have a large area through which gas exchange can occur. This could be accomplished in a relatively small space by folding or dividing the surface into many smaller tubes, plates, or sacs—like the alveoli of the lungs or lamellae of fish gills. The surface must be moist so that gas molecules will dissolve readily, and it must be thin so that gases can diffuse through it easily and rapidly. The respiratory surface should be protected in some way—perhaps folded inside the robot's body—to keep it from drying out or being damaged. Finally, there should be some mechanism for ventilating the respiratory surface—similar to breathing or movement of a fish's gill covers—to move fresh air over the surface to maximize gas absorption. Maybe a fan?

3. The carbon monoxide molecules would attach to hemoglobin and reduce its ability to carry oxygen. Cells cannot use carbon monoxide as they use oxygen, so they might be starved for oxygen.

4. 20 cigarettes/day × 11 minutes/cigarette × 365 days/year × 40 years = 3,212,000 minutes. There are 365 × 24 × 60 = 525,600 minutes in one year, so this will shorten the smoker's life by 3,212,000/525,600 = 6.1 years.

5. The breathing centers in the brain normally regulate breathing rate by monitoring blood pH. CO_2 combines with water in the blood to form carbonic acid, so if the concentration of CO_2 in the blood increases, the blood becomes more acidic, and pH drops. The brain detects this change and speeds up breathing, which removes carbon dioxide and returns blood pH to the acceptable range. In the submarine, CO_2 removal was functioning normally, so there was no buildup of CO_2 (and no drop in pH) in the blood of the sailors—and no response on the part of their breathing centers—even though oxygen was getting dangerously low.

Chapter 23

Review the Concepts

Exercise 1: 1. G   2. F   3. H   4. J   5. I   6. A   7. K   8. B   9. D   10. C   11. L   12. E

Exercise 2: A: a. 1   b. 3   c. 4   d. 2
B: a. 1   b. 3   c. 2   d. 6   e. 5   f. 4

Exercise 3: *Labels:* a. superior vena cava   b. capillaries of upper body   c. pulmonary artery   d. pulmonary artery   e. capillaries of right lung   f. aorta   g. capillaries of left lung   h. pulmonary vein   i. pulmonary vein   j. right atrium   k. left atrium   l. inferior vena cava   m. aorta   n. right ventricle   o. left ventricle   p. capillaries of lower body. *Colors and blood flow:* See Figure 23.3A in the text.

Exercise 4: 1. superior vena cava   2. right atrium   3. right ventricle   4. pulmonary arteries   5. pulmonary veins   6. left atrium   7. left ventricle   8. aorta   9. capillaries   10. inferior vena cava   11. right   12. pulmonary   13. left   14. systemic   15. artery

Exercise 5: 1. AV   2. SL   3. SL   4. AV   5. SL   6. AV   7. AV   8. SL   9. SL   10. SL

Exercise 6: Answers will depend on your heart rate and fitness level—typically on the order of 5,000 mL/minute.

Exercise 7: 1. cholesterol and saturated fat   2. pacemaker (SA node)   3. heart attack or myocardial infarction   4. AV node   5. angioplasty   6. atherosclerosis   7. bypass   8. brain   9. epinephrine   10. defibrillator (AED)   11. CT and MRI   12. coronary artery   13. stent   14. artificial pacemaker   15. Aspirin   16. ECG or EKG   17. cardiovascular disease   18. C-reactive protein (CRP)

Exercise 8: 1. arteries   2. heart   3. capillaries   4. veins   5. capillaries   6. heart   7. thin   8. thick

9. in-between   10. epithelium (and basement membrane)   11. epithelium, smooth muscle, connective tissue   12. epithelium, smooth muscle, connective tissue   13. no   14. no   15. yes

Exercise 9: 1. G   2. C   3. B   4. A   5. A or B   6. D   7. D   8. A or B   9. C or D   10. A   11. A or B   12. F   13. B   14. F or G   15. F and G

Exercise 10: Some possible changes that might cause increased blood flow through muscle capillaries are increased temperature, increased CO_2 concentration, increased concentration of other waste products, decreased O_2 concentration, and nerve impulses from the brain.

Exercise 11: The red arrow is oxygen-rich blood flowing into the capillary; the blue arrow is oxygen-poor blood leaving. At the arterial end of the capillary, the blood pressure arrow is bigger than the osmotic pressure arrow, so the net flow is out. At the venous end, the osmotic pressure arrow is bigger than the blood pressure arrow, so the net flow is in.

Exercise 12: 1. in-between   2. blood clotting   3. erythrocytes   4. medium-sized   5. most numerous   6. largest   7. defense and immunity

Exercise 13: 1. high   2. pulse   3. arterioles   4. erythrocytes   5. oxygen   6. leukocytes   7. phagocytes   8. plasma proteins   9. platelets   10. capillary   11. epithelium   12. interstitial   13. connective tissue   14. Platelets   15. fibrinogen   16. fibrin   17. coronary arteries   18. heart attack   19. venule   20. vein   21. valve

Exercise 14: 1. E   2. G   3. H   4. L   5. D   6. I   7. N   8. C   9. J   10. A   11. B   12. K   13. O   14. M   15. F

Exercise 15: 1. atrium (The others are all blood vessels.)   2. fibrin (The others are blood cells.)   3. fibrinogen (The others relate to blood pressure.)   4. aorta (The others are parts of the pulmonary circuit.)   5. leukocyte (The others are involved in blood clotting.)   6. semilunar valve (The others are parts of the heart's electrical system.)   7. leukocyte (The others relate to heart contraction and relaxation.)   8. epithelium (The others are concerned with delivery of oxygen to cells.)   9. pulmonary artery (The others relate to the systemic circuit.)

Test Your Knowledge

Multiple Choice: 1. d   2. e   3. e   4. a   5. d   6. b   7. b   8. e   9. a   10. a   11. a   12. b   13. b

Essay: 1. A fish has a single circuit of blood flow. The fish's two-chambered heart receives and pumps only oxygen-poor blood from body tissues. After leaving the heart, the blood passes through capillaries in the gills, where it picks up oxygen. In the gills, the blood slows down considerably, then it continues on to capillaries in various body tissues. A mammal, a terrestrial vertebrate, has two blood circuits instead of one. The right side of the heart receives oxygen-poor blood returning from the tissues and pumps it through capillaries in the lungs. The left side of the heart receives this oxygen-rich blood and pumps it to capillaries in body tissues. Blood passing through the lung capillaries loses speed and pressure, but the second trip through the heart gives it a boost, so the oxygen-rich blood is rapidly and efficiently delivered to the tissues. This supports the higher metabolic rate characteristic of many land vertebrates.

2. When you are exercising, cardiac output can be increased by increasing (1) the heart rate and (2) the amount of blood the heart pumps with each beat.

3. High blood pressure causes the heart to work harder, which may result in an enlarged and weakened heart. It can also cause small tears in the lining of blood vessels and increase the tendency for vessel-blocking plaques and clots to form there. High blood pressure increases the risk of heart attacks, heart disease, stroke, and kidney failure.

4. Blood pressure is highest in the aorta and only slightly lower in the arteries because in these vessels the heart is pushing on the blood with full force. As blood passes through the narrower arterioles, pressure drops sharply, mainly due to increased resistance to blood flow caused by friction between the blood and the extensive inner surface of the many arterioles. Pressure continues to drop as blood flows through the capillaries, due to the resistance and huge surface area of the capillary walls. By the time blood gets to the veins, its pressure is near zero. Reduced pressure in the chest due to breathing, squeezing action of skeletal muscles, and venous valves help keep blood moving back to the heart.

5. The large number of capillaries and their minute size give them a tremendous combined surface area through which materials can be exchanged between blood and interstitial fluid. Capillary walls consist of a single thin layer of epithelial cells, and there are gaps between these cells, also facilitating exchange.

Apply the Concepts

Multiple Choice: 1. c   2. a   3. c   4. b   5. b   6. a   7. d   8. e   9. d   10. a

Essay: 1. Cardiac output is equal to the amount of blood pumped by each heartbeat times the number of beats per minute. The runner's cardiac output is 90 mL per beat × 160 beats per minute = 14,400 mL, or 14.4 L, per minute.

2. In an artery, blood pressure increases during ventricular systole, when the heart is contracting and pushing blood into the artery. Pressure drops during diastole, when the heart relaxes. An arterial pressure of 130/80 tells you that the systolic pressure is 130 mm of mercury, and the diastolic pressure is 80 mm of mercury. The numerous arterioles and capillaries offer much resistance to blood flow. By the time blood gets to a vein, the force of the beating heart no longer propels it, so there are no pulses of pressure, just a slow, steady flow at a low, constant pressure.

3. At the capillaries, blood pressure tends to force fluid out of the blood and into the tissues. The osmotic pressure of blood causes fluid to leave the tissues and enter the blood. Blood pressure tends to win out at the arterial end of a capillary, and osmotic pressure is greater at the venous end, but overall these forces are just about in balance. In a person with high blood pressure, blood pressure would overcome osmotic pressure, and there would be a net movement of fluid into the tissues, causing swelling.

4. If the blood vessels were reversed as described, the right ventricle would pump blood to the tissues, and this blood would soon return to the right atrium and then to the right ventricle, where this oxygen-poor blood would again be pumped to the tissues, without being sent to the lungs to pick up oxygen. Similarly, the left atrium and left ventricle would keep sending the same blood to the lungs, without ever pumping this oxygen-rich blood to the tissues that need it. Once the baby is born and the lungs are functioning, there would be no way for oxygen-rich blood to get to the tissues, which would soon die from lack of oxygen.

Chapter 24

Review the Concepts

Exercise 1: 1. H 2. M 3. E 4. I 5. P 6. A 7. O 8. N 9. G 10. B 11. J 12. D 13. K 14. C 15. L 16. F

Exercise 2: A. bacteria B. histamine C. swelling D. fluid E. leakiness F. pathogens G. cell debris

Exercise 3: 1. Innate immunity—defenses such as the skin, inflammation, phagocytes, and antimicrobial proteins—is nonspecific, always ready to act against any threat. Acquired immunity—carried out by the immune system—fully develops only after exposure to a specific pathogen.   2. An antigen is a foreign molecule that elicits an immune response. An antibody is a protein that attaches to and counters a particular antigen.   3. T cells mature in the thymus and carry out cell-mediated immunity. B cells mature in bone marrow and carry out humoral immunity.   4. B cells carry out humoral immunity, sending out antibodies to attack bacteria and viruses in body fluids. In cell-mediated immunity, T cells attack body cells invaded by bacteria and viruses.   5. The lymphatic system consists of lymph vessels, lymph nodes, and other organs such as the thymus and spleen. It returns a fluid called lymph to the blood vessels and helps defend the body. The circulatory system consists of the heart, blood vessels, and blood.   6. Active immunity occurs when the body is stimulated by antigens and secretes antibodies in its own defense. In passive immunity, the body does not secrete its own antibodies but rather "borrows" antibodies temporarily.   7. Both blood capillaries and lymphatic capillaries are narrow tubes with thin walls, but blood capillaries are open at both ends, whereas lymphatic capillaries end in dead ends. 8. Plasma containing hepatitis antibodies could be injected into a person to give temporary passive immunity. A vaccine contains antigens from the hepatitis pathogen and elicits long-term active immunity.

Exercise 4: 1. innate 2. immune 3. lymphocytes 4. bone marrow 5. B cells 6. T cells 7. thymus 8. protein 9. antigens 10. antibodies 11. antigen-binding sites 12. determinants 13. lymph 14. selection 15. clone 16. effector 17. plasma 18. antibodies 19. primary 20. days 21. immune 22. secondary 23. memory 24. decades

Exercise 5: 1. antigens 2. B cells 3. primary immune response 4. antibodies 5. effector cells 6. memory cells 7. antigens 8. secondary immune response 9. antibodies

Exercise 6: 1. activation of complement 2. precipitation of dissolved antigens 3. neutralization 4. agglutination of microbes 5. cell lysis 6. phagocytosis

Exercise 7: For example, monoclonal antibodies might be used to identify human blood proteins in criminal investigations. Wildlife managers might use them to differentiate between wild and hatchery fish. Monoclonal antibodies might also be used to identify particular materials in archaeological digs—skins, animal remains, or food, for example.

Exercise 8: 1. H, V 2. D, P 3. F, T 4. A, S 5. G, R 6. J, Y 7. C, W 8. I, U 9. E, Q 10. B, X

Exercise 9: 1. M 2. E 3. C 4. J 5. B 6. I 7. A 8. F 9. L 10. K 11. O 12. G 13. P 14. H 15. D 16. N

Exercise 10: 1. In SCID and other immunodeficiency diseases, components of the immune system are defective and the body is susceptible to infection.   2. An allergy is an overreaction by the immune system to normally harmless antigens, such as those carried by cat hair. This triggers the inflammatory response, which produces allergy symptoms.   3. Radiation kills or impairs certain immune cells and may depress the immune system's ability to fight disease.   4. The immune system of a heart transplant recipient recognizes the MHC markers of the transplanted heart as foreign and attacks the organ.   5. In rheumatoid arthritis, some of the patient's own antibodies attack the joints, where they cause inflammation and damage.   6. HIV, the AIDS virus, attacks helper T cells, disabling the immune system and leaving the victim defenseless against infections and certain cancers.   7. Crohn's disease may be caused by an autoimmune reaction against normal bacteria in the intestine, causing inflammation of the lining of the digestive tract.

Test Your Knowledge

Multiple Choice: 1. d   2. e   3. d   4. e   5. a   6. d
7. e   8. c   9. c   10. b   11. a   12. c   13. e   14. d
15. a   16. e   17. b

Essay: 1. A vaccine contains a harmless form of a disease-causing microbe. It cannot make you sick, but it stimulates B and T cells to become effector cells and mount a primary immune response. At the same time, it stimulates some B and T cells to become memory cells, which are responsible for active immunity and are able to mount a rapid and massive secondary immune response if you are ever exposed to the "real" harmful microbe, even years later.

2. B cells carry out humoral immunity against microbes in body fluids. T cells carry out cell-mediated immunity against body cells infected by microbes or altered by cancer. When B cells are activated, they produce antibodies that circulate in body fluids and attach to microbe antigens. This clumps and inactivates the microbes and tags them for destruction by complement and phagocytes. Receptors on activated cytotoxic T cells enable them to bind to antigens on infected body cells. The T cells discharge a substance called perforin that destroys infected cells.

3. Immune system cells do not analyze antigens and produce antibodies to fit. Instead, the body contains small numbers of many types of immune cells, each type capable of making antibodies that fit a certain antigen. When an antigen appears, only the clone of cells capable of responding to that antigen are selected—a mechanism known as clonal selection. The cells that respond have the ability to make an antibody that fits the antigen before the antigen appears. Antibodies are not made to order.

4. They are lymph nodes. Lymph carries the microbes responsible for your sore throat to the lymph nodes, where T and B lymphocytes are activated. The activated B cells proliferate and produce antibodies that are carried by body fluids to sites of infection. T cells also travel to sites of infection. The rapid production of lymphocytes makes your lymph nodes swollen and tender.

5. When tissues are damaged, cells release chemical alarm signals such as histamine, which trigger various defensive mechanisms. Histamine causes blood vessels to dilate and become leakier. Blood flow increases, and plasma passes out of the blood vessels and into the damaged tissues. Other chemicals attract white blood cells, which engulf bacteria and debris. Blood-clotting proteins and platelets wall off the damaged area and allow repair to begin. Local increases in blood flow, fluid, and cells produce the redness, heat, and swelling characteristic of inflammation. Inflammation disinfects and cleans injured tissues and allows healing to begin.

Apply the Concepts

Multiple Choice: 1. a   2. b   3. d   4. c   5. e   6. b
7. b   8. a   9. a   10. d

Essay: 1. The first time a person is exposed to an allergen such as bee venom, B cells make special antibodies that attach to mast cells—cells that make histamine and other chemicals that trigger an immune response. By the time this slow sensitization process occurs, the venom is gone. The next time the individual is stung, the antibodies are already in place, and the mast cells suddenly release large amounts of inflammatory chemicals. This triggers an abrupt dilation of blood vessels, causing a sudden drop in blood pressure—anaphylactic shock—that can be fatal.

2. It is possible that each time you catch a cold it is a different virus. You get sick for a while, but eventually the primary immune response fights off the cold virus. Memory cells make you immune to that virus but not to other cold viruses, which probably have different antigens. A 2-year-old gets lots of colds because a 2-year-old has not yet developed immunity to many of the cold viruses. By the age of 80, an individual has been exposed to and developed immunity to most cold viruses, so colds are less frequent.

3. The injection contained antibodies against hepatitis, so it conferred passive immunity to the disease. The antibodies soon disappeared from John's blood. Because he was not exposed to hepatitis

antigens, his body never developed memory cells and active immunity against the disease.

4. A virus may have several different antigens on its surface. In addition, antigens are large molecules, and antibodies usually recognize localized regions, called antigenic determinants, on the surface of antigen molecules. It is possible that the same virus could activate lymphocyte clones that respond to different antigens or antigenic determinants in different people.

5. An antibody can attach to two different antigen molecules or microbes at once, and more than one antibody could attach to a given antigen or microbe. In this way, antibodies stick antigens and microbes together—processes called precipitation (antigens) and agglutination (microbes). The clumps of antigens or microbes formed in this way are easily engulfed by phagocytes.

6. The body directs the cell-mediated immune response, carried out by T cells, at transplanted organs, whose MHC antigens the body senses as foreign (non-self). Transplant recipients are given drugs that suppress immunity, particularly cell-mediated immunity. This makes recipients more prone to infections, especially viral infections, because cell-mediated immunity is primarily directed against virus-infected cells.

Chapter 25

Review the Concepts

Exercise 1: 1. ectotherms   2. endotherm
3. ectotherms   4. ectotherm   5. endotherms
6. endotherms   7. endotherm   8. ectotherm
9. endotherms   10. endotherm

Exercise 2: 1. D, S   2. B, Q, T   3. B, Q   4. C, T
5. D, S, T   6. A, P   7. D, R   8. A, T   9. D, T
10. B, R   11. A, P   12. B, R   13. D, S, T   14. C, T
15. E, T   16. B, P

Exercise 3: 1. osmoconformer   2. osmoregulator
3. osmoregulator   4. osmoregulator   5. neither
6. gain   7. lose   8. lose   9. both   10. lose
11. gain   12. lose   13. actively transports certain ions into and out of cells   14. kidneys excrete dilute urine; digestive system and gills take up ions
15. drinks seawater; expels salts through gills and concentrated urine   16. drinks water and consumes salts; behavior and waterproof skin save water

Exercise 4: 1. most aquatic animals, including many fishes   2. mammals, amphibians, some fishes   3. birds, insects, many reptiles, some amphibians, land snails   4. readily diffuses into water; no energy expended to make it   5. highly soluble in water, less toxic than ammonia

6. not toxic, requires little water for disposal; can be stored in shelled egg   7. highly toxic, cannot be stored in body, requires much water for disposal
8. energy expended to make it; requires some water for disposal   9. energy expended to make it

Exercise 5: See Figure 25.6A in the text.

Exercise 6: See Figure 25.6D in the text.

Exercise 7: 1. A   2. F   3. B   4. E   5. D   6. C

Exercise 8: 1. B   2. B   3. C   4. D   5. A   6. B
7. A   8. D   9. B

Exercise 9: 1. Dialysis is better if kidney damage is expected to be temporary. There are no problems with rejection or infection. Over the long term, dialysis is expensive, time-consuming, and not always available. It is no substitute for a real kidney, because dialysis restricts diet and lifestyle, and wastes build up between treatments.

2. A transplanted kidney works continuously, with no buildup of wastes as between dialysis treatments. The donor can live with one kidney, and over the long run a transplant is less expensive than dialysis. There may be a long wait for a transplant, and there is a chance of transplant rejection and infection due to drugs that suppress rejection.

Exercise 10: 1. ammonia   2. urea   3. renal
4. nephrons   5. glomerulus   6. Bowman's capsule
7. filtrate   8. reabsorption   9. secretion
10. excreted   11. active transport   12. water
13. loop   14. renal vein   15. secretion   16. urine
17. ureters   18. urinary bladder

Test Your Knowledge

Multiple Choice: 1. a   2. d   3. a   4. c   5. b   6. a
7. c   8. e   9. c   10. e   11. b   12. e   13. c   14. c
15. a

Essay: 1. When an animal is too warm, nerves signal blood vessels in the skin surface to dilate, so more blood flows from the warm body core to the surface, and its heat can escape to the surroundings. Heat loss can be further increased by evaporative cooling, such as sweating or panting. When the body is too cold, sweating and panting are reduced, and nerves signal blood vessels in the skin to constrict, decreasing movement of heat from the warm body core to the surface and reducing heat loss.

2. A goose's legs must contain countercurrent heat exchangers. In such a heat exchanger, warm blood from the core of the body and cold blood from the foot flow in opposite directions in adjacent blood vessels. Heat from the body warms the blood from the foot. Cold blood is thus warmed as it moves up the leg, so it does not chill the body, and warm blood is cooled as it flows to the foot, so there is little heat left to lose when it gets there.

3. The internal fluids of a freshwater fish have a higher solute concentration than the surrounding water. This causes water to enter the fish by osmosis, mainly through the gills. The fish does not drink water (except with its food). To get rid of the excess water, the fish's kidneys produce large amounts of dilute urine. Some solutes are lost with the urine. To replace them, the digestive system absorbs ions from food, and the gills take up salt from the surrounding water. A saltwater fish has the opposite problem. The solute concentration of its body fluids is less than that of the surrounding water, so it tends to lose water by osmosis. To compensate, it drinks salt water and disposes of the excess salts via its gills and its kidneys, which excrete small amounts of concentrated urine.

4. An animal can bask in the sun or huddle together with other animals to warm up. It can turn in a direction that exposes less surface to the sun, move into the shade, hide in a damp burrow, or bathe to cool off.

5. In filtration, water and other small molecules are forced by blood pressure from the porous glomerulus into Bowman's capsule, the first part of the nephron tubule. The process of reabsorption is reclaiming valuable solutes and water from the filtrate. Solutes are reabsorbed by active transport mainly in the proximal and distal tubules, and water is reabsorbed by osmosis in the loop of Henle and the collecting duct. Some substances, such as excess H^+ ions, are secreted by active transport from the blood into the filtrate in the proximal and distal tubules. After the filtrate is refined by the processes of absorption and secretion, it is called urine. It leaves the kidney and is expelled from the body—a process called excretion.

Apply the Concepts

Multiple Choice: 1. a   2. c   3. c   4. e   5. d   6. c   7. a   8. b   9. e   10. d   11. c

Essay: 1. The wind chill factor takes into account not only the temperature but also the movement of air, which accelerates heat loss. The body loses heat by conduction, convection, radiation, and evaporation. In comparison with still air, moving air increases heat loss by convection and evaporation. Thus, on a windy day, the body loses heat faster. The wind chill factor reflects the fact that a windy day seems colder than a day without wind.

2. A horned lizard is an ectotherm. It warms itself by absorbing heat from its surroundings, so it expends a lot of effort moving in and out of the sun to maintain the proper body temperature. It does not generate much heat from metabolism, so it does not eat much. A mouse is an endotherm. It

gets most of its heat from its own metabolism, so it does not need to expend much effort changing its position or location relative to the sun or outside energy sources. But because it maintains a constant high temperature, it needs to expend more effort than the horned lizard hunting for food to "burn."

3. Mammals and birds are endotherms, which generate body heat via their own metabolism. They are able to stay warm enough to function even when there is little heat available from the outside environment. Crocodiles and frogs are ectotherms, which obtain heat from their surroundings. There is little heat available in the polar regions; an Antarctic crocodile would not be able to obtain enough heat to function in this environment.

4. A large amount of water is filtered from the blood as a normal part of nephron function. As solutes are reabsorbed by the nephron tubules, water follows by osmosis. If you have been sweating and the body is short on water, blood solute concentration rises above a set point, and a control center in the brain triggers an increase in the blood level of antidiuretic hormone (ADH). ADH signals the nephrons to increase water reabsorption above normal levels. (ADH actually works by making the nephron tubule walls more permeable to water, allowing more water to leave the filtrate by osmosis and be retained by the body.) As a result, the body excretes a smaller amount of concentrated urine.

Chapter 26

Review the Concepts

Exercise 1: 1. neurosecretory cell   2. blood vessel   3. target cell   4. endocrine cell   5. secretory vesicles   6. blood vessel   7. target cell   8. nerve cell   9. nerve signals   10. nerve cell. For hormones and neurotransmitter molecules, see Figure 26.1 in the text.

Exercise 2: 1. both   2. N   3. both   4. N   5. both   6. E   7. N   8. E   9. N   10. E   11. E   12. N

Exercise 3: 1. steroid hormone   2. plasma membrane   3. target cell   4. receptor protein   5. nucleus   6. hormone-receptor complex   7. DNA   8. new protein   9. cellular response   10. epinephrine   11. receptor protein   12. target cell   13. plasma membrane   14. relay molecules   15. signal-transduction pathway   16. enzyme   17. cellular response

Exercise 4: See Figure 26.3 and Table 26.3 in the text.

Exercise 5: 1. thyroid gland   2. T_3 3. T_4 4. lowers blood calcium   5. estrogens   6. stimulate uterus; female sex characteristics   7. promotes uterine lining growth   8. adrenal medulla

9. epinephrine   10. norepinephrine   11. adrenal cortex   12. increase blood glucose   13. mineralocorticoids   14. raises blood glucose   15. insulin   16. lowers blood glucose   17. parathyroid glands   18. parathyroid hormone   19. pineal gland   20. involved in body rhythms   21. anterior lobe of pituitary   22. growth hormone   23. stimulates thyroid gland   24. stimulates milk production   25. ACTH   26. stimulates ovaries and testes   27. follicle-stimulating hormone   28. stimulates contraction of uterus and mammary gland   29. antidiuretic hormone   30. thymosin   31. stimulates T cell development   32. testes   33. support sperm formation; development of male characteristics   34. hypothalamus

Exercise 6: 1. P   2. H   3. A   4. H   5. P   6. A   7. A   8. P   9. A (or H)   10. H   11. P   12. A   13. H   14. H   15. A   16. P

Exercise 7: 1. E   2. C   3. B or D   4. D or B   5. A   6. E   7. A   8. D   9. B

Exercise 8: 1. blood clotting   2. parathyroid glands   3. parathyroid hormone   4. increased   5. increased   6. release of calcium from   7. calcitonin   8. thyroid gland   9. decreased   10. deposition of calcium in

Exercise 9: 1. increase   2. increases   3. increase   4. increase   5. decrease   6. decrease   7. decreases   8. increase   9. increases   10. increase   11. decreases   12. increases   13. decrease   14. decrease

Exercise 10: 1. E   2. E   3. M or G   4. M or G   5. M   6. E   7. E   8. G   9. M or G   10. M or G   11. G   12. M   13. E   14. E   15. G

Exercise 11: There is no single correct answer for a concept map. You learn by constructing it; many arrangements are possible.

Exercise 12: 1. A   2. D   3. E (and D)   4. B   5. C   6. E (and D)   7. B

Exercise 13: 1. All regulate reproduction.   2. Both are produced by the posterior lobe of the pituitary.   3. All act to raise blood glucose.   4. ACTH stimulates the adrenal cortex to secrete glucocorticoids.   5. All target the kidney.   6. All are produced by the thyroid gland.   7. Both are secreted by the pancreas and have opposing effects on blood glucose.   8. All are secreted in response to stress.   9. All are secreted by the anterior lobe of the pituitary.   10. They have opposing effects on blood calcium.   11. Both affect milk secretion.

Test Your Knowledge

Multiple Choice: 1. d   2. e   3. e   4. d   5. b   6. c   7. a   8. b   9. c   10. e   11. e   12. c   13. b   14. c   15. d

Essay: 1. The pituitary gland controls many other glands, but not all of them, and the pituitary itself is controlled by the hypothalamus, which is the real "master gland."

2. The adrenal medulla secretes epinephrine and norepinephrine, which trigger the fight-or-flight response to short-term stress. The adrenal cortex secretes mineralocorticoids and glucocorticoids in response to prolonged stress.

3. Hormones and neurotransmitters are both chemical signals between cells. Hormones are carried by the blood to distant cells, while neurotransmitters affect nearby cells directly.

4. For example: Gigantism—abnormally large size—is caused by oversecretion of growth hormone during development. Type 1 diabetes results from undersecretion of insulin. In type 2 diabetes, insulin levels are normal, but cells are less able to respond to it. In either type of diabetes, cells are unable to take up and use glucose and so burn fats and proteins instead. Glucose builds up in the blood, and numerous complications result.

5. Insulin lowers blood sugar, and glucagon raises it. Parathyroid hormone raises blood calcium, and calcitonin lowers it.

Apply the Concepts

Multiple Choice: 1. c   2. c   3. d   4. b   5. c   6. d   7. b   8. e   9. a   10. e   11. c   12. d

Essay: 1. When a person with hypoglycemia eats sugar, hyperactive beta cells in the pancreas secrete too much insulin, which causes blood sugar to drop too low. Eating sugar would just make this problem worse.

2. In diabetes, cells break down fat for energy because they have not received the insulin signal to take in and use glucose.

3. The enlarged thyroid gland might secrete more T_3 and T_4 than normal, boosting metabolic rate and body temperature. A pituitary tumor might cause the pituitary to secrete too much TSH, causing the thyroid gland to secrete too much T_3 and T_4.

4. An organ is affected by a hormone only if it has the proper receptor. Many cells have insulin receptors, but only kidney cells have receptors for ADH.

5. Dioxin probably acts as a steroid hormone mimic, like the estrogen mimics described in the chapter introduction. Number or responsiveness of receptors could vary in various tissues in different animals. An animal with a small number of receptors that respond to dioxin might not be affected at all. In another animal with a large number of receptors in the liver, cells might sustain serious liver

damage, while a different animal whose immune cells have some receptors might suffer slight immune system impairment.

Chapter 27

Review the Concepts

Exercise 1: 1. hermaphroditism 2. sperm 3. asexual reproduction 4. zygote 5. gamete 6. regeneration 7. external fertilization 8. ovum 9. copulation 10. budding 11. internal fertilization 12. fragmentation 13. infertility 14. fission

Exercise 2: 1. oviduct 2. ovary 3. uterus 4. vagina 5. follicles 6. corpus luteum 7. wall of uterus 8. endometrium 9. cervix 10. seminal vesicle 11. ejaculatory duct 12. prostate gland 13. bulbourethral gland 14. epididymis 15. testis 16. scrotum 17. vas deferens 18. erectile tissue 19. urethra 20. glans 21. prepuce 22. penis

Exercise 3: 1. seminiferous tubules of testes 2. follicles in ovary 3. daily and continuously 4. before birth 5. $2n = 46$ 6. $2n = 46$ 7. two secondary spermatocytes 8. one secondary oocyte and a first polar body 9. four immature sperm cells 10. one ovum and a second polar body 11. four 12. one 13. $n = 23$ 14. $n = 23$ 15. millions per day 16. one per month

Exercise 4: 1. hypothalamus 2. follicle-stimulating hormone (FSH) 3. luteinizing hormone (LH) 4. follicle 5. ovary 6. estrogen 7. endometrium 8. uterus 9. ovum 10. follicle 11. negative 12. FSH 13. LH 14. preovulatory 15. pituitary 16. FSH 17. LH 18. ovulation 19. fourteenth 20. oviduct 21. postovulatory 22. corpus luteum 23. progesterone 24. follicles 25. progesterone 26. fertilized 27. endometrium 28. menstruation

Exercise 5:

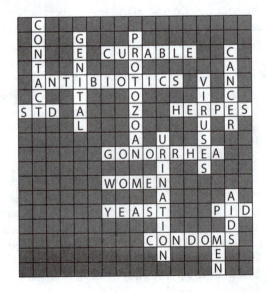

Exercise 6: 1. F 2. I or J 3. I or J 4. L 5. B 6. G 7. D 8. K 9. B 10. A 11. H 12. K 13. H 14. C 15. E 16. B 17. A

Exercise 7: 1. A sperm swims to the egg, propelled by a flagellum powered by ATP. 2. The sperm wriggles in between remaining follicle cells, and when it gets to the egg, the acrosome in the sperm head bursts on contact, releasing enzymes that digest an opening in the egg's coat. 3. Species-specific protein molecules on the surface of the sperm head bind with proteins on the egg. 4. If an egg were fertilized by more than one sperm, the resulting zygote nucleus would contain too many chromosomes and could not develop normally. 5. Fusion of sperm and egg membranes causes the egg membrane to quickly become impenetrable to other sperm, and then another barrier, the fertilization envelope, forms, which also keeps other sperm out. 6. Cellular respiration and protein synthesis speed up, and the egg and sperm nuclei fuse, forming the diploid zygote nucleus.

Exercise 8: A. 8 B. 5 C. 6 D. 4 E. 11 F. 1 G. 2 H. 9 I. 7 J. 10 K. 3 L. blastocoel M. digestive cavity N. blastocoel O. blastula P. neural tube Q. coelom R. somite S. zygote T. digestive cavity U. gastrula V. blastopore W. notochord X. ectoderm Y. mesoderm Z. endoderm

Exercise 9: 1. endoderm 2. mesoderm 3. endoderm 4. ectoderm 5. mesoderm 6. mesoderm 7. endoderm 8. ectoderm

Exercise 10: 1. induction 2. pattern formation 3. cell migration 4. programmed cell death 5. changes in cell shape 6. effects of homeotic genes 7. differentiation

Exercise 11: 1. I 2. M 3. J 4. V 5. A 6. K 7. P 8. T 9. R 10. S 11. B 12. U 13. N 14. Q 15. D 16. O 17. E 18. F 19. C 20. L 21. H 22. G

Exercise 12: There are many possible combinations. For example, a man's sperm could fertilize a donated egg in vitro. This egg could then be planted in his partner's uterus. A woman's eggs could be fertilized with her partner's sperm and the embryo implanted in the uterus of a surrogate mother. Or a donated egg could be fertilized in vitro by ICSI and implanted in a surrogate mother.

Test Your Knowledge

Multiple Choice: 1. e 2. c 3. d 4. c 5. b 6. c 7. b 8. a 9. b 10. d 11. d 12. c 13. a

Essay: 1. When food is abundant and the environment is unchanging, asexual reproduction would enable a single aphid to produce many offspring quickly. The populations could expand rapidly and exploit available resources. Sexual reproduction increases genetic variability among offspring. This might make the species more adaptable when the environment is changing and food supplies are less predictable.

2. The outermost extraembryonic membrane, the chorion, forms the embryo's part of the placenta. Extensions of the chorion, called chorionic villi, absorb oxygen and nutrients from the mother's blood. The amnion forms a fluid-filled sac around the embryo. The fluid absorbs shock and keeps the embryo from drying out. The yolk sac produces the embryo's first blood cells and germ cells, which later give rise to gamete-producing cells in the gonads. The allantois forms part of the umbilical cord and part of the embryo's urinary bladder.

3. Primary oocytes form in a female embryo and begin dividing to form eggs before birth. At maturity, FSH and LH cause a single follicle containing an egg to mature and rupture each month, and fertilization triggers the final meiotic cell division that completes the egg's development. Sperm production does not begin until a male reaches maturity, but after that point primary spermatocytes divide and mature at a rate of millions per day, stimulated by FSH. A primary oocyte divides to form a single ovum and some polar bodies, which are discarded. A primary spermatocyte divides to form four sperm.

4. See Figures 27.10–27.11 in the text.

Apply the Concepts

Multiple Choice: 1. c   2. d   3. c   4. b   5. e   6. a   7. a   8. d   9. e   10. e   11. a

Essay: 1. The fish that reproduce sexually would probably adapt most successfully to a changing environment, because sexual reproduction produces more variation among offspring than does asexual reproduction. Among a variety of offspring there is more likelihood of some individuals being suited to changed environmental conditions.

2. Sperm cannot develop at body temperature. The testes are located in the scrotum, a sac outside the abdominal cavity, where the sperm-forming cells can be kept cool enough to function normally. Extended periods of time soaking in a hot tub might interfere with sperm development.

3. The eggs of aquatic animals are usually fertilized externally, in the water. The male might court the female for her to lay her eggs, but it is not necessary for a male and a female to be close to one another for fertilization to occur. Land animals usually use internal fertilization. The male and the female must copulate so that sperm are deposited inside the body of the female. This requires closer cooperation between the male and the female, which is accomplished by more elaborate courtship behaviors.

4. What apparently happened was induction. The underlying blue fish tissue sent out a chemical signal that caused the transplanted red fish tissue to differentiate into a structure—a dorsal fin—appropriate to its new location. Normally, it would have become red fish belly skin. The red fish tissue grew into a red fish fin because it followed its own genetic instructions to form a fin at its new location. The blue fish signal only told the red tissue what it had to do, not how to do it..

5. Oogeneis, the formation of eggs, actually happens prior to birth. If a woman suffers from starvation, her own eggs have started to form long before, so they may be unaffected. But, if a woman is subjected to starvation while pregnant, when eggs in the developing ovaries of her daughters are starting to develop, the *daughters' eggs* might be affected. The changes will show up in *their* children, the *grandchildren* of the woman who starved!

Chapter 28

Review the Concepts

Exercise 1: 1. I, b   2. K, m   3. N, f   4. O, k   5. J, c   6. B, h   7. M, e   8. D, n   9. H, l   10. A, i   11. L, a   12. E, j   13. F, o   14. G, p   15. C, d   16. P, g

Exercise 2: See Figure 28.2 in the text.

Exercise 3:

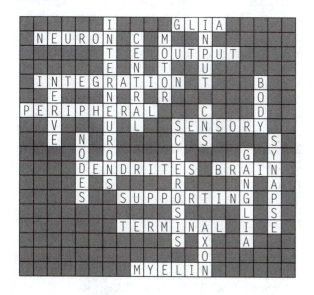

Exercise 4:   A. 3, T   B. 1, P   C. 2, Q   D. 4, R   E. 5, S

Exercise 5: 1. transmitting   2. electrical   3. neurotransmitters   4. cleft   5. axon   6. plasma membrane   7. receptor   8. receiving cell   9. channels   10. action potentials   11. enzyme

Exercise 6: 1. Y 2. N 3. Y 4. Y 5. N 6. N
7. N

Exercise 7: 1. Reduced level of serotonin would result in lower level of stimulation and depressed mood. 2. Increased inhibition would result in lower level of stimulation and depressed mood. 3. This would simulate stimulation by serotonin, elevating mood. 4. Increased inhibition would result in lower level of stimulation and depressed mood. 5. Decreased inhibition by GABA would elevate mood. 6. Increased stimulation by serotonin would elevate mood. 7. This would increase serotonin, elevating mood. 8. This would increase serotonin, elevating mood.

Exercise 8: 1. E 2. D 3. C 4. A 5. F 6. B
7. A, B, C, E 8. D, F 9. A, B, C, E

Exercise 9: 1. brain 2. somatic nervous system
3. autonomic nervous system 4. ventricles
5. meninges 6. cranial nerves 7. white matter
8. cerebrospinal fluid 9. peripheral nervous system 10. gray matter 11. central nervous system
12. spinal cord 13. spinal nerves 14. blood-brain barrier

Exercise 10: 1. stimulates activity 2. inhibits activity 3. constricts 4. dilates 5. stimulates erection 6. promotes ejaculation 7. slows
8. accelerates 9. stimulates saliva production
10. inhibits saliva production 11. constricts
12. dilates

Exercise 11: 1. cerebral cortex, E 2. thalamus, F
3. hypothalamus, G 4. pons, B 5. midbrain, C
6. cerebellum, A 7. medulla, D

Exercise 12: 1. rapid eye movement (REM)
2. memories 3. learning 4. pons 5. medulla
6. reticular formation 7. midbrain 8. motor
9. frontal 10. motor 11. cerebellum
12. limbic 13. right 14. sensory 15. spinal cord
16. thalamus 17. somatosensory 18. parietal
19. left 20. corpus callosum 21. short 22. long
23. amygdala 24. hippocampus 25. left
26. association 27. left 28. medulla oblongata
29. pons

Exercise 13: 1. E 2. B, C 3. B 4. A , D 5. C
6. E 7. A 8. D 9. A 10. C 11. A 12. A
13. E 14. A 15. C 16. D 17. B

Test Your Knowledge

Multiple Choice: 1. b 2. d 3. d 4. d 5. a 6. c
7. c 8. d 9. a 10. c 11. e 12. e

Essay: 1. See Figure 28.6 in the text. The transmitting and receiving neurons are separated by a narrow synaptic cleft. A chemical neurotransmitter is contained in small vesicles in the synaptic terminals at the end of the axon of the transmitting cell. When an action potential arrives at the end of the transmitting cell's axon, the vesicles fuse with the plasma membrane. Neurotransmitter molecules are released into the synaptic cleft, rapidly diffuse across, and bind to receptor molecules in the membrane of the receiving cell. This opens ion channels, allowing ions to diffuse through the membrane and initiating new action potentials in the receiving cell. Then the neurotransmitter is broken down by an enzyme, and the ion channels close.

2. The parasympathetic and sympathetic divisions are both parts of the autonomic nervous system, which exerts control over many involuntary (mostly internal) body functions. In general, the two systems oppose each other. The sympathetic system prepares the body for intense activities by speeding up the heart, dilating breathing passages, and stimulating glucose release by the liver, but slowing down activities of the digestive tract. The parasympathetic system primes the body for rest and digestion by slowing the heart, constricting the breathing passages, and stimulating the digestive system.

3. The relative size of the vertebrate brain increased in birds and mammals. The forebrain and hindbrain became subdivided into subregions with specific functions, such as the medulla, cerebellum, and cerebrum. The third trend was an increase in the power of the forebrain, specifically the cerebrum, which is much larger in birds and mammals than in other vertebrates.

4. The three functions of the nervous system are sensory input, integration, and motor output. As you read the question, sensory receptors in your eyes send signals to processing centers in the brain—sensory input. The brain interprets the signals, comparing them with memory and determining their symbolic and conceptual meaning—integration. Finally, the brain sends signals to the muscle cells in your arm and hand, directing their movements as you write an answer—motor output.

5. Some transmitting neurons deliver excitatory neurotransmitters, which open Na^+ channels and thus trigger action potentials in the receiving cell. Other transmitting cells deliver inhibitory neurotransmitters, which open channels for Cl^- or K^+ and decrease the receiving cell's tendency to transmit action potentials. Only if excitatory signals outweigh inhibitory signals will the receiving cell transmit action potentials.

Apply the Concepts

Multiple Choice: 1. e 2. a 3. e 4. c 5. a 6. d
7. b 8. a 9. c 10. d 11. b 12. b

Essay: 1. a. Acetylcholine stimulates receiving neurons to transmit action potentials. If the receptors for acetylcholine are blocked, acetylcholine will not be able to stimulate receiving neurons, and the action potentials will be stopped at the synapse.   b. If acetylcholine release is inhibited, transmitting cells will not be able to signal receiving cells, and action potentials will be stopped at the synapse. c. If DF prevents acetylcholine from being broken down, acetylcholine will remain in receptors in the receiving cell, and the receiving cell will continue to transmit signals, without being stimulated by the sending cell.

2. When dominoes topple, each domino affects the next one in the row. The dominoes do not travel along the row, but the topple is relayed along the row, one domino to the next. Similarly, when sodium channels open in a neuron, movement of sodium ions into the neuron triggers channels in the next portion of the membrane to open, letting sodium in at that point, which triggers channels to open farther down, and so on. Sodium does not travel along the neuron, just the change in the neuron membrane. Thus, the action potential travels along the neuron, much the way the topple travels along the row of dominoes. The biggest difference between the dominoes and the neuron is that the dominoes can only topple once. The neuron membrane can repolarize after each action potential and thus quickly transmit many action potentials in succession.

3. Action potentials from the brain pass through the spinal cord on their way to the nerves leading to the arms and legs. Apparently, Andrew's injury damaged the spinal cord between the brain and the place where nerves to the legs originate, so that signals to the legs were interrupted. His injury must have been below the connection between spinal cord and nerves to the arms, so action potentials to and from the arms were not affected. The brain is not involved in the knee-jerk reflex. Signals travel to the spinal cord via a sensory neuron, across a synapse, and back to leg muscles via a motor neuron. His spinal cord is injured above these neurons, so signals from sensory neurons do not reach the brain, and Andrew cannot feel his legs. But he doesn't have to feel his legs for the reflex to function.

4. The cerebrum is responsible for higher mental functions such as interpreting sensory information, language, and control of voluntary movements. Centers in the brainstem and midbrain, such as the medulla and hypothalamus, control involuntary vital functions such as heart rate, breathing, and body temperature. These centers may continue to do their jobs even if the cerebrum is damaged.

Chapter 29

Review the Concepts

Exercise 1: 1. C, M   2. A, L   3. B, N

Exercise 2: A. 2, M   B. 4, P   C. 3, N   D. 5, L   E. 1, Q   F. 6, O

Exercise 3: 1. electromagnetic receptor   2. near eyes of rattlesnake   3. mechanoreceptor   4. skin and hairs   5. electroreceptor   6. electromagnetic receptor   7. skin of electric fish   8. photoreceptor   9. electromagnetic receptor   10. light   11. pain receptor   12. hair cell   13. sound   14. thermoreceptor   15. heat   16. skin   17. stretch receptor   18. mechanoreceptor   19. skeletal muscle   20. chemoreceptor   21. chemicals in external environment   22. electromagnetic receptor   23. Earth's magnetic field

Exercise 4: 1. stirrup, H   2. anvil, H   3. hammer, H   4. auditory canal, G   5. semicircular canals, F   6. auditory nerve, D   7. utricle (and saccule), I   8. eardrum, E   9. pinna, C   10. oval window, J   11. cochlea, A   12. Eustachian tube, B   13. outer ear   14. middle ear   15. inner ear

Exercise 5: 1. sclera, K   2. ciliary body, G   3. ligament, I   4. cornea, B   5. iris, E   6. pupil, L   7. aqueous humor, C   8. lens, F   9. vitreous humor, J   10. choroid, N   11. retina, D   12. fovea, H   13. optic nerve, A   14. artery and vein, M   15. blind spot, O

Exercise 6: See Figure 29.8 in the text.

Exercise 7: 1. farsightedness (hyperopia)   2. nearsightedness (myopia)   3. astigmatism   4. distant   5. nearby   6. varies   7. nearby   8. distant   9. varies   10. eyeball too short or lens less elastic   11. eyeball too long   12. misshaped lens or cornea   13. corrective lens thicker in the middle than at the edges   14. corrective lens thinner in the middle than at the edges   15. lens that compensates for asymmetry in eye

Exercise 8: *Rods:* 3, 5, 6, 8, 9, 10, 12, 13, 15, 16, 17, 19 *Cones:* 1, 2, 4, 5, 7, 8, 11, 13, 14, 15, 18

Exercise 9: 1. B   2. F   3. E   4. G   5. H   6. C   7. I   8. A   9. D   10. E   11. E   12. D   13. F

Test Your Knowledge

Multiple Choice: 1. c   2. b   3. a   4. e   5. c   6. d   7. b   8. b   9. a   10. a

Essay: 1. Rattlesnakes have infrared receptors near their eyes that enable them to detect the heat given off by warm-blooded prey. Some animals may be able to detect the magnetic field of the Earth and use this information to find their way when migrating. Certain fishes find their prey by sensing the electrical fields produced by the muscles of the prey.

2. Sensory receptors send action potentials to the brain, and the brain is made aware of sensory stimuli. This is called sensation. Perception is meaningful interpretation of these sensory data.

3. Farsightedness is a vision defect that occurs when the eyeball is too short, and the lens cannot bend light rays sharply enough to bring an image into focus on the retina. It can also occur when the lens loses its elasticity and cannot curve enough to bring an image into focus. It can be corrected with a lens that is thicker in the middle than at the edges, making light rays converge slightly before entering the eye.

4. A stimulus usually interacts with a receptor molecule on or in a sensory receptor cell. This activates a signal transduction pathway that opens ion channels in the cell membrane. Positively charged ions flow into the cell, altering the membrane potential, increasing it to a level called the receptor potential. If the receptor potential is big enough, it triggers action potentials that are sent to the central nervous system.

5. The senses of smell and taste both rely on chemoreceptors. There are different kinds of smell and taste receptors, each stimulated by molecules of particular types. The particular combination of receptors stimulated gives rise to our perception of different tastes and smells. Smells are carried by the air, and their molecules attach to proteins on olfactory receptors in the nasal cavity. Taste receptors are usually stimulated by direct contact between chemicals in solution and proteins on receptor cells in taste buds. There are five kinds of taste sensations, but thousands of smells.

Apply the Concepts

Multiple Choice: 1. d   2. c   3. e   4. c   5. a   6. c   7. a   8. c   9. c   10. d   11. b

Essay: 1. Receptors for both the sense of hearing and balance are located in the inner ear. Sound receptors are in the cochlea, and balance receptors are in the semicircular canals, saccule, and utricle. Ménière's disease must impair the inner ear in some way.

2. Blue cones are stimulated when we see blue; red cones are stimulated when we see red. White stimulates all three types of cones, and black stimulates none of them.

3. One method that has been tried is to blindfold people, drive them around an unfamiliar area, spin them around to make sure they are completely mixed up, then ask them to point toward their starting point. Two groups of people were tested: One group wore helmets containing powerful magnets

(which would presumably interfere with a magnetic sense); the other group wore nonmagnetic helmets. (The results were tantalizing but inconclusive.)

4. Changes in the environment tend to be more important than ongoing "background" stimuli. Natural selection has enabled the brain to sense a sudden change more readily than a continuing stimulus. Pain is an important message about danger, injury, or disease. Natural selection would not favor an animal that gets used to pain. It is adaptive for the animal to be reminded of pain and deal with its cause.

5. Hair cells are the receptors for hearing; if they were damaged, they would not be able to respond to vibrations in the cochlea. A brain injury that affects the auditory portion of the cerebral cortex would impair the brain's ability to sense or perceive sounds. Arthritis of the middle-ear bones might interfere with their movement and prevent them from transmitting vibrations to the cochlea. A torn eardrum might not be able to pick up vibrations from the air and transmit them to the ear bones. Earwax in the auditory canal would block sound waves from setting the eardrum in motion.

6. Paired sensory structures allow an animal to locate a stimulus—to determine its direction. The animal compares its strength on either side and perceives that the stimulus originates on the side where sensation is the strongest. For example, if infrared feels strongest on the left, the rattlesnake knows the source of heat is on its left. The snake turns its head until the stimulus is equally strong on both sides; then it knows the source of heat (dinner!) is straight ahead.

Chapter 30

Review the Concepts

Exercise 1: 1. friction   2. gravity   3. three   4. blood   5. cannon   6. momentum   7. peristalsis   8. friction   9. gravity   10. streamlined   11. up and down   12. airfoils 13. greater   14. lift   15. gravity   16. muscles   17. skeleton

Exercise 2: 1. earthworm, *Hydra*, jellies   2. insect, crab, clam   3. sponge, sea star, human   4. fluid under pressure in a body compartment   5. armorlike covering or shell over outside   6. hard supporting elements embedded inside soft tissues   7. fluid (water)   8. insect: chitin; clam: calcium carbonate   9. sponge: calcium carbonate or silica spicules and protein fibers; human: bone and cartilage   10. cushions, gives body shape, peristaltic movement   11. protection, support, movement   12. support, movement, protection   13. cannot support locomotion in which body is held off ground; offers little protection   14. insect exoskeleton does not grow with animal; must be molted, leaving animal vulnerable   15. does not offer strong protection of exoskeleton

Exercise 3: See Figure 30.3A in the text.

Exercise 4: 1. A   2. D   3. B   4. C   5. B   6. A   7. B   8. E   9. B   10. B   11. A   12. B   13. C   14. E   15. B

Exercise 5: 1. G   2. L   3. H   4. B   5. M   6. O   7. A   8. E   9. D   10. K   11. C   12. F   13. N   14. I   15. J

Exercise 6: 1. B, E   2. C, F   3. A, D

Exercise 7: See Figures 30.7 and 30.8 in the text.

Exercise 8: 1. capillaries   2. tendon   3. nerves   4. motor unit   5. neuromuscular junction   6. fiber   7. acetylcholine   8. action potential   9. tubules   10. myofibrils   11. Z   12. sarcomeres   13. thin   14. actin   15. thick   16. myosin   17. slide   18. calcium   19. heads   20. ATP   21. shortening   22. calcium   23. endoplasmic reticulum

Exercise 9: 1. C   2. H   3. E   4. G   5. J   6. A   7. F   8. G   9. I   10. D   11. C   12. E   13. B   14. C

Exercise 10: 1. N   2. M   3. P   4. S   5. P   6. S

Test Your Knowledge

Multiple Choice: 1. c   2. b   3. e   4. d   5. d   6. a   7. b   8. d   9. e   10. b

Essay: 1. A wing is an airfoil, a shape that is thicker at the leading edge and more curved on top than underneath. When a bird flaps its wings, the air passing over the wings has to travel a longer distance than the air moving under the wings. Air molecules are spaced farther apart above a wing than below, resulting in greater air pressure below than above, which lifts the wing.

2. A skeleton supports the body against gravity; this function is illustrated by the vertebral column. The skeleton protects organs, as the ribs protect the lungs and heart. The skeleton works with muscles in movement; this is illustrated by the leg bones and their opposing sets of muscles.

3. Arthropods—animals such as insects, crabs, and spiders—have jointed exoskeletons. An exoskeleton is strong and effective at protecting the organs inside. A drawback of an exoskeleton is that it does not grow with the animal and must be molted periodically. This is wasteful and leaves the animals temporarily vulnerable.

4. The two main forces that must be overcome by a moving animal are friction and gravity. Aquatic animals are buoyed up by the water and are not affected much by gravity, but the frictional resistance of water is a problem for them. Land animals must expend energy supporting themselves against the pull of gravity, but air offers much less resistance than water, so friction is not a serious problem for them.

5. Check your drawings against Figure 30.9A in the text.

6. A motor neuron forms a synapse, called a neuromuscular junction, with the plasma membrane of a muscle fiber. When an action potential reaches the end of the neuron, vesicles release the neurotransmitter acetylcholine, which diffuses across the synapse and docks with receptors on the muscle fiber membrane. This causes a change in the permeability of the membrane, which triggers action potentials across the muscle fiber membrane. The action potentials penetrate deep into the muscle fiber via membranous transverse tubules. In the muscle cell, the action potentials cause the release of calcium ions from the endoplasmic reticulum. Calcium binds to troponin molecules on the thin filaments, causing tropomyosin strands on the thin filaments to shift, exposing myosin binding sites on the thin filaments. This allows myosin heads to attach to actin, initiating the sliding of thick and thin filaments that results in muscle contraction.

Apply the Concepts

Multiple Choice: 1. c   2. e   3. d   4. a   5. a   6. c   7. b   8. d   9. b   10. c   11. a

Essay: 1. The frog's long fingers and toes support webs that enable the frog to swim. Its hind legs, much longer than its front legs, are used for hopping. The frog's short, rigid backbone stiffens its body for hopping. Flexible cartilage in its joints and sternum absorb shock on landing.

2. An earthworm is segmented, and each of its segments is filled with fluid. Muscles in different segments can perform different movements simultaneously. For example, circular muscles can contract in anterior segments, lengthening them and pushing the front of the worm forward. At the same time, longitudinal muscles can shorten posterior segments, drawing the rear of the worm up behind. This enables the worm to make varied crawling and turning movements. The whole body of the roundworm is one big fluid-filled compartment, so the whole worm has to move the same way at once.

3. The short wings of a songbird are easy to flap and enable the bird to fly fast and make tight turns. An eagle uses its broad wings to lift its prey from the ground. The elongated wings of a gull are more suited to gliding long distances with minimal effort.

4. A motor unit is a neuron and the muscle fibers it controls. In the hand, the motor units are small; one neuron might control only one or a few muscle fibers, allowing for fine movements. In the leg, the motor units are bigger; one neuron might control hundreds of muscle fibers. Thus, even though the hand is smaller than the leg, it takes more neurons to control the hand, because of its one-to-one arrangement of neurons and muscle fibers. Therefore, more area in the brain is devoted to controlling the hands than the legs.

5. Walking with the legs splayed to the sides is hard work. A lot of energy is devoted simply to fighting gravity and keeping the body off the ground. Animals that walk this way are slower than those whose legs are directly under the torso. If dinosaurs' legs were under the body, like those of horses and elephants, they probably were a lot faster than alligators or lizards.

(Crossword puzzle grid — Exercise 3 answer — containing the following filled answers: GROUND, TAPROOT, HAIR, BULB, CHOR(oid/CHLOR), CELLULOSE, CUTICLE, COMPANION, SHOOT, PETIOLE, CORTEX, LAMELLA, APICAL, NODE, MESOPHYLL, WALLS, PARENCHYMA, FIBER, VEIN, TUBER, LIGNIN, STOLON, SCLEREID, PITH, SPINE, STOMA, GUARD, ROOT, SCLERENCHYMA, and related terms)

Chapter 31

Review the Concepts

Exercise 1: There are as many possible answers to this question as there are useful plants. As I write this, I can see landscape plants such as Oregon white oaks, rhododendrons, and a magnolia that shade the house, produce oxygen, hold soil in place, and beautify my surroundings. There is a pack of chewing gum on my desk; chicle for making gum comes from a tropical tree. I am wearing a cotton shirt and cotton jeans, and rubber-soled shoes. Rubber is made from the sap of a tropical tree. Even the ink on the papers on my desk is soy-based.

Exercise 2: *Monocots:* D—complex array of vascular bundles, E—floral parts usually in multiples of three, F—veins usually parallel, H—one cotyledon, I—fibrous root system. *Eudicots:* C—vascular bundles in ring, G—flower parts usually in multiples of four or five, B—veins usually netlike, J—two cotyledons, A—taproot usually present

Exercise 3: See crossword at the top of the next column.

Exercise 4: See Figures 31.5 in the text. a. monocot stem b. root c. leaf d. eudicot stem

Exercise 5: 1. Q 2. I 3. R 4. S 5. M 6. L 7. N 8. A 9. D 10. F 11. O 12. E 13. G 14. K 15. B 16. H 17. C 18. J 19. P

Exercise 6: See Figure 31.7B in the text.

Exercise 7: 1. E A D C F I 2. C 3. J G H C B F I 4. G C 5. J G H 6. B 7. E A D

Exercise 8: 1. bark 2. secondary phloem 3. cork 4. cork cambium 5. vascular cambium 6. sapwood 7. heartwood 8. secondary xylem (colors from inside out: red, pink, green, light blue, dark blue)

Exercise 9: See Figure 31.9A in the text. The carpel (stigma, style, and ovary) should be red and the stamens blue.

Exercise 10: 1. water 2. germination 3. seed coat 4. embryo 5. cotyledons 6. endosperm 7. sporophyte 8. meiosis 9. spores 10. gametophytes 11. sepals 12. petals 13. anthers 14. stamens 15. stigmas 16. carpels 17. male 18. sperm 19. pollination 20. ovary 21. sperm 22. an ovule 23. embryo sac 24. gametophyte 25. egg 26. zygote 27. endosperm 28. fertilization 29. seed 30. embryo 31. cotyledons 32. ovary 33. fleshy fruit 34. sugar 35. seed coats 36. dormancy

Exercise 11: 1. E 2. A 3. D 4. B 5. C

Test Your Knowledge

Multiple Choice: 1. e 2. c 3. a 4. d 5. e 6. b 7. b 8. a 9. b 10. c 11. e 12. c

Essay: 1. Roots anchor the plant, absorb water and minerals, transport water and minerals to other parts of the plant, and store food. The epidermis, especially root hairs, functions mainly in absorption; the vascular cylinder carries out transport; and the cortex stores food.

2. See Figure 31.5 in the text. The epidermis makes up the dermal tissue system, xylem and phloem the vascular tissue system, and pith and cortex the ground tissue system. The vascular cylinder is in the center of the root, surrounded by a thick cortex and the epidermis. The center of a stem is filled by pith, with the vascular tissues forming a ring, and a thin cortex on the outside covered by epidermis.

3. The diploid sporophyte stage is the full-grown flowering plant, with roots, stems, and leaves. Its flowers contain anthers and ovules, in which cells undergo meiosis to produce haploid spores. These spores divide by mitosis to develop into pollen grains (the male gametophytes) and embryo sacs (female gametophytes)—both haploid. The gametophytes produce haploid gametes—sperm and eggs—by mitosis.

4. A meristem consists of unspecialized parenchyma cells that divide and generate new cells and tissues. Apical meristems occur in the tips of roots and the apical and axillary buds of shoots. These are responsible for primary—lengthwise—growth. An apple tree also has lateral meristems—cylinders of vascular cambium and cork cambium just inside its trunk—which enable the tree to grow in diameter. This is secondary growth.

5. Sclerenchyma cells have thick, rigid secondary walls hardened with lignin, which enables them to form a sturdy framework for the stem. The water-conducting xylem cells in a root die when mature and form long, hollow pipes that carry water. Epidermal cells are thin and transparent, letting light into the leaf. They secrete a waxy cuticle, which keeps the leaf from drying out.

6. The vascular cambium forms a cylinder just underneath the bark of a tree, starting out between the primary xylem and the primary phloem. The vascular cambium is a meristem; its cells divide and give rise to xylem cells and phloem cells. As cambium cells divide and differentiate, new xylem cells are added to the inside of the cambium and new phloem cells are added to the outside of the cambium. This causes the tree to grow in diameter.

7. Pollination is transfer of a pollen grain from a stamen to the stigma of a carpel. The pollen grain germinates, and its tube cell gives rise to a pollen tube that grows down into the ovary to an ovule. Its generative cell divides to form two sperm cells, which enter the embryo sac. One of the sperm cells fertilizes the egg, forming the plant zygote. The other sperm fertilizes a large diploid cell, which divides to form the triploid endosperm, food stored in the seed for the embryo. Fertilization of the haploid egg by one sperm and a diploid cell by another sperm is called double fertilization, and it occurs only in flowering plants. Fertilization in animals involves only a single haploid egg and a single haploid sperm.

8. Some plants, such as turnips, have large roots that store food. Stems are modified in many ways. Horizontal runners and stolons enable grasses and strawberries to reproduce and spread asexually. Enlarged horizontal stems called tubers, such as potatoes, store food. A bulb is a shoot whose swollen leaves function in storage and reproduction by fragmentation. The leaves of some plants, such as celery, have enlarged petioles that store food and water. Leaves may be modified to form tendrils, which help plants climb. Cactus spines are modified leaves that protect the plant from grazing animals.

Apply the Concepts
Multiple Choice: 1. a   2. c   3. c   4. d   5. c   6. d   7. a   8. d   9. a   10. d   11. a

Essay: 1. Most of a tree trunk is xylem, and xylem cells are dead at functional maturity, so the cells that make up most of a tree are not alive.

2. The spike would pass through layers of cork, cork cambium, and secondary phloem, which make up the tree's bark. Then it would penetrate the vascular cambium, the functional secondary xylem of the sapwood, and the nonfunctional secondary xylem of the heartwood.

3. Garlic—bulb (modified shoot); walnut—seed; cabbage—leaves; carrot—root; cauliflower—flower buds; artichoke—leaves and flower bud; asparagus—stem; cucumber—fruit and seeds; broccoli—flower buds; rice—fruit (mostly seed); potato—tuber (underground stem).

4. Apical meristems at the tips of roots and shoots contain dividing, elongating, differentiating cells that are responsible for plant elongation (primary growth). After Grandpa nailed the horseshoe to the tree, the tips of its branches continued to grow, and other meristems caused the tree to expand in girth (secondary growth), but the part of the trunk where he nailed the horseshoe, 5 feet from the ground, no longer contained embryonic cells capable of elongation.

5. It would be difficult to kill a palm by girdling it. An oak tree is a eudicot. Its vascular tissues are arranged in a ring, with the phloem just under the cork layer of the bark, vulnerable to damage. The palm is a monocot, with its vascular bundles and meristems scattered throughout the diameter of the trunk, and therefore less vulnerable to damage.

6. The vascular cylinder is in the middle of a root. Lateral roots must start growing from the inside of the main root so that the vascular cylinders of the main root and lateral roots will connect.

7. Seeds contain all the nutrients needed to sustain the plant embryo until its leaves and roots are functional. Plant cells and human cells need much the same nutrients, so seeds are a good source of nutrition.

Chapter 32

Review the Concepts

Exercise 1: 1. carbon dioxide (CO_2) 2. water (H_2O) 3. oxygen (O_2) 4. inorganic ions (minerals) 5. CO_2 6. air 7. H_2O 8. soil 9. nitrogen 10. nitrogen 11. magnesium 12. nitrogen 13. phosphorus 14. minerals 15. O_2

Exercise 2: See Figure 32.2B in the text.

Exercise 3: A. 4, diffuse (evaporate) B. 1, xylem, osmosis C. 5, cohesion, adhesion D. 2, root pressure E. 6, hydrogen F. 3, guard, osmosis, stomata, transpiration G. 7, transpiration-cohesion-tension

Exercise 4: 1. source 2. sink 3. source 4. source 5. source 6. source 7. source 8. source 9. sink 10. source 11. sink 12. sink 13. sink 14. sink

Exercise 5: 1. C, H, O, N, S, P, Ca, P, Mg 2. C, H, O, N, S, P 3. Fe, Cl, Cu, Mn, Zn, Mo, B, Ni 4. N 5. Ca 6. K 7. Fe 8. P 9. Mg 10. N, P, K

Exercise 6: 1. B 2. H 3. K 4. J 5. D 6. I 7. C 8. A 9. E 10. F 11. G

Exercise 7: 1. Humus content of the soil can become depleted, reducing its fertility, and fertilizer runoff can pollute water. 2. Crop rotation maintains fertility. Different crops have different needs and contribute different qualities (nutrients, texture) to soil. 3. This could channel runoff, reduce penetration of water into the soil, and increase soil erosion. 4. Salts that are left behind when irrigation water evaporates can make soil too salty for crops. 5. This adds humus, restoring the natural texture and fertility of soil and reducing dependence on inorganic fertilizers and pesticides. 6. Drip irrigation uses less water, allows plants to absorb most of the water, reduces salt buildup, and reduces water and nutrient loss from evaporation and drainage. 7. These plants are rich in protein, but are expensive and demand a great deal of fertilizer.

Exercise 8: 1. Nitrogen 2. proteins 3. nucleic acids 4. air 5. ammonium 6. nitrate 7. nitrogen-fixing 8. ammonifying 9. nitrifying 10. amino acids 11. mycorrhizae 12. mutually 13. growth factors 14. antibiotics 15. sugar 16. dodder 17. mistletoe 18. epiphytes 19. Carnivorous 20. Venus flytrap 21. nitrogen 22. insects 23. legume 24. nodules 25. nitrogen-fixing 26. organic compounds 27. N_2 28. rotate 29. corn or wheat

Test Your Knowledge

Multiple Choice: 1. b 2. a 3. e 4. d 5. d 6. a 7. c 8. e 9. a 10. e 11. b 12. c 13. c

Essay: 1. Root cells actively pump mineral ions into the xylem. Water follows by osmosis and may exert a slight push, called root pressure, that may move water a few meters up a tree. The primary force is a pull exerted by the transpiration of water molecules from the tree's leaves. The water molecules cling to each other and cell walls by hydrogen bonds—phenomena known as cohesion and adhesion. As each water molecule evaporates from a leaf, it pulls on the next, which pulls on the next. This pull, exerted through a column of water molecules in xylem extending all the way to the roots, is what moves water from roots to leaves.

2. If you grew plants in hydroponic culture, bathing them in solutions of known composition, you could compare plants that received a complete nutrient solution with plants receiving the same solution minus molybdenum.

3. Phloem transports sugars, via a pressure-flow mechanism, from sugar sources to sugar sinks. In the summer, a leaf is a sugar source and a potato is a sugar sink. At the leaf, where sugar is being made, it is pumped into the phloem by active transport. Water follows by osmosis, raising the water pressure at this end of the phloem "pipeline" and pushing phloem sap toward the underground potato tubers. In a potato, sugar is transported out of the phloem and stored, and water follows. This lowers the water pressure at this end of the phloem tube, maintaining the pressure gradient that keeps phloem sap moving from leaves to potatoes. Early in the spring, when the potato is full of carbohydrates and the leaves are developing but not yet performing photosynthesis, their roles are reversed. The potato is a sugar source and the leaf a sugar sink, and phloem sap flows upward from potato to leaf.

4. Clay particles are negatively charged, and because opposite charges attract, positively charged nutrient ions tend to cling to them. Root hairs obtain these ions by cation exchange. The root hairs release hydrogen (H^+) ions into the soil solution. The H^+ ions take the place of the nutrient ions on the clay particles, and the root hair can then absorb the free positively charged ions.

5. Humus holds water but keeps the topsoil porous enough to allow aeration of roots. The nutrients in

humus are relatively insoluble and released gradually, so they are not washed from the soil and are not as likely to pollute water.

Apply the Concepts

Multiple Choice: 1. c   2. e   3. d   4. e   5. d   6. e   7. b   8. d   9. c   10. d   11. b

Essay: 1. Spraying with antibiotics probably killed beneficial nitrogen-fixing bacteria growing in nodules on the roots. Without their help, the bean plants became nitrogen deficient. The plants and root fungi probably formed a mutually beneficial mycorrhizal association. The fungi helped the plants absorb water and nutrients and may have protected them from disease. Without the fungi, the plants suffered. Bad move.

2. Bubbles in the xylem would interfere with transport of water to the leaves. Transpiration of water from the leaves pulls water through the xylem from the roots. Bubbles in the xylem tubes would disrupt the cohesion of water molecules and reduce this transpiration-adhesion-cohesion pull.

3. Most plants open their stomata during the day. This allows them to take in the CO_2 used in photosynthesis and transpire water, which pulls water and minerals up from the roots. In its hot, dry environment, the jade plant might lose too much water if its stomata remained open during the day. By opening its stomata at night, it reduces water loss.

4. Potassium is needed as an activator of several enzymes, and it is also the main solute that regulates water balance in a plant—the opening and closing of guard cells, for example. The plant does not build molecules containing potassium, but it needs potassium as a regulator of its activities.

5. Acid and heavy rain would be most likely to deplete the nitrogen in the soil. Organic matter decays slowly in acid soil, making little nitrogen available. Unlike phosphate or magnesium ions, nitrate is negatively charged. The rain would be most likely to wash these negative ions from the soil, because they would not cling to negatively charged clay particles. Denise and Roger could enrich the soil by planting and plowing under legumes, such as clover.

Chapter 33

Review the Concepts

Exercise 1: 1. left   2. left   3. straight   4. right (then left to light)   5. left   6. straight   7. left 8. straight   9. no growth   10. no growth 11. straight   12. left   13. straight   14. right 15. no growth

Exercise 2: 1. auxin   2. Cytokinins   3. IAA 4. apical meristem   5. roots   6. embryos   7. fruits 8. xylem sap   9. Auxin   10. cell walls   11. inhibits 12. stimulates   13. inhibits   14. axillary 15. bushy   16. Cytokinins   17. auxin 18. cytokinins   19. inhibit   20. stimulate 21. lower   22. cytokinins   23. root   24. shoot

Exercise 3: 1. ABA   2. G   3. ABA   4. ABA   5. G 6. G   7. ABA   8. ABA   9. G   10. ABA   11. G 12. G   13. ABA   14. G

Exercise 4: 1. ethylene   2. aging   3. softening 4. ripening   5. Tomatoes   6. CO_2 7. ethylene 8. deciduous   9. water   10. pigments 11. chlorophyll   12. shorter   13. cooler 14. auxin   15. abscission

Exercise 5: 1. gibberellins   2. ethylene   3. auxin 4. cytokinins

Exercise 6: 1. C   2. A   3. D   4. E   5. B

Exercise 7: 1. Growth of a shoot toward light 2. Auxin migrates from light to dark side of shoot tip, causing cells on dark side to elongate faster. 3. Gravitropism   4. Gravity may pull organelles downward, affecting auxin distribution and slowing the rate of growth on lower side of root. 5. Thigmotropism   6. Growth or movement in response to touch

Exercise 8: For example: Plants may exhibit circadian rhythms, daily cycles of stomata movement or opening. A rhythm may be controlled by a biological clock, perhaps a gene producing protein on a daily cycle.

Exercise 9: 1. yes   2. no   3. no   4. no   5. yes 6. no   7. yes   8. yes   9. yes   10. no

Exercise 10: 1. B   2. D   3. H   4. E   5. A   6. G 7. C   8. F

Test Your Knowledge

Multiple Choice: 1. e   2. a   3. e   4. b   5. c   6. b 7. c   8. b   9. a   10. e   11. c

Essay: 1. Responses called tropisms enable a plant to alter its pattern of growth to adjust to environmental changes. Phototropism is bending toward light. A hormone called auxin migrates from the light side to the dark side of a shoot tip, causing cells on the dark side to elongate faster. Gravitropism, response to gravity, also appears to be controlled by auxin. Gravity may pull organelles downward, redistributing auxin in the process. This causes the upper side of a root or the lower side of a stem to grow faster, even in the dark.

2. Circadian rhythms continue even when a plant is sheltered from environmental changes—kept in constant darkness, for example. A biological clock

does not keep perfect time, so circadian rhythms tend to drift somewhat in relation to the external environment. Environmental cues such as the daily cycle of light and dark enable plants to reset the clock and keep in sync with the environment.

3. As fruit ripens and ages, it produces ethylene gas, which can diffuse through the air from fruit to fruit. Ethylene accelerates ripening, so the ethylene from damaged or overripe fruit can speed up the softening of crisp apples.

4. The signals for leaf drop are the shorter, cooler days of autumn. Auxin prevents leaf drop, but as a leaf ages, it produces less auxin. Cells in the abscission layer, at the base of the leaf stalk, begin to produce ethylene. The ethylene causes enzymes to digest cell walls in the abscission layer. Eventually, the stalk breaks away, and the leaf falls.

5. Long-day plants flower when the night is shorter than a critical length. Short-day plants flower when the night is longer than a critical length. Phytochrome, a colored protein, absorbs light and helps the plant set its biological clock so that it can measure the length of the night. Light changes a form of phytochrome called *Pr* to *Pfr*. *Pfr* is converted back into *Pr* during the night. The biological clock may measure the timing of *Pfr*-to-*Pr* changes to measure the length of the night.

Apply the Concepts

Multiple Choice: 1. b 2. d 3. b 4. a 5. e 6. d 7. c 8. b 9. e 10. d

Essay: 1. The gas probably contained ethylene, which causes changes in the abscission layer at the base of a leaf that cause the leaf to drop from the stem.

2. Different concentrations of plant hormones can have quite different effects on the same plant. Relatively low auxin concentrations stimulate elongation of stems, but higher concentrations inhibit stem growth. See Figure 33.3B in the text.

3. Normally, production of auxin by stems would be balanced by production of cytokinins by roots. Auxin tends to inhibit axillary buds, while cytokinins stimulate them. Grafting a shoot to a large root system would alter this balance. The excess of cytokinins from the roots would stimulate development of axillary buds, which would grow into lateral stems. This would make the plant more bushy, producing a shoot system proportional to the root system.

4. Auxin, produced by the leaf, normally offsets the effects of ethylene, produced by the stalk. Ethylene causes the stalk to drop off. Smearing auxin on the stalk should offset the effects of ethylene and cause the stalk to remain attached to the stem.

5. Apparently, the age or size of soybean plants has little influence on whether or not they flower. Soybeans must be short-day (long-night) plants, which flower in the fall when the night exceeds a critical length. Once the plants are mature, they wait until the nights are a certain length to flower.

6. Abscisic acid (ABA) acts as a growth inhibitor. It causes seed dormancy by inhibiting germination. ABA might be a good place to start.

7. You could grow a shoot straight up, then turn the plant so that the shoot is horizontal. Place a light under the horizontal shoot. Will it grow toward the light or away from gravity?

Chapter 34

Review the Concepts

Exercise 1: 1. biosphere 2. landscapes 3. ecosystems 4. abiotic factors 5. biotic factors 6. nutrients 7. populations 8. habitats 9. ecology 10. interactions 11. groups belonging to same species 12. organisms

Exercise 2: 1. scarcity of water 2. snow leopard 3. high body temperature, fur 4. hot spring or hydrothermal vent organism 5. high temperatures 6. fire 7. sunscreen-like chemicals in overlapping leaves 8. cold and wind 9. energy source—light 10. low oxygen content of water 11. pronghorn 12. cold

Exercise 3: 1. directly 2. North Pole 3. tilt 4. summer 5. winter 6. warms 7. moisture 8. rises 9. doldrums 10. rain 11. descends 12. 30 13. desert 14. moisture 15. rotation 16. prevailing 17. trade winds 18. northeast 19. currents 20. winds 21. rotation 22. continents 23. Gulf Stream 24. oceans 25. rainfall 26. rain shadows 27. biomes 28. temperature 29. rainfall

Exercise 4: See crossword at the top of the next page.

Exercise 5: See Modules 34.8–34.16 in the text.

Exercise 6: 1. desert 2. temperate grassland 3. temperate forest 4. tropical forest 5. coniferous forest 6. tundra

Exercise 7: 1. E 2. C 3. G 4. F 5. B 6. A 7. B 8. D 9. G 10. H 11. H 12. B 13. E 14. G 15. C 16. H 17. D 18. E 19. F 20. C 21. A 22. C 23. D 24. F

Exercise 8: See Figure 34.9 in the text.

```
          P           W E T L A N D
      B E N T H I C         I
    M . S . Y             G
    I N T E R T I D A L     A P H O T I C
    X . U . O             T         O
    I . A . P                       N
    N . R . L     H . S             T
    G . Y . A     Y . E   P E L A G I C
            N     D . D   H         N
            K     R . I   O         E
            T     O . M   T         N
      P H O T O S Y N T H E S I S   T
            N     H . N   C         A
                  E . T             L
      P H O S P H O R U S
          X         M       C O R A L
          Y         A       L
  D U M P I N G     S A L T I N E S S
          E                 A
  Z O O P L A N K T O N   M A R S H E S
```

Exercise 9: 1. sun   2. sea   3. water vapor   4. sea   5. wind   6. less   7. evaporation, transpiration   8. water vapor   9. surface   10. groundwater   11. sea   12. sediments

Test Your Knowledge

Multiple Choice: 1. d   2. b   3. a   4. a   5. c   6. c   7. d   8. b   9. d   10. c   11. b   12. a

Essay: 1. During the postwar era, new technologies such as chemical fertilizers and pesticides came into widespread use. These increased food production and promised to wipe out insect-borne diseases such as malaria. In the 1960s it became clear that technology had a downside. Pesticide residues threatened wildlife, and pests became resistant to chemicals. Now many people are aware that human activities have a big impact on nature, even changing the climate, and are more concerned about such negative effects of human activities on the biosphere.

2. The sun's rays strike the equator directly, warming the air and evaporating moisture. Warm air rises, and as it rises, it cools. Its moisture condenses and falls as rain, watering tropical forests. Cool air is heavier than warm air, so the dry air descends, 30° north and south of the equator, creating deserts.

3. In North America, the temperate forest and temperate grassland have been most extensively altered, mostly by agriculture and urban development.

4. The intense cold, relatively light precipitation, and the short growing season of the far north and high mountains prevent the growth of trees. The ground is permanently frozen—permafrost. This kind of environment is characterized by the tundra biome—shrubs, grasses, mosses, and lichens—species that can live under these extreme conditions. Tropical rain forests are characteristic of warm, humid equatorial areas where there is abundant rainfall and little seasonal variation temperature.

5. The sun drives the global water cycle by evaporating water and creating water vapor, mainly over the surface of the ocean. Winds (also driven by solar heating) move the water vapor around. Most of the water vapor in the air returns to the sea as precipitation, but some blows over land, where water falls as rain and snow. Some of the water evaporates or is transpired by plants. Some pools as surface water or groundwater, but eventually it all returns to the sea. Some areas near the sea, such as the temperate rainforests of the Pacific Northwest coast of North America, are shaped by high rainfall. Other areas farther from the sea or behind mountain ranges (such as the areas farther inland from the Pacific coast) are dominated by drier biomes such as grasslands and deserts

Apply the Concepts

Multiple Choice: 1. e   2. d   3. a   4. d   5. d   6. b   7. d   8. b   9. b   10. c

Essay: 1. Temperature, rainfall, soil nutrients, wind, and fire are all abiotic factors that would affect a tree. Predation by herbivorous insects and competition with other trees are biotic factors that might affect the tree. Day to day, the tree might adapt to changes in these factors by adjusting its rate and pattern of growth. For example, it might grow more slowly in a colder location, or it might drop its leaves in the winter. Over many generations, the tree has become adapted to a particular range of environmental factors via natural selection.

2. Water absorbs and stores heat, so the range of temperatures in the ocean is much narrower than on land. Since most of the ocean is within the same narrow temperature range, temperature is not as important a factor shaping the distribution of life in the sea as it is on land. Temperature is more variable on land, so it has a bigger impact on the distribution of land life. Most of the ocean is without light, but most places on land are light for about half of each day. Thus, light is not a major factor in determining where living things can exist on land, but it is an important factor in the sea.

3. The routes of sailing ships depend on the prevailing winds. On their way to North America, ships from Europe first sailed south to catch the northeast trade winds. On the return to Europe, hey first sailed north to catch the westerlies.

4. The biomes characteristic of warm environments, such as most deserts and grasslands, will probably expand toward the poles. Those characteristic of colder climates, such as coniferous forest and tundra, will probably shrink.

5. The photic zone, where light is available for photosynthesis, is only a thin layer at the surface of the sea. Here there is plenty of light but few nutrients needed for the growth of phytoplankton. Sediments at the bottom of the sea, the benthic realm, are rich in nutrients, but here there is no light for photosynthesis. Life is abundant in the shallow waters of the continental shelves, where sunlight reaches the bottom, but the continental shelves make up only a small portion of the oceans.

6. There are many possible examples. Beavers dam streams and create ponds. Coral animals build coral reefs, which create habitats for many other species. Overgrazing might create a desert. Trees create shade that might affect temperature and soil moisture.

Chapter 35

Review the Concepts

Exercise 1: 1. *Proximate:* Movement of a small object. *Ultimate:* Fitness of frogs has been enhanced by their ability to detect and capture insects. The fact that it is "automatic" suggests it is genetic.
2. *Proximate:* Presence of a potential mate causes each vole to sniff the other and mate. Mating causes voles to bond and remember each other. *Ultimate:* Bonding with a mate enhances care of young and thus reproductive success. Sniffing and bonding are genetic (due to genes for dopamine receptors), but bonding with and remembering a particular mate is learned (a result of experience). 3. *Proximate:* The appearance and location of the hive. *Ultimate:* Bees that could find their way back to the hive were more successful than those who couldn't. The routine of circling the hive each day is probably genetic. The location of the hive can vary, so finding it each day results from experience. 4. *Proximate:* Red wing patches. *Ultimate:* Those males that identify themselves and defend their territories have greater reproductive success than those who do not. Based on the information given, this behavior could be genetic or learned. 5. *Proximate:* The locked cage. *Ultimate:* Those monkeys able to use their intelligence to solve these kinds of problems have the most offspring. The monkey has to figure out the lock, so it has not seen it before. The ability to open the lock must be a result of experience. 6. *Proximate:* The appearance or sound of a bee. *Ultimate:* Toads that could not learn how to avoid harmful animals soon died out. The behavior of the toad is modified by experience. 7. *Proximate:* Colder weather or shorter days of autumn. *Ultimate:* Monarchs that flew south for the winter had greater reproductive success. Because the butterflies have not made the trip before, the behavior must be genetically programmed. 8. *Proximate:* Moving "hawk" silhouette. *Ultimate:* Chicks that run for cover live to reproduce. Because newly hatched chicks do it, the behavior must be genetic. 9. *Proximate:* Shape and pattern of egg. *Ultimate:* Reproductive success of gulls has been enhanced by an ability to recognize and retrieve their own eggs. The pattern of spots must vary, so the pattern on a particular egg must be learned by experience, but looking for it could be genetic.

Exercise 2: 1. edge of cliff 2. turns away from edge 3. chicken 4. runs for cover 5. bird 6. stuffs food into mouth 7. red-winged blackbird 8. threatens and attacks intruder 9. frog 10. moving insect 11. human infant 12. adult face 13. gull or graylag goose 14. rolls egg into nest

Exercise 3: 1. associative learning 2. imprinting 3. imprinting 4. habituation 5. cognitive map 6. problem solving 7. imprinting 8. problem solving 9. cognitive map 10. associative learning 11. social learning 12. habituation (probably associative learning, too) 13. imprinting 14. spatial learning 15. social learning

Exercise 4: See Chapter 3, Exercise 10, for a discussion of constructing concept maps.

Exercise 5: 1. Catch some turtles off Brazil and tag them in some way. Later catch some turtles laying eggs on Ascension Island and look for the tags.
2. See which way worms move when placed on the boundary between dark and light areas. 3. First see if the crabs retain their rhythmic behavior if moved into a laboratory away from the ocean. If so, move some crabs to a lab in a different time zone, where tides would be different. See if they stay on the schedule of their original habitat or that of the new location. 4. Blindfold a homing pigeon and see if it can still get home. 5. See whether birds in the lab, at a constant temperature, still become restless as fall approaches.
6. See whether the rat retains a rhythm of activity under constant light. 7. Keep newly hatched flies isolated until ready to mate, then see if they mate successfully. 8. Add or remove flowers or nectar and see if aggressiveness (time spent fighting) changes accordingly.

Exercise 6: 1. specialists 2. dominant 3. search image 4. efficient 5. optimal foraging 6. territory 7. scent 8. food 9. offspring

10. dominance   11. control   12. fighting
13. communication   14. scent   15. visual
16. sounds   17. monogamous   18. pair bond
19. paternity   20. agonistic   21. ritual
22. courtship   23. agonistic   24. altruism
25. kin   26. genes   27. reciprocal

Exercise 7: Honeybee—near the center, but closer to movements. Ring-tailed lemur—in the center. Wolf—in the center. Cricket—near sounds. Human—between sounds and movements, but closer to sounds. Domestic dog—in the center, but closer to chemicals. Common loon—between movements and sounds, but closer to movements. White-crowned sparrow—between movements and sounds, but closer to sounds.

Exercise 8: 1. C   2. M   3. Q   4. L   5. D   6. N
7. B   8. O   9. S   10. F   11. H   12. P   13. R
14. G   15. J   16. E   17. I   18. K   19. A

Test Your Knowledge

Multiple Choice: 1. c   2. c   3. e   4. a   5. d
6. d   7. b   8. b   9. b   10. e   11. a   12. b   13. d

Essay: 1. Newly hatched geese are genetically programmed to learn, during a short critical period after hatching, to follow a moving object (usually their mother). Similarly, a songbird is programmed to learn its species' song during a critical period, then sing that song later. Other birds imprint on their own eggs or offspring. Imprinting provides potential mates, parents, and offspring with a quick way to identify each other, saving time and energy. The energy saved can be used in finding food and rearing young, increasing the chances of reproductive success.

2. A dominance hierarchy assures access to food, water, mates, and nesting or roosting sites for higher-ranking individuals. It establishes order within the group, so individuals can devote energy to finding food, defending against predators, locating mates, or caring for young, instead of fighting each other. This contributes to fitness.

3. Courtship rituals signal that potential mates are not threats to each other. They also confirm that individuals are of the correct species and sex and are in condition to breed. Also, courtship rituals allow individuals to assess each others' health and thus the contribution of a partner to healthy offspring and potential reproductive success.

4. Social behavior provides organization within animal populations, but for animals to coordinate their activities, they must be able to signal each other about their intentions and actions. Bees communicate the location of nectar by "dancing." Wolves and cheetahs mark their territories with scents. Loon courtship involves a complex series of movements. Lemurs wave their scented tails to communicate aggression. A frog attracts a mate by croaking.

5. Some birds use the position of the sun as a reference point, adjusting for its motion across the sky. Other birds that migrate at night use the fixed North Star as a directional reference.

6. Fixed action patterns enable animals to perform behaviors correctly the first time, with no time required for learning. FAPs seem to be most important in animals with short life spans, such as insects. In birds and mammals, FAPs are important in mating and parental care, where there is little chance for practice and little margin for behavioral error.

7. Wolves, chimpanzees, and bees cooperate in social groups. Social cooperation aids in hunting and foraging for food, mating, care of the young, and defense.

Apply the Concepts

Multiple Choice: 1. d   2. b   3. c   4. c   5. b   6. d
7. a   8. e   9. e   10. b   11. c   12. b

Essay: 1. Group hunting allows animals to capture prey that are too fast, too wary, or too big for one predator to bring down alone. The biggest disadvantage of group hunting is that each predator must share food with others in the group.

2. The proximate cause of the diving behavior is an environmental cue—the sound of a bat's sonar. The ultimate cause of the diving behavior has to do with natural selection. The fitness (reproductive success) of moths is enhanced by diving to the ground when they hear bat sonar.

3. This is a question of optimal foraging. Alone, a wagtail does not catch as many insects, so it does not obtain as much energy, but it does not go as far, so it does not expend as much energy either. In a flock, a wagtail gets more insects and energy, but it must expend more energy covering more territory to get them. Foraging alone or in a flock depends on which provides the most energy gain for energy expended.

4. This is an example of associative learning. The patients learn to associate particular foods eaten just before treatment with the negative experience of chemotherapy, and subsequently they want to avoid those foods. This kind of learning would help an animal quickly learn to avoid harmful foods.

5. Newly hatched songbirds, such as sparrows, cannot feed themselves and need a constant food supply. In this situation, the male might achieve more reproductive success by helping a single

mate raise her young than by seeking more mates. As soon as they hatch, baby chicks can feed on their own. The rooster does not need to help; he achieves more reproductive success by mating with a number of hens.

6. Altruism might be fatal for an individual bird, but if it saves the lives of its relatives, its genes will live on through them. An individual might enhance its fitness by helping its relatives. This idea is called kin selection. Alternatively, reciprocal altruism might be involved here. A favor extended today might be repaid by the recipient tomorrow. In this way, unrelated birds might enhance each other's fitness.

7. A sociobiologist might say that this behavior is at least partially a consequence of our evolution. Adults are genetically programmed to be attracted to and take good care of babies. An individual is likely to be around his or her own babies the most. Individuals who like their babies enhance their reproductive success, and liking babies is passed on via their offspring.

Chapter 36

Review the Concepts

Exercise 1: 1. 2 hectares   2. This figure will vary but will probably be about 30 shrubs.   3. This figure will vary but will probably be about 15 shrubs per hectare.   4. This figure will vary but will probably be about 250 shrubs.   5. One could take more samples or count every shrub, but this would take more time and cost more money.   6. The shrubs are clumped. This could be due to variations in moisture or soil nutrients or perhaps because offspring sprout close to parent plants.   7. Mice are mobile; they do not hold still to be sampled.   8. Using the method in the Web Activity, $N = (40 \times 40)/10 = 160$ mice.   9. $160/16.8 =$ about 10 mice per hectare.   10. You have to assume that a marked mouse has the same chance of being caught as an unmarked individual. You could be wrong. If the mice are territorial, and you put the traps in the same locations both times, you are more likely to recapture the same mice, making the population look smaller. If the captured mice learn to avoid the traps, you are then less likely to recapture them, making the population look bigger.

Exercise 2: 1. B   2. A   3. E   4. A   5. E   6. D   7. D   8. C   9. C

Exercise 3: 1. 6 aphids   2. 20 aphids   3. 8, 28   4. 11, 39   5. 55, 77, 108, 151, 211, 295   6. The graph should look a bit like Figure 36.4A in the text.   7. It is J-shaped.   8. This kind of growth could not

continue. Eventually, some environmental factor, such as limited food, would slow the growth of the aphid population.   9. The graph is S-shaped.   10. It looks as though the population briefly exceeded carrying capacity ("overshoot"), then dropped a bit.   11. The population leveled off at carrying capacity—the number of individuals the environment could support over the long term as food or space became limited.   12. around 700 deer   13. logistic growth

Exercise 4: 1. DD   2. DD   3. A   4. DD   5. A   6. DD   7. DD   8. DD   9. A   10. A   11. DD   12. A   13. A   14. DD   15. A   16. DD   17. DD   18. DD   19. A   20. DD   21. DD   22. DD   23. A

Exercise 5: 1. Rapid population growth when conditions are favorable   2. Small   3. Large   4. Many   5. Few   6. Older   7. Quantity   8. Yes; competition for resources   9. Density-dependent factors   10. Many pests   11. Many predators   12. Many insects and weeds   13. Many large land vertebrates, such as polar bears

Exercise 6: 1. 6.5   2. 450   3. 8   4. r 5. N 6. 0   7. birth   8. death   9. Europe   10. decreased   11. birth rate   12. health care   13. survivorship   14. birth control   15. transition   16. low   17. low   18. equal   19. 73   20. family   21. delay   22. fewer   23. contraception   24. age   25. developing   26. momentum   27. 30   28. 17   29. developed   30. baby boom   31. retirement   32. Social Security   33. 9.5   34. double   35. water   36. desert   37. extinction   38. faster   39. ecological footprint   40. 1.7-2.0   41. exceeds   42. 20   43. birth   44. death

Test Your Knowledge

Multiple Choice: 1. e   2. c   3. d   4. b   5. d   6. a   7. a   8. c   9. c   10. d   11. d   12. b

Essay: 1. The exponential growth curve is J-shaped and describes a population growing at an ever-accelerating rate under conditions that do not restrain population growth. The rate of growth depends only on how big the population is, and as the population increases, it grows faster. The logistic growth curve is S-shaped and describes the growth of a population that is limited by its environment. When the population is small, it grows slowly. Growth accelerates as the population increases, but growth slows down and eventually ceases as the population approaches and reaches carrying capacity, the number of organisms that the environment can support.

2. Carrying capacity depends on species and habitat. The carrying capacity of a desert for foxes might be quite different from the carrying capacity of a forest for mice. The animals are different sizes,

they require different kinds and amounts of food, and desert and forest habitats produce different amounts of food.

3. Density-dependent factors include food supply and predation. Abiotic factors include temperature, rainfall, fire, and floods. Abiotic factors are not affected by the density of the population. Density-dependent factors affect a greater fraction of the population as the population increases.

4. Human technology, not biology, is responsible for our rapid and sustained population growth. Nutrition, health care, and sanitation have enabled us to maintain a birth rate that exceeds our death rate. Agricultural and industrial technology have enabled us to more effectively exploit the Earth's resources and inhibit the mechanisms that would normally slow population growth. In effect, our population has continued to grow because we have increased Earth's carrying capacity.

5. Some populations, such as the lynx and the snowshoe hare, undergo regular ups and downs in population density, called population cycles. Predator population cycles probably depend on the ups and downs of their prey. Prey population cycles are probably caused by changes in the availability and quality of their plant food or overexploitation of prey by predators. It is also possible that stress from overcrowding may reduce fertility, leading to population fluctuations.

6. Organisms with life histories shaped by *r*-selection are usually small in size, mature quickly, produce many offspring, and do not invest much parental care in their offspring. Examples are dandelions and aphids. They usually live in unpredictable environments and multiply exponentially when conditions are favorable. They emphasize producing large numbers of offspring quickly. Organisms whose life histories are shaped by *K*-selection are larger in size, mature later, and produce fewer offspring, but give their offspring more care. Their populations may be quite stable, held near carrying capacity by density-dependent factors. They emphasize production of better-endowed offspring that can compete in a well-established population. Polar bears and chimpanzees are examples of such organisms.

Apply the Concepts

Multiple Choice: 1. e 2. b 3. e 4. a 5. d 6. a 7. c 8. b 9. d 10. c 11. b

Essay: 1. Using the mark-recapture method described in the Web Activity for Chapter 36, there are 2,000 smallmouth bass in the lake.

2. The total size of the population is different from population density. Density is the number of indi-

viduals of a species per unit of area. The population of the United States is bigger than the population of Japan, but the people of Japan are packed into a much smaller area, so Japan has a higher population density. (U.S. population: 301,140,000. Area: 9,826,630 sq km. U.S. population density: 30.6 people per sq km. Japan population: 127,433,000. Area: 377,887 sq km. Japan population density: 337.2 people per sq km.)

3. A population that grows logistically grows slowly when the population is either large (limited by resources such as food supply) or small (limited by "breeding stock"), and grows fastest when the population is at an intermediate level. (The S-shaped growth curve is steepest at around half of carrying capacity.) The salmon population will grow fastest when it is "fished down" to about 50,000 (where the logistic curve is steepest.).

4. Temperature is an abiotic limiting factor. Its effect alone on the robin population is not influenced by the size of the population. One would expect the same percentage of robins to be killed whether the population is large or small. Competition for shelter is a biotic factor that has density-dependent effects on birth and death rates. It affects a bigger percentage of the robins when the population is larger. This shows how a population is regulated by a mix of factors.

5. These populations were introduced to environments where they are not subjected to their normal limiting factors, such as predators and parasites. Without these limits, the populations grew larger and more quickly than would otherwise be the case.

6. Populations on islands or in lakes are easier to define than in other areas. The organisms are limited by natural geographical boundaries, and there is less mixing, interbreeding, and confusion with organisms from elsewhere. They might also be easier to find and sample than populations with less well-defined boundaries.

7. China has a much larger population, but its *per capita* ecological footprint is only about one-fourth as big as the *per capita* footprint of the U.S. In other words, because of the greater level of resource consumption in the U.S., each American has four times the impact of each resident of China.

Chapter 37

Review the Concepts

Exercise 1: 1. parasitism + −   2. herbivory +− 3. herbivory +−   4. parasitism + −   5. mutualism + +   6. competition − −   7. mutualism + +

8. parasitism + −   9. competition − −
10. predation + −   11. competition − −
12. mutualism + +

Exercise 2: 1. Spider   2. Coral animal   3. Polar bear   4. *Heliconius* caterpillar   5. Redwood tree 6. Poison-arrow frog   7. Nitrogen-fixing bacterium 8. Virginia's warbler   9. Squirrel (or bear)

Exercise 3: 1. P   2. 1C   3. 1C   4. 2C   5. 2C   6. P 7. 1C   8. 1C   9. 2C   10. 2C   11. 2C   12. 3C 13. 3C   14. 3C   15. 3C   16. 4C   17. 4C   18. S 19. D   Your food web should look something like Figure 37.9 in the text, with producers on the bottom, quaternary consumers on the top, and arrows pointing up to show direction of nutrient transfer.

Exercise 4: 1. A and C (same number of species) 2. B   3. A   4. B   5. A   6. B   7. B   8. A   9. B 10. Kangaroos rats appear to be keystone predators (keystone herbivores?) in this community. They must eat species T (and possibly other species as well), and when they are not present, species T outcompetes other plants. Removing the keystone species drastically changes the makeup of the community.   11. Pocket mice have some effect on community diversity, but not as much effect as kangaroo rats. Pocket mice appear to eat species S (and kangaroo rats apparently do not). When pocket mice are removed, species S competes more strongly with the other plants, but S is not as successful a competitor as T.   12. T is the most successful competitor among the plants. Without grazing by herbivores, it dominates the community.

Exercise 5: 1. EP   2. L   3. ES   4. L   5. EP   6. ES 7. ES

Exercise 6: 1. The introduced organism might escape from any factor that might limit it in its natural environment: predation, herbivory, parasitism, competition, or native resistance to infection.   2. Biological control is the intentional release of a natural enemy (usually a predator or parasite) to attack a pest. It is hoped that the biological control agent will attack the introduced pest without affecting native species (and becoming a pest itself).   3. Coevolution may strengthen the pest and weaken the control agent. The imported enemy might not be as successful in the new environment as the introduced pest. It may not disperse widely enough, or it may not reproduce fast enough to keep up with the pest. Finally, the control agent might become as invasive as its target.

Exercise 7: A. 8   B. 2   C. 5   D. 4   E. 7   F. 3 G. 6   H. 1

Exercise 8: 1. production energy   2. tertiary consumers   3. energy used in cellular respiration 4. secondary consumers   5. energy in wastes

6. primary consumers   7. producers   8. sunlight

Exercise 9: 1. N_2 2. abiotic   3. nitrogen-fixing bacteria   4. biotic   5. biogeochemical   6. protein 7. bacteria   8. ammonium   9. nitrifying 10. nitrate   11. ammonium   12. nitrate 13. assimilation   14. amino acid   15. denitrifying

Exercise 10: 1. Burning of wood and fossil fuels adds CO_2 to the air faster than it is used, and deforestation slows removal of CO_2 from the air. This is causing a buildup of CO_2 in the atmosphere, contributing to the greenhouse effect and global warming.   2. When a forest is clear-cut, vegetation holds less moisture, and the increased runoff of water washes mineral nutrients from the soil. 3. Runoff from deforested land, fertilizer from farmland, animal wastes from pastures and stockyards, or sewage from cities can add nutrients to a lake, causing rapid eutrophication. The lake can become choked with vegetation and algae, and their decomposition can deplete oxygen, kill fish, and reduce species diversity.   4. Using large amounts of inorganic fertilizers increases crop yields, but excess fertilizer runs off, overfertilizing lakes and rivers with nitrogen and phosphorus and causing "blooms" of algae and cyanobacteria. 5. Untreated runoff from feedlots enters fresh water, fertilizing blooms of algae and cyanobacteria and causing potential human diseases from intestinal microbes in the waste.

Test Your Knowledge

Multiple Choice: 1. e   2. a   3. c   4. c   5. a   6. b 7. d   8. c   9. b   10. d   11. c   12. e   13. e   14. a

Essay: 1. At each trophic level in an ecosystem, organisms fail to ingest some food, and some of the food they do eat passes through their systems undigested. Much of the food they do assimilate is burned in cellular respiration, releasing energy that is used to carry out body activities and is eventually given off as heat. This leaves little energy—an average of 10%—available to the next trophic level. Ninety percent of the energy is lost to the next level. If humans eat grain, they get 10 to 100 times the energy that would be available if they fed the grain to cattle and then ate beef because most of the energy in the grain is lost in passing through the additional trophic level.

2. Decomposition is the breakdown of organic materials—plant litter, animal wastes, dead organisms, and so on—into inorganic compounds. Bacteria and fungi are an ecosystem's most important decomposers. They recycle organic material and release inorganic nutrients such as ammonia, nitrate, and phosphate, which are then available for plants to use for growth.

3. The speed of antelope is an adaptation to pursuit by fast predators. The cooperative hunting of a lion pride is an adaptation to their speedy prey. Predator and prey have shaped each other's evolution. Symbiotic relationships are also shaped by coevolution. For example, plants and their pollinators often coevolve. The shape and color of a flower might adapt to the behavior of a particular insect, and the life cycle and mouthparts of the insect might adapt to the flower.

4. Squirrel—primary consumer; oak tree—producer; mosquito—secondary or higher consumer; great white shark—secondary or higher consumer; moose—primary consumer; cheetah—secondary or higher consumer; mushroom—decomposer; spider—secondary or higher consumer; phytoplankton—producer; grass—producer; vulture—detritivore

5. A fern captures a CO_2 molecule in photosynthesis and uses the carbon atom it contains to build a glucose molecule, which ends up in a leaf. An insect eats the leaf and burns the glucose molecule in cellular respiration, releasing CO_2 to the air. A tree captures the CO_2 and uses it to build a protein molecule in its bark. The carbon atom is released in CO_2 when the tree burns in a forest fire. A tree a hundred miles away again captures the CO_2 in photosynthesis and uses it to build a cellulose molecule. When the tree topples and dies, decomposers such as bacteria could break down the cellulose in cellular respiration and release the carbon in CO_2, but this tree does not decompose. It is buried and forms coal. Millions of years later, the coal is burned in a power plant, and the carbon atom again ends up in CO_2.

6. A cornfield or pear orchard is essentially a monoculture, an ecosystem consisting of only one (plant) species. There is little in the way of habitat for other species, even potentially helpful predators, and in a monoculture it is easy for pests to spread from plant to plant. This leaves the community open to devastation by insect pests and pathogens. Polyculture—planting multiple species together—mimics more natural communities, offering shelter for useful species and slowing spread of pests and disease.

7. Phosphorus and nitrogen tend to be recycled locally. When an organism dies, decomposers in the soil break it down and make the nutrients available for uptake by plants. If either of these nutrients is washed away or depleted, it may be a while before a supply can build up again in the soil (although nitrogen-fixing bacteria may obtain nitrogen from the air). Plant producers obtain carbon from carbon dioxide in the air.

Apply the Concepts

Multiple Choice: 1. e 2. d 3. e 4. c 5. b 6. c 7. e 8. a 9. b 10. d 11. b 12. b 13. e

Essay: 1. You could remove A—the predator species—from some of the ponds and see if this has any effect on the abundance of other species. If A is a keystone predator, you would expect its removal to alter the relative sizes of its prey populations. For example, species B might crowd out other prey species.

2. *P. aurelia* and *P. caudatum* have similar niches; their needs and abilities are similar, so they compete with each other. *P. bursaria* apparently has a different niche from *P. aurelia;* they do not compete as much. If *P. aurelia* and *P. caudatum* are similar and *P. aurelia* can coexist with *P. bursaria*, then *P. caudatum* could probably coexist with *P. bursaria* if they were cultured together.

3. Much of the energy in the food consumed by the inhabitants of the shelter is used in cellular respiration to power body activities; this energy is lost as heat. Thus, the body wastes of the inhabitants contain only a fraction of the energy that was originally in their food. The mushrooms, like the shelter inhabitants, use much of the energy in these wastes in their own cellular respiration, storing little of it in their tissues. As energy is lost from the system, the mushroom crops would rapidly dwindle, and the inhabitants would quickly starve. Energy flows through an ecosystem; it is degraded to heat and be dissipated, not recycled. Without energy from outside—light (from the sun or electric lights)—to grow crops, the system quickly runs out.

4. If 10% of the energy on one trophic level is transferred to the next level, 20 kilograms of algae will grow 2 kilograms of crustaceans, and this will grow 0.2 kilograms of fish. If 1 square meter of pond grows 0.2 kilograms of fish and you want 1,000 kilograms of fish, the pond will have to have an area of 1000/0.2 = 5,000 square meters.

5. The least chipmunk can live in a broader range of habitats—both the sagebrush and the sagebrush-piñon pine zones. The yellow pine chipmunk is narrower in its requirements and is confined to the sagebrush-piñon pine zone. The two species have somewhat similar niches because they compete. If both species are present, the yellow pine chipmunk is apparently a better competitor, and it excludes the least chipmunk from the sagebrush-piñon pine zone.

6. Coevolution shapes host-parasite relationships. The deadliest strains of viruses do not spread as readily because they die with their hosts. Less deadly strains are more successful and displace the more deadly strains. The parasites thus become

less deadly over time. Among the hosts, those individuals with the greatest natural resistance to the viruses survive the epidemic and pass their resistance on to their children, increasing the resistance of the host population over time. This is what happened to the rabbits in Australia, and it will probably happen among the indigenous peoples of the Amazon, but at great human cost.

Chapter 38

Review the Concepts

Exercise 1: A. 1. Moral and esthetic values
2. Products: food, medicines, wood products
3. Recycling of nutrients and wastes
4. Stabilization of climate (greenhouse effect)
5. Purification of air and water   6. Flood control   7. Pollination of crops   8. Control of pests   9. Protection from UV rays (ozone)
10. Protection from disease   11. Bioremediation (cleanup of toxics).   B. 1. Destruction of habitat (deforestation, damaging coral reefs)
2. Hunting and overexploitation (tigers, food fish)   3. Introduction of invasive species (Nile perch in Lake Victoria, tree snake on Guam)
4. Pollution (PCBs in gulls; CFCs, ozone depletion, and phytoplankton)   5. Climate change (shrinking polar bear and frog habitats, expansion of mosquitoes and malaria)

Exercise 2: 1. P   2. M   3. L   4. I   5. A   6. H
7. O   8. E   9. J   10. N   11. F   12. D   13. G   14. B
15. C   16. K

Exercise 3: 1. over   2. carbon dioxide (CO_2)
3. 0.8   4. 2–4.5   5. Ocean   6. land   7. northern
8. shrinking   9. flooding   10. earlier
11. Precipitation   12. hurricanes   13. 385
14. fossil fuels   15. methane (CH_4)   16. human
17. atmosphere   18. photosynthesis   19. reservoir
20. decomposition   21. Fossil fuels   22. carbon dioxide (CO_2)   23. Plants   24. ocean   25. tundra
26. deserts   27. wildfires   28. bloom
29. symbiotic algae   30. north   31. Polar bears
32. mountain   33. malaria   34. life cycles
35. plasticity   36. earlier   37. adaptation
38. variability   39. longer   40. 30

Exercise 4: See crossword at the top of the next column.

Exercise 5: 1. Closer to equator   2. Sensitive to habitat degradation   3. Many endangered species
4. High plant and animal diversity   5.Likely to be tropical forest   6. Many endemic species   7. High potential extinction rate

Exercise 6: 1. On a global scale, the Rocky Mountain area does not rank as a biodiversity hot spot. (Most

are in the tropics.) The area is of interest because it is large, relatively unsettled and undeveloped, and already contains many parks and reserves.   2. Landscape ecology is the application of ecological

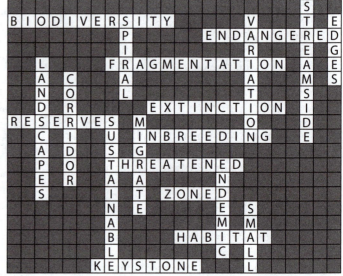

principles to the study of the structure and dynamics of a collection of ecosystems, such as the Yukon-Yellowstone region. One of the priorities of landscape ecology is to study human land use and to make biodiversity conservation a priority, as is being done in the initiative area.   3. One of the goals of the initiative is to connect existing parks and reserves in the region, so that wide-ranging species such as wolves and grizzlies can move about safely.   4. The gray wolf is a keystone species, with many effects on other species in its community.   5. Radio-tracking studies have shown that wolves range over a huge area. They do not stay within the boundaries of existing parks and reserves and are exposed to dangers such as traffic and hunting when they leave protected areas.
6. When wolves were removed, populations of large herbivores such as elk and deer grew unchecked, damaging vegetation and habitat for smaller animals.
7. Wolves don't stay inside reserves and are hit by cars. Some recreationists and ranchers fear wolf attacks. Some wolves have been shot.   8. Elk, moose, and deer populations dropped, and grasses and trees returned. Stream banks recovered, and birds and water-dwellers such as beavers made a comeback.

Exercise 7: 1. E   2. A   3. A   4. B   5. D   6. D
7. C   8. F   9. A   10. E

Test Your Knowledge

Multiple Choice: 1. e   2. b   3. d   4. e   5. c
6. d   7. e   8. b   9. b   10. d   11. c   12. b
13. e   14. a

Essay: 1. It is much less difficult and expensive to take a proactive approach and protect organisms and their habitats before species face extinction. Very small populations face problems of reduced variation due to genetic drift and inbreeding, and it may be very difficult for them to make it in the wild once their numbers are reduced. Besides, literally thousands of species face extinction; how do we choose, and how do we pay for their rescue?

2. True. Tropical forests harbor a disproportionately large number of the world's species of plants and animals. Most biodiversity hot spots are in the tropics. Besides, most of the forest land in the United States has already been cut.

3. Forests produces oxygen, use CO_2, and affect the water cycle. Destroying forests could affect climate. Altering a wetlands could exacerbate natural disasters such as hurricanes or floods. Wetlands are often the "nurseries" for fish species that we use as food. We may be able to obtain other useful products from forest or wetland species.

4. Climate scientists agree that global climate change is real and human activities are the cause. The scientific debate is over. Any delay at this point is political and economic. It might cost us to start now, but that will be less than the ultimate cost of not acting.

5. Not true. It is pollution of the atmosphere by CFCs from air conditioners and refrigerators that destroys the ozone layer. Combustion of fossil fuels intensifies the greenhouse effect and cause global warming, but this has little effect on ozone.

6. Many species suffer from population and habitat fragmentation. Many animal species need large areas to roam and migrate. Most existing reserves are too small. Setting aside biodiversity hot spots is the most effective way to protect the most species; many biodiversity hot spots are not yet identified and protected.

7. The rate of extinction this time may be much greater than in the past. And never before has a mass extinction of so many species been caused by a single species—in this case, humans—and we're supposed to be "intelligent"!

Apply the Concepts

Multiple Choice: 1. a 2. d 3. e 4. e 5. a 6. d 7. a 8. a 9. c 10. e

Essay: 1. Top predators depend on all the species below them in the food web. If a disturbance alters the vegetation or affects a primary or secondary consumer, the disturbance usually affects the top predators, who depend on the lower trophic levels for food. A good example is biological magnification of pesticides. Additionally, many top predators, such as the wolf, are keystone species, with disproportionate influence on ecosystem dynamics

2. a. Conversion to solar and hydroelectric power would decrease global warming, because this would reduce CO_2 emissions. b. Growing trees use CO_2, so this would reduce global warming. c. Burning and decomposition of logs add CO_2 to the air, and trees remove it, so deforestation would increase global warming. d. More efficient cars would burn less gasoline and produce less CO_2 per mile, so this would reduce global warming. e. Fewer people would use less fossil fuels and cut forests at a slower rate, slowing CO_2 buildup and global warming. (Note, though, that the biggest greenhouse gas producers are the developed nations, whose populations are not growing very fast, if at all.) f. Decomposition produces CO_2, so this would increase global warming.

3. Given enough iron, the algae might grow faster, speed up photosynthesis, and take CO_2 from the atmosphere at a faster rate. Since CO_2 traps heat in the atmosphere, the drop in CO_2 would reduce the rate of global warming. But trying this might have unintended effects, such as disrupting ocean habitats and food chains.

4. Restoration ecology is our overall approach—using ecology to repair a damaged ecosystem. Mature forest species such as owls might be helped by retaining or establishing corridors to link areas of big trees and reduce habitat edges. If necessary, we might introduce organisms to restore depleted soil, or use organisms to remove toxics (bioremediation). Ultimately we could protect the trees and the birds in a zoned reserve—undisturbed forest surrounded by forest that could be harvested in a controlled manner.

5. 1,000 pounds of forest life are destroyed for each hamburger; this amounts to about 55 square feet of forest. If you eat 100 hamburgers a year, you could destroy $55 \times 100 = 5,500$ square feet of rain forest, roughly the area of three suburban houses. (Explanation: 50 pounds of beef per year for 8 years is 400 pounds. Half is usable meat, so that totals 200 pounds. There are 4 4-ounce burgers in a pound, so 200 pounds of meat yields 800 burgers per acre. 800,000 pounds of forest/800 = 1,000 pounds. 43,793 square feet of forest/800 = 55 square feet.) Do you think the hamburger is worth it?